FUNCTIONAL
NEUROMARKERS
FOR PSYCHIATRY

FUNCTIONAL NEUROMARKERS FOR PSYCHIATRY
Applications for Diagnosis and Treatment

JURI D. KROPOTOV
N.P. Bechtereva Institute of the Human Brain
of the Russian Academy of Sciences
Saint Petersburg, Russia;
Norwegian University of Science and Technology
Trondheim, Norway;
Andrzej Frycz Modrzewski Krakow University
Krakow, Poland

AMSTERDAM • BOSTON • HEIDELBERG • LONDON
NEW YORK • OXFORD • PARIS • SAN DIEGO
SAN FRANCISCO • SINGAPORE • SYDNEY • TOKYO

Academic Press is an imprint of Elsevier

Academic Press is an imprint of Elsevier
125 London Wall, London EC2Y 5AS, UK
525 B Street, Suite 1800, San Diego, CA 92101-4495, USA
50 Hampshire Street, 5th Floor, Cambridge, MA 02139, USA
The Boulevard, Langford Lane, Kidlington, Oxford OX5 1GB, UK

Copyright © 2016 Elsevier Inc. All rights reserved.

No part of this publication may be reproduced or transmitted in any form or by any means, electronic or mechanical, including photocopying, recording, or any information storage and retrieval system, without permission in writing from the publisher. Details on how to seek permission, further information about the Publisher's permissions policies and our arrangements with organizations such as the Copyright Clearance Center and the Copyright Licensing Agency, can be found at our website: www.elsevier.com/permissions.

This book and the individual contributions contained in it are protected under copyright by the Publisher (other than as may be noted herein).

Notices
Knowledge and best practice in this field are constantly changing. As new research and experience broaden our understanding, changes in research methods, professional practices, or medical treatment may become necessary.

Practitioners and researchers must always rely on their own experience and knowledge in evaluating and using any information, methods, compounds, or experiments described herein. In using such information or methods they should be mindful of their own safety and the safety of others, including parties for whom they have a professional responsibility.

To the fullest extent of the law, neither the Publisher nor the authors, contributors, or editors, assume any liability for any injury and/or damage to persons or property as a matter of products liability, negligence or otherwise, or from any use or operation of any methods, products, instructions, or ideas contained in the material herein.

British Library Cataloguing-in-Publication Data
A catalogue record for this book is available from the British Library

Library of Congress Cataloging-in-Publication Data
A catalog record for this book is available from the Library of Congress

ISBN: 978-0-12-410513-3

For information on all Academic Press publications
visit our website at https://www.elsevier.com/

Publisher: Nikki Levy
Acquisition Editor: Nikki Levy
Editorial Project Manager: Barbara Makinster
Production Project Manager: Nicky Carter
Designer: Matthew Limbert

Typeset by Thomson Digital

Contents

Acknowledgments   xv
Introduction   xix

PART 1
METHODS OF ASSESSING NEUROMARKERS

1.1. Theory of Measurement

True and Observed Scores, Errors	3
Reliability	4
Validity	5
Distribution Across Population	5
Percentiles and z Scores	5
Sensitivity and Specificity	6
Effect Size	7
Requirements for Introducing a Neuromarker into Clinical Practice	8

1.2. Psychometrics and Neuropsychological Assessment

Psychological Models	9
Neuropsychological Testing	10
Supervisory Attentional System Model	10
Operations of Attentional Control	10
Dual Mechanisms of Cognitive Control	11
General Factor	11
Reaction Time	12
Reaction Time Variability	12
Continuous Performance Tasks	15
Infraslow Fluctuations in Performance	15
Big 5 Model	15

1.3. Functional Magnetic Resonance Imaging

Talairach Atlas	17
Montreal Neurological Institute Atlas	18
Physical Basis of Magnetic Resonance Imaging	18
Functional Magnetic Resonance Imaging	19
Bold Response	19

Bold Infralow Fluctuations	20
0.1-Hz Hemodynamic Oscillations	20
Processing Steps in Functional Imaging	22
Activation Maps of fMRI	22
Model-Dependent Correlational Methods	23
Model-Free Correlational Methods	23
Task-Negative and Task-Positive Networks	23
Functional Connectivity and Diffuse Tensor Imaging	24
Test–Retest Reliability	25
fMRI in Neurological Practice	25
Challenges for Clinical fMRI	25

1.4. Positron Emission Tomography

Physical Basis of Positron Emission Tomography	27
Neuroreceptors	28
Test–Retest Reliability	28
Neurotransmitters and Receptor Imaging in Clinics	29

1.5. Spontaneous Electroencephalogram

How an Electroencephalogram is Measured	31
Montages	31
Electrical Events in the Cortex	33
10–20 International System	34
Frequency Bands	35
Electroencephalograms as a Reflection of Cortical Self-Regulation	37
Voltage-Gated Ion Channels	37
Nonbrain Events (Artifacts) in Electroencephalograms	37
Spectral Analysis of Electroencephalograms	40
Interindividual Differences	41
Wavelet Transformation	42
Coherence	44
Neuronal Sources of Electrical Currents	44
Intracortical Connectivity	45
Cortical Focus and Spikes	47
Volume Conductance	47
Inverse Problem: Dipole Approximation	49
Inverse Problem: Nonparametric Solutions	49
Current Source Density	51
Blind Source Separation	52
Independent Component Analysis	53
Individual Electroencephalogram Decomposition Into Independent Components	54
Group ICA Decomposition	56
Test–Retest Reliability	57

1.6. Event-Related Potentials

Definition	59
Information Flow	59
Montages	60
Averaging	61
Number of Trials	62
Information Flow in Visual Pathways	63
Information Flow in Local Network	64
Two Packets of Information Flow	65
Canonical Visual Event-Related Potential	66
Event-Related Potential Paradigms	68
Multiple Sources of Event-Related Potentials	68
Separating Components: Subtraction Approach	69
Separating Components: Single Trial Independent Component Analysis	71
Separating Components: Group Independent Component Analysis in Multiple Tasks	72
Separating Components: Group Independent Component Analysis in a Single Task	72
Separating Components: Joint Diagonalization of Covariance Matrixes	74
Test–Retest Reliability	76
Interindividual Variability	77
A Roadmap for the Development and Validation of Event-Related Potential Neuromarkers	78

PART 2
NEUROMARKERS OF CORTICAL SELF-REGULATION

2.1. Infraslow Electrical Oscillations

Arrhythmic Electroencephalograms	81
Power-Law Function of Electroencephalogram Spectra	81
Infraslow Electrical Oscillations: History	83
Infraslow Fluctuation in Thalamic Neurons	83
Nonneuronal Origin of 0.1-Hz Oscillations	84
Responses to Tasks	85
Preparatory Slow Fluctuations	85
Neuronal Mechanisms	86
Functional Meaning	87

2.2. Alpha Rhythms

Historical Introduction	89
Types of Alpha Rhythms	90
Alpha Rhythms in the Somatosensory Cortex	90
Alpha Rhythms of the Visual System	91

Functional Reactivity	91
Parietal Alpha Rhythm	94
Negative Correlation with BOLD Signals	95
Age Dynamics	97
Frontal Alpha Asymmetry	98
Alpha Rhythms in the Dysfunctional Brain	99
No Alpha Rhythms: Low-Voltage Fast Electroencephalograms	100
Heritability	101
Neuronal Mechanisms	101
Model	104

2.3. Beta and Gamma Rhythms

Historical Introduction	107
The Mystery of Multiple Beta Rhythms	109
Rolandic Beta Rhythms	109
Correlations With BOLD fMRI	111
Frontal Beta Rhythms	111
Vertex Beta Rhythms	113
Occipital Rebound Beta Rhythms	114
Arrhythmic Beta Activity as an Index of Cortical Activation	116
Neuronal Mechanisms	117
Gamma Activity	117
Abnormal Beta Rhythms	119

2.4. Frontal Midline Theta Rhythm

Historical Introduction	121
Functional Features	122
Localization	123
Prevalence	124
Genetic Factors	125
Age Dynamics	125
Personality Traits	126
Cortical Metabolism	126
Working Memory	127
Conflict Monitoring and Anxiety	129
Model	130
Abnormal Theta Rhythms	132

PART 3
INFORMATION FLOW WITHIN THE BRAIN

3.1. Sensory Systems and Attention Modulation

Introduction	137
Separation of Ventral and Dorsal Visual Streams by fMRI	139
Attention Modulation Effects in fMRI	140
Vision as an Active Process	142

Bottom–up and Top–Down Selection Operations	142
Bottom–up Operations in the C1 Wave of Event-Related Potential	145
N1 Wave as Index of Visual Discrimination	145
Visual Mismatch Negativity as Index of Regularity Violation	146
Visual N170 Reflects Activation of Personal Memory	148
Visual N250 Repetition Effect	151
Visual P2 Discrepancy Effect	153
Latent Event-Related Potential Components of Visual Processing	154
A Neuronal Model	157
Principles of Information Flow in the Visual System	158
What and Where Streams in the Auditory Modality	159
Auditory N1/P2 Wave	160
Independent Components	161
Auditory Mismatch Negativity	162
Orienting Response	165
Role of Dopamine in Orienting Response	168
Loudness Dependence of Auditory N1/P2 Waves	168

3.2. Executive System and Cognitive Control

Introduction	171
Operations of Cognitive Control	172
Modes of Cognitive Control	173
Prepotent Model of Behavior	174
Behavioral Paradigms	174
Stroop Tasks	176
Models of Cognitive Control	177
Representations in Working Memory	178
Preparatory Cortical Activities	179
Frontal Lobe Functions	183
Basal Ganglia-Thalamo-Cortical Loops	183
Neuronal Correlates of Cognitive Control in the Basal Ganglia	186
fMRI of Cognitive Control	188
ERP Correlates of Cognitive Control	190
Independent Components of Cognitive Control	192
Lesion Studies	193
Correlation with Neuropsychological Parameters	195
Latent ERP Components of Reactive Cognitive Control	197
Functional Meaning of Latent Components	198
Target P3 (P3b) in Oddball Tasks	202
P3b and Noradrenaline	204
Cortical Dopamine and Working Memory	205
Striatal Dopamine as Regulator of Flexibility	205

3.3. Affective System, Emotions, and Stress

Introduction	207
Emotions as a Separate Dimension	209
Emotions as Habitual Responses	209
Classification of Emotions	209

Three Dimensions of Temperament 211
Brain Model 212
Model of Left–Right Asymmetry in Emotions 214
Big Five Model 216
Eysenck's and Gray's Models 217
Behavioral Paradigms 218
Amygdala as Detector of Fearful Stimuli 218
Anxiety is a State of Preparing to Fear 220
Hypothalamus is Involved in Expression of Emotions 220
Orbitofrontal Cortex as a Map of Rewards and Punishers 221
Ventral Anterior Cingulum and Anxiety 222
Connections to Cognitive Control System 223
fMRI of Emotions 224
Stages of Reactions of Affective System 225
Event-Related Potentials to Emotional Stimuli 226
Anxiety Enhances Visual N1 Wave 228
Neuromodulators of Affective System 228

3.4. Memory Systems

Introduction 231
Temporal Aspects of Memory 231
Working Memory Representations 233
Types of Long-Term Memory 233
Hippocampus as a Reference to Episodic Trace 235
Functional Neuromarkers of Episodic Memory 237
Neuronal Model of Episodic Memory 238
Retrieval Operations 238
Acetylcholine as Neuromodulator of Declarative Memory 239
Procedural Memory System 240
Neuromodulators of Procedural Memory 242

PART 4
METHODS OF NEURO-MODULATION

4.1. Pharmacological Approach

Historical Introduction 245
Current Crisis of Psychopharmacology 246

4.2. Neurofeedback

Definition 247

4.3. Electroconvulsive Therapy

Historical Introduction 267
Parameters of Electroconvulsive Therapy 268
Neuronal Model 269

Mechanisms	269
Efficacy	270
Relapse	270
Contraindications	270
Side-Effects	271

4.4. Transcranial Direct Current Stimulation

Historical Introduction	273
Procedure	274
Difference From Electroconvulsive Therapy	275
Neurophysiological Basis	275
Nonlinear Collective Short-term Effects of tDCS	277
Long-Term Post-tDCS Effects	277
NMDA Involvement in Long-Lasting After-Effects	279
Safety and Side-Effects	280
Limitations	280

4.5. Transcranial Magnetic Stimulation

Introduction	281
Physical Principles	281
Physiological Effect	281
rTMS at Low and High Frequency	283
Model	283
Safety	283

4.6. Deep Brain Stimulation

Introduction	285
Procedure	286
Neuronal Mechanism	286
Advantages and Limitations	287

PART 5
NEUROMARKERS IN PSYCHIATRY

5.1. Attention Deficit Hyperactivity Disorder

Historical Introduction	291
Symptoms	292
Latent Classes in ADHD Symptoms	293
Prevalence	295
Age Onset	295
Persistence in Adulthood	295
Outcome	295
Comorbidity	296
Environmental Factors	296
Genetic Factors	296

Rolandic Focus	297
Executive Functions	297
Heterogeneity of Neuropsychological Profile	298
Inhibition Deficit	299
Delay Aversion	300
Reaction Time Variability	300
Interference With Default Mode	301
State Regulation and Energization Function	302
Hypoarousal Hypothesis	302
Maturation Delay in Neurodevelopment	303
Theta/Beta Ratio	304
QEEG Endophenotypes in ADHD	307
Frontal Beta Synchronization in Childhood ADHD	309
Magnetic Resonance Imaging Correlates	310
fMRI Correlates	310
Decreased P3b Wave	310
ERP Correlates of Cognitive Control in Children	311
Event-Related Potential Correlates of Cognitive Control in Adult ADHD	313
Pharmacological Treatment	315
Event-Related Potential Predictors of Response to Psychostimulants	317
Dopamine Hypothesis	318
Neurofeedback	319
tDCS	320
Transcranial Magnetic Stimulation	321

5.2. Schizophrenia

Historical Introduction	323
Symptoms	323
Prevalence	325
Timecourse	325
Neurodevelopment	326
Heterogeneity	328
Heritability	328
Environmental Risk Factors	329
Treatment	329
Neuropsychological Assessment	329
Volumetric Studies	330
Motor Abnormalities	330
Spontaneous Electroencephalography	330
Sensory-Related Neuromarkers	332
Automatic Predicting Ability Failure	335
Object Recognition Deficit in N170	336
P3b as Endophenotype	337
P3b as a Predictor of Psychosis	338
Proactive Cognitive Control Deficit	339
Reactive Cognitive Control in Schizophrenia	340
Hypofrontality—fMRI Studies	340
Hypofrontality as Predictor of Response to Medication	343

Neurotransmitters	343
Neuronal Model	344
tDCS	348
Transcranial Magnetic Stimulation	348

5.3. Obsessive–Compulsive Disorder

Historical Introduction	351
Symptoms	352
Prevalence	352
Development	353
Heterogeneity	353
Heritability	353
Comorbidity	354
Neuropsychological Profile	354
Lesions	355
Structural Magnetic Resonance Imaging	355
fMRI in Symptom Provocation	355
fMRI in Conflict Conditions	356
Quantitative Electroencephalography	357
Error-Related Negativity and N2 Event-Related Potential Waves	357
Latent Components of Cognitive Control	359
Neuronal Model	359
Neurotransmitters	361
First-Line Treatment	361
Psychosurgery and Deep-Brain Stimulation	362
Neurofeedback	362
Transcranial Magnetic Stimulation	363
Transcranial Direct Current Stimulation	364

PART 6
ASSESSING FUNCTIONAL NEUROMARKERS

6.1. Working Hypothesis

Reasons for Assessment	367
Conventional Diagnostic Categories as a Starting Point	368
Multiple Causes of ADHD	369
Theses to Test	370
Prognostic Power	370

6.2. Technical Implementation

Arrangement of the Working Space	371
QEEG/ERP Databases	371
Arsenal of the 21st Century Psychiatrist	373
Selecting the Behavioral Paradigm	374
Correcting Artifacts	374

6.3. Testing Working Hypotheses: Spontaneous EEG

Rolandic Spikes	377
Excessive Theta/beta Ratio	379
Excess of Frontal Beta Activity	380
Excessive Frontal Midline Theta Rhythm	382
Excessive Alpha Activity	383
Individual Independent Components for Neuromodulation Protocols	384

6.4. Testing Working Hypotheses: Event-Related Potentials

Independence from Other Functional Neuromarkers	387
Selective Deficit of Cognitive Control	388

6.5. Monitoring Treatment Effects

Pharmaco-Electroencephalography	391
Pharmaco-Event Related Potentials	392
Neurofeedback	392

PART 7
THE STATE OF THE ART: OVERVIEW

7.1. Objective Measures of Human Brain Functioning

7.2. Rhythms of the Healthy Brain

7.3. Information Flow in the Healthy Brain

7.4. Current Treatment Options in Psychiatry

7.5. Functional Neuromarkers in Diseased Brain

7.6. Implementation in Clinical Practice

Postscriptum	**421**
References	**423**
Further Readings	**431**
Subject Index	**447**

Acknowledgments

This book is dedicated to those I love. First of all this includes my family: my parents Anna and Dmitrii, my wife Nelia, my sons Maxim and Vania, and my stepsons Igor and Anatoly. All of them believed in me and helped me in different but loving ways.

My early interest in physics was inspired by Victor Kobushkin when he educated me at the famous Lyceum/school 239 in Leningrad with advanced teaching in mathematics and physics. My broad view of mathematics was shaped by Professor Boris Pavlov at Leningrad State University.

In 1972, being a postgraduate student at the university's Department of Theoretical Physics, I met Professor Natalia Bechtereva, a granddaughter of the famous Russian psychiatrist Bechterev. This meeting changed my life: I became involved in studies of the healthy and diseased human brain. In my days at the Institute of Experimental Medicine in Leningrad some severe cases of Parkinson's disease and epilepsy were treated with deep-brain stimulation and local lesions by means of implanted electrodes. This approach provided a unique opportunity of recording the impulse activity of neurons and other physiological parameters from the human brain. The main idea was to search for a "neuronal code of human mental activity" by recording responses of neurons in different behavioral paradigms. Those were years when sophisticated methods of EEG analysis were still to be invented and researchers hoped that the recording of neuronal reactions in various behavioral tasks would tell them how the brain processes information. Unfortunately, these hopes were never completely satisfied. However, the studies of human neuronal reactions performed by our group showed that neurons of the basal ganglia were involved not only in motor actions, but also in sensory and cognitive functions. For this research in 1985 together with Yury Gogolitsyn and Natalia Bechtereva I was awarded the highest award in the whole Soviet Union—the State Prize. Soon after, I created a Laboratory of Neuroinformatics at the Institute of Experimental Medicine.

In the 1990s, neuroscience horizons were widened by the invention of new neuroimaging methods: positron emission tomography (PET) and magnetic resonance imaging (MRI). The Institute of the Human Brain (director Sviatoslav Medvedev) was the first in the Soviet Union to build up a PET center. My laboratory moved to this brand new institute in the naive hope that new imaging methods would answer all our questions. But again, the initial euphoria was replaced by deep disappointment:

no qualitatively new data were obtained and no clinical applications for psychiatry were found.

All this happened before *perestroika*, proclaimed by Mikhail Gorbachev, became a real disaster and the Soviet Union collapsed. The funding of science ceased as well. To earn a living, people in my laboratory started bargain tea packages and did a lot of other "business-like" enterprises. Most of them left the Soviet Union and went to the West for a better life. Yury Gogolitsyn immigrated to England, Andrey Sevastianov and Michael Kuznetzov went to the United States, Aleksander Popov to Australia, Oleg Korzukov and Olga Dubrovskaya to Finland. In 1992–93, I was working with Peter Kugler, Helen Crowford, and Karl Pribram at the Brain Research Center at Radford University in Virginia on mathematical simulation of realistic neural networks. I am very grateful for their help in those hard years.

However, removal of the iron curtain opened new opportunities for my laboratory for collaboration with other universities in the West. Here I want to mention our joint research projects with Nobel Prize winner Ilia Prigogine and the outstanding Finnish psychologist Risto Naatanen. Risto introduced me to the field of ERPs. We published with him several papers on intracranial correlates of mismatch negativity—a small negative fluctuation in response to deviance in repetitive auditory stimulation. It was the first time I saw a neuronal marker of a psychological operation that could be potentially used in clinical practice. To continue ERP research in the laboratory, we needed the corresponding equipment. But where would we get the funding?

In cooperation with the Institute of Television we set up Potential, a company, with the aim to manufacture EEG machines for Russian clinics and for research. At the beginning we were working in collaboration with Don Tucker, a professor at the University of Oregon and the founder of EGI, a company. Here I would like to mention Valery Ponomarev, a senior researcher in my lab, who started writing the software for the EEG amplifiers developed by Potential. Highly talented in mathematics he was able to introduce into the software a lot of advanced methods for EEG/ERP analysis. So, instead of buying the equipment for our research, we developed it ourselves.

We still needed the money to keep people in the lab. So, I decided to provide services for ADHD children in whom, as Western studies showed, EEG could be used for diagnostic purposes. However, no medical treatment existed in Russia because ADHD was not considered by Russian pediatricians as a real disorder and psychostimulants were forbidden. We needed a new approach and the classical method of operant conditioning provided the answer. It should be noted that EEG operant conditioning had a long history in Russia with animal experiments undertaken by Professor Nikolai Vasilevskii and with clinical research undertaken by Professor Natalia

Chernigovskaya. In the United States, similar studies were being carried out by Joe Kamia and Barri Sterman. However, the first neurofeedback protocols for ADHD were suggested to us by Joel Lubar and Siegfried and Sue Othmer from the United States. The quantitative EEG (QEEG) diagnostic procedures were inspired by Roy John, Barry Sterman, and Robert Thatcher. These new technologies were implemented in my lab by Olga Dubrovskaya and Vera Grin-Yatzenko. Traditional neurofeedback equipment was developed in our laboratory.

By teaching these technologies at a workshop in Norway, I earned enough money to buy tickets to the 2001 ISNR conference in Monterey (USA). ISNR stands for International Society for Neuronal Regulation—one of the few research communities that were using QEEG/neurofeedback technologies in clinical practice. At the conference I met Jay Gunkelman, the president of the society in those days, and I was elected president of the newly formed European chapter of ISNR. In 2002 the workshop organized by Jonelle Villar for Jay and me in Portugal inspired Dr. Andreas Mueller to set up an ERP database. I constructed the tasks and sent my PhD student Katja Beliakov to Andy's practice in Switzerland to collect the data together with Gian Gadrian. After 4 months they had recorded data from 250 healthy children and after 6 months Valery Ponomarev wrote the software for comparing individual QEEG/ERP parameters with the database. For the first time, group independent component analysis was applied for extracting ERP functional components. Andy immediately started using the database in his practice.

It was approximately at this time that Jan Brunner from the Norwegian University of Science and Technology, Trondheim began using the technology in his clinical practice. He was the first to prove that ERP components were reliable and that they selectively correlated with scores in certain neuropsychological domains. Another approach for applying the methodology to neuropsychological practice was suggested by Professor Maria Pachalska from Krakow. She organized a series of lectures and workshops for me at Krakow Academy in which we tested many patients with different psychiatric conditions. The results of the work were published in a series of papers and the 2014 book *Neuropsychologia kliniczna*.

It took 10 years of hard work using different approaches to find the functional neuromarkers of the healthy and diseased brain. Here I would like to mention Valery Ponomarev, Marina Pronina, Sergei Evdokimov, Ekaterina Tereschenko, Vera Grin-Yatzenko, Elena Yakovenko, Inna Nikishena, Olga Dubrovskaya (Kara), Leonid Chutko, Galina Poliakova, and Yury Poliakov from the Institute of the Human Brain in Saint Petersburg (Russia); Andreas Muller, Gian Gadrian and Gian-Marco Baschera from the Brain and Trauma Foundation in Chur (Switzerland); Bernhard Wandernorth from BEE Systems in Germany and Switzerland; Jan Ferenc Brunner, Ida Emilia Aasen, Anne Lise Hoyland, and Knut Hestad from the

Norwegian University of Science and Technology in Trondheim; Venke Arntsberg Grane from the Helgeland Hospital in Norway; Geir Ogrim from the Østfold Hospital Trust in Norway; Antonio Martins-Mourao and Tony Steffert from the Open University in England; Beverly Steffert from the British Psychological Society; Mirjana Askovic from STARTTS in Sydney (Australia); Nerida Saunders and Rustam Yumash from the Brain Mind & Memory Centre & Research Institute in Australia; Maria Pachalska, Anna Rasmus, and Andrzej Mirski from the Andrzej Frycz Modrzewski Cracow University in Poland. The list of my coworkers is not complete.

The book I am presenting here is the final result of a lot of hard work of a multidisciplinary team of extremely dedicated and highly qualified people, and I am very thankful to all of them. Studies in my laboratory in Saint Petersburg were supported by grants from different agencies such as the Soros Foundation, Russian Foundation for Fundamental Research, Russian Humanitarian Science Foundation, US National Science Foundation, and Austrian Academy of Sciences. The latest grants are Grant 14-06-00937a of RGNF (Russian Humanitarian Science Foundation) and Grant 16-15-10213 of RSF (Russian Science Foundation).

Juri (Yury) D. Kropotov
April 2016, Saint Petersburg

Introduction

TWO RECENT EVENTS THAT MAY CHANGE PSYCHIATRY

Two crucial events occurred recently in psychiatry: (1) the appearance of the fifth edition of Diagnostic and Statistical Manual of Mental Disorders (DSM) published by the American Psychiatric Association and (2) approval by the US Food and Drug Administration (FDA) of the first electroencephalogram (EEG)-based biomarker of Attention Deficit Hyperactivity Disorder (ADHD). **DSM-5** was published on May 18, 2013. It was criticized by various authorities for lacking empirical support and for the influence of the psychiatric drug industry (see the response letter of the Coalition for DSM-5 Reform at http://dsm5-reform.com/). This event evoked feelings of **disappointment** in psychiatrists awaiting biomarkers of mental dysfunctions.

The second event happened on Jul. 15, 2013 when the FDA allowed marketing of the first medical device based on EEG to help assess ADHD in children and adolescents 6–17 years old. The device, the **Neuropsychiatric EEG-Based Assessment Aid (NEBA) System**, records EEG for 15–20 min by computing the theta/beta ratio, which has been shown to be higher in children and adolescents with ADHD than in children without it.

MENTAL ILLNESS AS A CHALLENGE TO HEALTHCARE SYSTEMS

Mental disorders are a global problem and represent one of the biggest challenges for **healthcare systems**. Worldwide there are some 500 million people suffering from mental disorders. In the European Union, mental disorders are considered one of the leading causes of the disease burden. What makes the situation worse is that the prevalence of mental disorders is expected **to grow** for a variety of reasons including aging of the whole population and increasing economic problems such as reduced job security, work intensification, and enhancement of stress.

DSM AND ICD AS DICTIONARIES OF PSYCHOPATHOLOGY

Although the book is titled *Functional Neuromarkers for Psychiatry*, its main goal is to introduce **neuroscience methodology** into psychiatric practice. So, historically this book can be considered as a reflection of a neuroscience age of psychiatry.

In the past, mental disorders were defined by the **absence of organic lesions**. Mental disorders became neurological disorders at the moment a brain lesion was found. Because no obvious brain damage was usually found in mental disorders by the conventional clinical methods of the 19th and 20th centuries, the psychiatry of the past did not rely on brain damage markers and was confined by **description of symptoms and signs**.

In this description approach, DSM and the International Classification of Disorders (ICD) in the corresponding chapters of mental and behavioral disorders were a clear **step forward**. Psychopathology was decomposed into separate disorders. DSM-5 is supposed to serve (at least in the United States) as a universal authority, a "bible" for **psychiatric diagnosis**. It has practical importance in that psychiatric diagnosis determines treatment recommendations for clinicians and the payment strategy for healthcare providers.

However, in contrast with other fields of the advanced medical science such as heart or cancer diseases, these manuals are based not on **objective laboratory measures** but rather on some vague **consensus** about clusters of clinical symptoms. But, as we learn from other medical fields, symptoms alone rarely indicate the best choice of treatment, only knowing **the cause of the symptoms** on the basis of laboratory measures provides the optimal treatment.

THE HURDLES TO OVERCOME

If a cardiologist sees a patient with chest pain symptoms he or she would ask for an **electrocardiogram (ECG)** recording the heart's electrical conduction system. An ECG picks up electrical impulses generated by the polarization and depolarization of cardiac tissue in the form of specific waves. This information is used to measure directly the rate and regularity of heartbeats, and indirectly the size and position of the chambers and the presence of any damage to the heart. The ECG may provide a clue as to whether a patient is having an ischemic event.

However, a psychiatrist seeing a patient with, say, schizophrenia symptoms would not ask for an **EEG** recording and assessment as the first choice. This is despite the solid fact that an EEG registers electrical signals of the brain and theoretically could tell us much about the course of the symptoms. This begs the question: why do we use recordings of electrical signals for diagnosis in heart disease but do not use recordings of electrical signals in psychiatry? There are several reasons for that.

First, there is the **difference in complexity**. The brain is much more complex than the heart. Heart physiology provides a relatively limited number of ECG parameters that can be used for diagnosis. ECG parameters include only five waveforms which are easily identified in each subject and have a clear functional meaning in the heart cycle. In contrast, the number of parameters provided by multichannel EEG in the resting state is enormous. If event-related potentials (ERPs) are included in the analysis, the amount and quality of information increases tremendously.

Second, ECG waveforms do not show big intraindividual differences and have clear diagnostic value. In contrast, EEG and especially ERP parameters show **large interindividual variation.** Moreover, EEG/ERP parameters are very sensitive to fluctuations in the state of subjects. It appears that the healthy population **is not homogeneous** in terms of brain functioning. If we take into account the heterogeneity of a given diagnostic category of mental illness, the difficulties in separating patients with the given diagnosis from healthy controls by means of a biomarker becomes evident.

Third, the recording and measuring procedures in ECG are fully standardized. In contrast, in EEG only electrode placement is standardized. The other parameters, especially conditions of ERP recordings, **are not standardized**. As a consequence, setting up databases in clinical practice is limited in the EEG field and practically absent in the ERP field.

Fourth, the heart is a relatively simple organ of the body (a pump) whereas the brain is the **most sophisticated system in the world**, one that is intended not only to self-regulate but also to process external and internal information unconsciously and consciously. A cardiologist in his or her clinical practice relies on the well-established theory of the heart. No theory of brain functioning has so far been built up.

Similar reasons can explain why other functional neuromarkers such as magnetoencephalography (MEG), functional magnetic resonance imaging (fMRI), and positron emission tomography (PET) have not been used in clinical practice so far.

FROM SYMPTOMS TO NEURONAL CIRCUITS

The diagnostic categories of mental illness described in the DSM and ICD **are not organized according to impairment of brain systems**. However, as neuropsychology shows, similar symptoms can be induced by local damage in different parts of a given widely distributed neuronal circuit of the brain or even in functionally different systems. This statement is supported by studies of functional brain activity (such as ERP, fMRI, PET) recorded under various psychiatric conditions in comparison with functional brain activity in healthy controls. More and more studies clearly demonstrate that the **patterns of brain activation** during particular functional tests may be diagnostic, just as cardiac imaging during a stress test is now used to diagnose coronary artery disease. A promising approach for "biomarker" discovery has been based on pattern-recognition methods applied to neuroimaging data. In a 2015 paper Thomas Wolfers from Radboud University, the Netherlands, and coworkers reviewed the literature on MRI-based pattern recognition for making diagnostic predictions in psychiatric disorders and evaluated the recent progress made in translating such findings toward clinical application. In 2014, Gráinne McLoughlin, Scott Makeig, and Ming Tsuang from University of California, San Diego showed convincing evidence that new computational approaches in EEG research such as independent component analysis provided powerful tools for identifying distinct cortical source activities that are sensitive and specific to the pathophysiology of psychiatric disorders.

FROM "DECADE OF THE BRAIN" TO "DECADE OF TRANSLATION"

The 1990s were called the **"decade of the brain"** in which new concepts of brain functioning were formulated on the basis of accumulated data (Fig. I.1). This was also when new methodological approaches were developed such as blind source separation techniques for separating local sources of brain activity. The early part of this century (2000–2010) may be recognized as the **"decade of discovery"** during which brain circuits of normal and abnormal brain functioning have been identified. This is the time when new nonpharmacological methods of treatment such as transcranial direct current stimulation, transcranial magnetic stimulation, and other forms of neuromodulation have been tested in clinical practice. The "decade of

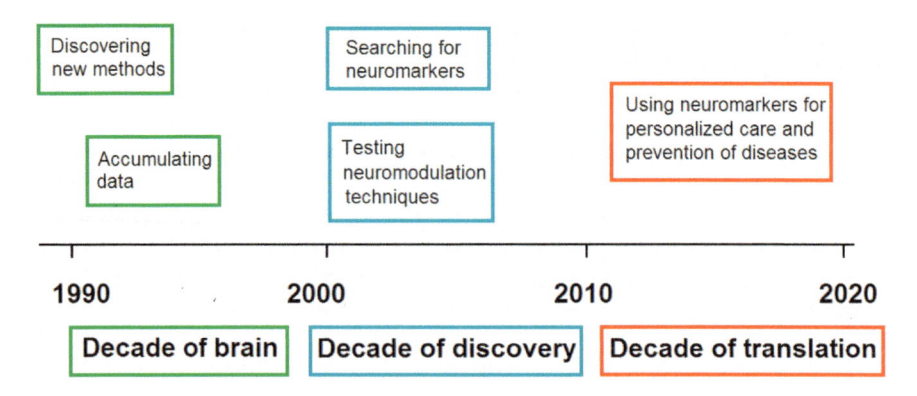

FIGURE I.1 Developing neuroscience of mental illness.

discovery" will be followed by the **"decade of translation,"** which will focus on application of neuromarkers to each of the major mental disorders for providing early detection and prevention as well as personalized care for a particular patient. The early detection of neuromarkers of mental illness in its turn will require development of preventive interventions.

In the United States the **BRAIN Initiative** (Brain Research through Advancing Innovative Neurotechnologies, also referred to as the **Brain Activity Map Project**) with the goal of mapping the activity of every neuron in the human brain was proposed by the Obama administration on Apr. 2, 2013. In Europe a large 10-year scientific research program named the **Human Brain Project (HBP)** directed by the École polytechnique fédérale de Lausanne and largely funded by the European Union was established in 2013. The project aims to simulate the complete human brain on supercomputers to better understand how it functions and to simulate drug treatments.

CONCEPT OF BIOMARKER

Intuitively, we think of a **biomarker** (or biological marker) as a characteristic that can be objectively measured and evaluated as an indicator of normal biological processes, pathogenetic processes, or pharmacologic responses to therapeutic intervention (Biomarkers Definitions Working Group, 2001). According to the type of information that they provide, biomarkers for CNS disorders can be classified as clinical, neuroimaging, biochemical, genetic, or proteomic markers. Expectations toward the development of biomarkers are high since they could lead to significant improvement in diagnosing and possibly preventing neurological and psychiatric diseases.

According to a 2012 consensus report of the World Federation of ADHD (Johannes Thome and coworkers from the University of Rostock), an ideal biomarker by analyzing ADHD must have (1) a diagnostic **sensitivity** >**80%** for detecting ADHD, (2) a diagnostic **specificity** >**80%** for distinguishing ADHD from other disorders with ADHD-like symptoms. In addition, the biomarker must be (3) **reliable, reproducible, and inexpensive** to measure, noninvasive, and simple to perform, (4) confirmed by at least **two independent studies** conducted by qualified investigators with the results published in peer-reviewed journals. The definition of the ideal biomarker can be extended for all psychiatric conditions.

NEUROMARKERS AND NEUROIMAGING

In this book we consider **neuroimaging** a set of neuroscience methods including MRI, fMRI, PET, and measuring parameters of EEG and MEG such as quantitative EEG (QEEG), event-related de/synchronization (ERD/ERS), and ERPs. Many objective measures of brain anatomy and physiology can be obtained in clinical practice using neuroscience methods. These objective measures are called **neuromarkers**. The term "neuromarker" was first introduced by Evian Gordon from the University of Sydney in 2007. It includes any neuropsychological parameter of behavior and any structural or functional index of the brain. Structural parameters include anatomical measures of brain anatomy and axonal pathways taken from postmortem brains or in vivo by MRI and **diffuse tensor imaging (DTI)**. PET is used for in vivo imaging of neurotransmitter systems within the brain. Functional parameters include dynamical measures of brain metabolic activity in the **second/decisecond time frame** by means of fMRI, PET, as well as dynamical measures of brain electrical activity at the **millisecond time resolution** by means of EEG/MEG including ERPs.

The term neuromarker is narrower that the term biomarker of disease. Biomarker in general is any gene, biochemical substance, structural index, physiological characteristic, or behavioral parameter indicating the presence of disease. Biomarkers are used to measure the **start and evolution of disease** or the **effects of treatment**. Although the term biomarker is relatively new, biomarkers have been used in clinics for a considerable time. For example, body temperature is a well-known biomarker for fever. Blood pressure is used to determine the risk of stroke. Cholesterol values are a biomarker and risk indicator for coronary and vascular disease. In epilepsy, spikes in EEG are considered as biomarkers of the focus in the cortex.

The current diagnostic categories of mental disorders were formulated 100 years ago by a small number of psychiatrists such as Bleuler (1911) and Jaspers (1923). To formulate these categories they relied on similarities in behavioral syndromes and clinical outcomes of patients they encountered. The founders of psychiatry were aware that such categories reflected only **observable behaviors** rather than dysfunctions in distinct anatomical–physiological substrates. By analogy, dysfunctions in a number of different automobile mechanisms (the electrical system including spark plugs, battery cables, etc. and the gas-distributing system including many different elements) might lead to similarly perceived symptoms—a car fails to start. It would be difficult to distinguish which mechanisms are dysfunctional to fix them without the ability to look under the hood of the car. Finding a neuromarker of a disease allows us to look under the hood of the mind and discover which mechanisms may be dysfunctional for a given disorder.

ENDOPHENOTYPE IN PSYCHIATRY

In psychiatry the term **endophenotype** is quite popular. It was proposed by Irving Gottesman of the University of Minnesota Medical School and J. Shields in 1973 as a response to the failure to find a strong association between genes and the most common psychiatric conditions. The purpose of the concept is to divide psychiatric behavioral symptoms into more stable phenotypes associated with certain neurophysiological systems, which is turn can be more directly connected to genes. In this context endophenotypes are considered a subset of biomarkers.

Endophenotypes are **intermediate phenotypes**, often undetectable by the unaided eye, that link disease-promoting sequence variations in genes (such as alleles) to lower level biological processes, and further link lower level biological processes to the observable syndromes that constitute diagnostic categories of disorders.

There is common agreement that a useful endophenotype should (1) **cooccur with the disorder;** (2) **be reliably measured;** (3) **be heritable;** and (4) **show familial overlap with the disorder.**

The issue of familial overlap is important because, without such evidence, we could find genes for a biologically based phenotype, but they may not be genes for the disorder of interest. Because an endophenotype is conceptualized as an expression of the genetic liability for a disorder, it should appear in individuals who carry genes for a condition but do not express the disorder itself, that is, the unaffected relatives of diagnosed individuals.

Deficits found in affected but not unaffected relatives raise the possibility that impairments are a result of the disorder itself or of unique environmental factors.

EXTENDED ENDOPHENOTYPE

There are several levels of endophenotypes varying from **anatomy** (structure), to **function at the neuronal level** (neurophysiology), to **function at the psychological level** (behavior). For example, schizophrenic patients show decreased volumes of the dorsolateral prefrontal cortex (DLPFC) (anatomy), low levels of metabolic activation in tasks on working memory (neurophysiology), and poor performance on working memory tasks (behavior). For such a functionally linked set of endophenotypes the term "extended endophenotypes" was proposed by Konasale Prasad and Matcheri Keshavan in 2008 from the University of Pittsburgh School of Medicine.

GENETIC AND EPIGENETIC FACTORS

Most common psychiatric conditions run in families, which presumes a genetic factor in their development. **Familial studies** show high-risk factors among parents and siblings of patients in a given diagnostic category. However, familial studies cannot distinguish between the contribution of genetics and environmental effects in the aetiology of a disorder.

Adoption and twin studies can help to separate, although not completely, genetic from environmental factors observed in family studies.

Another genetic approach is called the **candidate gene approach** that selects genes of interest based upon knowledge of the disorder. For example, in the case of ADHD we know that drugs which block reuptake of dopamine and noradrenaline are effective in treating ADHD, so that genes responsible for regulation of these neurotransmitters are good candidates for genetic research of ADHD.

Linkage and association studies allow the identification of genes that cosegregate with the disorder within families. The principle of a linkage study is the following: if a disease runs in a family, one could look for genetic markers that run exactly the same way in the family (eg, from grandma, to dad, to child). If we find one, we assume the gene that causes the disease is somewhere in the same area of the genome as the marker. In practice, a popular design is to genotype affected siblings, and use the following logic—for a given bit of chromosome, each sibling gets two copies, one from mom and one from dad. If the two have inherited the same bits from each parent, the area is more likely to be involved in the disease than if each sibling inherits different bits.

Association studies come from the other direction. The principle of an association study is also simple—gather some people with a disease and some people without a disease, and see if a certain genotype is present more often in the affected cases than the controls. If the allele plays a role in causing the disease, or is correlated with a causal allele, it will have a higher frequency in the case population than the control population. By comparing the frequency of mutations in a gene in a sample of patients with controls, we can determine whether a gene is associated with the corresponding psychiatric condition.

Genome-wide association designs are now available because of the Human Genome Project which was actually aimed at identifying sources of genetic variation between individuals that could be used to map different diseases including psychiatric conditions. This association design is hypothesis free, meaning that no a priori knowledge about a gene is needed for it to be linked to a disease.

COMMON PSYCHIATRIC DISEASES ARE NOT MENDELIAN DISORDERS

Two types of genetically inherited brain disorders can be separated: **Mendelian and non-Mendelian disorders**. An example of a Mendelian disorder is Huntington's disease. The most common psychiatric disorders are non-Mendelian. Each Mendelian disorder is the end product of the inheritance of only one or two mutations (rare alleles). In contrast, common mental disorders are hundreds of times more common than Mendelian disorders, and result from the complex interaction of multiple genes with environmental factors affecting each disorder. In other words, common mental disorders, such as schizophrenia, bipolar disorder, and depression, are caused by numerous genetic and environmental factors, each of which have individually small effects and which only result in overt disease expression if their combined effects cross a hypothetical **threshold of liability**.

Genes influencing this liability affect broadly defined neural systems including the sensory, memory, affective, and executive systems. Each of these systems involves a combination of neurotransmitters and neuromodulators such as glutamate, GABA, dopamine, and serotonin. The idea behind the concept of endophenotype is that dysfunctions of these higher order brain systems might be more directly connected to phenotypes associated with mental disorders than neurotransmitters and neuromodulators themselves (Fig. I.2). Several genes can be suspected in the disease

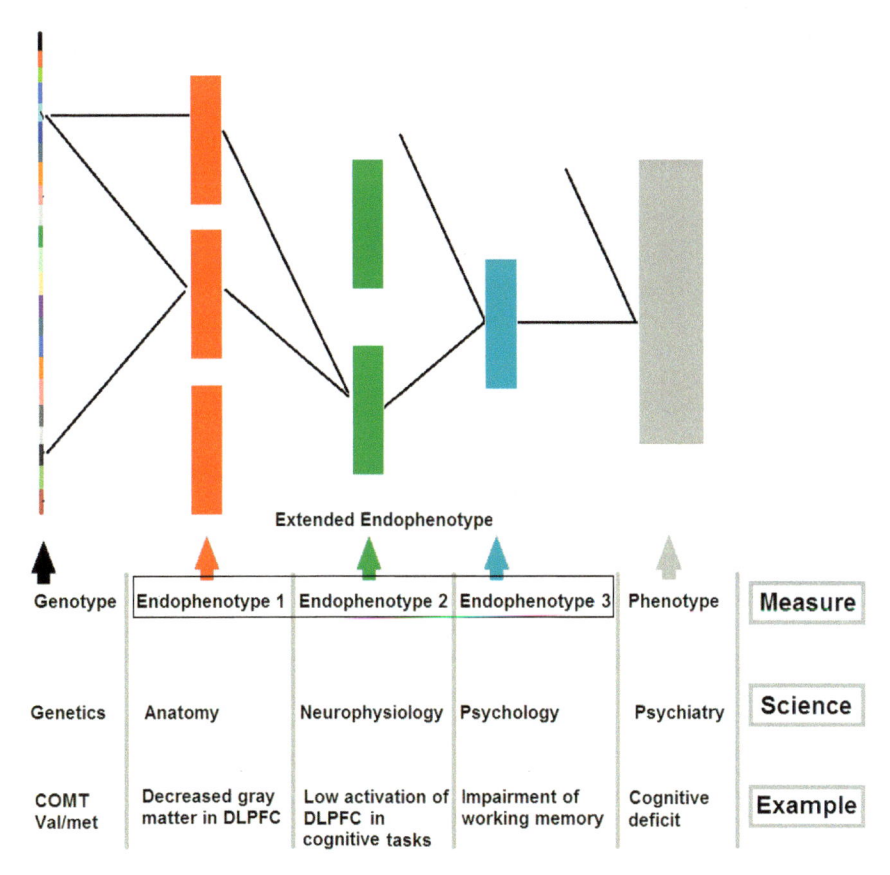

FIGURE I.2 From genes to phenotype through low-level (L) and high-level (H) endophenotypes.

(eg, COMT in schizophrenia). Genes are responsible for protein production and together with environmental factors define anatomical structures (such as gray matter in the DLPFC measured by MRI)—endophenotype 1. The interaction of genetic, anatomical, and environmental factors is expressed in metabolic activity of the DLPFC during cognitive tasks (such as GO/NOGO task) measured by fMRI—endophenotype 2. The level of DLPFC activation correlates with behavioral parameters (such as reaction time to GO targets)—endophenotype 3. Finally, poor performance in cognitive tasks is associated with cognitive deficits found in the schizophrenic patient phenotype. Endophenotypes 1, 2, and 3, studied by different sciences, form an extended endophenotype.

RISK FACTORS

Most common mental disorders lie on a **continuum of severity** that ranges from nonaffected (healthy) individuals to those with extreme forms of the disorder. The subject is diagnosed by a psychiatric disorder when his behavioral pattern (phenotype) "exceeds" some **threshold** defined by a diagnostic manual. Setting such a diagnostic threshold is a subjective procedure.

An endophenotype **quantitatively** indexes some brain structure or function. When measured in a human population it continuously varies from subject to subject. However, the measure of endophenotype in a clinical population (such as with schizophrenia) by definition must be deviant from a corresponding measure in the healthy population. Fig. I.3 schematically depicts the relationship between endophenotype and phenotype. Continuous measures allow for scaling liability in the nonaffected population.

SEARCH FOR NONPHARMACEUTICAL METHODS OF TREATMENT

Historically, the golden age of psychopharmacology was inspired by **serendipitous discoveries** of new drugs, such as the antipsychotics, antidepressants, and anxiolytics. However, attempts to find an effective **pharmacological treatment** for a separate diagnostic category, such as schizophrenia, have so far been **unsatisfactory** as a result of the limited efficacy of pharmacological treatment, major side-effects, and a lack of novel mechanisms or compounds. One of the basic problems is drugs are widely distributed within the brain by blood circulation and their effect is not local and may cause undesirable **side-effects**. As a result of these failures many pharmaceutical companies shut down their drug discovery programs in mental illness.

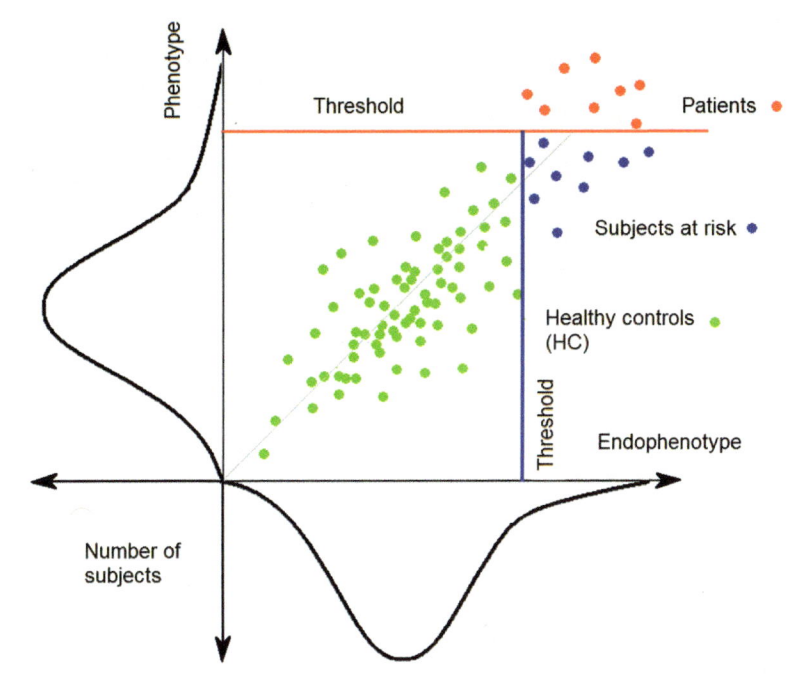

FIGURE I.3 **A combination of endophenotypes determines the likelihood of the subject having a psychiatric disorder.** The endophenotype is distributed in the population and correlated with the phenotype (behavioral pattern) of the disorder. Increased phenotypic severity is marked by the vertical *arrow*. The corresponding diagnostic category is determined by a threshold (disorder threshold, marked by *red* horizontal line) beyond which it is agreed (according to diagnostic criteria) that behavioral difficulties are impairing and require intervention. "Healthy" subjects according to the diagnostic criteria are marked by *green* circles; "diseased" subjects are marked by *red* circles; the subjects at risk for developing the disorder are marked by *blue* circles.

Instead, a substantial research effort is now devoted to nonpharmaceutical methods of treatment including those that use electrical intervention such as **transcranial magnetic stimulation (TMS), transcranial direct current stimulation (tDCS), deep-brain stimulation (DBS)** (Fig. I.4). These methods are supposed to modulate neuronal circuits of the brain by **local injection** of electrical currents.

Another new approach is **optogenetics**. It presumes injection of neurons with a benign virus containing the genetic information for light-sensitive proteins. These cells in turn can be controlled with flashes of light sent through embedded fiber optic cables. Although the method has only been used in animals there is a hope that it could be used in humans in the next few years.

The nonpharmaceutical methods of treatment are based on the idea that the core of a psychiatric condition is not a neurotransmitter but rather a

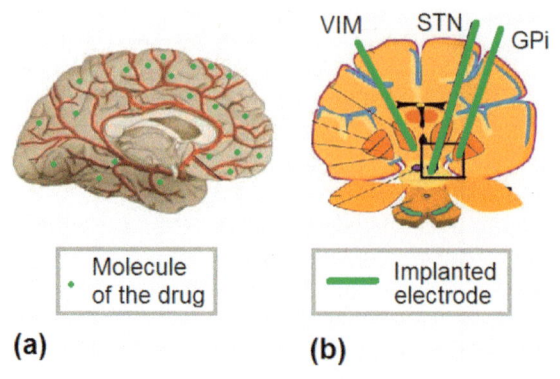

FIGURE I.4 Distributed effect of a drug treatment of (a) mental illness and (b) a local effect of DBS in the case of Parkinson's disease.

dysfunction of a specific neuronal system. This idea corresponds to the neuroscience concept that we can better understand human emotion and cognition by understanding **neural circuits**. To study such systems we need to use imaging techniques such as **PET, fMRI,** and **ERPs** under conditions when patients are required to make some functional tasks. One of the basic achievements of modern neuroscience is decomposing the brain into anatomically and functionally segregated neuronal circuits such as the **sensory, memory, affective, and cognitive control systems**. In its turn **neuropsychology** as a part of neuroscience demonstrated that similar symptoms can be caused by damage to different nodes of a particular system.

TRAINING IN CLINICAL NEUROSCIENCE

The psychiatrists (and neurologists) of the future more likely will be considered **clinical neuroscientists** as a result of applying the revolutionary insights from neuroscience to the care of patients with brain disorders. Consequently, the psychiatrists of the future will need to be educated as brain scientists. For example, in the past unconscious processes and motivation were the sole province of psychoanalysis. Now these operations (together with attention and cognition) are studied by cognitive neuroscience and in the future they will change our understanding of mental illness.

HUMAN BRAIN INDICES DATABASE AS A SOURCE OF REFERENCE

This is a personal book based on my attempts to find the neuromarkers of healthy and diseased human brain. During the last 15 years I analyzed the EEGs and ERPs of **thousands of healthy subjects and thousands of**

patients with different diagnostic categories (sometimes of unknown origin). To compare patient groups with healthy controls, the **Human Brain Indices (HBI) reference database** was developed during 2003–2005. This database was built up following the methodology developed in the Human Brain Institute of the Russian Academy of Sciences and the Institute for Experimental Medicine of the Russian Medical Academy of Sciences. Recordings were done in Switzerland and Russia on about 1000 healthy subjects without any history of neurologic and psychiatric conditions both individually and as families.

The collection of the normative data was initiated by the Swiss psychologist **Dr Andreas Mueller** from the Brain and Trauma Foundation. He was inspired by the QEEG/ERP methodology which Jay Gunkelman and the author of the book were teaching in a workshop in Cascais, a beautiful resort in Portugal, in 2002. This workshop was organized by Jonelle Villar and Patricia Bellinghausen—enthusiasts of the European chapter of the International Society of Neuronal Regulation (ISNR) which was later transformed into European Society of Applied Neuroscience (SAN).

In the data collection almost every subject was recorded by **19-channel EEG during two resting-state conditions with the eyes open and the eyes closed and under six task conditions** (the passive three-stimulus tasks presenting standard, deviant, and novel sounds; the active auditory oddball task; the two cued GO/NOGO tasks with animal/plants and happy faces/angry faces as GO/NOGO stimuli; the mathematical task with arithmetic operations on visually presented numbers; and the reading task with matching auditory and visually presented stimuli). The age of subjects was between **7 and 90 years** so that all subjects were able to complete the tasks with relatively small numbers of errors (Fig. I.5). To improve the signal-to-noise ratio of ERPs the number of trials in each category was selected to be 100, except the passive auditory three-stimulus task where the number of standard trials was 2000.

Data processing was carried out by means of the original software written by **Dr Valery Ponomarev**, senior research fellow at my laboratory in Saint Petersburg. The software included almost all methods of EEG/ERP analysis described in this book including two variants of the blind source separation approach.

Original data presented in the book are the results of **teamwork** of several groups of researchers in different countries over three continents (Fig. I.6).

Normative data of European origin were collected in Switzerland, Russia, and Norway (Andreas Mueller, Juri Kropotov, Jan Brunner, Anne Lise Høyland). Normative data of Asian origin were collected in South Korea (Seung Wan Khan). ADHD data were collected by

Methods of analysis
Automatic artifact correction
Automatic spike detection
Spectral analysis
Wavelet analysis, ERS/ERD
Blind source separation
ERP components

Comparing neuromarkers

Separating neuromarkers

Patient database
~1000 patients
ADHD, dyslexia and LD,
autism, schizophrenia, OCD,
depression, PTSD, TBI

HBI methodology

Reference database
~1000 heathy subjects
Ages from 7–90
Two resting-state
and six task conditions

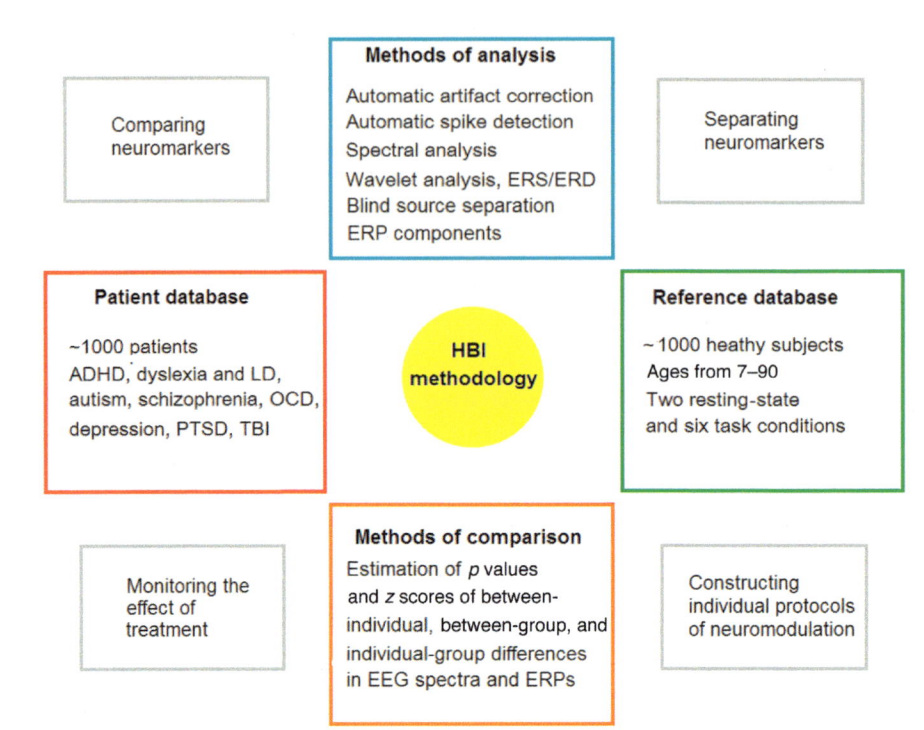

Monitoring the effect of treatment

Methods of comparison
Estimation of p values
and z scores of between-
individual, between-group, and
individual-group differences
in EEG spectra and ERPs

Constructing individual protocols of neuromodulation

FIGURE I.5 **The HBI database of behavioral, QEEG, and ERP data.** The database consists in two parts: raw EEG recordings of about 1000 healthy subjects and 1000 patients with different resting-state and task conditions, as well as processed data. Processed behavioral and electrophysiological parameters are obtained by the methods described in this book. The database includes the software that enables the user to compute for an individual subject or a group of subjects the parameters (such as reaction time, variance of RT, omission and commission errors in the tasks, EEG spectra and coherence, ERPs and latent components of ERPs) and to compare them either with the reference and patient data stored in the database or with the individual or averaged data obtained by the user.

the international team for Switzerland (Andreas Mueller), Norway (**Venke Arntsberg Grane**, Geir Ogrim), and Russia (Juri Kropotov). Data from a group of schizophrenic patients were collected in Russia (Marina Pronina). Data from a group of obsessive–compulsive disorder (OCD) patients were collected in Russia and England (Marina Pronina, Antonio Martins-Mourao). Data from the group of patients with depression were collected in Russia, Belgium, and England (Galina Poliakova, Vera Grin-Yatsenko, Colin Robertson). Data from the group of patients with posttraumatic stress disorder (PTSD) were collected in Australia (Mirjana Askovic, Russell Downham, Nerida Saunders, and Rustam Yumash).

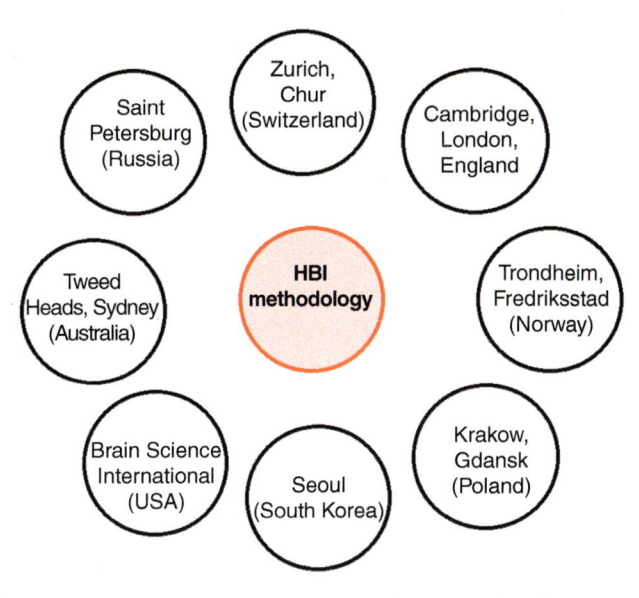

FIGURE I.6 Places that have participated in the collection of data for the patient database and that are using the HBI methodology in clinical practice.

It should be stressed here that the present book does not include data collected from patients of different diagnostic categories in the Brain Centers in Chur and Zurich (Switzerland) (Andreas Mueller), in the Department of Neuropsychology of Krakow University (Maria Pachalska), and in the Laboratory of Correction of Psychic Development (Leonid Chutko). Experience of working with these patients has been presented in a number of books: (1) Mueller, A., Candrian, G., Kropotov J.D. *ADHS: Neurodiagnostik in der Praxis*; (2) Pachalska, M., Kachmarek, B., Kropotov J.D. *Neuropsychologia kliniczna*; (3) Kropotov, J.D., *Current Diagnostics and Correction of Attention Deficit* published, respectively, in German, Polish, and Russian.

The beginning of the work dates back to the 1990s when **Risto Näätänen** from Helsinki University invited my lab to participate in a joint project on the intracranial correlates of mismatch negativity (MMN)—an automatic brain response to deviant stimuli elicited in the auditory oddball paradigm. In this project we recorded MMN-like responses from intracranial electrodes implanted for diagnostic and therapeutical purposes into patients with different neurological and psychiatric disorders. We found that scalp-recorded MMN is not a single entity but rather presents a sum of several sources with different locations and different functional properties.

Against the background of intracranial studies of local field potentials, we developed a cued auditory and visual GO/NOGO tasks and started the long journey of extracting functional neuromarkers from ERPs elicited in this paradigm. Intracranial correlates of hypothetical operations of comparison with working memory, response inhibition, and conflict detection were found. **Kimmo Alho** from Helsinki University and Drs **Anrei Sevastianov, Oleg Korzyukov,** and **Olga Dubrovskaya** from my laboratory were involved in these studies. A few years later, scalp-recorded ERPs and ERS/ERDs in the cued GO/NOGO task were used by **Olga Dubrovskaya** and **Elena Yakovenko** in differentiating groups of ADHD children from healthy controls. Research into the efficacy of neurofeedback and tDCS in children with ADHD and autistic spectrum disorder was initiated in studies by **Vera Grin-Yatsenko, Inna Nikishena, Nadezhda Kozushko,** and **Leonid Chutko**. These methods of neuromodulation are now effectively applied in the two diagnostic treatment centers of the Institute of the Human Brain headed by **Leonid Chutko** and **Nadezhda Kozushko.**

In 2003, I was invited by **Knut Hestad** from the Institute of Psychology of Norwegian University of Science and Technology (NTNU) in Trondheim to teach neuroscience courses and to organize an EEG lab. The HBI methodology inspired neuropsychologists **Jan Ferenc Brunner** from the Department of Physical Medicine and Rehabilitation of St. Olav's University Hospital in Trondheim and **Venke Arntsberg Grane** from the Neuropsychological Service, Helgeland Hospital, Mosjøen (Norway). Jan performed a **test–retest reliability** study and a study to associate the extracted ERP components in the cued GO/NOGO task with indices of **neuropsychological domains**. Venke participated in the multicenter study on ERPs in adult ADHD subjects. Recently, Ida Emilia Aasen became a member of the team and a PhD student at NTNU with the ultimate goal of studying the ERP correlates of **traumatic brain injury**. Psychologist **Geir Ogrim** from the Neuropsychiatric Unit, Østfold Hospital Trust, Fredrikstad (Norway) completed his PhD study on a large group of ADHD children using the QEEG/ERP methodology with the goal of **predicting the effect of medication** on the basis of neurophysiologically extracted parameters. Psychiatrist **Anne Lise Høyland** from NTNU is currently doing her PhD on a group of high-functioning autistic spectrum disorder children.

The European team have strong connections with Australia and the United States. In Australia **Mirjana Askovic** from the STARTTS Neurofeedback Clinic in Sidney collected HBI-oriented data from a large group of PTSD patients. **Nerida Saunders** and **Rustam Yumash** from Tweed Heads are using the HBI methodology for neurophysiological assessment of their clients and for constructing individually tailored neurofeedback and tDCS protocols. Jay Gunkelman from Brain Science International in the USA is using the HBI spectra database to assess the brainwaves of clients of the company. I have to stress here that meeting Jay in 2001 in Monterey, USA made a difference in my life: his enthusiasm inspired creation of the European chapter of ISNR and started research activities in Europe, in general, and in my laboratory in particular. My first book on QEEG, ERPs, and neurofeedback, published by Elsevier in 2009, was inspired by Jay.

Antonio Martins-Mourao from the Open University, London collected the QEEG/ERP data from a large sample of OCD patients. **Tony and Beverly Steffert** from Cambridge University accumulated the QEEG/ERP data on a large group of children with dyslexia and learning disabilities.

In 2008, I was invited by Professor **Maria Pachalska**, head of the Department of Neuropsychology in Andrzej Frycz Modrzewski Cracow University (Poland) to teach courses on QEEG, ERPs, and neurotherapy and to assess neurological and psychiatric patients. As a result of this cooperation a series of papers and two books on application of the HBI methodology for diagnosis and treatment of different categories of patients were published.

PERSONAL MOTIVES

My lab in Saint Petersburg is a **multidisciplinary** one and includes neurophysiologists, psychologists, physicists, mathematicians, and psychiatrists. Psychiatrist **Yury Poliakov** is a senior research fellow in my laboratory and head of the Psychiatric Clinic of our institute. For many years, I tried to persuade him to use the ERP methodology in clinical practice. At the beginning, he, much like any other psychiatrist, was very skeptical because his patients usually did not have any neurological abnormalities and application of conventional **clinical EEG seemed of little use** in such cases. Using his clinic, we started several projects on the application of cognitive ERP for ADHD, schizophrenia, OCD, and depression. The results were so convincing that now he does not see any patient without looking at the patient's ERPs. I believe that most psychiatrists are in a similar position to Yury Poliakov's 10 years ago. One of the goals of this book is to show the **advantages of the ERP methodology in psychiatric practice**.

PART 1

METHODS OF ASSESSING NEUROMARKERS

CHAPTER 1.1

Theory of Measurement

TRUE AND OBSERVED SCORES, ERRORS

Neuromarkers are measured quantitatively in each person selected from the general population. From our experience we know that any measurement is made with errors, not talking about mistakes. The question is how those errors can be estimated. Charles Spearman at the beginning of the 20th century was the first who figured out how errors can be estimated and who laid down foundations of the theory of measurement (Spearman, 1904).

In brief, classical test theory assumes that each person has a **true score**, T, that would be obtained if there were no errors in measurement. A person's true score is defined as the score over an infinite number of independent applications of the test. Because the infinite number is never reached, a person's true score is never obtained. So in any test only an **observed score** X is measured. It is assumed that observed score = true score plus some **error** E:

$$X = T + E$$

The error E is composed of two types of error: random E_r and systematic E_s.

$$E = E_r + E_s$$

The random error E_r is defined as a randomly varying effect on the same subject across different testing sessions. Sources of E_r are subject variations in mood, fatigue, stress, etc.; subject bias in motivation, cheating, etc.; variations of the environment in noise, temperature, lighting, seat comfort, etc.; test administration bias such as nonstandard instruction, scoring errors. When we are talking about brain activity a source of E_r could be **spontaneous variations in the local brain state** as well as errors of the device measurement. Spontaneous variation in the parameter can be decreased by **averaging**. In such case the signal-to-noise ratio increases as the square root of numbers of averaged trials. Errors in the device measurement, such as 1-μm error in most electroencephalogram (EEG) devices, is defined by manufacturers and cannot be decreased.

E_r affects the **repeatability** and **reproducibility** of measurements. **Repeatability** of measurements refers to variation in repeat measurements made on the same subject under **identical conditions** and over a short period (several hours or days) of time. Variability in measurements made under these conditions can then be ascribed only to measurement errors E.

Reproducibility refers to variation in measurements made on a subject under **changing conditions**. Changing conditions may be due to different measurement methods or instruments being used, measurements being made by different observers or raters, or measurements being made over a long (from several weeks to several months) period of time.

The systematic error E_s is that error which is consistent across test sessions such as poorly written or verbalized directions, inclusion of items unrelated to the content, etc. E_s affects **validity**, but not repeatability and reproducibility. In general, validity refers to how well a test measures what it is supposed to measure.

RELIABILITY

The classical test theory deals with relations between T (true score), X (observed score), and E (error) measured in a population of subjects. The **reliability** of observed test scores X, which is denoted as ρ^2_{XT}, is defined as the ratio of true score variance σ^2_T to observed score variance σ^2_X:

$$\rho^2_{XT} = \frac{\sigma^2_T}{\sigma^2_X};$$

which after simple transformations can be presented as:

$$\rho^2_{XT} = 1 - \frac{\sigma^2_E}{\sigma^2_X}$$

From this equation one can see that (1) the reliability of test scores is **lower than one**, (2) the reliability becomes higher as the **error variance** σ^2_E **becomes lower**, (3) the heterogeneity of subjects in the population, measured by σ^2_X, affects the value of reliability: **the higher the heterogeneity, the higher the reliability**. It can be mathematically shown that the square root of reliability (called the reliability index or ρ_{XT}) equals the **correlation between observed and true scores in the population of persons**.

In practice the correlation between the observed X scores measured at two different times t_1 and t_2 on a selected population of subjects is used for assessing the reliability of the measured X value. This correlation measures **test–retest reliability**. The other measure of reliability is half-split reliability in which the whole sample is randomly divided in two equal sets and the Pearson coefficient is computed between these sets.

VALIDITY

While reliability of the test is necessary for appropriate measurement, alone it is not sufficient. For any measurement to be reliable, it also needs to be valid. As was mentioned previously, validity is the extent to which a measurement is well founded and corresponds accurately to the real world. The word "valid" is derived from the Latin *validus*, meaning strong.

DISTRIBUTION ACROSS POPULATION

Any parameter when measured in a given population varies from one person to another. EEG spectra (both absolute and relative) are not distributed normally. In all known databases EEG spectra are subjected to a normalization procedure by taking the **log** of corresponding values. In contrast to EEG spectra, the amplitudes of event-related potentials are **normally** distributed and are not subjected to any normalization procedure.

PERCENTILES AND z SCORES

When describing the results of measurement within the population **percentiles** and **z scores** are used (Fig. 1.1.1). Converting measured scores to percentiles has the important advantage that percentiles directly express the rarity of an individual's score. In addition, percentiles are easily comprehended. In statistics, a **percentile** is the value of a variable below which a certain percent of observations fall. For example, the 20th percentile is the value (or score) below which 20% of the observations may be found (Fig. 1.2.1a).

z **Scores** are an alternative method of expressing scores. For the normally distributed parameter X with mean value M and standard deviation σ the z score is defined as:

$$z = \frac{(X-M)}{\sigma}$$

The relationship between z scores and percentiles is presented in Fig. 1.1.1a.

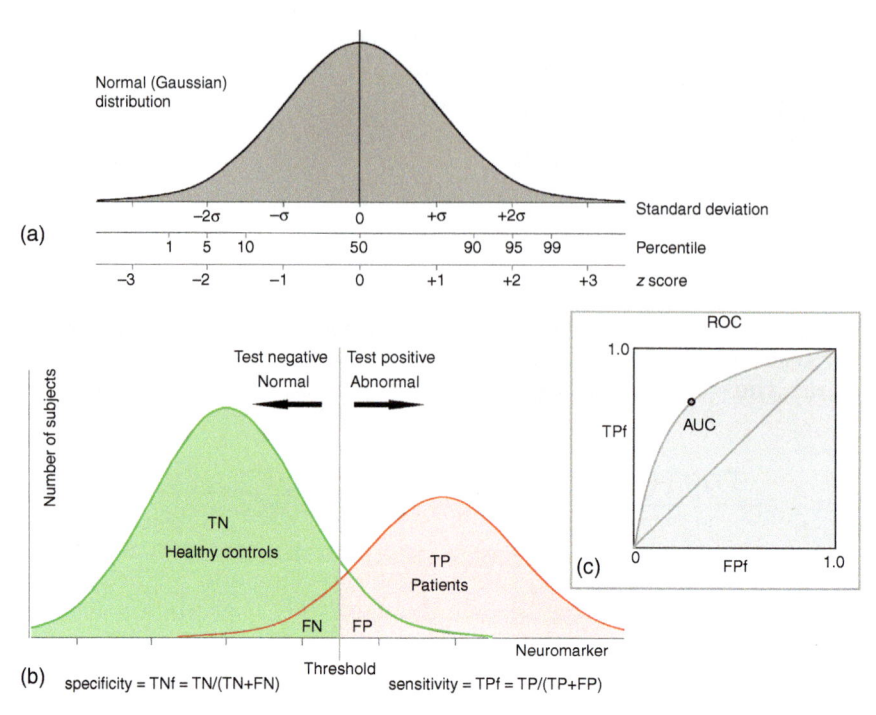

FIGURE 1.1.1 **Specificity–sensitivity measures for discriminating patients from healthy controls.** (a) Normal (Gaussian) distribution in terms of standard deviations, percentiles, and z scores. (b) For a given neuromarker (x-axis), a distribution over a group of healthy subjects *(green)* and a group of patients with a certain categorical dysfunction *(red)* are shown. The observer defines a binary decision threshold *(vertical line)* so that the cases to the right of the threshold (toward abnormality) are diagnosed as abnormal, and the cases to the left are diagnosed as normal. The two distributions and the threshold define cases of: (1) true positive *(TP)* when the patient is correctly positively diagnosed, (2) true negative *(TN)* when the healthy subject is correctly negatively diagnosed, (3) false positive *(FP)* when the healthy subject is wrongly positively diagnosed, and (4) false negative *(FN)* when the patient is wrongly negatively diagnosed. These values are used to compute sensitivity and specificity. (c) The ROC curve is defined as the sensitivity (TP_f) versus $FP_f = (1 - \text{specificity})$ when the threshold changes across the x-axis. The *diagonal line* shows an ROC curve for the case when the two distributions practically coincide with each other. The area under the curve *(AUC)* is used as a metric for the overall performance of diagnostic procedure. AUC values change from 0.5 (pure guessing) to 1.0—a perfect diagnostic procedure.

SENSITIVITY AND SPECIFICITY

Suppose a test for a neuromarker screens people for a disease. Each person taking the test either has or does not have the disease. The test outcome can be positive (predicting that the person has the disease) or negative (predicting that the person does not have the disease). The test results for each subject may or may not match the subject's actual status.

Sensitivity of neuromarker is defined as the probability of the positive test given that the subject is ill. **Specificity** is the probability of the negative test given that the subject is healthy (Fig. 1.1.1b). For any test, there is a **tradeoff** between sensitivity and specificity. This tradeoff can be represented graphically as a **receiver operating characteristic (ROC) curve** (Fig. 1.1.1c).

The choice of the **gold standard** or reference test is an essential component for defining specificity and sensitivity. This is the standard against which the test will be measured. The currently accepted gold standard is the **best estimate diagnosis**. It is reached by agreement among a number of experts relying on multiple sources of information.

EFFECT SIZE

Suppose we have two groups recruited from healthy subjects and from a group of patients with a given psychiatric disorder. Intuitively, a measure of difference between the two groups could be estimated as the difference between means in terms of the standard deviation:

$$\text{ES} = \frac{M_1 - M_2}{\sigma};$$

where M_1 and M_2 are mean values for the first and second group, and σ is the standard deviation of either group.

This is what Cohen (1988) defined as the **effect size (ES)** of a given measure between two independent groups.

In practice, the pooled standard deviation σ_{pooled} is commonly used:

$$\sigma_{\text{pooled}} = \sqrt{\frac{\sigma_1^2 + \sigma_2^2}{2}};$$

where σ_1^2 and σ_2^2 are squared standard deviations for the first and second groups.

It should be stressed here that the ES can also be derived from the value of the t test of the differences between the two groups. Following Cohen, in this book we define ESs as small for $d < 0.2$, medium for $0.2 < d < 0.5$, and large for $d > 0.8$.

One can see that theoretically (when the sample size is very large) the ES between the two groups is determined only by mean differences and standard deviations and does not depend on the sample size. However, in practice the ES is measured by relative small sample sizes such as 20–30. In such cases the ES is computed within the confidence interval defined by the standard error (SE) of the ES, which is larger when the sample size is smaller.

Especially in metaanalysis, where the purpose is to combine multiple ESs, the **SE of ES** is of critical importance. The SE of ES is used to weight ESs when combining studies, so that large studies are considered more important than small studies in the analysis.

REQUIREMENTS FOR INTRODUCING A NEUROMARKER INTO CLINICAL PRACTICE

A neuromarker is designed to **aid** psychiatrists in diagnostic as well in prognostic procedure. To be of real help the neuromarker must be a (1) **reliable**, (2) **sensitive**, and (3) **specific index** of brain functioning and/or dysfunctioning. Due to the heterogeneous nature of psychiatric disorders, there is a minor chance that any neuromarker is able to identify all patients classified into a certain category according to the gold standard. Consequently, it is important to define the **clinical characteristics of the whole group** of patients that are identifiable by a particular test. On the other hand, testing several neuromarkers simultaneously is necessary. This set of neuromarkers defines a **profile of abnormality** of a certain diagnostic category in the corresponding multidimensional space.

The testing procedure for a neuromarker requires **standardization.** Standardization includes unequivocally defining laboratory settings, task parameters, instructions, etc. as well as the methodology for quantifying the neuromarker.

In order to be used in a clinic for a single patient, larger **normative databases** must be collected. The larger the number of subjects in the normative database the more precisely can the parameters of "normal" distribution be estimated. Having patient databases collected for different diagnostic categories would allow developing optimal **discrimination procedures**. Development of such large databases is challenging and requires collaboration among **multiple research groups**.

CHAPTER

1.2

Psychometrics and Neuropsychological Assessment

PSYCHOLOGICAL MODELS

Behavior is the subject of psychology. Behavioral parameters are measured by special instruments and procedures unified under the general concept of **psychometrics**. The parameters include behavioral indexes of human abilities, skills, and knowledge. The parameters also include qualitative descriptions of specific human social behavior known as personality traits. As a science, psychometrics deals with the construction and validation of measurement instruments such as questionnaires, tests, and personality assessments.

The first psychometric instruments were designed to measure intelligence. Psychometrics has been applied in education to measure abilities in reading, writing, and mathematics. A well-known instrument of psychometrics is the **Minnesota Multiphasis Personality Inventory** (**MMPI**). The other broadly accepted instrument is the **Five-Factor Model (Big 5)** that has been used in assessment of personality traits.

Measures are usually built upon a **psychological model**. Psychological models evolve with time. At the beginning of the 20th century some people thought that models should look like theories from physics with postulates and equations. In contrast, some people suggested that internal representations or cognitive maps could be an explanation basis of basic functions, such as learning and memory. An influential model of brain–behavior interrelations was suggested by Alexander R. Luria after the Second World War. In the 1960s computer metaphors emerged for explanation of psychological experimental findings. In general, the psychological model defines specific procedures of behavioral measurement. Then, the model is tested to determine whether or not it explains or predicts behavior as intended. The collected data are statistically assessed so that the model may be tested and further adjusted.

NEUROPSYCHOLOGICAL TESTING

The relationships between psychological processes, on the one hand, and brain function and structure, on the other hand, are studied by **neuropsychology**. Neuropsychology is seen as a clinical and experimental field of both psychology and neurology that aims to measure **behaviors directly related to brain lesions** or other neurological problems.

There are many neuropsychological tests invented to measure sensory, affective, memory, and cognitive (executive) dysfunctions of the damaged brain. Although computerized variants of the tasks have been recently developed still many neuropsychological settings use the paper-and-pencil approach. The measures are **numbers** of correctly recalled objects, numbers of omission and commission errors, **time** needed to accomplish the task, speed of the task, reaction time (RT).

SUPERVISORY ATTENTIONAL SYSTEM MODEL

In this book we will be using the **supervisory attentional system (SAS) model** suggested by Donald Norman and Tim Shallice (1986). The model suggests that information processing in the frontal lobes may be modulated by attention (or cognitive control) in order to handle situations involving planning, novel sequences of action, and the need to overcome strong habitual responses in decision making.

OPERATIONS OF ATTENTIONAL CONTROL

Inspired by the SAS model, the the Rotman–Baycrest Battery for Investigating Attention (ROBBIA) approach suggested by Donald T. Stuss and coworkers from the Rotman Research Institute in 1995 argues for fractionating the attentional system into three anatomically and functionally independent processes: energization, monitoring, and task setting (Stuss, Shallice, Alexander, & Picton, 1995).

Energization refers to a process that **facilitates** and boosts other SAS processes, especially those necessary for making decisions through initiation and maintenance of optimal response patterns. The **monitoring** process is thought to provide **quality control** of behavior by checking task performance and outcome over time, which is a prerequisite for appropriate adjustment of behavior. The **task-setting** process refers to the formation of a criterion of how to respond to a defined target and to organize the schemata to complete a specific task. Although these processes are described as being independent, they act in concert to control lower order processes and optimize behavior.

These processes can be reflected in specific neuropsychological parameters. The effect of energization is evident in any speeded behavior, as slowing in **RT** tasks is the core behavioral parameter of a reduced level of energization. The effect of energization can also be observed in verbal fluency performance tasks. The task-setting process can be assessed behaviorally by measures reflecting poor criterion setting, such as increases in **false positive (commission) errors.** Monitoring is reflected in all types of errors, including **false negatives (omission) errors**. The **detectability parameter (d')** is based on both commission and omission errors and indexes how well a subject is able to respond differentially to targets and nontargets.

DUAL MECHANISMS OF COGNITIVE CONTROL

To emphasize the dynamics of cognitive control processes Todd Braver from Washington University in St. Louis in 2012 described a model of dual mechanisms (Braver, 2012). In this framework, **proactive control processes** involve early selection, in which goal-relevant information is actively maintained in a sustained manner, prior to the occurrence of cognitively demanding events. **Reactive control** processes, on the other hand, are late correction mechanisms mobilized only as needed, in a just-in-time manner, such as after a high interference event is detected.

GENERAL FACTOR

For describing cognitive abilities a general factor (g **factor**) has been introduced. It is a variable that summarizes positive correlations among different cognitive tasks, reflecting the fact that an individual's performance at one type of cognitive task tends to be comparable with his or her performance at other kinds of cognitive tasks. In Russia there is a saying that a talented man is talented in everything. The g factor typically accounts for 40–50% of between-individual variance in **IQ** test performance. The terms IQ, general intelligence, general cognitive ability, or simply intelligence are often used to refer to the g factor.

The idea of the g factor was first formulated by the English psychologist Charles Spearman at the beginning of the 20th century. He observed that children's performance ratings across seemingly unrelated school subjects were positively correlated and reasoned that these correlations reflected the influence of an underlying general mental ability that eventually defined performance on all kinds of mental tests. Spearman suggested that all mental performance could be described in terms of a single general ability g factor and a large number of specific ability factors. Behavioral genetic research has shown that the construct of *g* is highly heritable. The *g* factor correlates with a number of behavioral parameters including RTs.

REACTION TIME

RT is associated with the ability of the subject to perform fast. Using a stimulus–response design, instances of intermittent long RTs were found to be associated with **decreased** metabolic activity in the frontal cortex prior to stimulus presentation as well as with **increased** activity in the posterior cingulate, precuneus, and middle temporal gyrus.

REACTION TIME VARIABILITY

Response time distributions are not Gaussian distributions. They rise rapidly at short RT and have a long positive tail at long RT. RT distributions can be approximated by **ex-Gaussian and log-normal** distributions (Fig. 1.2.1). The ex-Gaussian distribution is a convolution of a Gaussian and an exponential distribution. In this case it is described by a **mean (μ)** and **standard deviation (SD) (σ)** for the Gaussian part and an **exponential**

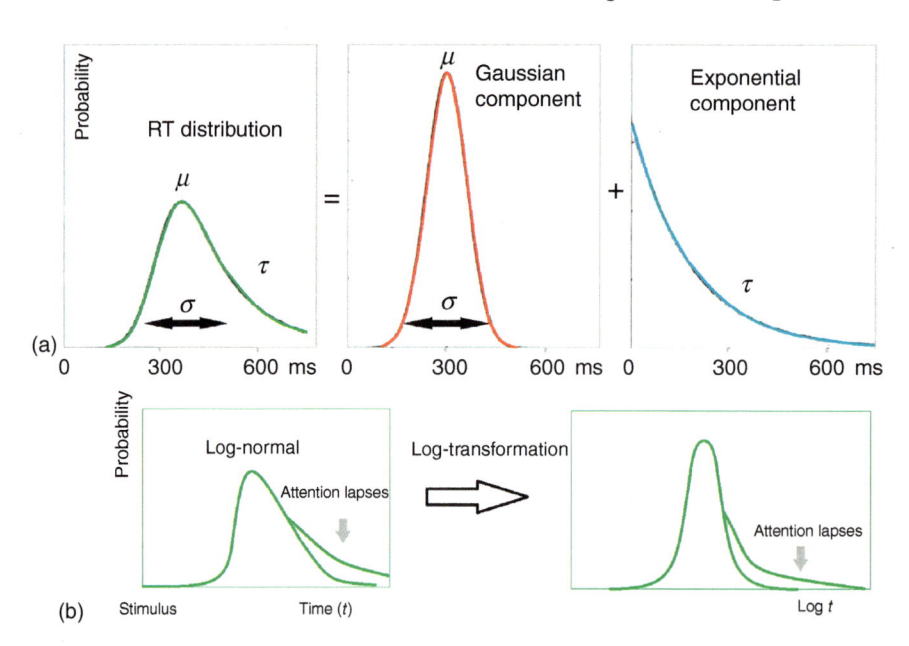

FIGURE 1.2.1 **Two models of RT distributions.** (a) In the ex-Gaussian approach μ and σ describe the mean and SD of the Gaussian component, and τ reflects both the mean and SD of the exponential component. (b)The RT distribution can also be approximated by a log-normal distribution. Attention lapses (eg, in an ADHD subject) add a right "wing" to the distribution at large latencies.

parameter (τ) for the exponential component. The logarithm of RT can be modeled by a Gaussian distribution.

In spite of non-Gaussian distribution, in practice, mean RT, **reaction time variability (RTV)**, and coefficient of variability (CV) are usually estimated as:

$$\text{Mean RT} = \frac{1}{N}\sum_{I=1}^{N} RT_i$$

$$RTV = \sqrt{\frac{1}{(N-1)}\sum_{i=1}^{N}(RT_i - \text{Mean RT})^2}$$

where i indexes the trial in a task, and N is the number of trials.

The majority of studies of RT variability in attention disorders have used SD to quantify RT variability. Although mean RT and RT variability are assumed to reflect different measures of cognitive control they are highly correlated with each other. To compensate for this correlation some studies calculate the **coefficient of variation (CV)**:

$$CV = RTV/\text{Mean RT}$$

There is consistent evidence that RTV distinguishes individuals with Attention Deficit Hyperactivity Disorder (ADHD) from typically developing populations with medium to large-effect sizes for children/adolescents and small to medium-effect sizes for adults. Most studies of RTV implicitly assume that heightened RTV reflects occasional **lapses in attention**. An example of an excessive RTV in a 14-year-old ADHD patient is shown in Fig. 1.2.2. In this patient the number of omission and commission errors

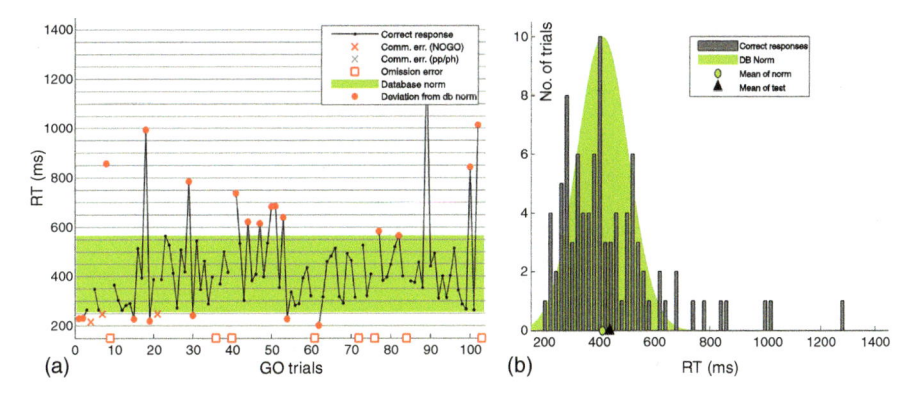

FIGURE 1.2.2 **Attention lapses in an ADHD patient.** (a) Dynamics of RT over consecutive GO trials in the cued GO/NOGO task; confidence intervals of RT variation in healthy populations are depicted in green; (b) RT distribution of the patient (black) in comparison to the distribution averaged across the healthy subjects (green).

and RT do not significantly differ from the mean reference values. Only the RTV is larger in comparison to healthy controls at p < 0.02.

RTV has been suggested as a potentially important index of the stability/instability of a subject's nervous system. However, it **lacks specificity** as increased RTV characterizes populations ranging from ADHD, schizophrenia to dementias and traumatic brain injury.

fMRI-related interpretation of lapses of attention suggests that they are caused by intrusions of **task-negative brain network** activity during task performance. The network is a part of the **default mode network**, a network of interconnected brain regions activated during rest states such as a resting condition with the eyes closed. The default mode network is considered a task-negative network because it must be suppressed concomitant with activation in the **task-positive network** (ie, the neural network required to perform the cognitive task at hand). Brain activation in the default mode network is negatively correlated with activation in regions of the task-positive network and positively correlated with RT fluctuations. In other words, an inability to suppress the default mode network leads to inconsistent performance and increased RT.

The inconsistency in RT across the age obeys the U-law with higher inconsistency throughout childhood and advanced age (Fig. 1.2.3). Differences

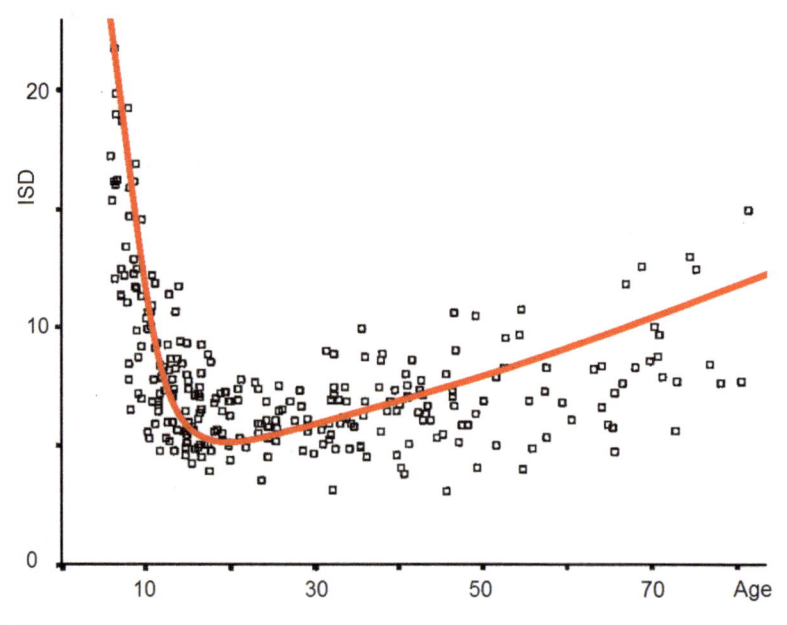

FIGURE 1.2.3 **Scatterplot of inconsistency (intraindividual standard deviation—ISD of choice RT) across the life span.** *Adapted with permission from Gordon, Cooper, Rennie, Hermens, and Williams (2005).*

in inconsistency are independent of practice, fatigue, and age-related differences in mean level of performance.

CONTINUOUS PERFORMANCE TASKS

To obtain reliable parameters of behavior, **continuous performance tasks (CPTs)** are usually used where the subject performs a constant-difficulty task for minutes or tens of minutes without interruptions. The first version of the CPT was developed in 1956 by Haldor Rosvold and coworkers and demonstrated that patients with brain damage had difficulties in task performance compared with healthy controls. The most commonly used CPTs are the **Test of Variables of Attention (TOVA)**, the **Integrated Visual and Auditory CPT (IVA)**, and Conners' CPT (Rosvold, Mirsky, Sarason, Bransome, & Beck, 1956). These tests are often used for helping to support or rule out a diagnosis of ADHD. In addition there are some CPTs, such as the QbTest and Quotient, that combine attention and impulsivity measures with motion tracking analysis.

INFRASLOW FLUCTUATIONS IN PERFORMANCE

At the beginning of research of CPTs it was suggested that the trials of a task were independent and that averaging over trials gave a robust estimation of behavior. However, the following research showed that consecutive trials in continuous performance tasks were not purely independent but weakly autocorrelated so that similar behavioral parameters appeared in clusters. Estimated frequency from autocorrelation histograms for hit rates, RTs, and other behavioral parameters lies between **0.01 and 0.1 Hz**. This frequency band is often labeled as **infralow** and oscillations in performance are named **infraslow fluctuations (ISF)**. These behavioral parameters correlate with corresponding ISFs at the electrophysiological level as measured by broad band EEG (see chapter: Spontaneous Electroencephalogram) and at the metabolic level as measured by fMRI (see chapter: Functional Magnetic Resonance Imaging).

BIG 5 MODEL

We are all different in our ability to perceive, feel, think, and act. This diversity is the foundation of richness of our culture, science, and technology. In modern psychology the diversity of the human population is reflected in factor models of personality traits. The factor model presumes that the whole population can be characterized by a few general factors. The widely accepted model is the **Five-Factor Model (Big 5)**.

In the neuroscience of human behavior, interindividual differences are often wrongly treated as a source of "noise" and therefore discarded through averaging data from a group of participants. Note that the interindividual variability lies in foundation of the methods of group independent and latent components of spontaneous and evoked EEG [see chapters: Spontaneous Electroencephalogram; Event-Related Potentials (ERPs)]. Moreover, university students of industrialized Western countries are typically the participants in many psychology and neuroscience studies. Despite the narrow selection of human diversity in such experiments, it is wrongly assumed that the conclusions drawn from a small sample generalize to the entire population. To overcome this problem a wider selection of participants in large groups of healthy subjects and patients is recommended.

CHAPTER 1.3

Functional Magnetic Resonance Imaging

TALAIRACH ATLAS

For functional magnetic resonance imaging (fMRI) the measured parameters are **maps**—brain activation patterns superimposed on the 3D images of the brain. To measure the parameters of the maps one needs an atlas. Talairach and Tournoux (1988) designed such a stereotaxic atlas of the human brain. This invention has been enormously influential in functional imaging.

Talairach and Tournoux introduced two keystone ideas: (1) a coordinate system for identification of a particular brain location relative to anatomical landmarks; (2) a spatial transformation to match a given brain to a standard brain, with anatomical and cytoarchitectonic labels.

The **coordinate system** of Talairach and Tournoux postulates that the brain should be aligned according to the two relatively invariant small subcortical structures: **anterior commissure (AC)** and **posterior commissure (PC)**. A coordinate system with the AC as the origin can be defined as follows. The y-axis represents the AC–PC line. The z-axis is a vertical line that passes through the interhemispheric fissure and the AC. The x-axis is a horizontal line at right angles to the y- and z-axes that passes through the AC. Any point in the brain can be defined in the Talairach coordinate system.

The atlas describes the Talairach proportional grid normalization to match a given brain to the atlas—a series of labeled diagrams of transverse (axial), sagittal, and coronal brain slices from the postmortem brain of a 60-year-old French woman—the **Talairach brain**. This is a simple set of scalings that can be used to transform one brain to give a rough match to the atlas in overall brain size and shape.

MONTREAL NEUROLOGICAL INSTITUTE ATLAS

In order to define a brain that is more representative of the population the **Montreal Neurological Institute (MNI)** created a new template in a two-stage procedure. First, they took **250 normal MRI scans**, and manually defined various landmarks to identify the AC–PC line. Each brain was scaled to match the landmarks to equivalent positions on the Talairach atlas. They then took an extra 55 images, and registered them to the 250 atlas using an automatic linear registration method thus creating the **MNI305 atlas**. The current standard MNI template is the **ICBM152**, which is the average of 152 normal MRI scans that have been matched to the MNI305 using a nine-parameter affine transform. The International Consortium for Brain Mapping adopted this as the standard template.

PHYSICAL BASIS OF MAGNETIC RESONANCE IMAGING

The MRI method is based on a physical phenomenon known as **magnetic resonance**. The point is that many organic elements like hydrogen atoms are elementary magnets. In their common state, these tiny magnets are oriented randomly. However, if an external magnetic field is applied, all the magnets will be arranged along that field just like a compass needle is oriented along the magnetic field of the Earth. Then, if external magnetic waves of the **radiofrequency band** pass through an area magnetized in the earlier described way, these waves make the elementary magnets rotate in a certain direction. When the radio waves are turned off, the atoms return to their original state and generate output waves that are registered by magnetic detectors on a tomograph placed around the patient's head.

Consequently, the measured parameter in MRI is **amplitude of radio waves** registered by magnetic detectors. Density and magnetic features of elementary magnets define the power of the signal. To restore a 3D density distribution pattern for these magnets, special mathematical image reconstruction methods are used. Many neurologists and neurosurgeons, when studying MRI images, do not even suspect that the source data looks quite different from those shown on the MRI images. To obtain brain structure images, source data are processed by complex **mathematical algorithms** requiring computational facilities of large (even from today's point of view) capacity.

For a clinical practitioner and researcher the measured parameter is a structural image of the brain in a selected plane. Spatial resolution reached by means of MRI is **really astonishing**. On MRI scans one can see separate convolutions, corpus callosum, caudate nucleus, and even smaller structures like mammillary bodies or thalamic nuclei.

FUNCTIONAL MAGNETIC RESONANCE IMAGING

An advanced modification of MRI, called functional MRI (fMRI), has been developed for studies of **vascular/metabolic reactions** of the brain tissue in response to different tasks. The primary contrast mechanism exploited for fMRI is the **blood oxygenation level dependent (BOLD) contrast**. The discovery and exploitation of the BOLD effect has revolutionized many areas of neuroscience research. Technically speaking, the BOLD effect should be called the "blood deoxygenation level–dependent" effect, because it is the deoxyhemoglobin that alters the blood's magnetic susceptibility. Physiologically speaking BOLD contrast reflects a complex interaction between **neuronal activity, blood flow, blood volume**, and **hemoglobin oxygenation**.

Deoxyhemoglobin is a **paramagnetic** molecule whereas **oxyhemoglobin** is **diamagnetic**. The presence of deoxyhemoglobin in a blood vessel causes a susceptibility difference between the vessel and its surrounding tissue. Such susceptibility differences cause dephasing of the MR proton signal so that the presence of deoxyhemoglobin in the blood vessels causes a darkening of the image in those voxels containing vessels. Since oxyhemoglobin is diamagnetic and does not produce the same dephasing, changes in oxygenation of the blood can be observed as the signal changes.

Task-specific BOLD signal changes are not directly quantifiable in physiological units, but rather are expressed as a **percentage signal change** or as a statistical significance level based on a particular statistical model. Absolute or resting function cannot be easily assessed, and for clinical studies it may be difficult to know whether any observed abnormalities are due to baseline or task-specific effects. A typical BOLD response consists of a **0.5–5% change in regional image intensity**, which develops over **3–8 s** following task initiation. The peak latency of several seconds represents a major limiting factor in the temporal resolution of fMRI.

BOLD RESPONSE

There is growing evidence that prior to the increase in regional cerebral blood flow, there is a more localized decrease in hemoglobin oxygenation, presumably due to a more rapid increase in oxygen utilization than in blood flow. In healthy human subjects, the increase in cerebral blood flow is dominant over the other changes, with the near paradoxical consequence that increased neural activity leads to an increase in the BOLD signal. The BOLD response to a short task-related stimulus may show three distinct phases (Fig. 1.3.2c): (1) a small negative initial response that attains its minimum value at **2–3 s** poststimulus followed by (2) the main

BOLD response that is conventionally used in fMRI experiments, with a time to peak of about **5 s** and a response width of roughly **4 s**, followed by (3) a poststimulus undershoot that may take **60 s** to return to baseline.

BOLD INFRALOW FLUCTUATIONS

In the resting state, BOLD signals exhibit spontaneous fluctuations with **infralow** frequencies between **~0.01** and **0.15 Hz**. They have a relatively featureless $1/f$-type spectrum and are tightly coupled to neuronal activity. The BOLD oscillations in different brain regions correlate with each other thus forming so-called resting state functional connectivity mapping—a popular approach in fMRI. These spatially distinct correlations in the timecourse of spontaneous hemodynamic fluctuations are interpreted as the results of **large-scale functional and spatial network organization**.

0.1-Hz HEMODYNAMIC OSCILLATIONS

Distinct from $1/f$-type fluctuations, slow near-sinusoidal hemodynamic oscillations at a frequency of about 0.1 Hz are found in the human brain. They can be also observed by near-infrared spectroscopy. These oscillations were confused with sinusoidal, around 0.1 Hz, oscillations in **systemic blood pressure** discovered by a German physiologist Siegmund Mayer and his coworkers at the end of the 19th century and named after him as **Mayer waves**. The possible influence of Mayer waves on local cortical hemodynamics has not been resolved until recently. Mayer waves might be associated with 0.1 oscillations in **heart rate variability**.

Rayshubskiy et al. (2014) from Columbia University reported the direct observation of large-amplitude, **local sinusoidal ~0.1-Hz hemodynamic oscillations** in the cortex of an awake human undergoing surgical operation. They applied intraoperative multispectral optical intrinsic signal imaging to show that these oscillations were spatially localized to distinct regions of the cortex, exhibited wave-like propagation, and involved oscillations in the diameter of specific **pial arterioles** (Fig. 1.3.1), suggesting that the effect was not the result of systemic blood pressure oscillations. These oscillations were reported to be similar to fMRI oscillations detected around the same region.

These new studies remind me of my own studies back in the 1970s. In those years I recorded the local concentration of **extracellular oxygen by polarographic method** in neurological patients with gold electrodes

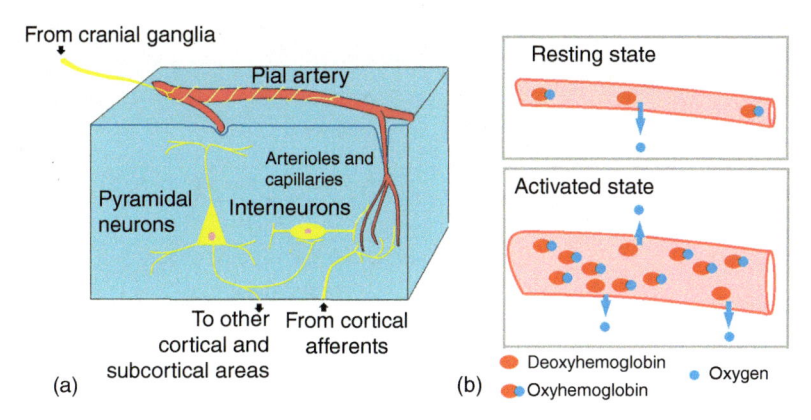

FIGURE 1.3.1 **Physiological basis of fMRI (model).** (a) Pial artery is shown at the surface of the cerebral cortex. It gives rise to a penetrating arteriole that enters the brain parenchyma and divides to form a capillary network. Peripheral nerve fibers of cranial ganglia innervate cerebral arteries through the smooth muscle cell coat of the arterial wall. Cortical neurons are in close contact with small arterioles and capillaries which are in turn innervated by cortical afferents from ascending activation system of the brain stem. (b) Brain local blood flow in a resting state is compared with that of an activated state. In the activated state there is a slight increase in oxygen extraction from the blood which is accompanied by a much larger increase in cerebral blood flow bringing more oxyhemoglobin.

implanted for diagnostics and therapy. I also recorded **impulse activity of neurons** from the same electrodes in the resting state and in response to psychological tasks. The polarographic method is based on the fact that the voltage of −0.63 V applied to a polarizable electrode (such as a gold electrode) inserted into a brain tissue creates an electric current that is proportional to the concentration of oxygen in the surrounding tissue. In our studies we showed that the level of oxygen in the local brain tissue does not remain constant but oscillates at **infralow frequencies** (Fig. 1.3.2a).

These oxygen oscillations varied in frequency from **0.5** to **0.01 Hz** and showed a rather local distribution. In some areas they were accompanied by up and down states in neuronal activity so that shifts from a lower to higher level of activations induced clear responses in the infraslow band of oxygen. Similarly, the evoked responses were observed in response to the activation tasks such as repetitions of arithmetic calculations or motor movements (Fig. 1.3.2b). The most thrilling feature of these oscillations was that at the cortical level they were associated with slow oscillations of electrical potentials measured from the skin of the head. Similar decisecond oscillations were recorded in Moscow by Nina Alexandrovna Aladjhalova (Aladzhalova & Kol'tsova, 1971).

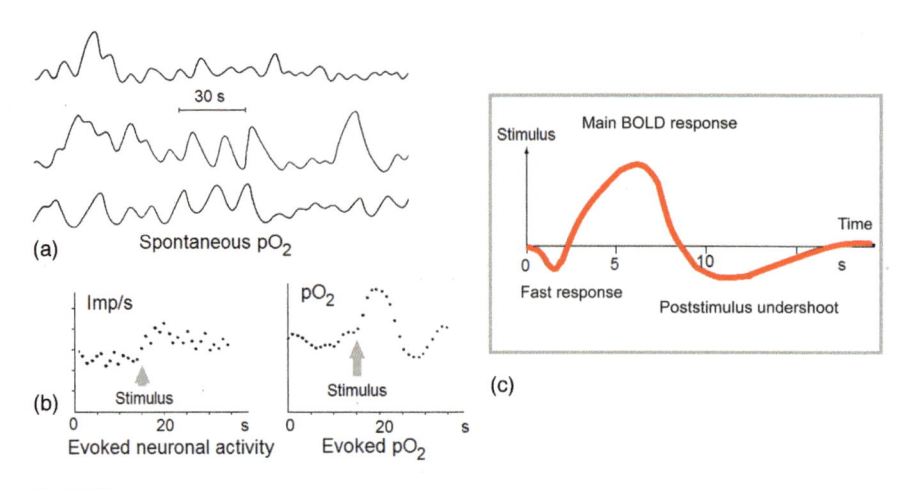

FIGURE 1.3.2 **Physiology of fMRI infralow oscillations.** (a) Spontaneous oscillations of local oxygen in basal ganglia and thalamic nuclei measured by polarographic method. (b) Poststimulus time histogram of multiunit impulse activity in the ventral thalamus and simultaneously recorded response in local oxygen averaged over 10 trials of an arithmetic task of neurons. (c) Scheme of the BOLD response to an activation event. *Part (a, b): adapted from Gretchin and Kropotov (1979).*

PROCESSING STEPS IN FUNCTIONAL IMAGING

Any fMRI study involves the collection of one or more functional scans for each subject, which show signal changes in regions where neuronal activity associated with the function under study has been changed in relation to **the baseline.** To compensate for **subject movement**, it is usual to realign the functional images to one of the images in the series. The next step is **transformation** of the images for each subject to match a template brain. Statistical analysis may involve any of a number of methods, most of which result in an **index at each voxel** that reflects the response of this voxel to the experimental manipulation. This activation 3D map can be **thresholded** to identify activated brain regions. If we have scanned several subjects, then we can do intersubject averaging to find regions that respond on average across subjects. The final step of the analysis is labeling the activated areas in terms of stereotaxic coordinates, macroanatomy, microanatomy, or function.

ACTIVATION MAPS OF fMRI

In spite of the facts that fMRI measures are indirect, reflecting incompletely understood mechanisms of coupling blood flow and neural activity, and are limited by a spatial resolution of a few millimeters, they provide a powerful tool for measuring functional response properties of voxels within the human cerebral cortex. In the **sensory modalities** the method allows direct mapping of the complexity of receptive field properties from primary to secondary visual, auditory, and somatosensory cortical areas, suggesting serial hierarchical processing in the human sensory cortex.

Neuroimaging studies in **cognitive tasks** revealed simultaneous activation in multiple regions in the parietal, frontal, temporal, and cingulate cortices, suggesting a distributed nature of brain activation in cognitive control. However, the main limitation of fMRI is poor temporal resolution which does not allow studying separate but temporally overlapping subcomponents of cognitive control such as context updating, action inhibition, and conflict detection.

MODEL-DEPENDENT CORRELATIONAL METHODS

A straightforward way to examine the functional connections of a particular brain region (called **seed**) is to correlate the resting state time series of this region against the time series of all other regions. The seed can be a priori defined region on the basis of previous research or it can be selected from a task dependent activation map acquired in a separate fMRI experiment. For example, a seed in the left primary motor cortex can be selected from the task in which the subject is instructed to move their right hand. However, this approach is limited to the functional connections of the selected region, making it difficult to examine the whole-brain connectivity pattern.

MODEL-FREE CORRELATIONAL METHODS

Several model-free methods have been suggested and successfully applied to resting-state time series, including principal component analysis, **independent component analysis (ICA)**, and hierarchical and normalized cut clustering. ICA-based methods have been reported to show a high level of consistency. They are designed for separating spatial sources of resting state signals that are maximally independent from each other. Interpretation of the obtained independent component is more difficult to make in comparison to the straightforward seed methods. However, seed-based and ICA-based methods tend to show strong similarity in their results.

TASK-NEGATIVE AND TASK-POSITIVE NETWORKS

When a subject is not focused on the outside world in a **resting state** or a state of introspection, some brain areas are highly active. These brain areas are characterized coherent by infralow oscillations of the BOLD signal.

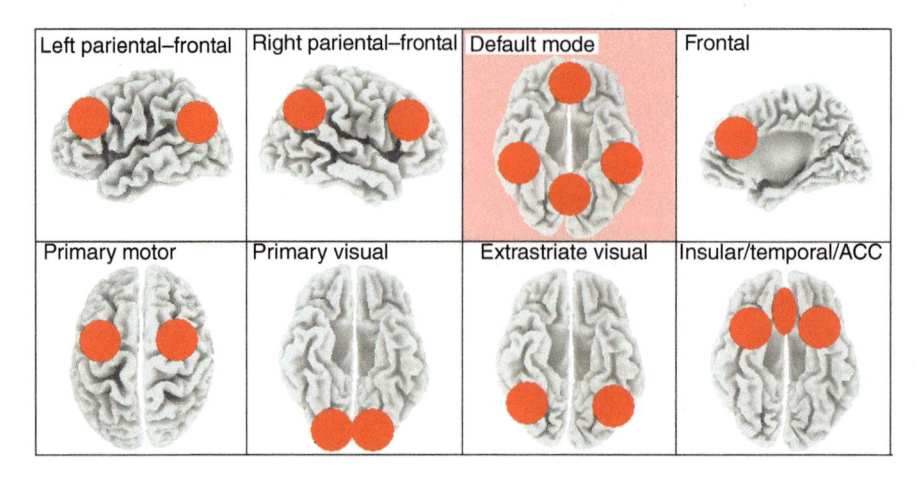

FIGURE 1.3.3 fMRI networks. Coherently occillating brain areas during resting state (default mode) and during sensory, motor, emotional, and cognitive tasks are schematically depicted. The nodes of the networks are depicted as *red circles*.

These coherently oscillating brain areas are labeled as the **default mode network (DMN)** (Fig. 1.3.3). The name was coined by neurologist Marcus E. Raichle and colleagues from Washington University School of Medicine in 2001. The DMN includes part of the medial temporal lobe, part of the medial prefrontal cortex, and the posterior cingulate cortex.

During a certain task (such as GO/NOGO) the DMN is deactivated and another network, the **task-positive network** is activated. Several task-positive networkshave been discovered (Fig. 1.3.3).

FUNCTIONAL CONNECTIVITY AND DIFFUSE TENSOR IMAGING

Functional connectivity MRI (fcMRI) and diffuse tensor imaging (DTI) have recently emerged as promising tools for mapping the connectivity of the human brain, each with distinct strengths and weaknesses. DTI measures the diffusion of water, thus allowing direct noninvasive mapping of white matter pathways. fcMRI measures intrinsic functional correlations between brain regions. Although not a direct measure of anatomical connectivity, the functional couplings detected by fcMRI are sufficiently constrained by anatomy to provide insights into properties of circuit organization.

TEST–RETEST RELIABILITY

Numerous fMRI test–retest studies have quantified fMRI reliability for a range of paradigms, from sensory stimulation to complex cognitive tasks and concluded that **group activation maps** are highly reproducible across measurement sessions and across different scanners, whereas **intrasubject reliability** measured as an intraclass correlation coefficient is **low with a mean of 0.50**. The low intrasubject reliability of fMRI imposes a clear limitation for using this methodology in psychiatric practice.

fMRI IN NEUROLOGICAL PRACTICE

When talking about clinical application of fMRI we separate **between patient group analysis** (such as comparison of a group of patients on medication A with a group of patients on medication B) from application for assessment of **individual patients**. To date there has been much more progress on the first application than the second.

In spite of poor test–retest reliability fMRI are applied in neurological practice for **localization of motor cortex and lateralization of language**. This method is used to complement invasive methods rather than to replace them entirely. For example, the absence of activation near a left frontal tumor during a language task might imply a reduced risk of aphasia following its resection. Unfortunately, the false-negative rate for fMRI is much less well specified in current image analysis schemes, and it is hard to assign a significance or confidence to this interpretation.

CHALLENGES FOR CLINICAL fMRI

Full clinical implementation of fMRI faces many challenges including: (1) low signal-to-noise ratio, (2) uncertainties in BOLD mechanisms in a healthy brain, (3) uncertainty in BOLD modulation by pathophysiological states, (3) lack of standardization in experimental paradigms as well in data acquisition and analysis, and (4) lack of fMRI testing against other gold standard measures. It seems clear that further research and some modifications to analysis packages will be needed before fMRI assessment can be properly and routinely applied in psychiatric studies.

CHAPTER 1.4

Positron Emission Tomography

PHYSICAL BASIS OF POSITRON EMISSION TOMOGRAPHY

Positron emission tomography (PET) is based on the physical properties of **isotopes** – radioactive forms of simple atoms (like hydrogen, oxygen, fluorine, etc.)—emitting **positrons** when they decay. In PET centers, isotopes are obtained by means of cyclotrons. Simple atoms are arranged into more complex molecules like molecules of oxygen, water, glucose, etc. These chemical reactions are carried out in special radiochemical laboratories (called "hot cells") of PET centers. During measurements using a PET scan, the **radioactive substance** is administered into the patient's blood and reaches his/her brain through **circulation**. In the brain, cells **consume** the radioactive substance (for example, glucose or oxygen) or cellular receptors **bind** the corresponding ligand. In the first case, the higher the activity of a specific brain area, the larger the quantity of glucose (or oxygen) consumed and the more radioactive substance is accumulated. In the second case, the higher the density of the receptors, the larger is the quantity of the ligand bound to these receptors.

The radioactive substance when it is accumulated in a certain area of the brain emits positrons. As positrons encounter electrons they annihilate, emitting **two gamma quantums** per event. Special detectors placed around the subject's head register the gamma quantums, and the number of encounters registered is directly proportional to the metabolic activity of the brain area being investigated. The source parameter for PET is the **number of events** when two gamma quantums are emitted. In a PET scanner there are many detectors registering gamma quantums, and they surround the subject's head in a sort of ring—or rather in layers, rings, or cylinders.

To restore a 3D pattern of radioactive substance distribution density, special mathematical reconstruction methods are applied, similar to those used for MRI. We should note that the **spatial resolution** of PET method is significantly **lower** than that of the MRI.

PET is a rather expensive and invasive method. To apply it, a cyclotron and a special radiochemical laboratory are needed. To reduce the dose of radiation the subject is exposed to, relatively small quantities of isotopes are administered leading therefore to a poorer quality PET image. However, for some scientific tasks (such as studying the density receptors of dopamine reuptake) PET seems to be the only method available at the moment.

NEURORECEPTORS

Receptors have a prominent role in brain function, as they are the effector sites of neurotransmission at the postsynaptic membrane, have a regulatory role on presynaptic sites for transmitter reuptake and feedback, and are modulating various functions on the cell membrane. In general, a **neuroreceptor** is a **membrane protein** that is activated by a neurotransmitter. The protein is embedded into the phospholipid bilayer that encloses the cell. A neurotransmitter binds to the receptor thus allowing cells in the brain to communicate with each another through chemical signals. In postsynaptic cells, neurotransmitter receptors regulate activity of ion channels. There are two types of neurotransmitter receptors: **ligand-gated receptors** or **ionotropic** receptors and **G protein–coupled receptors** or **metabotropic** receptors. Ligand-gated receptors can be activated by neurotransmitters (ligands) like glutamate (fast-acting excitatory receptors), gamma-aminobutyric acid (GABA; fast-acting inhibitory receptors), and dopamine (slow-acting modulatory receptors) (Fig. 1.4.1a,b). G protein–coupled receptors are neither excitatory nor inhibitory, they modulate the actions of excitatory and inhibitory neurotransmitters. Most neuroreceptors are G protein coupled. Receptors can be characterized by their affinity and density. Receptors can also be located on the presynaptic membrane as a mechanism for negative feedback and for the reuptake of transmitters.

Ligand-gated receptors are the focus of pharmacology. In PET many **radioligands**—5-HT2 receptor, serotonin transporter, D2 receptor, dopamine transporter (DAT), GABA receptors—have been synthesized for studying the receptor systems of the brain.

TEST–RETEST RELIABILITY

Long-term test–retest reliability of PET has been assessed for different radioligands. In many studies the reliability was shown to be good with intraclass correlations around 0.8. However, the number of subjects participating in the studies was relatively low (usually not exceeding 10), which provides large confidence intervals for the estimated parameters.

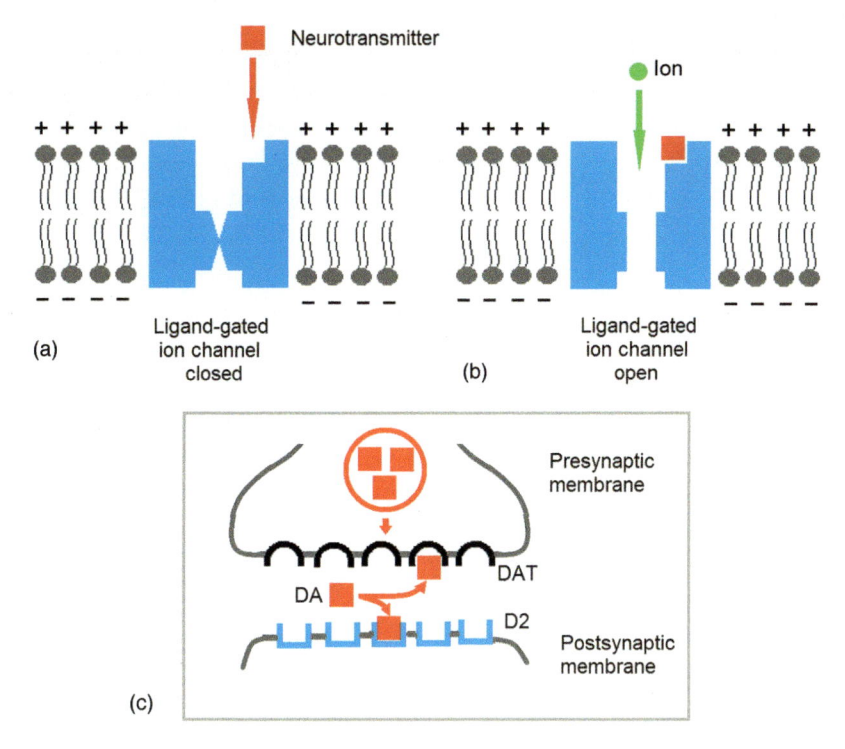

FIGURE 1.4.1 **Neuroreceptors.** Ligand-gated receptor in closed (a) and open (b) state. (c) The dynamics of dopamine *(DA)* in the synapse: DA diffuses into the synaptic cleft and binds to the dopamine receptor (D2 as an example); DA is "washed out" by the reuptake of DA by a presynaptic DAT receptor.

NEUROTRANSMITTERS AND RECEPTOR IMAGING IN CLINICS

Distribution, density, and activity of receptors in the brain can be visualized by radioligands labeled for PET, and the receptor binding can be quantified by appropriate tracer kinetic models, which can be modified and simplified for particular applications. As mentioned above selective radioligands are available for the various transmitter systems, by which the distribution of these receptors in a normal and diseased brain can be visualized.

Quantitative imaging for several receptors showed potential clinical importance. Dopamine (D2) receptors can be used for differential diagnosis of **movement disorders** and for assessment of receptor occupancy by **neuroleptics drugs**. Serotonin (**5-HT**) receptors and the 5-HT transporter can be used in affective disorders for the assessment of activity of antidepressants. **Nicotinic receptors** and **acetylcholinesterase** can be

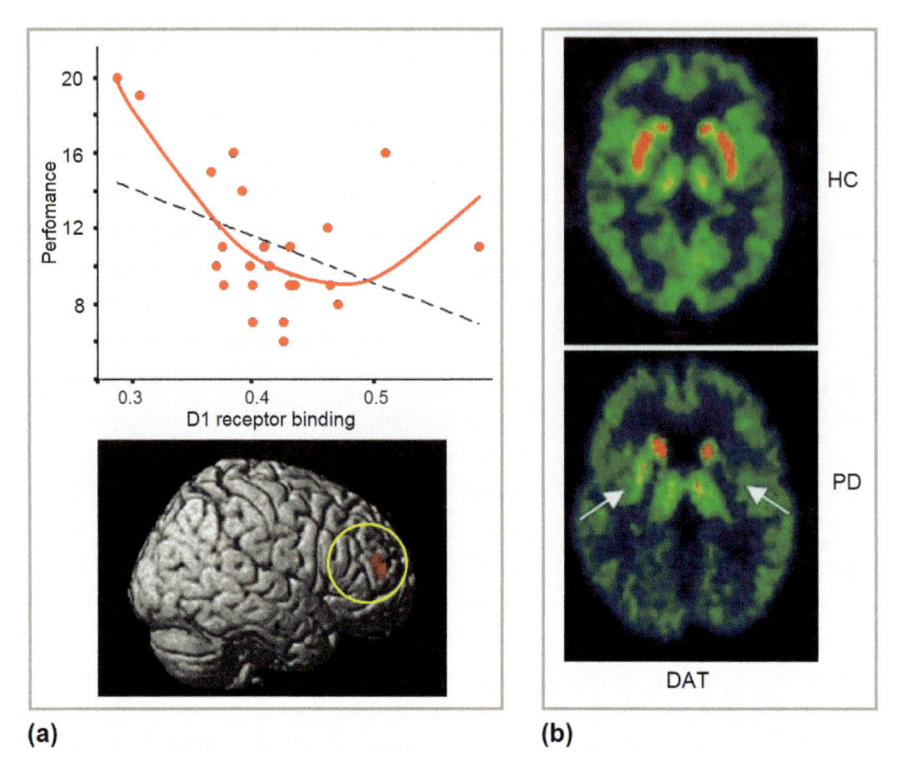

FIGURE 1.4.2 **PET as neuromarker in healthy and diseased brain.** (a) Inverted U-law between D1 receptor binding in the prefrontal cortex (x-axis) and performance of the Wisconsin Card Sorting Test in healthy controls. (b) Striatal uptake of dopamine transporter (DAT) measured by $_{11}$C-RTI 32 in a healthy control subject (HC) and in an early Parkinson's disease (PD) patient. *Part a: adapted with permission from Takahashi (2013); Part b: adapted with permission from Brooks and Piccini (2006).*

used as neuromarkers of **cognitive** and **memory impairment**. Central benzodiazepine-binding sites at the **GABA$_A$** receptor complex can be used as markers of neuronal integrity in **neurodegenerative** disorders, **epilepsy**, and **strokes**. **Opioid receptors** can detect increased cortical excitability in focal epilepsy, in emotional response to **pain**. Fig. 1.4.2 illustrates applications of PET for: (a) indexing the D1 receptor binding in the prefrontal cortex during the Wisconsin Card Sorting Test in healthy subjects, and (b) for measuring the density of dopamine transporter in a patient with Parkinson's disease in comparison to a healthy subject.

CHAPTER 1.5

Spontaneous Electroencephalogram

HOW AN ELECTROENCEPHALOGRAM IS MEASURED

The electroencephalogram (EEG) as an **electric field** is measured by electronic devices called **differential amplifiers**. By definition a differential amplifier amplifies the difference between two input voltages but suppresses any voltage common to the two inputs. In EEG applications the amplifier has two inputs: one from a site of interest and the other from a reference site. This means that (1) an EEG measures the potential difference between the two sites thereby yielding **relative** rather than absolute measures, and (2) the properties of the **reference** have a fundamental impact on the signal of interest.

MONTAGES

According to the laws of physics, electrical fields are **passively transmitted** through a biological tissue. The EEG as an electrical field is **volume-conducted** throughout the brain, skull, and body so that a local source generates a potential distributed practically over the whole brain. For isolating localized EEG abnormalities, clinical applications implement **bipolar** derivations when two sites of the head are located close to each other.

For research purposes, **unipolar** or **referential** (ie, relative to the selected reference) recordings are used, which are by definition reference dependent. Reference electrodes can be placed on the mastoids, earlobes, tip of the nose, even the vertex of the head or below the neck. But **no neutral reference location** exists on the body.

One attempt at approximating neutral reference was made by using the **common average montage**. In theory the common average of all recorded

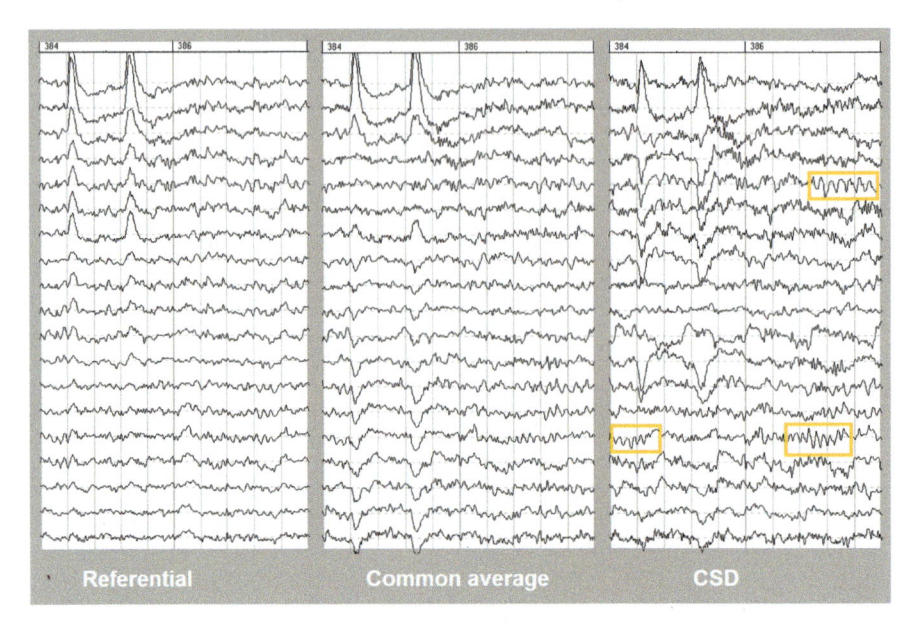

FIGURE 1.5.1 Comparison of three different montages: *Referential, Common average,* and *CSD*. The three montages are applied to the same EEG fragment. Note that only the *CSD* montage reveals bursts of local theta and alpha rhythms *(boxes)* that are difficult to see in the other two montages.

EEG activity approximates zero if there is a sufficient number of electrodes and they cover the whole head. However, this is never practically achieved even with 256 electrodes because no electrodes are placed on the ventral side of the head.

Another attempt to suggest a reference-free montage was made by Francis Perrin and coworkers in the end of 1980s in the French National Institute of Health and Medical Research by computation of **current source density (CSD)**. In this method **spherical spline interpolation** of recorded potentials was used (Perrin, Pernier, Bertrand, & Echallier, 1989).

It should be stressed here that the absence of a universal reference standard raised issues of **validity and across-study comparability** in EEG research and makes it hard to translate the results of these studies into clinical practice.

Fig. 1.5.1 illustrates three different montages applied to the same EEG fragment which contains bursts of frontal theta and parietal alpha rhythms (highlighted by yellow boxes) . One can see that only the CSD montage enables the observer to see these local features in an EEG. The other two montages make it very difficult to detect rhythmicities by the naked eye.

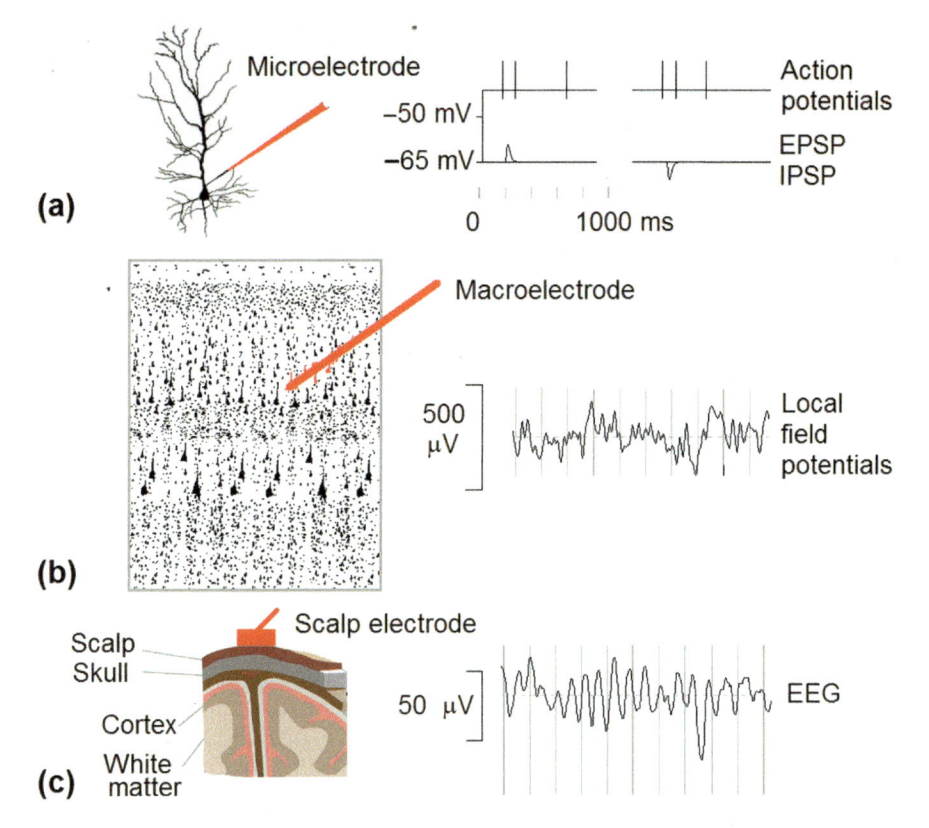

FIGURE 1.5.2 **Structural organization of electrical events in the brain.** (a) Action potentials, excitatory and inhibitory postsynaptic potentials *(EPSP and IPSP)* are recorded from microelectrodes within the neuron (around 30 μ) and correspond to changes of a single-neuron membrane potential. (b) LFPs are recorded with macroelectrodes within the cortex and can be different at 3-mm distances. (c) Electroencephalogram *(EEG)* signals are recorded by metal electrodes placed on the skin of the head and are volume-conducted over the whole head. Note the large amplitude differences between the electrical events.

ELECTRICAL EVENTS IN THE CORTEX

From the structural point of view, electrical activity can be measured (1) in a **single neuron** in the form of **action potentials, inhibitory and excitatory postsynaptic potentials** by microelectrodes inserted into or near the neuron, (2) in a neuronal module in the form of **local field potentials** recorded by macroelectrodes implanted into the gray matter, and (3) at the large systemic level in the form of electrical potentials recorded from **scalp electrodes** (Fig. 1.5.2).

The local field and large systemic potentials are manifestations of **collective activity of synaptic and other transmembrane potentials**. From the classical view, neurons process information by generating spikes. Consequently, the local field potentials might be simply a byproduct of neuronal firing without any functional meaning. However, growing experimental evidence in animals challenges this classical view by observations that the local oscillatory environment causally affects neural firing via ephaptic coupling. Here we refer the reader to a recent review by György Buzsaki, Costas Anastassiou, and Christof Koch from the State University of New Jersey, published in 2012.

10–20 INTERNATIONAL SYSTEM

In 1959 the first standards in EEG recordings were laid down. Among them was the system of placement of electrodes. This system is called the **10–20 International System** of Electrode Placement or simply the 10–20 System (Fig. 1.5.3a). The name comes from the fact that any electrode is **10 or 20% of some distances from another**. In the direction from nose to back of the head this distance is defined by the circumference of the head measured from the nasion to the inion. In the direction from one ear to another the distance is defined by the circumference of the head measured from the small fossa just anterior to one ear canal over the central vertex to the opposite ear. The distance between the peripheral electrodes is defined as 10% of the circumference of the head measured just above the eyebrows and ears.

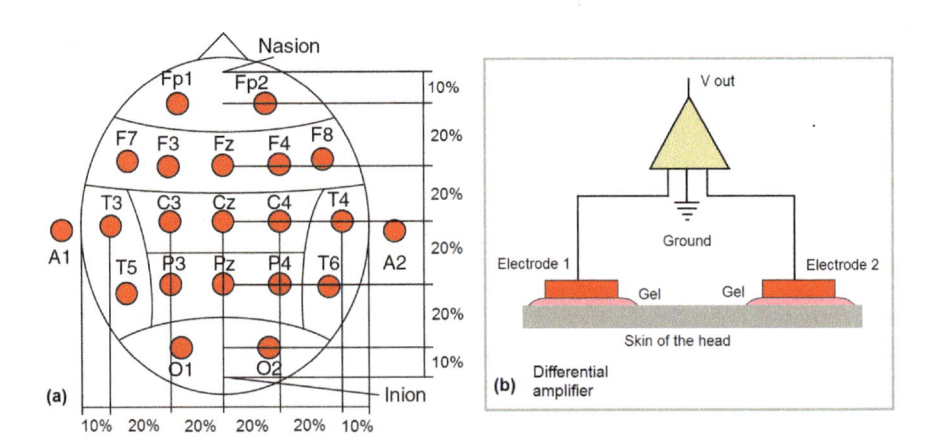

FIGURE 1.5.3 **Scheme of recording potentials from the head.** (a) 10–20 International system of placement of electrodes on the head. Each electrode position has a letter to identify the lobe and a number to identify the location within the lobe. All electrodes are 10 or 20% distant from one another. (b) Scheme of differential amplifiers for recording the voltage difference between two electrodes.

The names of electrodes include the first letter associated with the area where the electrode is placed, and the number indicates the side and placement within this area: Fp1, Fp2 = prefrontal; F3, F4 = frontal; Fz = frontal midline; C3, C4 = central; Cz = central vertex; P3, P4 = parietal; Pz = parietal midline; F7, F8 = anterior temporal; T3, T4 = mid temporal; T5, T6 = posterior temporal; A1, A2 = earlobes. Odd numbers indicate the left hemisphere. Even numbers indicate the right hemisphere.

The potentials recorded from the scalp of a healthy adult are about 50 µV and must be amplified before processing. Amplification is done by **differential amplifiers** (Fig. 1.5.3b).

FREQUENCY BANDS

According to the specifications of EEG recording in clinical settings defined 50 years ago the bandwidth of a "conventional" EEG is **0.5–50 Hz**. One of the challenges was to get rid of slow shifts of potentials of both brain and nonbrain (artificial) origins that saturated the amplifier's dynamic range. This was solved by equipping amplifiers with a high-pass filter which provided stable recordings. The other challenge was to get rid of muscle artifacts and main noise. These frequencies are conventionally filtered out by low-pass filters at around 50–70 Hz.

In contrast to clinical settings, for research purposes both infralow and high-frequency bands were investigated in a few laboratories. Here I would like to mention the methodology of **infraslow (<0.1 Hz)** recordings developed at the Institute of Psychology (N.A. Aladjalova) and the Institute of Experimental Medicine (V.A. Ilukhina) in the Soviet Union, and at Vienna University (H. Bauer) in Austria. My first PhD dissertation was about the close relationship between electric infraslow oscillations with metabolic fluctuations such a local blood flow and extracellular oxygen.

Recently interest in **infraslow fluctuations (ISF)** has been renewed partly because of the discovery of ISFs in the **blood oxygenation level–dependent (BOLD) signal**, partly because of the success of applying ISFs to **neurofeedback technology**, and partly because **direct-coupled (DC) amplifiers** became commercially available. The currently available technology for DC recording includes: (1) nonpolarizable electrodes (eg, Ag/AgCl), (2) a sufficiently stable electrode gel, (3) subcutaneous tissue–gel interfaces, (4) a genuine DC amplifier with sufficiently high input impedance and DC stability. As suggested by a research group from Helsinki University (Sampsa Vanhatalo and coworkers in 2005) the **full-band EEG (fbEEG)** is likely to become the standard approach for a wide range of applications in both basic science and in the clinic (Vanhatalo, Voipio, & Kaila, 2005).

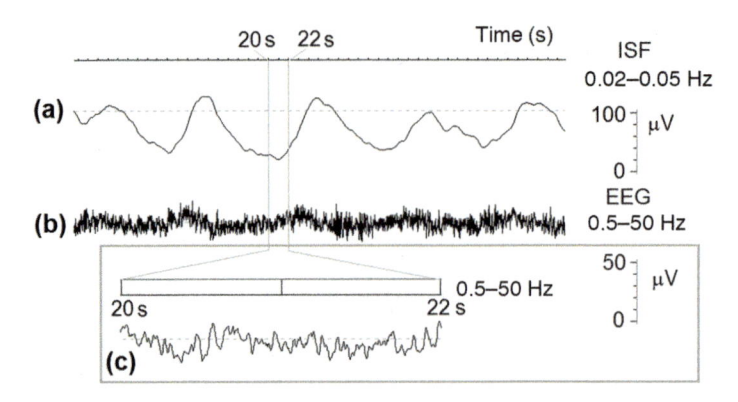

FIGURE 1.5.4 **ISF and conventional EEG.** (a) *ISF* in the frequency band 0.02–0.5 Hz recorded from Cz of a healthy adult under the eyes-closed condition. The horizontal scale is time in seconds (the 20-s and 22-s points of recording are marked). The vertical scale is amplitude of the signal in microvolts. (b) Electroencephalogram *(EEG)* in the frequency band 0.5–50 Hz at the same time and amplitude scale. (c) Conventional EEG at larger time and lower amplitude scales.

There is no consensus regarding the physiological mechanisms of ISFs in the EEG. Some people suggest that these slow shifts of potentials might be associated with **depolarization of apical dendrites of pyramidal cells** accompanied by long-lasting expectation periods in cognitive control, others suggest that they are generated across the **blood–brain barrier** and have nonneuronal origin. Both views are supported by experimental findings so that it might be that they deal with different phenomena expressed in different types of slow shifts of scalp-recorded potentials.

It has also become evident that monochromatic ultraslow oscillations in **a narrow band around 0.1 Hz** constitute a separate phenomenon (such an example can be seen in the oscillations in Fig. 1.5.4). These electrical oscillations correlate with similar rhythms recorded by near-infrared spectroscopy and arterial blood pressure, but show a complex topography which cannot be explained by one source generator. A group from University Hospital Charité (Vadim Nikulin and coworkers) in 2014 suggested these oscillations might be of extraneuronal origin reflecting cerebral vasomotion similar to the origin of the 0.1-Hz oscillations in the BOLD signal described in the previous section (Nikulin et al., 2014).

From the temporal point of view, electrical activity can be measured in different frequency bands: (1) **direct current** and **infraslow fluctuations** below 0.1 Hz, (2) **slow electrical potentials** of 0.1–1 Hz, (3) **delta** waves of 1–4 Hz, (4) **theta** waves of 4–8 Hz, (5) **alpha** waves (8–12 Hz), (6) **beta** waves at frequencies higher than 13 Hz, (7) the **gamma** band is considered a separate band with frequency higher than 30 Hz.

ELECTROENCEPHALOGRAMS AS A REFLECTION OF CORTICAL SELF-REGULATION

The main feature of an EEG is the oscillatory nature of potential dynamics. The **oscillations** are seen at different frequency bands: from infraslow to gamma. The oscillations are seen in raw EEGs as up and down fluctuations similar to sinusoidal waves and are seen in EEG spectra as peaks at the corresponding frequency. The generators of these rhythms are usually certain neuronal networks with different membrane and synaptic mechanisms. Nevertheless, in all these networks there is similarity in the goals: they all are designed by evolution to insure optimal conditions of brain functioning by interplay between **excitatory and inhibitory** processes. The interplay between these opponent processes leads to oscillations of potentials which in turn through **voltage-gated channels** might stabilize neuronal network functioning.

VOLTAGE-GATED ION CHANNELS

An example of oscillations at a single neuron level is a **burst of action potentials** induced by a profound membrane depolarization evoked by an external event. The membrane **depolarization** above a threshold causes Na^+ channels to open rapidly and to produce inward Na^+ current (Fig. 1.5.5). The inward current leads an avalanche of depolarization. After some delay this strong depolarization opens K^+ channels and produces an outward K^+ current that repolarizes the membrane (not shown in the figure). If external depolarization is kept, the cycle repeats and produces a burst of action potentials. This example shows the role of voltage-gated Na^+ and K^+ channels in the generation of oscillations at a single neuron. The duration of an action potential is around 1 ms. So, to be seen at the scalp, millions of cortical neurons must fire synchronously. This may happen in the case of epileptiform activity in the generation of interictal spike/slow wave complexes.

NONBRAIN EVENTS (ARTIFACTS) IN ELECTROENCEPHALOGRAMS

EEGs might be contaminated by voltages associated with noncortical events such as eye movements, muscle activity, heart electrical and mechanical artifacts. Eye movement artifacts are often the largest ones. They are generated by vertical and horizontal eye movements. The main source of artifacts is the potential of the eyeball. The eyeball acts as an electric dipole with the positive pole oriented anteriorly. **Eye blink** results in reflexive upward vertical eye movement that produces positive deflection

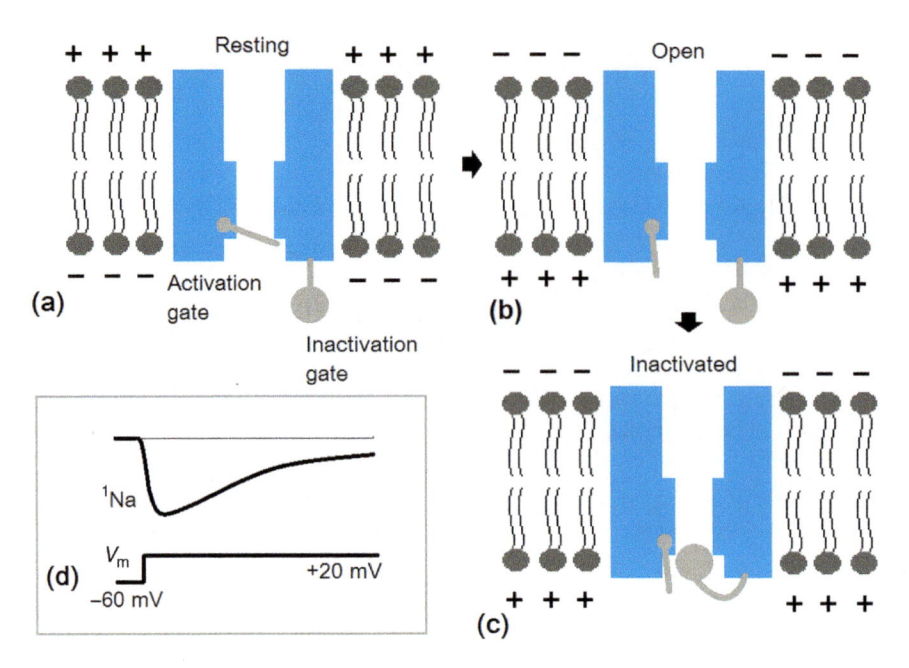

FIGURE 1.5.5 Na⁺ **electrical current in response to membrane depolarization is associated with opening and closing of the Na⁺ channel.** (d) The depolarizing external current first activates and then inactivates the Na current. (a)–(c) Illustration of the position of the activation and inactivation gates when the channel is at rest (a), when the Na channel has been opened (d), and when the channels have been inactivated (c). *Adapted from Fig. 7–13 of Chapter 7 [Kandel, E. R., Shwartz, J. H., Jessel, T. M., Siegelbaum, S. A., & Hudspeth, A. J. (2012). Propagating signaling: the action potential from principles of neural science (5th ed.)].*

at frontal areas with maximum at Fp1, Fp2 electrodes. Eye closing is associated with a similar artifact, while eye opening results in downward vertical eye movement and negative deflection at the Fp1, Fp2 electrodes. Horizontal eye movements (also called **saccades**) produce opposite changes of potentials at F7 and F8. The old way of correcting eye movements is to record the electrooculogram (EOG) and to compute the proportion of ocular contamination in each EEG channel. Then the EOG signals are scaled by the estimated proportion and subtracted from the original EEG signals (Fig. 1.5.6). The modern way of correcting eye movements is based of independent component analysis of the multichannel EEG in which the eye movement artifacts are associated with certain independent components and are corrected by eliminating the corresponding components.

Muscle artifacts arise from the electrical activity of muscles. In particular, frontalis and temporalis muscles are the most common source of myogenic activity, respectively, in frontal electrodes (mostly Fp1 and Fp2) and in temporal electrodes (mostly T3, T4). Usually, it is not difficult to separate muscle activity from beta cortical activity. Indeed, at the spectra the range

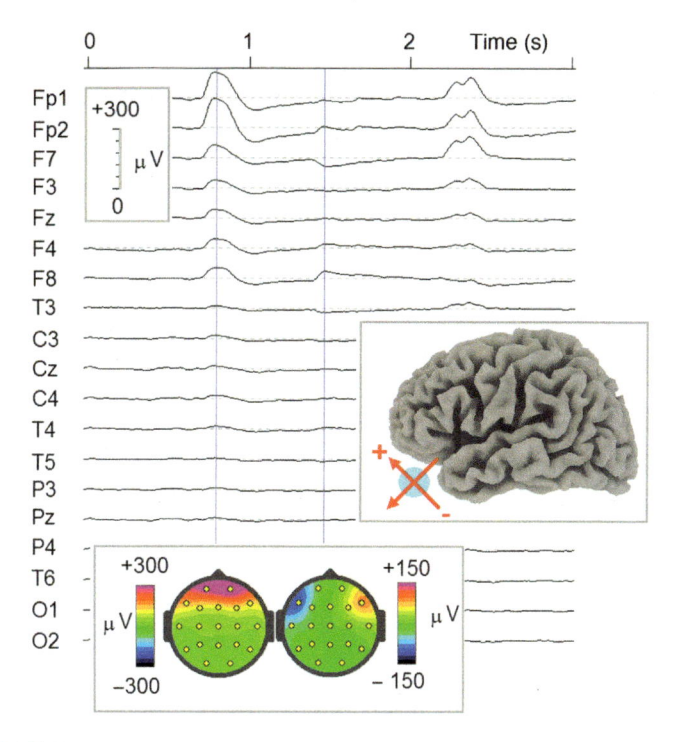

FIGURE 1.5.6 Eye movement artifacts. Eye blink (with topography maximal at Fp1 and Fp2) is followed by saccadic eye movement (with opposite potentials at F7 and at F8).

of muscle artifact is usually broader than the range of beta activity and because of that at the recordings, the muscle activity looks like a thicker line when compared with genuine EEGs. Single muscle discharges may look like epileptic spikes, but muscle "spikes" are shorter in duration and are limited to only one electrode.

In individuals with short necks and large hearts, electrical fields of the heart may be detected by ear or other basal electrodes. It is difficult to confuse electrocardiogram (ECG) artifacts with epileptic spikes because these artifacts are regular and usually seen with the same polarity in many electrodes. Simultaneous ECG recording usually helps to differentiate these artifacts (labeled as **ECG artifacts**), but an experienced practitioner can easily do so without such recording.

Another common type of nonbrain-related potential change is the **cardioballistic artifact**. This type of artifact is caused by a periodic (the same period as a heart beating) movement of an electrode located just above a blood artery of the head. Pulsation of the vessel moves the electrode and induces a periodic artifact. The cardioballistic artifact is usually observed under one electrode. This artifact can easily be detected on the map of EEG spectra as a local peak at about 1-Hz frequency (Fig. 1.5.7).

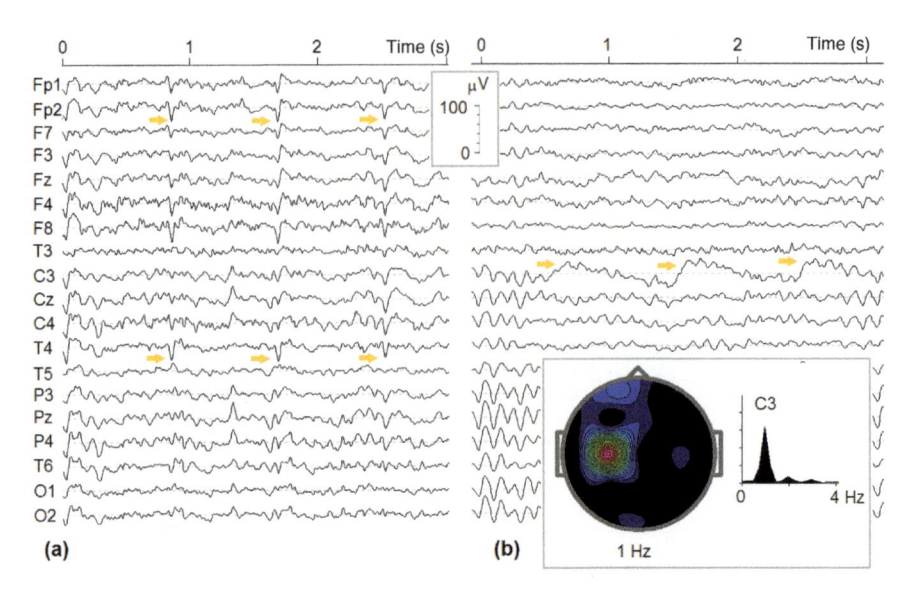

FIGURE 1.5.7 **Heart-related artifacts.** (a) ECG artifacts. Note regular (with 1-s intervals) spikes at frontal-central electrodes *(arrows)*. (b) Cardioballistic artifacts. Note local (C3) regular oscillations at 1 Hz expressed in the peak of EEG spectra.

SPECTRAL ANALYSIS OF ELECTROENCEPHALOGRAMS

One important feature of EEGs is their rhythmicity. In a conventional EEG (0.5–30 Hz) of healthy subjects in the resting state several types of rhythmicity are observed. They include **posterior alpha rhythms (8–12 Hz)**, **frontal and central beta** (13–30 Hz) **rhythms**, and **frontal midline theta rhythms** (6–8 Hz). The rhythms change in time by waxing and waning. Sometimes they are seen in bursts with long intervals between them so that it is not easy to observe the rhythms in the EEG recording by the naked eye. A powerful method of extracting and compressing information about rhythmicity over time is given by **Fourier analysis** which is based on the Fourier theorem introduced by the French mathematician Jean-Baptiste-Joseph Fourier in 1822.

The **Fourier theorem** states that any signal $X(t)$ within the time epoch from 0 to T can be decomposed into a set of simple sinusoidal functions, called Fourier series:

$$X(t) = A_0 + \sum A_n \sin(nft + \varphi_n)$$

where Σ is a sum over all frequencies (harmonics), f is a **fundamental frequency of the signal** $f = 1/T$, A_n and φ_n are the amplitude and phase of a sinusoidal component, n is a harmonic number. The fundamental frequency defines the resolution of spectra (eg, for $T = 4$ s the frequency resolution is $1/T = 0.25$ Hz).

A_n and φ_n define **amplitude and phase spectra**. The direct integral-based approach to compute A_n and φ_n by convoluting the signal with sinusoidal waves is computationally very laborious. In 1965, James W. Cooley of IBM and John W. Tukey of Princeton published a more efficient algorithm for the computation of Fourier series called the **fast Fourier transform (FFT)** algorithm (Cooley and Tukey, 1965).

To obtain the reliable spectra in a certain condition one needs at least 3 min of recording. This gives 45 nonoverlapping 4-s epochs ($T = 4$ s). The slicing of a continuous signal stream into finite epochs introduces contamination of the spectra by artifactual frequencies created by the abrupt transitions at the ends of the epoch. This type of distortion is minimized by multiplying potentials in each epoch by a window function. A variety of window functions have been described, including the triangle and **Hanning windows** and all of them equal zero at the end of the epoch. Applying a time window produces another problem: fragments near the beginning and the end of epochs of analysis give a smaller input to spectra in comparison with the middle fragments. To compensate for this distortion, epochs are selected with overlapping. The overlapping usually constitutes **one-half of the epoch**.

Spectra can be displayed in two different ways: (1) in a form of dynamics of spectra computed for each epoch sequentially, (2) in a form of spectra **averaged** for all epochs over the whole interval of recording. Examples of averaged amplitude spectra (amplitude vs. frequency) for a healthy subject are presented in Fig. 1.5.8. One can see bursts of theta, alpha, and beta rhythms in raw EEG together with an artifact of eye blink. The rhythms are presented in **peaks** on EEG spectra at corresponding frequencies.

INTERINDIVIDUAL DIFFERENCES

EEG patterns differ significantly between healthy subjects. Some people have strong alpha rhythms even in the eye open condition, some has frontal midline theta rhythm, some have no alpha and theta rhythmicities at all. The amplitude of an EEG recorded in a particular subject depends on many factors including the genetic, neurophysiological, anatomical and physical properties of the brain and surrounding tissues (skin, bone, dura mater, and pia mater). These parameters vary from one subject to another and are basically unknown. To compensate for this variation, **relative EEG power** is sometimes computed.

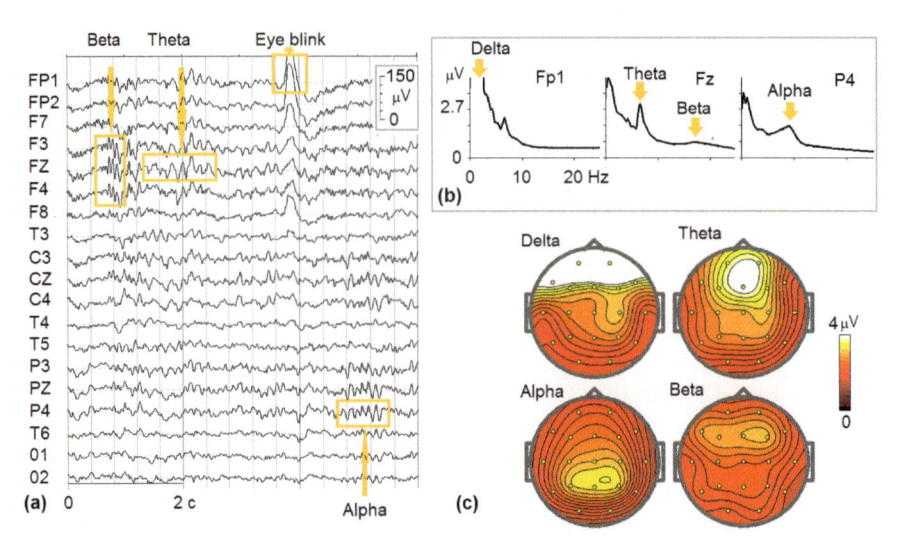

FIGURE 1.5.8 **Amplitude spectra of EEG.** (a) A 19-channel fragment of an EEG in the referential montage of a healthy subject during resting state with bursts of beta, theta, and alpha rhythms *(boxes)*; (b) EEG amplitude spectra for Fz and P4 electrodes with peaks corresponding to the rhythms, and for Fp1 electrode with delta activity corresponding to eye blinks; (c) Maps of EEG spectra at frequencies indicated on (b) by *arrows*.

WAVELET TRANSFORMATION

As indicated above in Fourier analysis an EEG signal is decomposed into a set of sinusoidal functions of different frequencies, amplitudes, and phases. These functions are infinite in the time domain. Consequently, Fourier analysis lacks time dynamics and cannot be applied for studying responses of brain waves to certain events. The method of **wavelet analysis** is used in such cases. In this method an EEG is decomposed into discontinuous (finite length) oscillating waveforms called **wavelets.**

The concept of wavelet transformation was first proposed by the geophysicist Jean Morlet who together with the physicist Alex Grossman from the Université de la Méditerranée Aix-Marseille invented the term "wavelet" in 1984 (Grossman and Morlet, 1984). In mathematics the wavelet transform refers to representation of a signal in terms of finite length oscillating waveforms. These waveforms are scaled and translated so that when summed up they match the input signal. The functions mostly used are complex Morlet wavelets, defined by:

$$w(t, f) = A \exp(-(t-t_o)^2/2s^2)\exp(2i\pi f t)$$

where $w(t,f)$ has a Gaussian shape both in the time domain (around time t_o) and the frequency domain (around frequency f). The wavelets have the

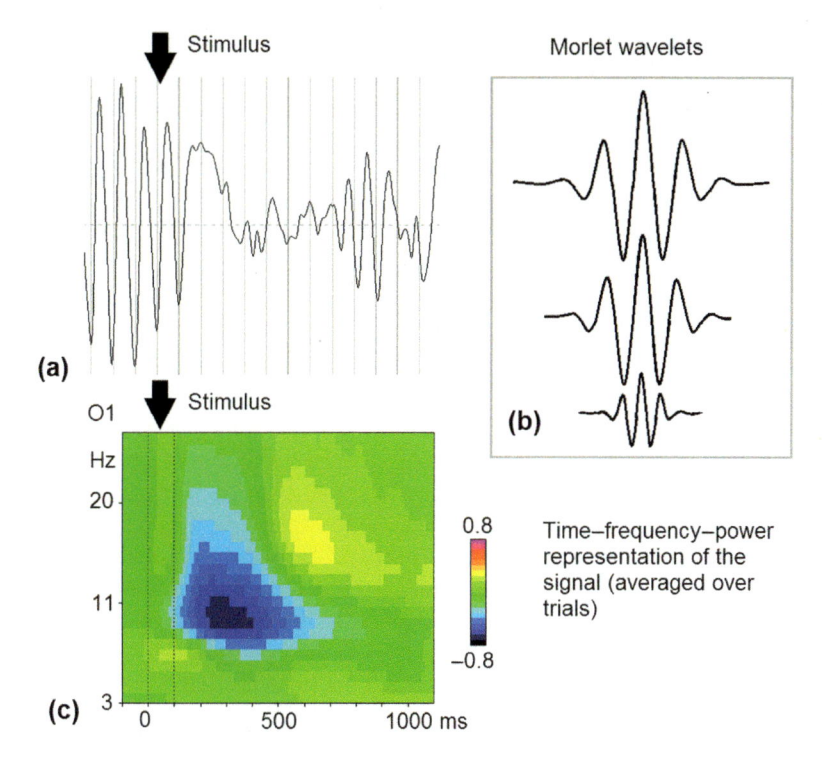

FIGURE 1.5.9 Wavelet analysis: computing the power–time–frequency representation. (a) EEG alpha desynchronization in response to visual stimulus at the O1 electrode. Note decrease of alpha wave amplitude in response to stimulus; (b) wavelets. The signal is decomposed into a series of wavelets; and (c) the time–frequency–power representation of the signal at the O1 electrode averaged over the trials. *Dotted vertical lines* represent a visual stimulus presentation. The *x*-axis time is in milliseconds. The *y*-axis is frequency. Color codes represent relative change in the power of the signal in the corresponding frequency and time in comparison with the prestimulus background. Note the 80% decrease in power around the alpha band and around 300 ms after stimulus.

same number of cycles for different frequency bands, resulting in different wavelet durations.

Morlet functions are presented in Fig. 1.5.9b. The signal $P(t)$ is convoluted by wavelets in a similar way as it is convoluted with sinusoidal functions. The result is squared and gives the time-dependent power of EEG around frequency f. Repeating this calculation for a family of wavelets having different frequencies f provides a time–frequency power representation of signal components. The results of wavelet analysis are presented in Fig. 1.5.9c, which shows how relative EEG power changes in O1 in response to visual stimulus in the time and frequency domains: time–frequency presentation. Note that visual stimulus presentation induces occipital alpha desynchronization as shown in Fig. 1.5.9a.

COHERENCE

From the anatomical point of view, the most striking feature of the brain is the abundant connectivity between neurons. From the functional point of view, this connectivity is reflected in synchronous activities within the brain: neurons in anatomically connected structures tend to fire synchronously. Electrophysiological data show that this synchronicity is performed in bursts repeating at different frequencies. The frequency of the synchronization seems to define the functional meaning of connectivity. For example, alpha frequencies are idling rhythms of sensory systems, and synchronization at 10 Hz frequency indicates the state of the sensory system when neurons do not relay sensory information but are ready to commence when a relevant stimulus appears. Oscillatory synchronization in the **gamma band** has been proposed as a **binding mechanism** for combining different features of an object into a single percept. In this case, synchronization at 40-Hz frequency indicates synchronous activation of neurons responsible for detecting different features of the same stimulus. The disruption of "normal" synchronization may be a sign of neurological or psychiatric dysfunction. For example, an abnormal pattern of synchronizations between different parts of the basal ganglia seems to be responsible for tremor and dyskinesia in Parkinson's disease.

In EEGs the measure of synchronization between two different scalp locations is given by a parameter called **coherence**. Coherence by definition is a measure of correlation between EEG power values computed in the same frequency band but in two different locations. For 19 electrodes the total number of pairs equals 171, which gives the number of all possible coherence measures. This number of values is difficult to map in one picture. Several ways of presenting coherence have been suggested.

NEURONAL SOURCES OF ELECTRICAL CURRENTS

Synaptic activity is the most important source of **local field potentials (LFPs)**. The dendrites and soma of a pyramidal neuron form a tree-like structure with basal and apical dendrites. They can be modeled as an electrically conducting interior that is surrounded by a relatively insulating membrane with thousands of synapses located along it.

The excitatory fast-acting neurotransmitter **glutamate** binds to **AMPA** or **NMDA** receptors and induces excitatory (depolarizing) currents, involving Na^+ or Ca^{2+} ions, respectively, which flow inward into the synapse. This influx of cations from extracellular into intracellular space gives rise to a local extracellular sink. The inhibitory fast-acting **GABA** binds to subtype A (**GABA$_A$**) or subtype B (**GABA$_B$**) receptors and induces inhibitory (hyperpolarizing) currents, involving inward flow of Cl^- ions—a local **source**.

The transmembrane currents that flow in or out of a neuron at active synaptic sites are **compensated** by currents that flow in the opposite

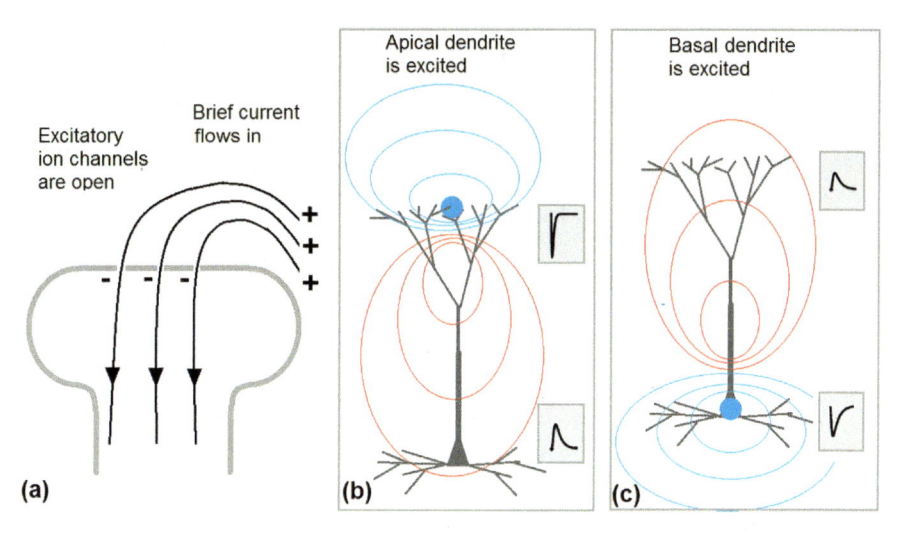

FIGURE 1.5.10 Electric currents in neurons (model). (a) Glutamate binds to AMPA receptors and opens the ion channels to induce a positive ion flow with the sink on the dendrite; (b) activation of the apical dendrites by glutamate generates a negative local field at this part of the neuron, accompanied by a positive local field at soma and basal dendrites. Local field contour lines are presented as *blue* (for negative) and *red* (for positive) values. LFP responses to the corresponding brief injection of current are presented in *boxes* to the right of the model; and (c) The inverse pattern is generated by activation of the basal dendrites.

direction elsewhere along the neuronal membrane. Consequently, in the case of an EPSP, besides the active sink at the level of the synapse, there are distributed passive sources along the soma-dendritic membrane (Fig. 1.5.10). The pyramidal cells are arranged in columns with the apical dendrites aligned perpendicularly to the cortical. When millions of these cells fire synchronously they produce open fields that are strong enough to be measured at the scalp.

In general terms, excitation of the postsynaptic neuron generates an extracellular potential near the neural dendrites that is more negative than elsewhere on the neuron. This kind of potential distribution when a region of positive charge is separated from a region of negative charge by some distance is named an **electrical dipole**. In a conducting brain tissue the potential gradients induce currents which are reflected in concept of a current dipole. So, the current dipole is a current from a **sink** to a **source** of the current. Current dipoles are used to model neuronal sources of EEG and MEG.

INTRACORTICAL CONNECTIVITY

Roughly, two types of neurons dominate the cortical structures (at least in the sensory cortical areas): (1) **principal cells** (about 80% of all neurons) that use the excitatory neurotransmitter glutamate and (2) **interneurons** (20%) that use the inhibitory neurotransmitter GABA (Fig. 1.5.11).

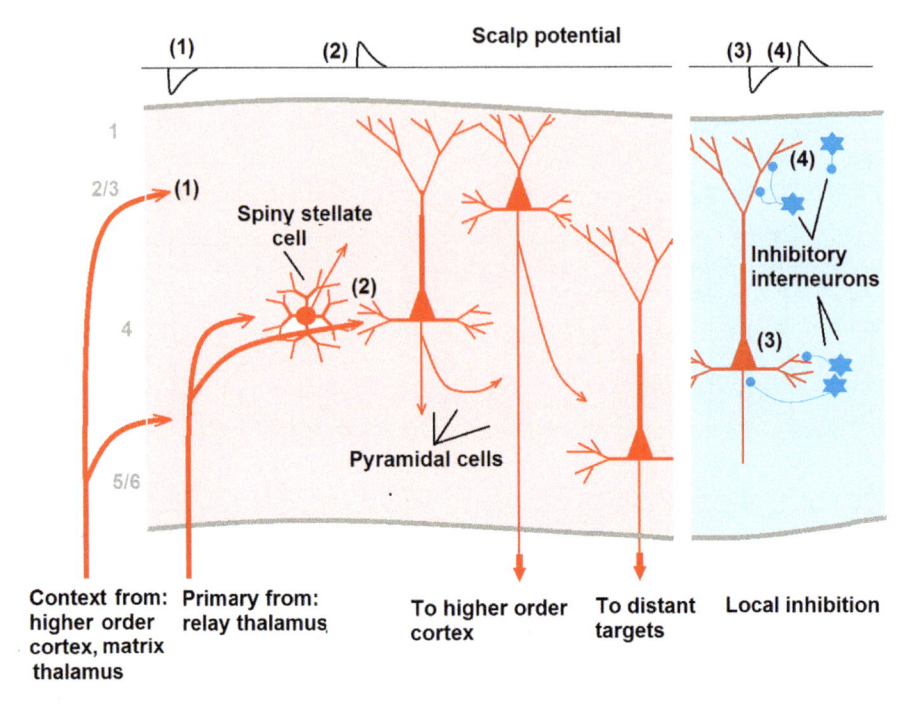

FIGURE 1.5.11 Intracortical connectivity. Left *(red)* shows input/output and interconnections in the excitatory network. Right *(blue)* shows output of inhibitory interneurons. Excitatory synapses produce a sink at the corresponding layer. Inhibitory synapses produce a source. Sinks and sources produce scalp potentials with polarities depending on their localization within the cortex.

Bottom–up primary sensory information arrives from the relay neurons of the thalamus most densely into **layer 4**. **Top–down contextual inputs** from the higher order cortex and the matrix neurons of the thalamus target **upper and lower cortical** layers but avoid layer 4. Principal cells in layer 4 comprise two morphological classes: **pyramidal and spiny stellate** cells. These cells most strongly project to the upper layers. Principal cells of **layers 2 and 3** project to **higher order** and contralateral cortices, as well as locally to layer 5. Principal cells in layer 5 include **subcerebral projection** neurons that integrate inputs from multiple cortical layers and send powerful outputs to distant targets. As can be seen the context excitatory input to **layer 2/3** generates a **negative potential** on the scalp located just above the source while the primary input to **layer 4** generates a **positive potential** on the scalp.

The cortex contains three major types of inhibitory interneurons: **parvalbumin-expressing interneurons (PVs)**, **somatostatin-expressing interneurons (SOMs)**, and **5HT3A receptor–expressing interneurons (5HTs)**. PVs consist of two main subgroups: basket cells that inhibit the soma and proximal dendrites of principal cells, and chandelier cells that inhibit the axon initial segment. PV cells receive strong excitatory inputs from the thalamus and cortex and play a key role in **stabilizing** cortical networks. A substantial decrease in PV cells leads to **epileptiform** activity, whereas a chronic dysfunction of these cells was suggested in **schizophrenia**. SOMs include Martinotti cells that inhibit the tuft dendrites of principal cells. In contrast to PVs, SOMs receive the majority of their input from local principal cells. These cells have been implicated in the **control of dendritic integration**. 5HTs are located in superficial layers and release GABA by volume transmission. They have been implicated in **learning** and **control of cortical circuits** by higher-order cortex.

CORTICAL FOCUS AND SPIKES

In the cortex, excitatory neurons are heavily interconnected so that without the mechanism of inhibition the activity would spread out over the whole brain. The spreading of excitation is limited by inhibitory neurons. However, in cases of genetically determined epilepsy or of local cortical damage (dysplasia) the inhibition can be locally compromised and these local areas start generating spike/slow-wave patterns—neuromarkers of a **focus** in the cortex. The focus does not mean that the subject has epilepsy—in many cases the power of the focus is not strong enough to spread over the whole brain and generate a seizure. When neurons in the focus fire synchronously they generate a sharp potential on the scalp called a **spike**, (don't mix with spikes of action potenitals generated by a single neuron).

Studies of animal models show that the spike itself is a **sum of excitatory postsynaptic potentials**. The spike is followed by a slow wave, which represents **GABA-mediated recurrent inhibition** as well as intraneuronal processes such as afterhyperpolarization.

Some software packages offer automatic spike detection procedures for detecting spikes. Fig. 1.5.12a shows the results of spike detection in the EEG of an epileptic patient in the resting state without epileptic seizures. One can see that the sharp negative spike with maximum at F8 is followed by a slow wave of a similar distribution. Fig. 1.5.12c presents the current source density dynamics measured by the implanted multielectrode (Fig. 1.5.12b) during the novel spike generation in an epileptic patient. One can see that the synaptic depolarization at the layer V of the cortex initiates the given paroxysmal event.

VOLUME CONDUCTANCE

A given single-current dipole generates a distribution of positive and negative potentials recorded on the scalp (Fig. 1.5.13). Volume conduction results in a **widespread, smeared potential topography** over the whole scalp with

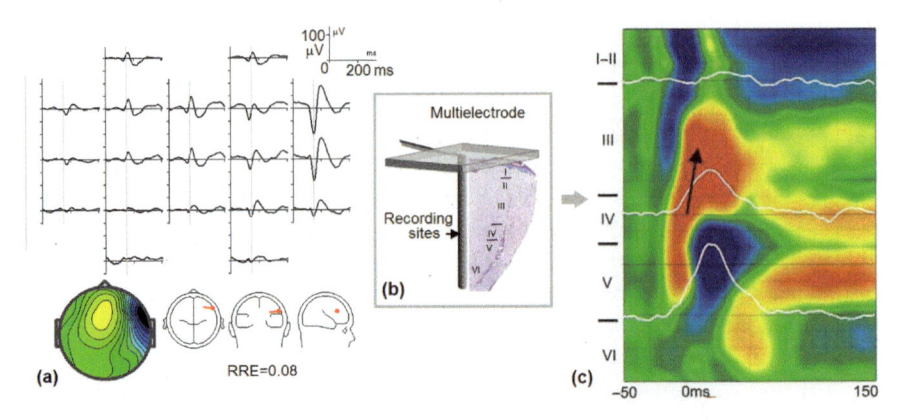

FIGURE 1.5.12 Interictal spikes in the human brain. (a) Averaged interictal spike/slow-wave complex in an epileptic patient in 10–20 montage with a map taken at the peak of the spike; (b) schematic drawing of a multielectrode implanted into a patient with medically intractable epilepsy with the aim to localize and further remove the seizure focus; and (c) current source density and multiunit analysis of a spike generated in the focus. *White curves show multiunit activity recorded from the electrodes. Adapted from with permission Ulbert, Heit, Madsen, Karmos, & Halgren (2004).*

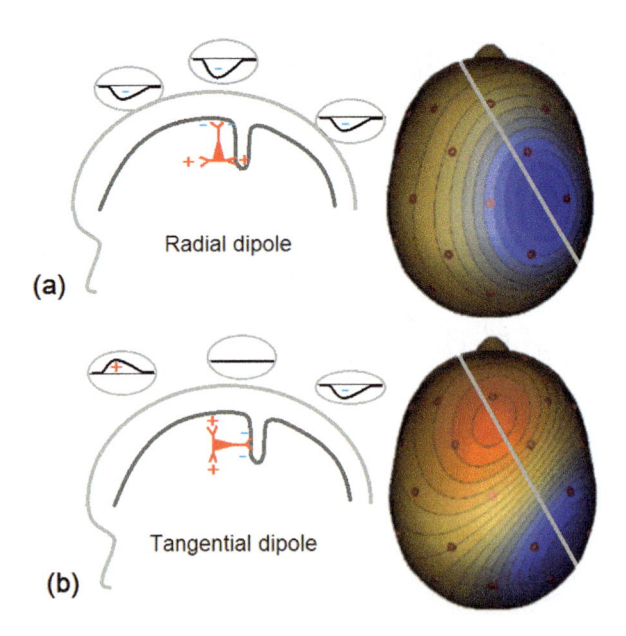

FIGURE 1.5.13 Dipole orientation defines the potential distribution. (a) Radially oriented dipole generates widely distributed negative potential with minimum just above the dipole position in the cortex; and (b) tangential dipole at the same location generates a negative–positive pattern of distribution with zero potential above the dipole.

a maximum that does not necessarily occur right over the activated cortical patch. By the laws of physics, the integral of the potential over the whole head is zero. Therefore, any negativity has a corresponding positivity somewhere else over the head. The polarity of scalp-recorded potential at a given electrode is defined by the orientation and position of the dipole. Alterations in the electrical dipole travel at the speed of light, so that the cortical potential at a given instant reflects essentially instantaneous change in the current source.

The orientation and polarity of the dipole are not usually possible to associate unequivocally with excitation or inhibition processes within the cortex because the same dipole could be a result of different combinations of excitatory and inhibitory postsynaptic potentials occurring in different cortical layers.

INVERSE PROBLEM: DIPOLE APPROXIMATION

Currents at the cortical convexity have a predominantly radial orientation, while currents in cortical fissures have a predominantly tangential orientation. Quite often a patch-activated cortex in a sensory, motor, or focus area generates a current of oblique orientation depending on the net orientation of the activated cortex. Areas up to **3 cm in diameter** can be very accurately modeled by a **single equivalent dipole**. There are six parameters that specify a dipole: three spatial coordinates and three dipole moment components (orientation angles and strength). The relative difference between the real potential and the single-dipole model that approximate the reality is usually presented in **relative residual energy (RRE)**. In the case of the spike-related averaged potential from Fig. 1.5.12a the RRE is only 8%. This low value of RRE indicates that the spike potential is well approximated by a single dipole. The orientation and location of this dipole are schematically presented on the right from the map of the spike-related potential.

INVERSE PROBLEM: NONPARAMETRIC SOLUTIONS

The modeling of real potential by a dipole is the simplest way of approaching the **inverse problem of EEG** (ie, finding the generators of the potential measured on the scalp). According to the laws of physics, opposite to the direct problem (ie, finding the potential for known generators) the indirect problem has **infinite number of solutions**. So, the EEG inverse problem is **ill posed** because for all realistic potentials the solution is nonunique and unstable (ie, highly sensitive to small changes in noisy data).

There are two main approaches to the inverse solution: the **nonparametric** and **parametric** methods. Nonparametric optimization methods are also referred to as **distributed source models** or imaging methods. In these models, dipole sources with fixed locations and possibly fixed orientations are **distributed in the whole cortical surface**. In the parametric approach, few dipoles are assumed in the model whose location and orientation are unknown.

Hämäläinen and Ilmoniemi (1994) in Helsinki University of Technology introduced a tomographic (a distributed source) solution to the EEG/MEG (magnetoencephalogram) inverse problem: the **minimum norm solution**. Several minimum norm estimates and their generalizations were suggested such as **LORETA, sLORETA**, VARETA, S-MAP, ST-MAP, Backus–Gilbert, LAURA, Shrinking LORETA, FOCUSS.

In this book we use a methodological approach called **low-resolution tomography (LORETA)**. LORETA was first introduced in 1994 by Roberto Pascual-Marqui from the KEY Institute for Brain-Mind Research in Zurich and his coworkers (Pascual-Marqui, Michel, & Lehmann, 1994). LORETA is based on certain electrophysiological and neuroanatomical constraints. It assumes that neighboring cortical areas produce similar local field potentials, which electrophysiologically could be the result of synchronous activation of neighboring neurons. From the mathematical point of view, these physiological constraints imply the smoothness of the spatially distributed potential. In essence, LORETA gives the smoothest of all possible inverse solutions. Its inverse solution corresponds to the 3D distribution of electric neuronal activity that has maximum similarity (ie, maximum synchronization),

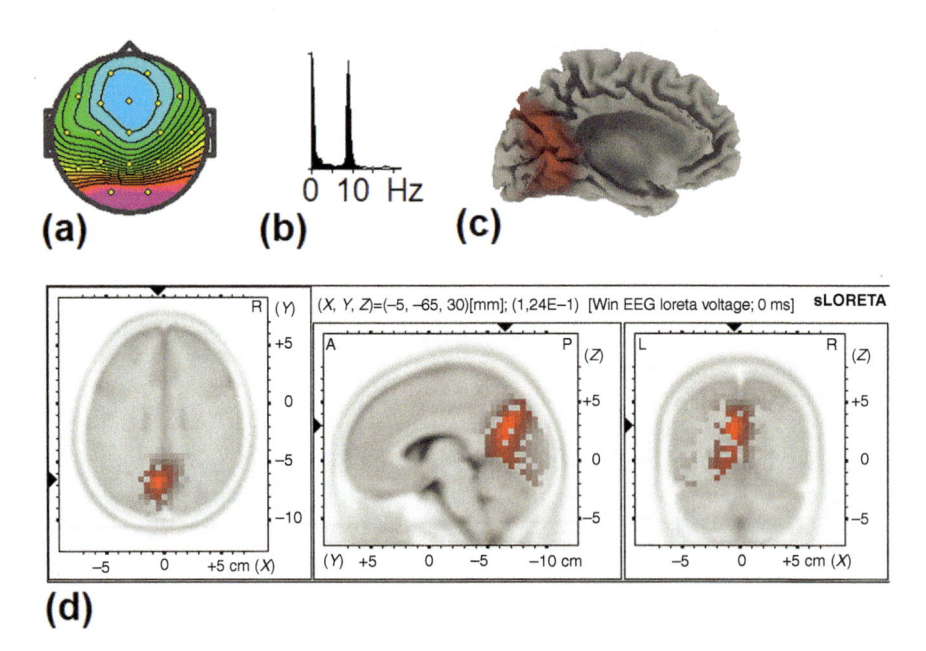

FIGURE 1.5.14 **sLORETA image of the topography corresponding to occipital alpha activity in a healthy subject.** (a) Map of the alpha rhythm; (b) EEG spectra; (c) 3D view of tomography; and (d) slice view of tomography.

in terms of orientation and strength, between neighboring neuronal populations (represented by adjacent voxels).

An example of application of sLORETA to the topography (a) of the occipital alpha rhythm with a 10-Hz peak at spectra (b) is presented in Fig. 1.5.14. 3D and 2D images are presented correspondingly in Fig. 1.5.14c,d.

CURRENT SOURCE DENSITY

Typically, LFPs such as those recorded from implanted multiple electrodes in the experiments in Fig. 1.5.12b,c are measured against a distant reference and are thus vulnerable to volume-conducted far-field effects. The **second spatial derivative** of LFPs eliminates this problem. Moreover, it has the ability to precisely localize transmembrane currents. From the point of view of physics, the method is based on quasistatic approximation of the **Maxwell equations** which can be expressed as:

$$I = -\nabla(\sigma \nabla P)$$

where P is the LFP, σ is the conductivity tensor, positive I represents an outgoing current (source), and negative I an incoming current (sink).

When multielectrodes are inserted radially to the cortical surface (as in Fig. 1.5.12b), the dendrites are oriented along the z direction and the dominant current flows along that direction. σ is homogeneous, so the equation can be rewritten as:

$$I = -\sigma \frac{\partial^2 P}{\partial z^2}$$

That is, the local current is proportional to the second derivative of the potential. In discrete values, I can be approximated as:

$$I = -\sigma \frac{P(z+h) - 2P(z) + P(z-h)}{h^2}$$

Fig. 1.5.12c shows the results of such computations made for multielectrode for the averaged interictal spike.

The formulas for current source density computations were extended to the case of multiple electrodes placed over the scalp by Francis Perrin and colleagues in the end of 1980s in the French National Institute of Health and Medical Research. They suggested a spherical spline interpolation method for modeling the potential distribution, from which the current density is easily computed. CSDs reflect sinks and sources of EEG currents and sometimes are used as a reference-free montage for EEG representation (Fig. 1.5.1) (Perrin et al., 1989).

BLIND SOURCE SEPARATION

The **blind source (or signal) separation problem** relates to separation of a set of sources from a set of mixed signals, without the aid of information about the source signals or the mixing process. The mathematical problem is illustrated by the **cocktail party problem** in which the goal is to separate individual voices from a mixture of acoustic recordings from several microphones installed in the cocktail room. This is actually a task for the KGB—to know who said what. In his novel *In the First Circle* (an allusion to Dante's first circle of Hell in *The Divine Comedy*) released in 1968, Aleksandr Solzhenitsyn (the famous Russian writer) depicts how a similar problem was solved by the occupants of a research and development bureau near Moscow made up of gulag inmates. One of the first solutions of the problem was presented in the *Proceedings of the Academy of Sciences of the USSR* in the 1950s.

Blind signal separation (Fig. 1.5.15) relies on the assumption that the source signals are **not related** to each other (ie, knowledge about one signal cannot tell you anything about the other signals). The particular definition type of being "not related" describes the particular method. For example, the signals can be defined as **mutually decorrelated** or **statistically independent**. A blind source separation method thus separates a set of recording signals into a set of source signals, such that the regularity of each resulting source signal is maximized and the regularity between the source signals is minimized.

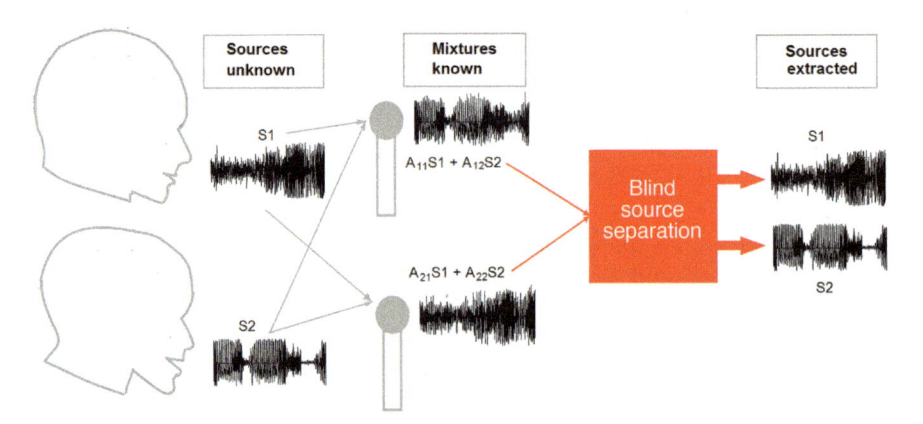

FIGURE 1.5.15 **Blind source separation.** The traces of two voices are mixed at two microphones. The voices are independent of each other: recognizing one voice does not tell us anything about the other voice. The blind source separation method restores the voices from their mixtures.

INDEPENDENT COMPONENT ANALYSIS

One of the approaches in blind source separation is **independent component analysis (ICA)**. We are now facing an era when new recording technologies such as PET (positron emission tomography), fMRI (functional magnetic resonance imaging), and EEG produce vast amounts of data. The extraction of essential features from the data therefore becomes crucial. The use of ICA for PET, fMRI, and EEG data is now very widespread since the ICA concept is easy to understand: it decomposes the data as a linear combination of statistically independent components.

The assumptions underlying ICA are: (1) the sources are **statistically independent**, (2) the mixing process is **linear and instantaneous**, (3) the sources have **non-Gaussian probability density functions**, (4) each source signal can be modeled as an **independent and identically distributed (IID) process**, (5) the number of independent sources is the **same as the number of sensors**.

Human spontaneous resting-state EEG seems to fit the first three assumptions because EEG can indeed be modeled as a collection of a fixed number of statistically independent brain processes, volume conduction of the electrical field in the brain tissue is **effectively instantaneous and almost linear**, and the EEG signal which is oscillatory by nature has a **non-Gaussian distribution**.

Assumption (4) is hard to prove because EEGs show nonzero **autocorrelation functions** and cannot be modeled as IID processes. Recall that IID by definition implies that an element in the sequence **is independent of the random variables that came before it**. However, this limitation could be overcome by taking a considerable number of sampling points for analysis (ie, by recoding EEG in a single subject for a very long time or by collecting many recordings in different subjects). Assumption (5) is also questionable since the EEG is temporally and spatially **nonstationary**, so that it is very likely that quite a large number of statistically independent sources may be necessary to accurately describe this complex process.

A **model of ICA** can be presented as follow. Let us suggest that there are N independent sources S_i (such as patches of synchronously activated dipoles within the cortex) that are linearly mixed to produce N potentials recorded from the scalp. Because of the volume conductance of the brain tissue, scalp, and skin any recorded potential $P_i(T)$ from the scalp is a **linear combination of the sources** (ie, $P_i(t) = \Sigma\ A_{ij}S_j(t)$, where A_{ij} are the constant weights that approximately correspond to distances from the source). This produces a system of linear equations which is the subject of a separate field of mathematics called **linear algebra**:

$$\begin{aligned} P_1(t) &= A_{11}S_1(t) + A_{12}S_2(t) + \ldots + A_{1N}S_N(t) \\ P_2(t) &= A_{21}S_1(t) + A_{22}S_2(t) + \ldots + A_{2N}S_N(t) \\ &\ldots \\ P_N(t) &= A_{N1}S_1(t) + A_{N2}S_2(t) + \ldots + A_{NN}S_N(t) \end{aligned}$$

In **matrix form** this equation will be simply presented as:

$$P(t) = AS(t)$$

where $P(t)$ is a matrix of potentials with N rows and M columns (in which M is the number of time samples), $S(t)$ is a matrix of potentials with N rows and M columns (in which M is the number of time samples), A is a mixing matrix with N columns (the **i column corresponds to the i source**) and N rows.

So the goal of ICA is to find the mixing matrix A such that $S_i(t)$ are maximally **statistically independent**. When A is found, S can be found as:

$$S(t) = A^{-1}P(t)$$

where A^{-1} is the inverse matrix.

To **back-project** the source S_i to the scalp (ie, to define the potential distribution which the source i generates on the scalp), a spatial filter can be used defined as:

$$F = AE^i A^{-1}$$

where E^i is a diagonal matrix in which only $\delta^{ii} = 1$ and the other elements are zero.

There are several ways to define the statistical independence of variables. Accordingly, the methods can be separated into two broad types: the first type of algorithms is intended to **minimize mutual information** and uses measures such as Kullback–Leibler divergence and **maximum entropy**; the second tries to maximize the non-Gaussianity of the sources and uses kurtosis and negentropy as the measures. Typical algorithms use centering (subtract the mean to create a zero mean signal) and whitening (ensures that all dimensions are treated equally for preprocessing the raw EEG signal). There are many algorithms of ICA including **Infomax**, FastICA, and JADE.

INDIVIDUAL ELECTROENCEPHALOGRAM DECOMPOSITION INTO INDEPENDENT COMPONENTS

For illustration of ICA possibilities, let us apply the Infomax algorithm to a 19-channel EEG recorded in a 20-year-old healthy subject (Fig. 1.5.16).

When applied to a raw EEG fragment of 20 minutes in the resting state, ICA gives 19 components (only 5 of them are shown). Each component is characterized by topography (a column in the unmixing matrix) and timecourse.

Physiological interpretation of independent components must be made with caution. They do not necessarily correspond to fMRI local areas of activation or to single dipoles approximating local patches of activation. By definition, independent components define **spatially fixed** and

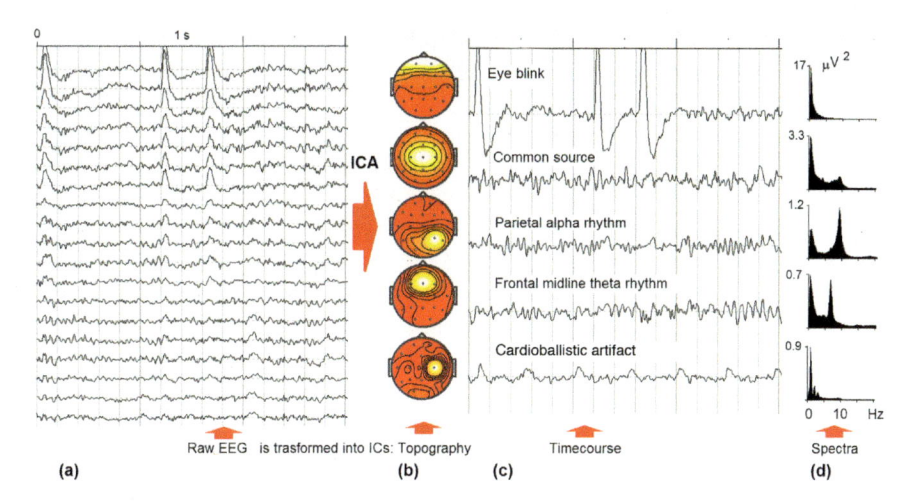

FIGURE 1.5.16 Decomposition of 19-channel *EEG* into independent components *(ICs)*. The 20-min resting-state EEG, a fragment of which is shown at (a) is decomposed by the Infomax algorithm of independent component analysis *(ICA)* into 19 components (5 of them are depicted); (b) topographies of the components; and (c) timecourse of the components; and (d) EEG spectra of back-projected components.

temporally independent potentials arising from cortical areas which may be **spatially distributed**. So the sources of independent components may be distributed networks rather than local patches of activity. In this way, ICA decomposition is different from CSD montages. ICA methods and software for imaging source dynamics of cortical activity from multi-channel EEG are freely available as open source software –EEGLAB. EEGLAB is an interactive Matlab toolbox for processing continuous and event-related EEG, MEG and other electrophysiological data. The toolbox incorporates ICA, artifact rejection, event-related statistics, and several useful modes of visualization of the averaged and single-trial data. For detail description of EEGLAB the reader is referred to the 2004 paper by Arnaud Delorme and Scot Makeig. The review of application of ICA for developing biomarkers in psychiatry is presented in a 2014 paper by Gráinne McLoughlin, Scott Makeig, and Ming Tsuang from University of California San Diego.

The use of ICA decomposition in clinical practice has certain limitations: (1) it does not presume identification of the EEG sources, (2) the topographies critically depend on the type of EEG so that (3) it is almost impossible to compare two different EEGs, (4) it is difficult to compile databases of these components to compare individual components with normative data. Indeed, the appearance of frontal midline theta rhythm and parietal alpha rhythm in the EEG presented in Fig. 1.5.16 is a relatively rare phenomenon, so that it would be difficult to compare independent components corresponding to these oscillations in this particular subject with a normative database. However, group ICA decomposition is a recently emerged methodology that seems to overcome these limitations.

GROUP ICA DECOMPOSITION

The group ICA methodology requires large and relatively homogeneous databases. The first attempt was made in 2010 by Marco Congedo and coworkers who used the normative database (N=57) of the Brain Research Laboratory of New York University School of Medicine headed by Roy John and colleagues, and the normative database of Nova Tech EEG, Inc (N=84) (Sherlin & Congedo, 2005; Sutton, Braren, Zubin & John, 1965). Both databases comprised EEGs collected in the eyes-closed condition only. In the same 2010 year our laboratory in the Institute of the Human Brain in Saint Petersburg published a paper on the group ICA decomposition of EEGs collected in the resting states of healthy controls from the HBI database (N=526) and of patients with depression (N=111) recruited from a private clinic in the Netherlands (Grin-Yatsenko, Baas, Ponomarev & Kropotov, 2010).

The main difference between the two group ICA methods was the technique for assessing the mixing matrix. The first team used the decorrelation technique based on second-order statistics in the frequency domain. This approach has a certain limitation: it cannot separate sources having an identical temporal correlation structure (ie, proportional power spectra). Our group used the InfoMax algorithm in the time domain, which is free from the abovementioned drawbacks but have some other limitations. In 2014 our group (Ponomarev, Mueller, Candrian, Grin-Yatsenko & Kropotov, 2014) made the next step in improving the method. In particular, the estimation of the optimal order of the ICA model of EEG (the number of components) was performed using three information theoretic criteria, and the reliability of the gICA model was assessed using five different methods. The gICA approach was applied for comparison the EEG spectra in a group of adult patients with ADHD (N=96) and a groups of healthy control subjects from the HBI database (N=376). It was demonstrated that the size effect (Cohen's d) of spectral differences for the independent components between these two groups of participants was considerably greater than in the case when spectral characteristics for single electrodes were compared. The topographies of the group-independent components with their sLORETA images obtained in the 2014 study are presented in Fig. 1.5.17.

All topographies except one *(red arrow)* reflect locally distributed sources. According to sLORETA images the **local components** are generated by the sources located in the cortex lying below the corresponding electrode. These local sources are weakly correlated with the signals from other local areas.

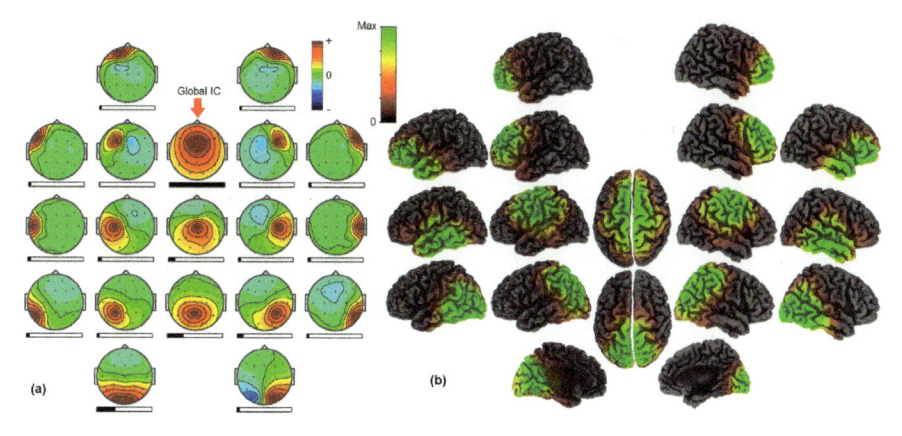

FIGURE 1.5.17 **Topographies and sLORETA images of group ICA.** ICA was performed on a collection of 3 min 19-channel EEG records of 376 healthy subjects (20–50 years of age) in the eyes-open condition. Horizontal bars below the maps partially filled with black indicate the relative power of components. *Adapted with permission from Ponomarev, Mueller, Candrian, Grin-Yatsenko, & Kropotov (2014).*

The **global component** integrates quite widely distributed sources. The nature of the global component is not clear. One possible explanation is that it reflects the activity of the reference electrode which in turn affects the potentials at all electrodes. Recall that the potential at the i electrode is $P_i = P_{itrue} - P_{ref}$, where P_{itrue} is a hypothetical true potential and P_{ref} is the potential at the reference electrode. But, this assumption does not explain the form of topography. The other possible explanation is the global component reflects a general source located, for example, in the brain stem which drives activity in widely distributed cortical areas with maximum in the frontal electrodes.

TEST–RETEST RELIABILITY

The test–retest reliability of spectral parameters of EEG in the resting state has been assessed in numerous studies. This parameter depends very much on the methods of measuring including the procedure of artifact correction, total time of recording, the epoch of spectra computation, and how well other parameters of recording—such as time of the day and interval since last meal—are controlled. Many studies report the lowest coefficient of reliability for delta and high-beta frequency bands. The reliability measures for theta, alpha, and low-beta bands are reported to be between 0.7 and 0.9.

# CHAPTER 1.6

Event-Related Potentials

DEFINITION

Event-related potentials (ERPs) are scalp-recorded voltage fluctuations that are **time-locked to an event**. The event can be a stimulus presentation followed by **sensory-related operations** (such as estimation of color, shape, or category of the visual stimulus), by **cognitive control operations** (such as selection of appropriate response or suppression of prepared action), and by **affective operations** (such as associated with positive or negative emotions) or **memory-related operations** (such as recalling an item or remembering a new item). The event can also be a **motor** or other type of subject response. ERPs as brain potentials are measured by the same amplifiers as EEG, and in this sense the measured parameter in EEG and ERPs is the same: electrical potential generated by the brain. However, from the functional point of view, there is a fundamental difference between these two measures of brain function.

INFORMATION FLOW

ERPs reflect stages of information processing in response to some event. The stages may include the flow of activation/inhibition patterns in sensory-related hierarchical neuronal networks induced by a sensory stimulus and the flow of activation/inhibition patterns in executive neuronal networks with a final output in the action. The hypothetical systems that deal with different patterns of information flow are presented in Fig. 1.6.1. Two "opposite" types of neuronal activity are shown in (b): **sensory-related activity** (high levels of activation just after the stimulus and low levels of activation before and during motor actions) and **motor-related activity** with an opposite pattern (low levels of activation just after the stimulus and high levels of activation before and during motor actions). In the sensory systems the first type of neuronal activity dominates. In the motor and cognitive control systems the second type of activity dominates.

Functional Neuromarkers for Psychiatry. http://dx.doi.org/10.1016/B978-0-12-410513-3.00006-1
Copyright © 2016 Elsevier Inc. All rights reserved.

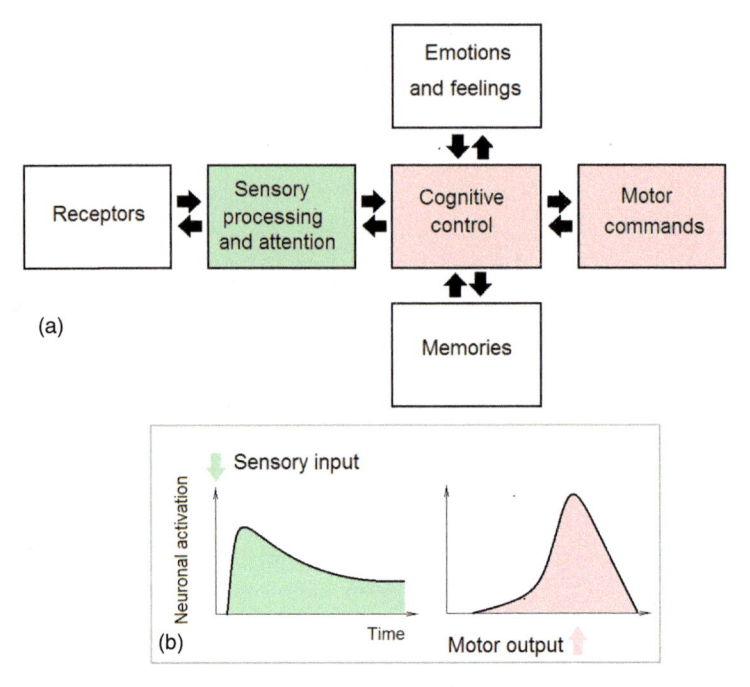

FIGURE 1.6.1 **Information flow in the brain.** (a) Brain systems that participate in "transcription" of sensory information into motor actions. The systems are heavily interconnected (only a few interconnections are shown for simplicity). (b) A schematic presentation of activity in sensory-related neurons and motor-related neurons.

The **sensory systems** extract sensory information from the input and select the most important part of the information by **attention mechanisms**. The **affective system** extracts the emotional content of the input and transfers it into emotions and feelings. The **memory systems** add the previous experience to the current situation and consolidate memories of the events for future use. The **cognitive control system** plans and initiates the relevant actions, suppresses irrelevant actions, and monitors behavior for quality control. These systems are distributed in different parts of the brain and are activated in different time intervals. The functioning of these different systems is studied by distinct behavioral paradigms and is reflected in different ERP components.

MONTAGES

ERPs as an **electric field** induced by information flow in neuronal networks are measured by the same electronic devices as EEGs (Fig. 1.6.2). So, everything written above about the methods of EEG recordings is equally true of ERPs. Like EEGs, ERPs are **volume-conducted** throughout

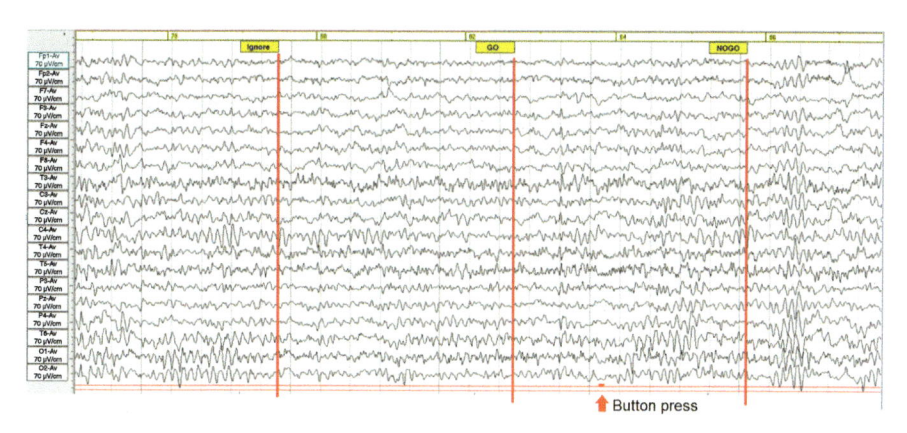

FIGURE 1.6.2 **ERPs are obtained from EEG recorded during the task.** An example of a 19 channel EEG fragment in a healthy subject performing a GO/NOGO task is shown. The first stimulus presentation in a trial is marked by the *red vertical line*. Time marks (seconds) are at the top. Note that it is impossible to detect by the naked eye the potential response evoked by stimulus presentation.

the brain, skull, and body so that a local source generates a potential distributed practically over the whole brain. In addition to the referential montage, other types of montages are used in ERP research such as the common average montage and current source density montage. As in EEG research the absence of a universal reference standard raised issues of **validity** and **across-study comparability** in ERP research and makes it hard to translate the results of these studies into clinical practice.

AVERAGING

The ERP amplitude is usually smaller than the amplitude of background EEG, so that it is quite difficult for the naked eye to separate the ERP waveform from ongoing EEG oscillations (Fig. 1.6.2). In other words, the ERP/EEG ratio (or the **signal-to-noise ratio**) in a single trial is rather small. To increase the ratio, EEG fragments are usually **averaged** over a large number of trials.

Let us suppose an EEG is recorded during a psychological task (eg, GO/NOGO task) in which trials of different categories (such as Ignore, GO and NOGO) are randomly presented many times (Fig. 1.6.3). Let us further suppose the Ignore condition is selected. In this condition visual stimuli are presented and no response is required. The whole EEG epoch is divided into small fragments corresponding to trials and consisting of **pre- and poststimulus** parts. An ERP is obtained by **averaging EEGs in these fragments**. In the optimal situation the EEG fluctuations preceding the stimulus should cancel each other in averaging and should make a

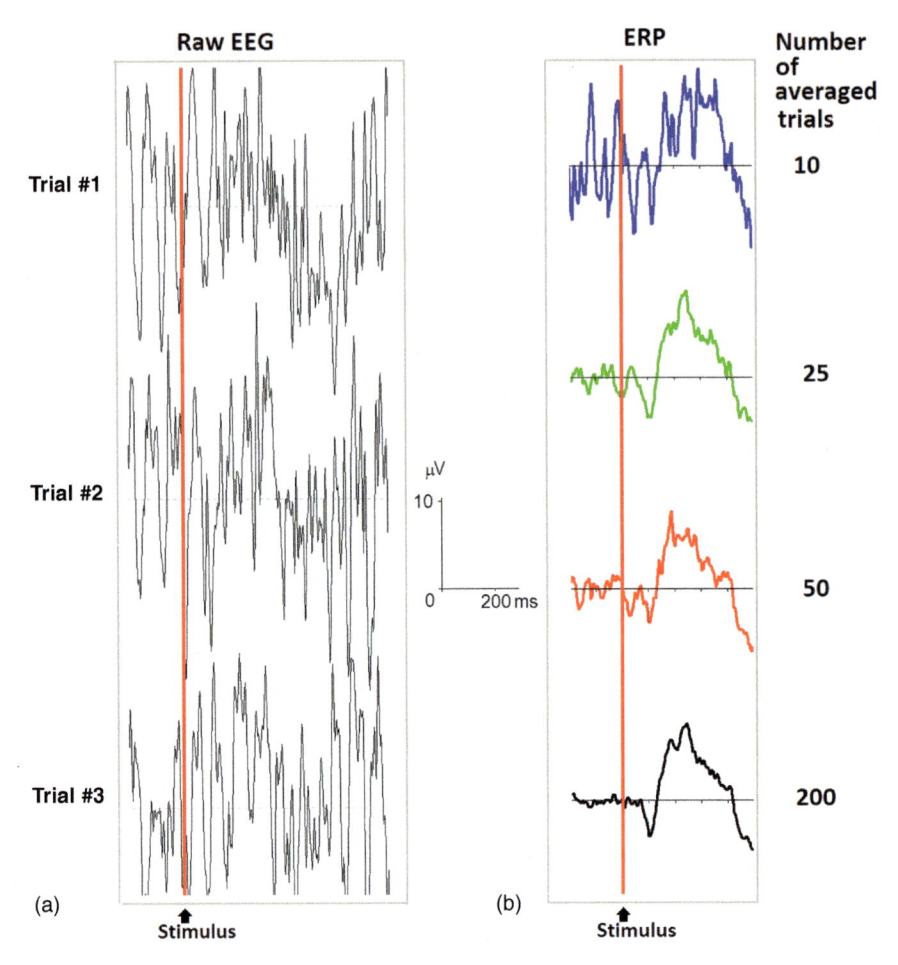

FIGURE 1.6.3 ERP is obtained by averaging raw EEG fragments. (a) An example of EEG fragments recorded in a healthy subject at the O1 electrode during a short presentation of a visual stimulus (no response is required) in three trials (*red line*, stimulus presentation). (b) ERPs in the same subject in response to visual stimulus presentation obtained by averaging different numbers of trials (10, 25, 50, and 200 trials).

prestimulus baseline close to zero. One can see that this is achieved by averaging over 200 trials in this particular subject and in this particular task.

NUMBER OF TRIALS

To obtain a reliable ERP the optimal number of averaged trials must be chosen. The as-many-as-possible general rule has a certain limitation: the subject may become tired during a long session such that ERP would change from the start to the end because of fatigue and adaptation.

In general, the optimal number of trials is defined by the **signal-to-noise ratio** in a single trial. For example, the amplitude of a mismatch negativity (MMN) wave measured at Fz, as the ERP difference between ERP to deviants and ERP to standards varies from one to just a few microvolts, makes the signal-to-noise ratio around 1/10. In this case the number of averaged trials should be large. To assess how large, one needs to rely on the fact that the signal-to-noise ratio increases with the number of averaged trials (N) as the square root of N ($\sqrt{N}$). So if the ratio after averaging needs to be as high as 10 in the case of MMN the number of averaged trials should be *10,000*! If we take into account that the probability of the deviant stimulus in the oddball task is 10%, then the number of overall trials must be *100,000*! If the interstimulus is 1 s, it takes more than a day to run the task. In practice, no one can afford it and researchers invent many techniques for increasing the signal to noise ratio during the limited time of testing.

A good estimation of the reliability of ERPs in a psychological task is given by assessing the variance of the averaged potential in the time interval preceding the onset of the trial. Because nothing usually happens before the trial, the averaged potential should be close to zero. A rule of thumb is the deviance of the background potential from zero must be much smaller than the component under study.

It should be noted here that the signal-to-noise ratio is defined not only by the amplitude of the measured wave but also by the amplitude of the background activity. For example, when measuring the visual N1 wave at occipital electrodes one needs to take into account the occipital alpha rhythm which could be rather high (at least in some individuals). In this case, to get a good measure of the C1 from a subject can take over 1000 artifact-free trials due to alpha noise, its small amplitude, and substantial overlap with the P1 component. However, large components such as P300 can be assessed with only about 40–60 trials.

INFORMATION FLOW IN VISUAL PATHWAYS

A simplified scheme of information flow in the visual system presented in Fig. 1.6.4 obeys the following general principles. First, there are two main streams: (1) the **ventral visual stream** that extracts information about shape, color, and category of the visual object, and (2) the **dorsal visual stream** that extracts information about position of the visual object in the 3D space needed for orientation and manipulation with visual objects. Second, information flow within each stream is **hierarchically organized**. Third, each cortical area receives numerous **feedback projections** from the higher cortical areas and from the higher order thalamic nuclei (not shown in the scheme). Fourth, the ventral and dorsal streams exchange information through **interconnecting pathways**.

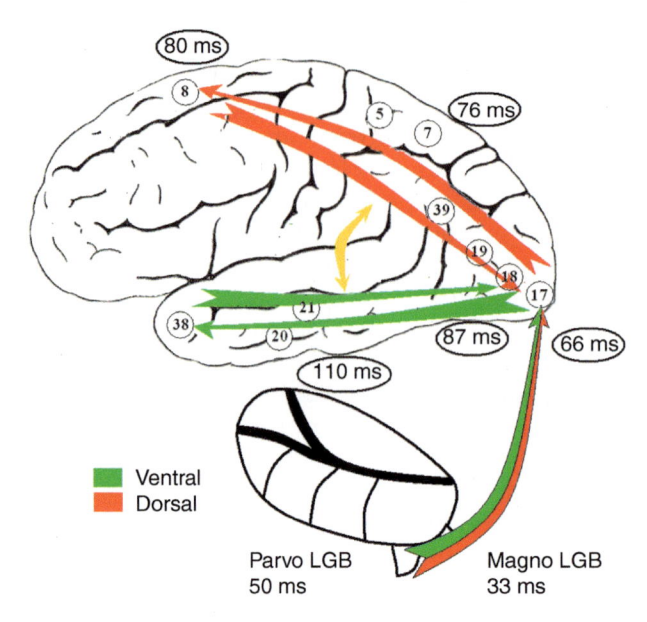

FIGURE 1.6.4 **Latencies of information flow in the visual system.** Schematic representation of dorsal *(red)* and ventral *(green)* streams in the visual system with onset latencies of neuronal responses (in milliseconds) to visual stimuli. Averaged latencies are obtained from monkey experiments of Schmolesky et al. (1998).

In a 1998 study by Mathew Schmolesky and his coworkers from Utah University onset latencies from many primate visual areas using the same experimental and analytical techniques were obtained (Fig. 1.6.4) (Schmolesky et al., 1998). The results demonstrate that neurons in the dorsal stream that arise from **magnocellular** neurons in the lateral geniculate body of the thalamus respond with shorter latencies than neurons in the ventral stream arising from parvocellular neurons of the thalamus. Note that it takes only 30 ms for information to reach the thalamus and 70 ms to reach the primary visual cortex. In the dorsal stream it takes only 14 ms for information to travel from the primary visual cortex to the frontal lobe.

INFORMATION FLOW IN LOCAL NETWORK

A stimulus presentation evokes a burst of action potentials in the thalamic neurons. Via local excitatory interconnections, initial activation from the lateral geniculate nucleus of the thalamus spreads over the local cortical neuronal network (Fig. 1.6.5). The spread of activation excites local short-axon intracortical neurons. Because of the short time constants of excitatory postsynaptic potentials the phase of excitation is rather fast. When inhibitory neurons become activated they inhibit adjacent excitatory neurons. The

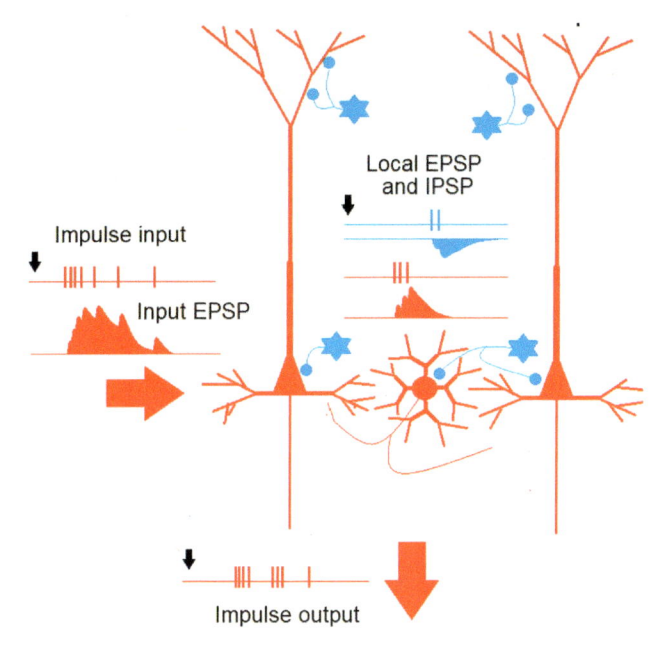

FIGURE 1.6.5 **Information flow in a local cortical network.** The action potentials arriving from external sources to layer 4 induce excitatory postsynaptic potentials in all neurons. When input EPSP reach the thresholds, they induce action potentials. Some of these action potentials leave the cortex and form the impulse output. Action potentials locally generate EPSP and IPSP depending on the type of neurons. Red color indicates excitatory neurons and EPSP. Blue color indicate inhibitory neurons and IPSP.

functional role of the intracortical inhibition is twofold: (1) preventing the spread of excitatory-driven information flow within the cortex and (2) shaping up the selectivity (such as orientation or spatial frequency selectivity in the visual cortex) of neuronal responses. Because of the slower time constants of inhibition, the phase of inhibition lasts longer than that of excitation.

TWO PACKETS OF INFORMATION FLOW

Excitatory neurons in visual streams process information in **two bursts of spikes** separated from each other by a relative silence interval of about 100 ms (Fig. 1.6.6a). Although feedforward and feedback information flows take part in shaping both packets, the first packet is dominated by **feedforward processing** (from primary to higher level cortical areas) whereas the second packet is dominated by **feedback** (from higher level to lower level cortical areas). Local field potentials recorded within the cortex show that neuronal responses in a distinct cortical area are reflected in positive/negative fluctuations repeated twice. In Fig. 1.6.6 the repeated

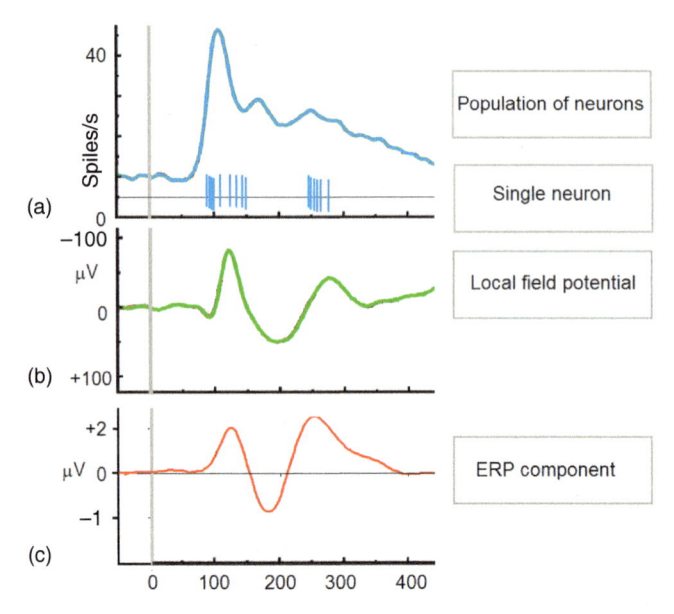

FIGURE 1.6.6 **Information in the ventral stream is processed in two packets.** (a) Responses to the discharge rate of the neuronal population and a single neuron. (b) Local field potential in the inferior temporal cortex of monkey in response to visual targets. (c) The temporally distributed independent component extracted from visual scalp recorded *ERPs* in human subjects. Note the difference in polarity and amplitude of intracranial and scalp recordings. *Part (a, b): adapted with permission from Anderson & Sheinberg (2008).*

response is presented by P1/N1 and P2/N2 waves in local field potentials and scalp-recorded ERP.

Fig. 1.6.7 illustrates cortical–cortical feedback connections between high- and low-level sensory cortical areas. Feedback connections seem to be responsible for the recurrent stage of information processing and for the P2/N2 waves of ERPs.

CANONICAL VISUAL EVENT-RELATED POTENTIAL

A visual stimulus activates both ventral and dorsal visual pathways. The potential waveform in response to a visual stimulus is schematically presented on Fig. 1.6.8a. It should be stressed that for simplicity all waveforms are presented on a single curve, but in reality they have different topographies. The waves are marked by a letter (**N for negative, P for positive**) and a number that approximately corresponds to the peak latency (eg, N1 or **N100** corresponds to a negative wave with peak latency around 100 ms).

The first wave is named **C1** (from central because it is mapped into the central posterior area). C1 peaks around 70–100 ms and is generated in

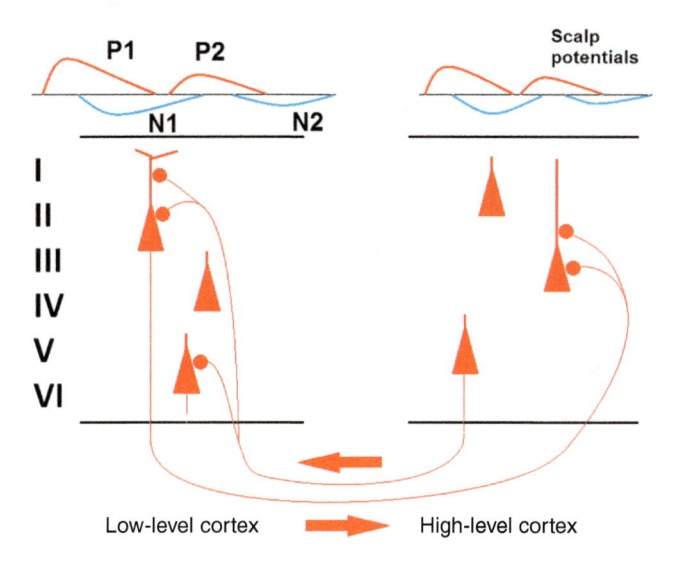

FIGURE 1.6.7 **Recurrent activation.** Low- and high-level cortical areas are interconnected. This interconnection enables recurrent activation at both cortical levels, so that the initial positive/negative potential wave is followed by the secondary (recurrent) positive/negative wave. For simplicity, no subcortical recurrent pathways are shown.

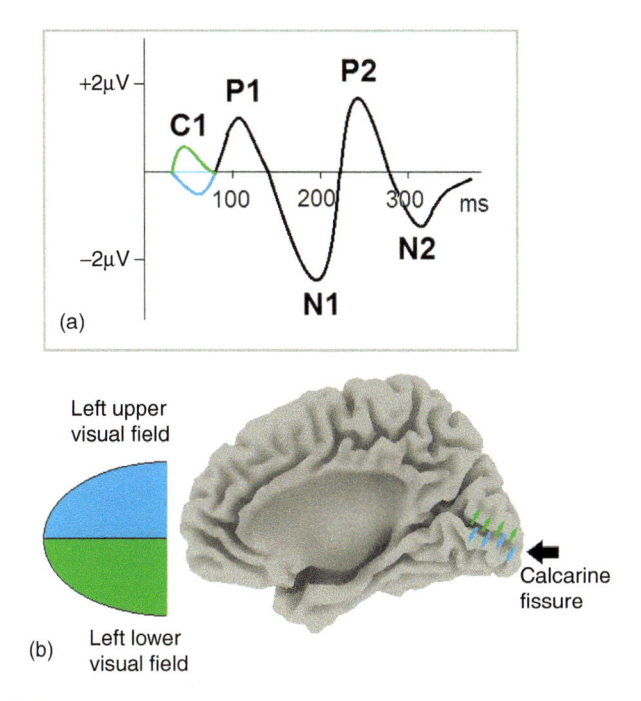

FIGURE 1.6.8 **Canonical visual ERP.** (a) The potential response consists of C1, P1, N1, P2, and N2 waves. Only the C1 wave fits the latencies of primary visual cortex activation presented in Fig. 1.6.4. The other waves appear to be a result of the complex interaction in feedforward, feedback, and parallel connections. (b) Stimulus in the upper (lower) visual field activates neurons in the lower (upper) part of the calcarine fissure (a primary visual cortex).

1. METHODS OF ASSESSING NEUROMARKERS

the striate cortex located in the calcarine fissure of the occipital lobe. The polarity (positive or negative) depends on spatial location of the visual stimulus because of the convolution of the striate cortex along the calcarine fissure (Fig. 1.6.8b). The C1 wave is a marker of the **earliest visual processing** within the cortex. When the visual stimulus is presented in the center of the visual field the C1 wave is quite small and difficult to detect.

EVENT-RELATED POTENTIAL PARADIGMS

Almost any psychological computerized test with consequently presented trials can be used for ERP recording. The ERP paradigms can be separated into large categories depending on the system under study: sensory, motor, attention, emotion, memory, and cognitive control tasks.

Sensory ERP components are modulated by modality (visual, auditory, somatosensory, etc.), by physical features within the modality (eg, position, spatial frequency, color, orientation), by stimulus category (eg, living objects vs. nonliving objects, faces vs. tools), by rate of stimulus presentation (eg, with 100- to 2000-ms interstimulus intervals), by stimulus intensity (eg, loudness of the acoustic stimulus). The paradigms in which these modulations have been studied include the **dichotic listening task** in the auditory modality; the **passive oddball task** and its variants in the auditory and visual modalities; random presentation of auditory stimuli with **different intensities**; and the **two-click paradigm**.

In studying the motor system, ERPs are locked to motor actions—not to sensory stimuli. The action can be self-initiated, or contingent on some sensory events. In studying attention, three stimulus paradigm and the Posner paradigm are used. Cognitive control tasks mimic behavioral situations when a **preportent model of performance** is formed by task setting and dominated during the task, but in some cases it is violated so that the subject has to **override** the prepared action with a different action or to **suppress** the prepared action.

In studying the affective system, emotionally competent stimuli are presented to induce positive or negative emotions. In studying the memory system, stimuli may be memorized before the experiment and subjects are instructed to differentiate remembered stimuli from new ones.

MULTIPLE SOURCES OF EVENT-RELATED POTENTIALS

Neuroscience shows that multiple cortical sources are activated even in response to a brief visual stimulation. The visual stimulus is represented more than **20 times** in spatially distinct cortical areas. Each of these areas

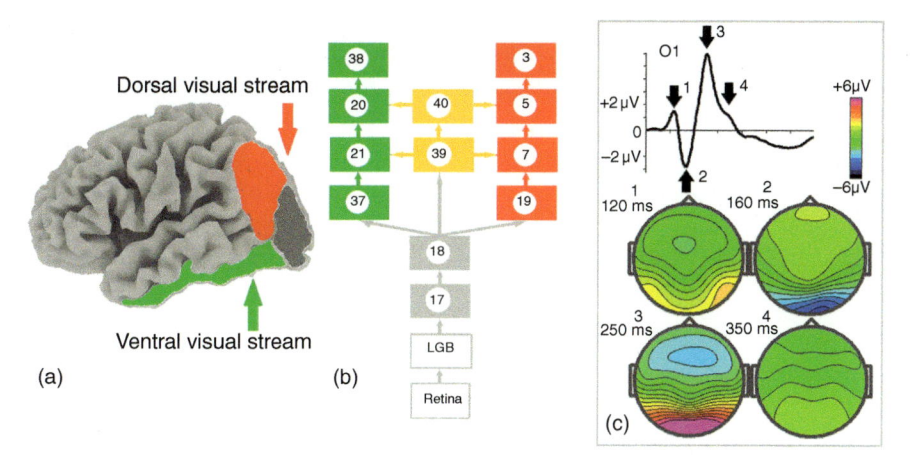

FIGURE 1.6.9 **Multiple sources of ERPs.** (a) Visual ventral and dorsal streams include (b) multiple cortical areas in which different properties of a visual stimulus are extracted. (c) A brief presentation of a visual stimulus induces ERPs that have quite different topographies at different latencies indicating the dynamics of spatial distribution of neuronal responses.

keeps a topographic presentation of the visual field and extracts specific features of the visual stimulus. The existence of multiple neuronal representations of the stimulus presumes that even a simple visual ERP is a sum of activation patterns in distinct cortical areas. This inference is illustrated by the 19-channel ERPs computed in response to a short (100-ms) presentation of a visual stimulus under the Ignore condition (no response is required) (Fig. 1.6.9).

One can see that a brief presentation of the visual stimulus evokes a sequence of electrical fluctuations. These fluctuations show quite different topographies indicating that they are generated by different cortical sources. These different sources are activated at different times and appear to have different functional meanings. This begs the question: How can these sources be separated?

SEPARATING COMPONENTS: SUBTRACTION APPROACH

In ERP research, functional sources generating ERP waves are called "components." Components are associated with hypothetical psychological operations. One way of separating components is to obtain a difference between the two ERPs performed under conditions which differ in only one respect: the presence or absence of a **hypothetical operation.**

As an example let us consider the cued GO/NOGO task used for studying operations of cognitive control. In this task, two categories of

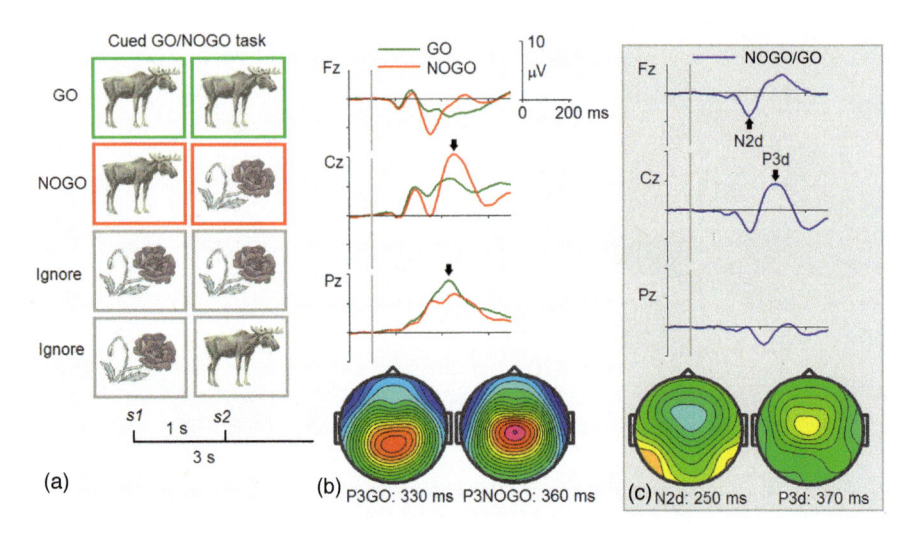

FIGURE 1.6.10 **Subtraction approach for separating ERP components.** (a) Scheme of the Cued GO/NOGO task; (b) Grand-averaged ERPs in the common reference montage for a group of healthy subjects (20–25 years old) for GO *(green)* and NOGO *(red)* conditions in the cued GO/NOGO task with maps made at maximums of the waves at the bottom. (c) NOGO/GO difference waves with maps at peaks of N2 and P3 NOGO difference waves.

stimuli – images of animals (*a*) and images of plants (*p*) – are randomly presented in pairs: *a–a*, *a–p*, *p–p*, *p–a* with the subject task to press a button as fast and precise as possible in response to *a-a* pairs (GO stimuli), to ignore *p–p* and *p–a* pairs (Ignore conditions), and to withhold from prepared action in *a–p* pairs (Fig. 1.6.10a). The inter-stimulus intervals within the pairs are of 1000 ms and inter-trial intervals are of 3000 ms. The duration of stimuli is 100 ms. The working hypothesis presumes that the brain builds up a prepotent model of behavior according to which the subject presses a button as soon as possible in response to the GO stimulus. Presentation of the NOGO stimulus evokes two hypothetical operations: detection of conflict (that is not what the subject has been prepared to see) and its monitoring, and inhibition of the prepotent action (the subject has to withhold from prepared action). So, the difference wave NOGO–GO must reflect these hypothetical operations. Fig. 1.6.10 depicts the GO and NOGO ERPs and their differences in the cued GO/NOGO task. One can see that the differences in ERPs include two waves: (1) a negative wave with peak latency of 250 ms and frontal distribution **N2d**, (2) a positive wave with peak latency at 370 ms and frontal-central distribution **P3d**. When these waves were first discovered they were associated with action inhibition and conflict-monitoring operations. No agreement about which wave reflects which operation has yet been achieved though.

SEPARATING COMPONENTS: SINGLE TRIAL INDEPENDENT COMPONENT ANALYSIS

The other approach is to apply blind source separation methodology for EEG fragments (eg, 1-s fragments) corresponding to GO and NOGO conditions recorded in an individual subject over the entire number of trials in the task. Independent component analysis (ICA), as a method of blind source separation, can be applied. This method was introduced into ERP research by Scott Makeig and his group from the Naval Health Research Center in San Diego in 1997.

As an example, let us consider 19-channel ERPs recorded in a 19-year-old healthy subject who performed the same cued GO/NOGO task twice with an interval of 11 days. The matrix P of single trial potentials of the ICA is constructed as a $19 \times 50{,}000$ matrix where the number of rows corresponds to the number of channels and 50,000 is the number of potential samples obtained as a product of 2 (number of categories, GO and NOGO), 250 (number of points for 1000-ms interval sampled at 250 Hz, the sampling rate of the EEG machine), and 100 (the number of trials). The results of application of the Infomax algorithm to the matrix P for the first and second recording in the same subject are presented in Fig. 1.6.11. One can see that in both recordings of this subject the NOGO stimulus elicits N2d/P3d waves in comparison with the late positive fluctuation elicited by the GO stimulus. This pattern is decomposed by single trial ICA into two components with different localizations and different latencies: the frontally distributed component shows peak latency of P3 NOGO at 380 ms while the centrally distributed component has peak latency at

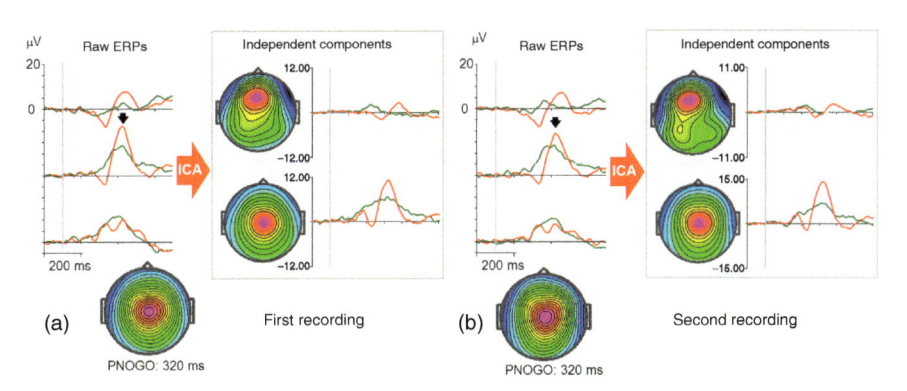

FIGURE 1.6.11 Single trial ICA. Single trial ICA applied to ERPs in the common reference montage of one healthy subject preforming the same cued GO/NOGO task two times with interval of 11 days. The P3 GO/NOGO waves (on the left) of the subject for the first (a) and second (b) recordings are decomposed into the two components (on the right in the box). Each component is characterized by topography and time course for GO (*green*) and NOGO (*red*) trials. Common reference montage. *Data from Brunner et al. (2013).*

320 ms. This decomposition is extremely robust showing almost the same temporal–spatial parameters for the two recordings separated by 11 days.

The components obtained by this approach are quite individual, so that it is impossible to directly compare two subjects or a subject with a database. However, for research purposes, it is important to know the common properties of a class of independent components. One way of solving this is to approximate each component by a single dipole and perform cluster analysis on the array of dipole localizations.

SEPARATING COMPONENTS: GROUP INDEPENDENT COMPONENT ANALYSIS IN MULTIPLE TASKS

Group ICA performed on a large array of ERPs lacks the weaknesses of the single trial ICA. However, this approach faces one important limitation: the number of individual ERPs in the group should be large enough to obtain enough training points to perform a robust ICA procedure. As Onton and Makeig (2006) paper written at the University of California, the number of training points must be much larger than $20N^2$, where N is the number of electrodes. In our studies the estimated lowest number of subject*conditions for 19 electrodes and 250 sampling points for one electrode and condition is more than 100.

In a recent study of our group the number of healthy subjects was 40 and the number of tasks was 4 thus making the number of subject*conditions 160. Schemes of the tasks used in this study are presented at Fig. 1.6.12.

Fig. 1.6.13 depicts two components that were separated by ICA from the collection of ERPs in the four tasks described above. The components are presented for the conventional a–a "r" GO task. One of the components (top) has more frontal distribution and 50-ms longer latency than the other one (bottom). The frontal component in the a–a 'r' GO task shows clear N2 fluctuation under the NOGO condition and is generated, according to sLORETA, in the cingulate cortex. The central component is generated near the supplementary motor cortex.

SEPARATING COMPONENTS: GROUP INDEPENDENT COMPONENT ANALYSIS IN A SINGLE TASK

The next approach would be to collect more subjects and use a single task. Fig. 1.6.14 depicts two components that were separated by ICA from the collection of 267 ERPs in the a–a 'r' task. One of the components (top) has more frontal distribution and 50-ms longer latency than the other one (bottom). The frontal component shows clear N2 fluctuation under the

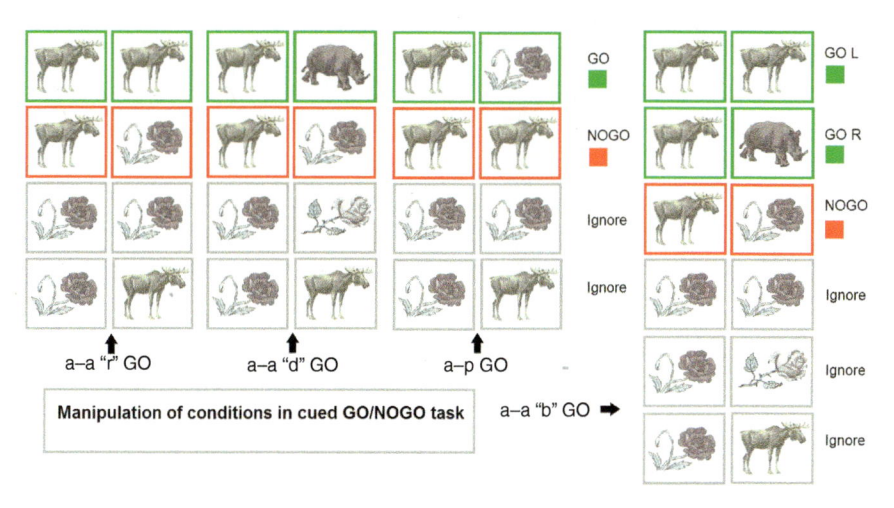

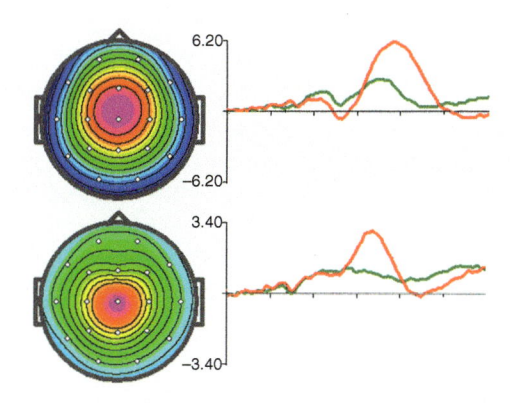

FIGURE 1.6.12 **Four task modifications used for group component analysis.** Two categories of stimuli animals *(a)* and plants *(p)* are used as stimuli. They are presented in different combinations in pairs *a–a, a–p, p–p, p–a*. The intratrial interval is 1000 ms, the intertrail interval is 3000 ms. In the first task (*a–a 'r' GO*) the images of stimuli repeated in *a–a* and *p–p* trials. In the second task the stimuli in the first and second stimuli in *a–a, p–p* pairs are physically different, although belonging to the same category. In both tasks the subjects were instructed to press a button to *a–a* pairs. In the third task, which used the same stimuli as the first, the subjects were instructed by press a button to *p–a* pairs. The fourth task used trials of the first two tasks with an instruction to use the index finger on the right hand to respond to identical *a–a* pairs, and use the middle finger on the right hand to respond to physically different *a–a* pairs. *Data from Kropotov et al. (in preparation).*

ICA components: 40 subjects*4 tasks

FIGURE 1.6.13 **Four tasks, 40 subjects: ICA decomposition of P3 GO/NOGO waves into two components (top and bottom).** (Left) Topographies of the components; (right) timecourses of the components for the task *a–a 'r' GO*. Common reference montage. *Data from Kropotov et al. (in preparation).*

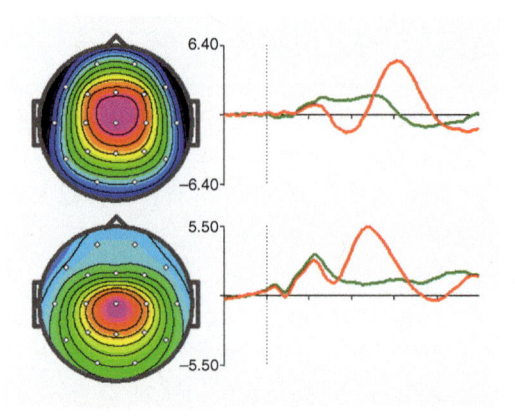

ICA components: 267 subjects*1 task

FIGURE 1.6.14 **One task, 267 subjects: *ICA* decomposition of P3 GO/NOGO waves into two components (top and bottom).** (Left) Topographies of the components; (right) timecourses of the components. Common reference montage. *Data from Human Brain Indices (HBI) database.*

NOGO condition. It should be noted here that the correspondence between the components obtained by the two ICA-based approaches is not perfect which can be explained by difference in the groups that participated in the two approaches (40 subjects of age 20–26 in the first approach and 267 subjects of age 17–45 in the second approach).

SEPARATING COMPONENTS: JOINT DIAGONALIZATION OF COVARIANCE MATRIXES

The two group ICA decomposition methods have limitations: (1) the ERP can hardly be presented as an **identically distributed process** with zero autocorrelation function, (2) the number of components is not necessarily 19 and must be estimated before the analysis. These limitations were overcome in the approach recently developed by Jury Kropotov and Valery Ponomarev at the N.P. Bechtereva Institute of the Human Brain in 2015 (Kropotov and Ponomarev, 2015). The ERP $x_i^j(t)$ at the i electrode ($i = 1, ..., M$) of the j subject ($j = 1, ..., P$) at time point t ($t = 1, ..., T$) is modeled as a linear combination of hidden (latent) signals $S_k^j(t)$, $k = 1, ..., K$, $K \leq M$, with some unknown coefficients a_{ik}:

$$x_i^j(t) = \sum_{k=1}^{K} a_{i,k} S_k^j(t) + \varepsilon_i^j(t),$$

where ε is the noise component that is assumed to be additive, stationary, isotropic, and independent of the signals $S_k^j(t)$. A is the $M \times K$

mixing matrix with elements a_{ik}, which is assumed to be the same for all subjects.

To separate the latent components hidden in ERPs, we used the blind source separation method based on second-order statistics described by Arie Yeredor from Tel-Aviv University in 2010 (Yeredor, 2010). The main assumption underlying the separation method is that the variability of deviations of individual ERPs from group average potentials can be modeled by linear combinations of low-correlated signals. To determine the optimal number of hidden signals K, we used a model selection method based on the Bayesian information criterion (BIC) first described by David MacKay from California Institute of Technology in 1992 (MacKay, 1992). An example of application of this approach to a collection of ERPs recorded in the a–a 'r' GO/NOGO task in 454 subjects is presented in Fig. 1.6.15. Note that the P3 NOGO wave was decomposed into two frontal-centrally distributed components. Comparison of these components with the previously selected components must be done with caution taking into account the difference between the groups (age of the last group is 18–84) and the difference in montage: referential montage for the last group and common reference montages for the first two groups.

In spite of all the differences between the individual and the three group approaches for decomposing ERPs into latent components for the cued GO/NOGO task, the consistent finding is the existence to two different components: (1) the frontally distributed component with long (about

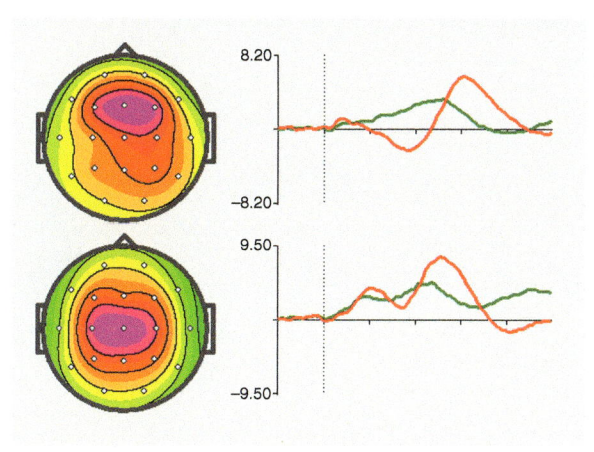

FIGURE 1.6.15 **One task, 454 subjects: decomposition based on joint diagonalization of covariance matrixes (JD of P3 GO/NOGO waves) into two components (top and bottom).** Referential (linked ears) montage. (Left) Topographies of the components; (right) timecourses of the components. *Data from Human Brain Indices (HBI) database.*

400-ms) P3 NOGO latency and clear N2 NOGO fluctuation and (2) the centrally distributed component with latency at about 50 ms shorter in comparison with the frontal component. sLORETA images often indicate localization of the frontal component in the cingulate cortex while the central component is localized in the supplementary motor cortex.

TEST–RETEST RELIABILITY

ERPs are quite reliable measures of brain functioning. Numerous studies in attempts to estimate the test–retest reliability of ERP waves mostly focused on the target P3 wave in the auditory oddball paradigm. Test–retest reliability of this wave varied in these studies from 0.50 to 0.86 for amplitude parameters and from 0.40 to 0.88 for latency parameters. In a 2013 study by Jan Brunner and coworkers from the Norwegian University of Science and Technology in Trondheim, ERPs in the cued GO/NOGO task were recorded in 26 subjects on two occasions with an interval of 6–18 months between recordings (Fig. 1.6.16) (Brunner et al., 2013). Interclass correlations between first and second recordings were around 0.8 for amplitude and around 0.9 for latency of the P3 NOGO waves.

The authors further extracted the NOGO components for each subject by means of spatial filtration. Spatial filters were obtained from the group ICA approach performed on a group of 102 healthy subjects. Interclass correlations between first and second recordings were around 0.8 for amplitude and around 0.9 for latency of the components (Fig. 1.6.17). The authors conclude that the long-term stability of the P3 NOGO wave and

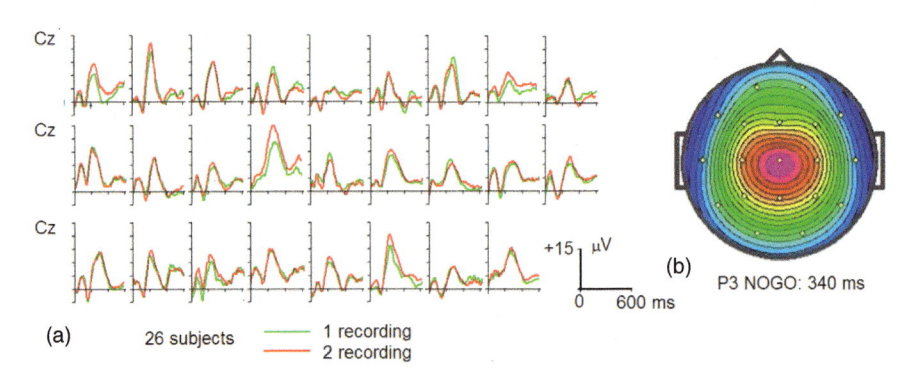

FIGURE 1.6.16 **Interindividual differences in ERPs.** (a) NOGO ERPs are computed at Cz for a group of 26 healthy subjects preforming the same cued GO/NOGO task at two different sessions with intervals of about a few months. *Green line*, first recording; *red line*, second recording. (b) map of P3 NOGO for the first recording at 340 ms. *Adapted with permission from Brunner et al. (2013).*

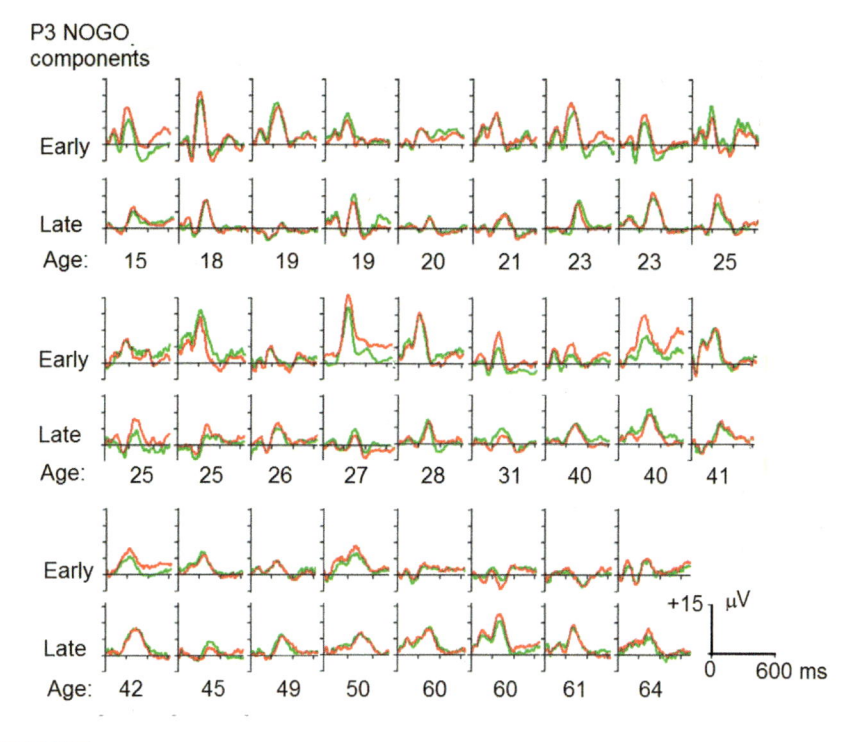

FIGURE 1.6.17 **Interindividual differences in ERP independent components.** Late and early independent components are extracted for each subject by means of spatial filtration on the basis of the group ICA presented above (see section "Separating components: group independent component analysis in a single task"). The components are back-projected and presented for Cz. *Green line*, first recording; *red line*, second recording. *Adapted with permission from Brunner et al. (2013).*

the independent components can be used in the clinical practice for reliable assessment of cognitive control function.

INTERINDIVIDUAL VARIABILITY

Although ERP waves, if recorded according to standard requirements, are quite stable in time, they vary substantially from subject to subject (Fig. 1.6.16). One source of interindividual differences could be variances in **anatomical folding patterns** of the cortex. This might be true for components generated in local cortical areas, but will hardly be correct for widely distributed components.

Another source of interindividual differences could be **fundamental differences in information processing** that grant each of us unique sensory, emotion, and cognitive abilities. This suggestion is supported by the

2015 study of Jan Brunner and coworkers from the Norwegian University of Science and Technology in Trondheim in which components of the NOGO wave in the cued GO/NOGO task were shown to be strongly correlated with neuropsychological parameters of individual performance (Brunner et al., 2015).

A ROADMAP FOR THE DEVELOPMENT AND VALIDATION OF EVENT-RELATED POTENTIAL NEUROMARKERS

In October 2009 the Cognitive Neuroscience Treatment Research to Improve Cognition in Schizophrenia Initiative convened a meeting to discuss the development of biomarkers of cognitive dysfunction in schizophrenia. A group of scientists headed by Steven J. Luck from University of California, Davis wrote a paper on the consensus about application of ERPs as biomarkers of cognitive dysfunction in schizophrenia (Luck et al., 2011). On the basis of the data available in those days they reached a number of conclusions. First, ERPs may serve as **endophenotype markers** and are relatively convenient to use in large-scale genetic studies. Second, ERPs can be used in **preclinical research** to **define potential treatment targets**. That is, if a given ERP component measures the operation of a given cognitive process, neural circuit, or transmitter–receptor system, then an abnormality of this component suggests that treatments targeting that process, circuit, or system might produce therapeutic benefits. Third, homologs of human ERPs can be found in both rodents and primates, providing opportunities for **across-species translational research**. Fourth, ERPs can be used in human **clinical research** to **determine whether a given treatment influences the specific process, circuit, or transmitter system of interest**. ERPs can also be used in healthy volunteers in whom ERP abnormality has been transiently induced by a pharmacological challenge. Fifth, ERPs might be useful in **defining subgroups within a disorder**. These subgroups might uniquely respond to new treatment methods and, consequently, can be used in corresponding clinical trials.

According to the authors of the consensus it is realistic to expect that ERPs could be used as biomarkers in the identification of **genes related to mental illness, in efforts to identify individuals at high risk** of developing disorders before illness onset, and in the development and assessment of new treatments.

PART 2

NEUROMARKERS OF CORTICAL SELF-REGULATION

I recall that at a scientific conference 10 years ago a psychiatrist announced that he did not use EEG in his clinical practice because no mechanisms of EEG rhythms were known because no quantitative measure existed, and because no evidence that EEG would aid in diagnosis had been proved. This chapter is intended to show that this statement is no longer correct. New insights into the field of EEG discovered during the last few decades have changed classical postulates which psychiatrists and neurologists used to learn in medical schools. In the majority of neuroscience textbooks, task-related alterations in neuronal impulse activity were considered the major indicators of information processing in the human brain. EEG rhythms were usually ignored or considered background activity.

We are now facing the renascence of EEG. This renascence is associated with the appearance of new methods in human EEG assessment and new experimental findings in animal and human research. These findings demonstrate that the oscillatory patterns of EEG play a critical role in maintenance of brain functions and consequently may be used as a powerful tool for the diagnosis of brain dysfunctions.

It should be stressed that EEG is a sensitive parameter of a subject's state and EEG rhythms change dramatically when the subject falls asleep and transfers from one stage of sleep to another. For example, at stage II specific oscillations called sleep spindles emerge. Sleep spindles disappear while theta and delta rhythms develop at further stages of sleep. During wakefulness, rhythms can be a sensitive measure of brain responses to different psychological tasks. For example, occipital alpha rhythms are suppressed (desynchronized) while frontal beta rhythms are enhanced (synchronized) in response to behaviorally meaningful visual stimuli.

In the diseased brain the normal mechanisms of EEG rhythms may be impaired and normal rhythms may (1) become slower in frequency (so-called EEG slowing), (2) may appear in unusual places (eg, alpha rhythms at temporal areas), (3) may become higher in amplitude (a phenomenon called hypersynchronization) and in greater synchronicity with other areas (a phenomenon called hypercoherence), (4) in some severe cases (eg, characterized by disconnection of cortical areas from subcortical structures due to stroke, trauma, or tumor) a separate slow rhythm in delta frequency (1–3 Hz) may appear. In some cases the local balance between excitation and inhibition may be destroyed to produce an epileptic focus which is manifested in interictal spike/slow-wave complexes recorded in spontaneous EEGs. Normative databases help a clinician to recognize those abnormal patterns and to assess the level of statistical significance of the abnormality.

CHAPTER 2.1

Infraslow Electrical Oscillations

ARRHYTHMIC ELECTROENCEPHALOGRAMS

In EEG, **oscillations** are usually emphasized. By definition, brain oscillations are repetitive patterns of brain waves with a particular temporal beat. On EEG spectra the oscillations are expressed in a form of local peaks (bumps). The peak frequency of these bumps corresponds to the frequency of the repetitive waves. There are several types of rhythmicities in the cortex with different underlying mechanisms and different roles in brain dysfunction.

However **irregular, arrhythmic patterns** can be clearly seen in human EEGs; they coexist with brain oscillations. Moreover, EEGs of approximately 10% of healthy subjects with low-voltage fast EEG type are dominated by irregular patterns. The arrhythmic patterns in EEGs are less studied. We know surprisingly little about their mechanisms and their implication for brain dysfunctions.

POWER-LAW FUNCTION OF ELECTROENCEPHALOGRAM SPECTRA

What we know is that EEG power spectra of arrhythmic EEG obey a simple rule: power dramatically decreases with frequency. Mathematically, the arrhythmic component of EEG spectra can be well approximated by **a power-law function**:

$$P(f) = \frac{1}{f^b}$$

where P is EEG power, f is frequency, and b is a parameter (typically between 0 and 3) named the "power-law exponent."

Note that **white noise** is a special case of arrhythmic activity, in which case b equals zero and power is constant across all frequencies. The power-law function indicates that no particular frequency (or time scale

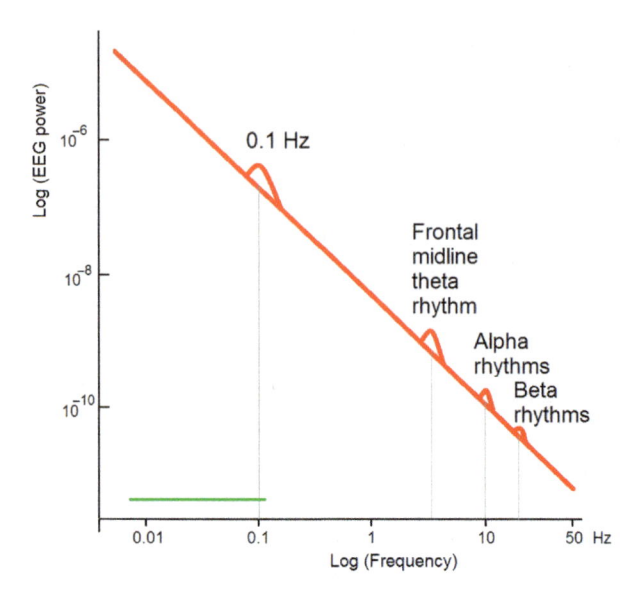

FIGURE 2.1.1 **Power-law function of EEG.** Schematic presentation of the power-law function of EEG. The mean power spectrum in the double-logarithmic coordinates is approximated by the straight line. The theta, alpha and beta rhythms in EEG are represented as bumps on the straight line

in general) dominates the dynamics of the parameter. This feature is called "scale invariance." It was described at many structural levels: in single-neuron membrane potentials, in local field potentials, and in EEG.

The power-law distribution has been observed in a variety of systems, such as earthquakes, finance, and solar flares. My old friend theoretical physicist Yury Pismak found such distributions in realistic neuronal network models suggested by me in the 1980s. So, there is a temptation to see in this equation a universal law of a wide range of complex systems.

In double-logarithmic coordinates the $1/f^b$ function appears as a straight line. Brain oscillations appear as bumps on top of this line (Fig. 2.1.1). For decades, brain activity contained in the $1/f$ slope was considered unimportant and was often removed from analyses to emphasize brain oscillations.

We now see increasing evidence suggesting that scale-free brain activity, in general, and EEG, in particular, is an important part of brain functioning. However, this field of research is still in its infancy. For more information we refer the reader to a 2010 review published in *Trends in Cognitive Sciences* by Biyu Jade He from National Institutes of Health, USA and coauthors.

INFRASLOW ELECTRICAL OSCILLATIONS: HISTORY

Infraslow spontaneous oscillations of human cortical potential were discovered in the middle of the 20th century when appropriate methods for their recording were designed. These oscillations became the subject of intensive research in the 1970s and 1980s in the former Soviet Union. The pioneers in this field were Nina Aladjalova from the Institute of Psychology in Moscow and Valentina Ilukhina from the Institute of Experimental Medicine in Saint Petersburg. In their studies they found two types of infraslow oscillations with periods of 10 and 30–90 s, respectively. In my own study described in the 1979 book by Gretchin V.B. and Kropotov J.D. *Slow Non-Electrical Oscillations in the Human Brain* similar spontaneous and evoked local field potentials were found in various cortical and subcortical regions including the basal ganglia and thalamus (Gretchin & Kropotov, 1979). These electrical potentials were correlated with infraslow metabolic oscillations expressed in spontaneous and evoked fluctuations of local oxygen.

In Austria similar research was carried out by Herbert Bauer and Gezilher Guttmann from Vienna University. In Germany Brigitte Rockstroh, Niels Birbaumer, and coworkers made a considerable contribution (Birbaumer et al., 1990). Research in these early years was mostly associated with empirical questions of how infraslow oscillations change in psychological tasks and how they modulate quality of performance in various psychological tasks. In particular, the Austrian group has shown that negative shifts of potentials recorded from the human skull were associated with improvement in performance (increases number of correct responses) while positive shifts of cortical potentials were associated with decline of performance.

Past and recent studies clearly show the existence of at least two types of spontaneous electrical infraslow oscillations: (1) **periodic oscillations** with a peak frequency on spectrograms of around 0.1 Hz and (2) **arrhythmic fluctuations** with no clear peak on EEG spectrograms. 0.1-Hz rhythms can be found in the resting state in a relatively large number of healthy subjects and when present show a striking monochromatic spectrum within the 0.07–0.14-Hz frequency band.

INFRASLOW FLUCTUATION IN THALAMIC NEURONS

Twenty first century studies on animals not only confirmed the initial findings in humans but obtained important data at the cellular level. In particular, Stuart Hughes from Cardiff University in the United Kingdom and his coworkers in 2011 described infraslow oscillations in the **impulse activity of thalamic neurons** in a form of robust firings interspersed with

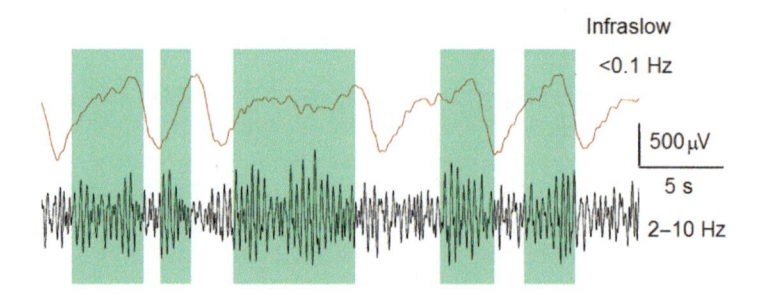

FIGURE 2.1.2 Local field potentials in the lateral geniculate body of the thalamus of a freely moving cat. Upper trace, infraslow fluctuations (filtration at <1 Hz); bottom trace, conventional EEG filtered at 2–10 Hz. Note that periods of increased alpha activity are associated with positive-going deflections in infraslow fluctuations. *Adapted with permission from Hughes et al. (2011).*

periods of neuronal silence. These oscillations coordinated bursts of local alpha rhythms. The authors conclude that oscillatory activity at <0.1 Hz is a fundamental property of cerebral functioning with still unknown neuronal mechanisms (Fig. 2.1.2).

NONNEURONAL ORIGIN OF 0.1-Hz OSCILLATIONS

The **nonneuronal nature of 0.1-Hz oscillations** of cortical potentials was recently discussed in a 2014 study of Vadim Nikulin and coworkers from the University Hospital Charité in Berlin. They simultaneously recorded full-band EEG, near-infrared spectroscopy, arterial blood pressure, respiration, and laser Doppler flowmetry (Fig. 2.1.3). They suggested that local cerebral vasomotion is involved in the generation of 0.1-Hz oscillations. This hypothesis seems to fit experimental data obtained by Aleksandr Rayshubskiy and colleagues from Columbia University in 2014. These authors reported direct observation of large-amplitude, local sinusoidal ~0.1-Hz hemodynamic oscillations correlated with **BOLD fMRI oscillations** detected around the recording region.

During the last few years the interest in infraslow electrical potentials has increased because of the observation of similar infraslow fluctuation in BOLD measures of fMRI and numerous applications of slow cortical potentials in **neurofeedback** methodology for modulating symptoms of Attention Deficit Hyperactivity Disorder (ADHD) and other dysfunctions.

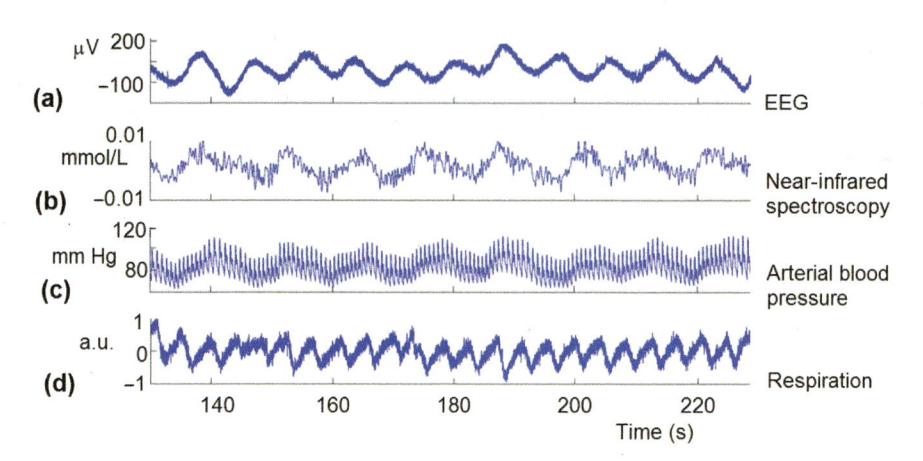

FIGURE 2.1.3 **0.1-Hz oscillations in different measurement systems.** (a) A 10-min unfiltered EEG at the C5 electrode in a healthy subject, (b) NIRS (source 8/detector 15, close to PO1), (c) blood pressure, and (d) respiration. *a.u.*, actual units; *EEG*, electroencephalogram. *Adapted with permission from Nikulin et al. (2014).*

RESPONSES TO TASKS

Infraslow changes in scalp-recorded potentials appear when subjects are preparing for a response over a relatively long period. An example of such a response is the *contingent negative fluctuation* (**CNV**) recorded between the cue and target in the cued two-stimulus paradigm. Fig. 1.2.4 shows a fragment of a spontaneous filtered EEG under the eyes-closed condition (<0.5 Hz) of a healthy 21-year-old subject. Robust 0.1-Hz oscillations are apparent in (a). Note that there is a shift of about 1 s between oscillations at Cz and Pz. Averaged spontaneous oscillations with maps at moments indicated by *red arrows* are shown at (b). Event-related potentials in the cued GO/NOGO task in the same subject with the same parameters of filtration are presented in (c). Note the clear **differences** in amplitude, distribution, and time dynamics of spontaneous and evoked potentials.

PREPARATORY SLOW FLUCTUATIONS

In the 1970s and 1980s, preparatory activities in the human brain were studied by means of electrophysiological methods at Tübingen University in Germany (Brigitte Rockstroh, Niels Birbaumer). The Tübingen theory of preparatory activity can be briefly formulated as follows: when the brain is prepared to receive a stimulus or make a motor action, **apical dendrites** of pyramidal cells receive excitatory inputs from higher cortical areas. The pyramidal cells are thus preset so that information processing in the cortical networks is facilitated. This facilitation was demonstrated

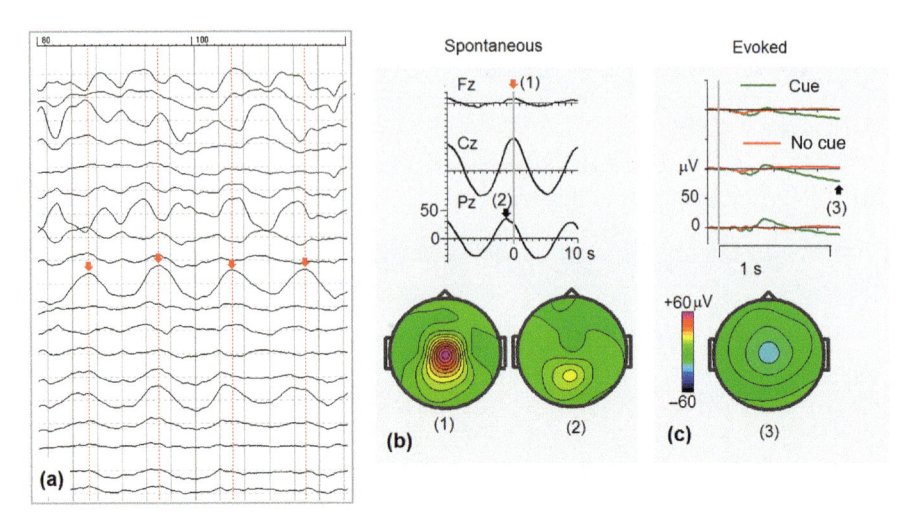

FIGURE 2.1.4 Resting-state spontaneous 0.1-Hz oscillations and evoked slow shifts in the cued GO/NOGO task. (a) A fragment of the spontaneous EEG (<0.5 Hz) of a healthy 21-year-old subject. *Red arrows* mark the peaks of 0.1-Hz oscillations at Cz, (b) averaged to the red-marked oscillations. Red error at 0 time corresponds to peaks in spontaneous EEG. Note that there is a shift of about 1 s between oscillations at Cz and Pz. Maps for *arrows* 1 and 2 are presented at the bottom, and (c) ERPs in the same frequency band in the same subject performed the cued GO/NOGO task. The subject is preparing to press a button after the cue. Note the relatively small CNV wave in the map below.

by showing improvement of the quality of performance (eg, the ability to recognize a stimulus presented with near-threshold expositions) during negative shifts of decisecond oscillations.

If the theory is correct, negative shifts of infraslow potentials have to **facilitate paroxysmal activity**, such as spikes and spike/slow-wave complexes, in the corresponding cortical areas, while the suppression of cortical negativity and increase of positivity might have an opposite effect—suppression of paroxysmal activity. In a 1990 study of the Tübingen group headed by Niels Birbaumer an attempt was made to use this regulatory function of DC potentials for the treatment of patients with epilepsy. Epilepsy patients were trained by means of neurofeedback technology to regulate their own slow-cortical potentials. The most important finding was that patients with complex partial and secondarily generalized seizures were more likely to experience seizure reduction if they demonstrated good control of their infraslow cortical potentials at the end of their training.

NEURONAL MECHANISMS

As mentioned above, there is strong evidence that 0.1-Hz oscillations are associated with local **hemodynamic regulatory oscillations** in the human brain. The neuronal source of aperiodic infraslow spontaneous

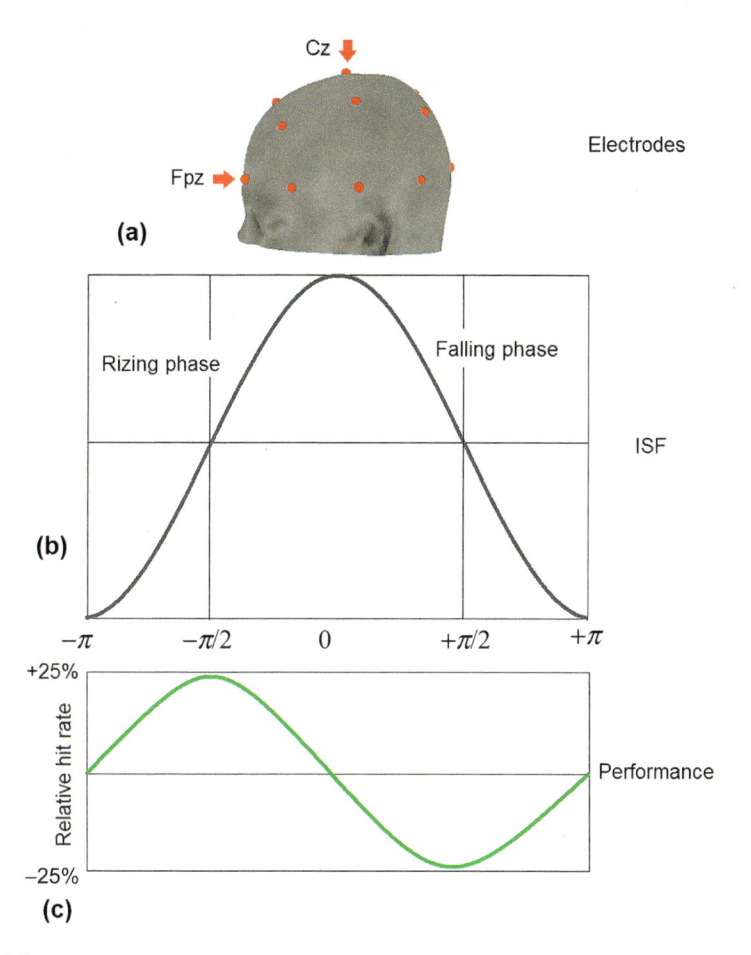

FIGURE 2.1.5 **Phase of infraslow EEG correlates with task performance.** (a) Scalp electrodes (Fpz and Cz) from which the full band EEG has been recorded. (b) Rising and falling phases of infraslow fluctuations (ISF) of the scalp potential. (c) Relative hit rates for somatosensory stimuli presented to the distal part of the right index finger at the threshold of sensation: the probability of a hit (but not reaction time) is larger on the rising phase of ISF and lower at the falling phase while no correlation with amplitude of ISF is found. *Schematic representation of experimental data from Monto et al. (2008).*

fluctuations is not so clear and definitely needs further research. The slow shifts of potential during long-lasting cognitive tasks have **different neuronal mechanisms** from those of spontaneous oscillations.

FUNCTIONAL MEANING

The functional meaning of infraslow oscillations is based on two facts: (1) the phase of these oscillations correlates with human psychophysical performance, (2) the phase also correlates with the amplitude of oscillations in higher frequency bands. The first fact was established in 1970 and later confirmed by a group from Helsinki University (Simo Monto, Satu Palva, Juha Voipio, & J. Matias Palva, 2008) (Fig. 2.1.5). Using full-band EEG technology they also demonstrated that the phase of infraslow oscillations correlates with the amplitude of EEGs in the faster frequency bands.

Although these findings are correlational, there is a temptation to suggest that the infraslow potentials of the human brain reflect fluctuations in cortical excitability which in turn modulate faster fluctuations in the brain state.

CHAPTER 2.2

Alpha Rhythms

HISTORICAL INTRODUCTION

The German psychologist **Hans Berger** from Jena University (Germany) was the first to observe electrical alpha rhythms from the scalp of human subjects who were sitting quietly with their eyes closed. He published his discovery in **1929** and named these electrical events "waves of the first order" or **α (alpha) waves**. He further showed that alpha waves were blocked upon eye opening or during certain types of mental effort, leading to the appearance of "waves of the second order" or "β waves." Berger's results were later confirmed by several other investigators, in particular **Edgar Adrian and Bryan Matthews**, who introduced the brilliant concept that different sensory brain areas possessed their own alpha rhythm representing a "resting" or **"idling"** state of that brain region.

It took more than 50 years for scientists to discover the neuronal mechanisms of alpha rhythms. Of specific importance is the discovery of a novel form of rhythmic burst firing, termed **high-threshold bursting**, which occurs in a subset of thalamocortical neurons under a depolarized state. Now we know that during relaxed wakefulness the human brain exhibits several types of distinct rhythmic electrical activity in the alpha frequency band (8–13 Hz) in the occipital, parietal, and central areas. These rhythms differ in topography, frequency, and sensitivity to tasks. Despite these differences alpha oscillations seem to have a general function: **active and adequate inhibition** of irrelevant sensory pathways.

An extensive review on **classical and modern studies of alpha rhythms in association with neuroscience and neuropsychology** was published in 2003 in the Elsevier book *The Brain's Alpha Rhythms and the Mind*, edited by J. C. Shaw.

TYPES OF ALPHA RHYTHMS

At least three types of alpha rhythms can be separated in EEG in the resting state of healthy subjects. During the resting state with the eyes closed the flow of visual information is completely shut down and the flow of somatosensory information is pretty much restricted, which may lead to **idling** (a passive concept that implies simply not working or sleeping) and may lead to **inhibition** of the corresponding sensory areas (an active concept which implies inhibition of irrelevant information). The first type of alpha rhythm includes a family of **posterior alpha rhythms** with maximums of power at the occipital or parietal electrodes. The second type includes a family of **mu-rhythms (μ)** with maximums of power at the central electrode—areas located above the somatosensory cortex.

In the auditory cortex, alpha rhythms have so far been observed only by means of magnetoencephalography (MEG). The first report about this rhythm was made in 1997 by Rita Hari and her coworkers from Helsinki Technical University (Lehtela et al., 1997). They found a clear decrease of MEG power in their studies of event-related desynchronization in response to auditory stimulation. The rhythm was localized in the auditory cortex and named the **tau (τ) rhythm**. The generators of the tau rhythm lie deep in the Sylvian fissure, occupy relatively small local areas, are kept in a desynchronized state by acoustic noise and, consequently, are difficult to record from the scalp.

In sum, the cortical areas of three main modalities (visual, somatosensory, and auditory) generate alpha rhythms in the resting state which are desynchronized during active information processing.

ALPHA RHYTHMS IN THE SOMATOSENSORY CORTEX

An example of mu-rhythms in a healthy subject is presented in Fig. 2.2.1. The rhythms pop out of arrhythmic activity only in the current source density (CSD) montage (compare Fig. 2.2.1a,b). The temporal pattern of the rhythm has a special shape which when inverted in polarity (negativity is up) is reminiscent of the Greek letter μ (mu) (Fig. 2.1e). On CSD spectra the rhythms are reflected in peaks around 11 Hz at the C3 and C4 electrode (Fig. 2.2.1c). The map of EEG spectra at this frequency is reminiscent of a monkey's face (Fig. 2.2.1d).

The rhythm is also named Rolandic alpha or sensory-motor rhythm (SMR) because of its localization. Mu-rhythms are considered indirect indications of functioning of the mirror neuron system and are used in studies with psychiatric patients.

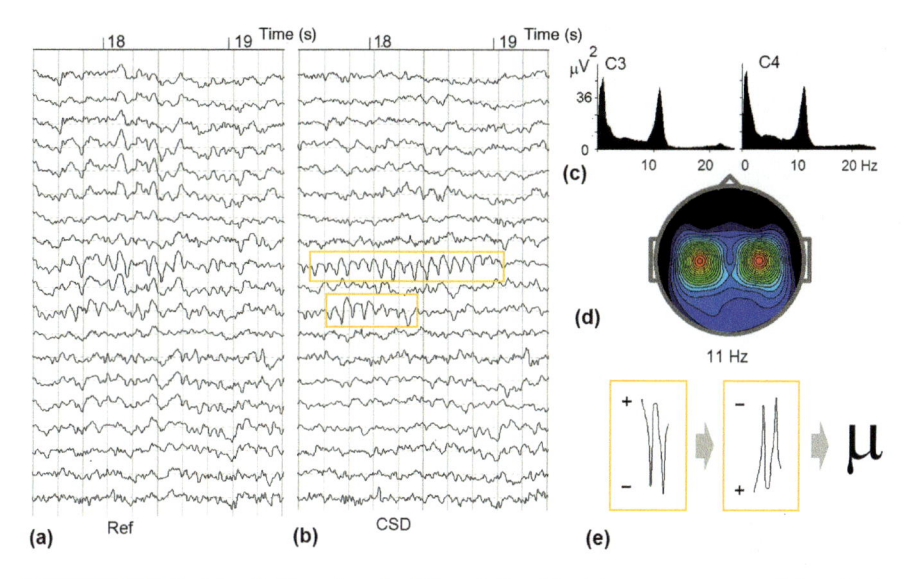

FIGURE 2.2.1 **Mu-rhythms in a healthy adult subject in eyes open condition.** (a) A fragment of EEG in the referential montage; (b) the same fragment in the CSD montage *(mu-rhythms are highlighted in boxes)*; (c) EEG power spectra at C3 and C4; (d) spectra map at 11 Hz; and (e) the mu-rhythm pattern in inverted polarity is reminiscent of the Greek letter μ.

ALPHA RHYTHMS OF THE VISUAL SYSTEM

An example of **posterior alpha rhythms** in a healthy subject is presented in Fig. 2.2.2. The rhythms pop out of arrhythmic activity only in the CSD montage (compare Fig. 2.2.2a and b). The temporal pattern of the rhythms is closer to a sinusoidal wave than to a mu-rhythm. On CSD spectra the rhythms are reflected in two peaks around 8.5 and 10 Hz, respectively. Independent component analysis (ICA) taken in a single subject usually reveals several independent components at the occipital–temporal–parietal electrodes. So, it is not uncommon for a single subject to have several posterior alpha rhythms with different frequencies and topographies. The largest component, however, usually occupies the occipital electrodes.

FUNCTIONAL REACTIVITY

About **90% of healthy subjects** show prominent **occipital alpha rhythms** in eyes -closed condition. Substantially smaller numbers of healthy subjects demonstrate **prominent mu-rhythms** in their EEG.

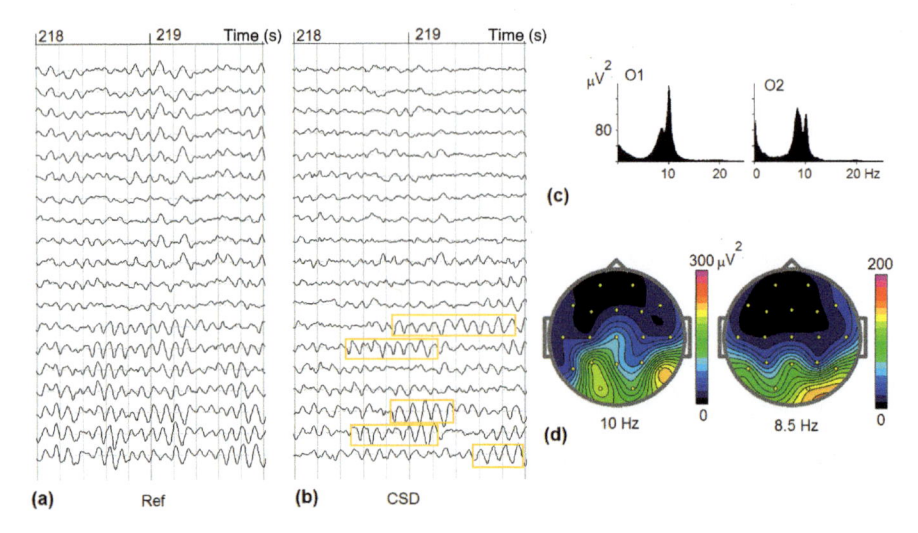

FIGURE 2.2.2 **Posterior alpha rhythms in a 27-year-old healthy subject in eyes closed condition.** (a) A fragment of EEG in the referential montage; (b) the same fragment in the CSD montage *(posterior alpha rhythms are highlighted in boxes)*; (c) EEG power spectra at O1 and O2. Note two peaks at 10 and 8.5 Hz., and (d) spectra maps at 8.5 and 10 Hz.

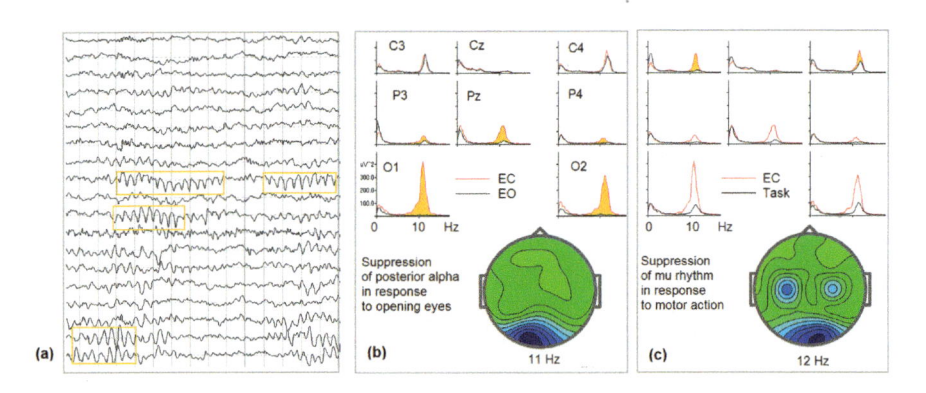

FIGURE 2.2.3 **Posterior and central alpha rhythms in EEG of a healthy subject.** (a) A fragment of EEG in eyes closed condition. The posterior (O1 and O2) and central (C3, C4) rhythms are marked; (b) EEG spectra in eyes closed *(red)* and eyes open *(black)* conditions and the map of the difference "eyes open–eyes closed"; and (c) EEG spectra in eyes closed *(red)* and Visual Motor task *(black)* and the map of the difference "task–eyes closed."

An example of an EEG of a healthy subject with both posterior alpha and mu-rhythms is presented in Fig. 2.2.3. Fig. 2.2.3a depicts a fragment of CSD montage of EEG of a healthy subject in eyes-closed condition. Note two types of rhythms: occipital (O1, O2) and central (C3, C4) rhythms. Both are in the alpha frequency band. The occipital alpha dominates in eyes closed condition and is suppressed when the subject performs a visual task or simply opens the eyes. The central rhythm does not respond to eyes opening but is suppressed in the motor task.

Standardized low-resolution brain electromagnetic tomography (sLORETA) images and different reactivity patterns of EEG independent components to opening the eyes in the same subject are illustrated in Fig. 2.3.4. The components are extracted from resting-state raw EEG by the Infomax algorithm of ICA. Fig. 2.2.4a,b correspond to two occipital alpha rhythms with different topographies. They are generated in BA 18 and 19 of the **occipital cortex**. Fig. 2.2.4c,d correspond to mu-rhythms generated in the postcentral gyrus in the somatosensory cortex corresponding to **representation of the hand on the homunculus map**. One can see that only the posterior alpha rhythms are suppressed by opening the eyes.

Fig. 2.2.5 demonstrates different reactivity of the occipital and Rolandic alpha rhythms in the same subject performing the **cued visual GO/NOGO task**. The task involves pressing a button using the right hand and consequently activating the left somatosensory cortex. The topographies of the components—occipital alpha 1 and Rolandic alpha left—are presented in Fig. 2.2.5a. The results of wavelet analysis for the two components for GO condition are presented in Fig. 2.2.5b. The occipitally generated component

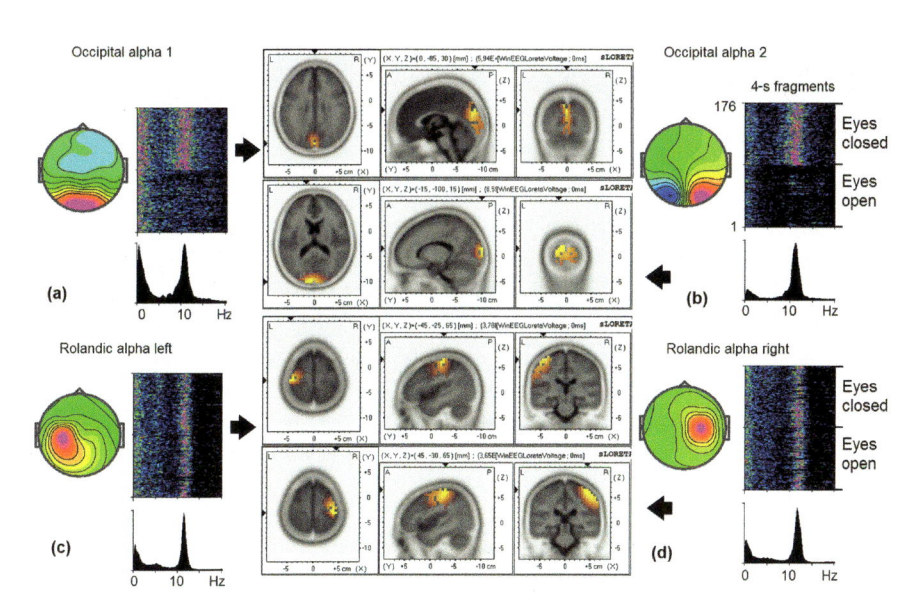

FIGURE 2.2.4 Independent components and their sLORETA images for occipital alpha and mu-rhythms in *EEG* of the healthy subject from Fig. 2.2.3. On a, b, c and d: map and spectra of the component; color coded spectra computed for sequential 4 s fragments of eyes-closed and eye-open condition, sLORETA image indicated by an *arrow*. Note that the occipital alpha rhythms are suppressed in response to eyes opening.

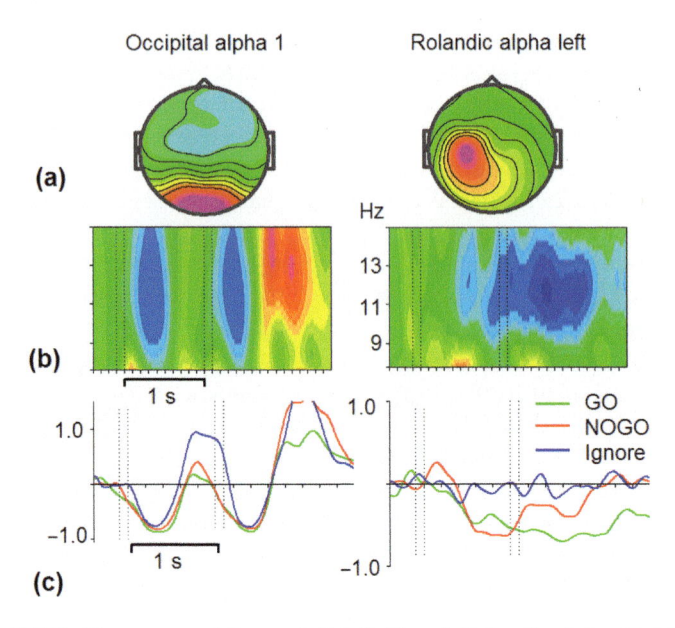

FIGURE 2.2.5 **Responses of the occipital (left) and Rolandic alpha (right) rhythms to visual stimuli and motor actions in the visual cued GO/NOGO task.** (a) Topography of independent component; (b) The results of wavelet analysis of responses to GO stimuli (*y*-axis, frequency in Hertz; *x*-axis, time). The percent of EEG power change in comparison to prestimulus background activity is color-coded (*blue*, decrease; *red*, increase). The GO trials are averaged; and (c) Event-related de/synchronization in the alpha band in GO *(green line)*, NOGO *(red)*, and Ignore trials *(blue)* (*x*-axis, time). *Dotted lines* in (b) and (c) indicate onset/offset of the stimuli.

responds with **transient suppression** (with a duration of about 700 ms) of the power in the alpha band in response to visual stimuli irrespectively of whether the stimulus is presented in the first position or second position or whether it is ignored or responded to. The mu-rhythm on the left side shows a completely different pattern: it does not respond to visual stimuli *per se*, but shows **slowly evolving suppression** when the subject is **preparing to make a movement**, strongly **suppressed** in response to **movement**, and is not suppressed when movement has to be withheld.

PARIETAL ALPHA RHYTHM

When healthy subjects are involved in visual tasks a clear parietal alpha rhythm is sometimes induced. The existence of **parietal alpha rhythms** was first demonstrated in 2002 by Ole Janson and his coworkers from Helsinki University of Technology. The phenomenon is in agreement with a suggestion that posterior alpha band oscillations suppress irrelevant visual information and thus gate the relevant visual information so that

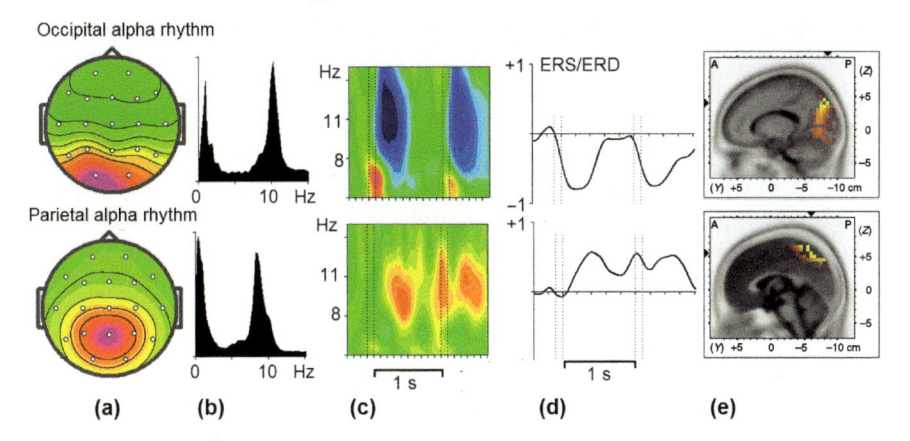

FIGURE 2.2.6 **Reciprocal pattern of response to visual stimulation of parietal and occipital alpha rhythms in a healthy subject performing the cued GO/NOGO task.** The responses are in Ignore condition to visual stimulus presentations. The occipital rhythm is suppressed while the parietal rhythm is synchronized. (a) Topographies of independent components; (b) spectra of the components; (c) the results of wavelet analysis of responses of the rhythms under the Ignore condition; (d) the dynamics of event-related synchronization *(ERS)* and event-related de-synchronization *(ERD)* in Ignore trials. Note the almost 100% changes in EEG power of the components; and (e) sLORETA images of the components.

emergence of the parietal alpha rhythm in response to the visual stimulus would indicate inhibition of the irrelevant parietal pathway while inhibition of the occipital alpha rhythm to visual stimulation would indicate engagement of the relevant occipital pathway. The **reciprocal activation/ inhibition pattern of the occipital and parietal alpha rhythms** is illustrated by Fig. 2.2.6. The two independent components (upper and lower rows) are extracted by ICA from 19-channel EEGs recorded in a healthy adult performing the cued GO/NOGO task. Fig. 2.2.6c,d depict, respectively, wavelet decomposition and event-related synchronization/ event-related desynchronization (ERS/ERD) curves in Ignore trials. sLORETA images show the occipital and parietal localization of the components. Note that the occipital alpha rhythm desynchronizes in response to visual stimuli while the parietal alpha rhythm synchronizes in response to visual stimulation. Note also that the frequency of the parietal rhythm is 1 Hz lower than that of the occipital alpha rhythm.

NEGATIVE CORRELATION WITH BOLD SIGNALS

The resting state with the eyes closed is reflected by a specific pattern of BOLD signal in fMRI recordings characterized by a **hotspot in the precuneous** and relative decrease of activity in the visual and somatosensory systems. The spatiotemporal pattern of the BOLD signal in the resting state is called the **default mode network**.

When EEG and BOLD signals are measured simultaneously the power of the posterior alpha rhythm negatively correlates with BOLD parameter in the occipital–temporal–parietal. In a 2005 study by Bernd Feige and coworkers from University of Freiburg (Germany), the EEG alpha rhythm in healthy subjects was related to BOLD signal change areas (Fig. 2.2.7 *left*). Topographical EEG was recorded simultaneously with fMRI under open/closed eyes conditions. The EEG was separated into spatial components by independence using ICA. EEG alpha component amplitudes were associated with BOLD signal decreases in occipital areas when the standard BOLD response (maximum effect at 6 s) was assumed. The area deactivated during increased alpha includes **the calcarine and fusiform gyri**.

In a 2009 study by Petra Ritter and colleagues from Germany and Norway simultaneous EEG and fMRI were recorded during a bimanual motor task (Fig. 2.2.3 *right*). A blind source separation algorithm was used for EEG data. Rhythm "strength" (ie, spectral power determined by wavelet analysis) was **inversely correlated** with the fMRI–BOLD signal in the postcentral cortex for the Rolandic alpha (mu)-rhythm.

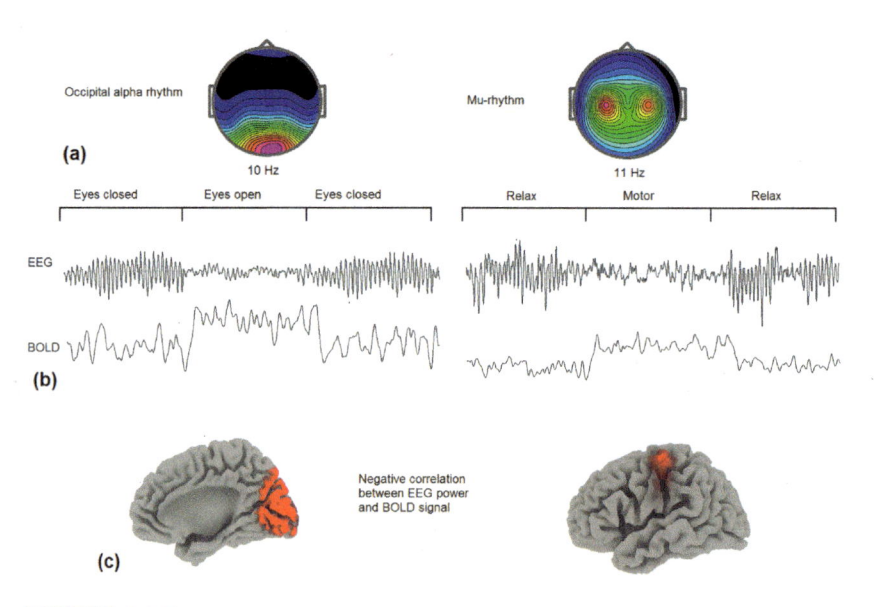

FIGURE 2.2.7 **fMRI correlates of posterior and Rolandic alpha rhythms.** fMRI correlates of occipital and Rolandic alpha rhythms. (a) Schematic presentation of topographies of occipital (left) and Rolandic (right) alpha rhythms; (b) Schematic presentation of two different paradigms with simultaneous recordings of EEG and fMRI BOLD signal: (left) sequential eyes open/eyes closed paradigm for occipital alpha rhythms, (right) repetitive bimanual motor task with relaxation periods; (c) negative (*red*) correlations of BOLD signal with the corresponding EEG alpha power: (left) for the occipital alpha rhythm and (right) for the Rolandic alpha rhythm. *Part (c, left): adapted from Feige et al. (2005). Part (c, right): adapted from Ritter et al. (2009).*

AGE DYNAMICS

The power of alpha rhythms in healthy subjects decreases with age in the interval from 7 to 80 years (Fig. 2.2.8). A different age dynamics is seen for frequency of alpha rhythms. On average, the frequency of mu-rhythms is about 1 Hz higher than posterior alpha rhythms. The frequency reaches the highest values at around 20 years of age and slowly declines. Variations with age are relatively small: around 1 Hz, so that any decrease in occipital alpha rhythms below 7.5 Hz should be considered as abnormal (Fig. 2.2.8).

Because the frequency of alpha rhythms decreases with age, this neuromarker was considered a candidate for a correlate of cognitive function regulation. The question still remains open: Do smarter brains run at higher alpha frequencies? Danielle Posthuma and colleagues from Vrije University in Amsterdam in 2001 examined 271 extended twin families (688 participants). IQ was assessed using the Dutch version of the Wechsler Adult Intelligence Scale (WAIS-IIIR), from which four dimensions (verbal comprehension, working memory, perceptual organization, and processing speed) were extracted. The data showed that that both peak frequency and the dimensions of IQ were highly heritable (ranging from 66 to 83%) but there was no evidence of a genetic correlation between alpha peak frequency and any of the four WAIS dimensions: smarter brains do not seem to run faster.

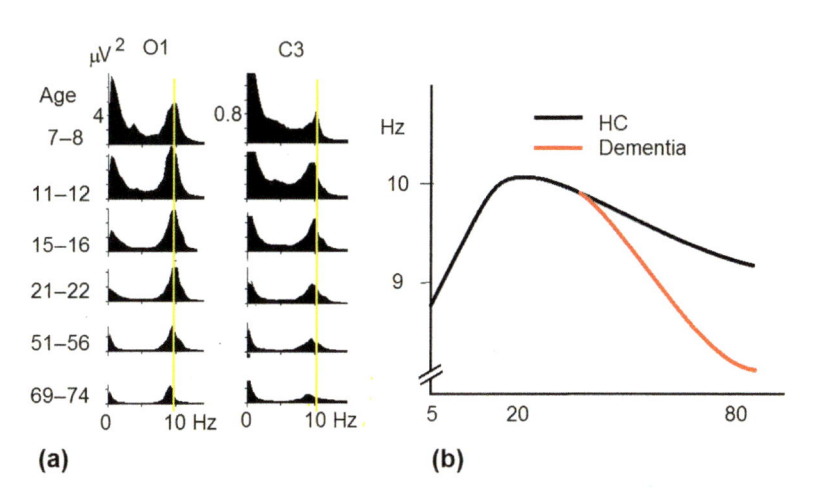

FIGURE 2.2.8 **Age dynamics of alpha rhythms.** (a) EEG spectra for different age groups under the eyes closed condition at O1 and C3; (b) schematic presentation of posterior alpha frequency as a function of age in healthy subjects *(HC)* and patients with dementia. *Data from Human Brain Indices (HBI) database.*

FRONTAL ALPHA ASYMMETRY

In mental health, it has been suggested that **frontal alpha asymmetry** may serve as a biomarker for depression. The idea was inspired by **Richard Davidson"s theory** regarding asymmetrical involvement of the orbitofrontal cortex in emotional reactions (Davidson, 1993). Roughly, according to this theory the left prefrontal lobe is biased to processing positive emotional stimuli, while the right prefrontal lobe processes mostly negative emotional stimuli. Taking into account a negative correlation between power of alpha rhythms and local metabolic activity the larger alpha activity in the left frontal–temporal electrodes in comparison to the right electrodes may indicate a depressive state.

Early research suggested that frontal alpha asymmetry in a resting state reflects a trait-like asymmetry of anterior cortical activity and predicts individual differences in affective/motivational behaviors. Unfortunately, recent studies have not been so decisive. The inconsistency of the results might be explained by several factors. One factor is the heterogeneity of depression, which suggests that there indeed might be a group of depressed patients with frontal alpha asymmetry, but this group is relatively small, The other factor is a methodological one. The asymmetry index is usually a small number with quite large standard deviation and, consequently, needs a long interval of recording for getting a robust measurement. The index also depends on the montage selected. In Fig. 2.2.9 the index is computed in the common reference montage.

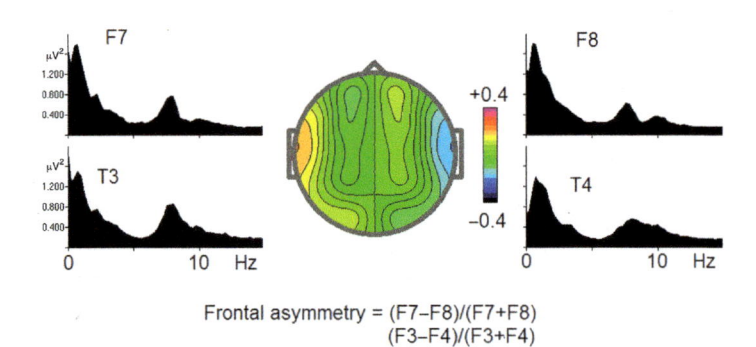

FIGURE 2.2.9 **Frontal alpha asymmetry.** An example of resting-state EEG spectra of a subject with a history of depression. Frontal alpha asymmetry is defined as the ratio: the difference of EEG power in the alpha band taken at symmetrical frontal leads such as F7 – F8 or F3 – F4 divided by the sum of the values.

ALPHA RHYTHMS IN THE DYSFUNCTIONAL BRAIN

In the healthy brain in the resting state, alpha rhythms are located only in posterior or central areas. No sources of alpha rhythms are usually found in the frontal lobes. However, in the dysfunctional brain the generators of alpha rhythms can be localized in **unusual cortical areas**. An example is presented in Fig. 2.2.10. The EEG was recorded in the resting state and during the cued GO/NOGO task in a subject with history of depression. Two independent components of alpha rhythms were decomposed from the EEG by independent component analysis. One rhythm (the posterior alpha rhythm) has a peak frequency of 9.8 Hz and is generated in the precuneous (Fig. 2.2.10, left). The other rhythm (the frontal-temporal alpha rhythm) has a peak frequency of 8.3 Hz and is generated in the inferior frontal/superior temporal cortex (Fig. 2.2.10, right). The two rhythms not only generated in different cortical areas but they also exhibit quite different pattern during the cued task: the posterior alpha rhythm is suppressed in response to stimulus presentations reflecting activation of the corresponding area in visual processing, whereas the frontal-temporal alpha rhythm is enhanced during GO and NOGO, but not Ignore trials, reflecting inhibition of the corresponding area in cognitive control.

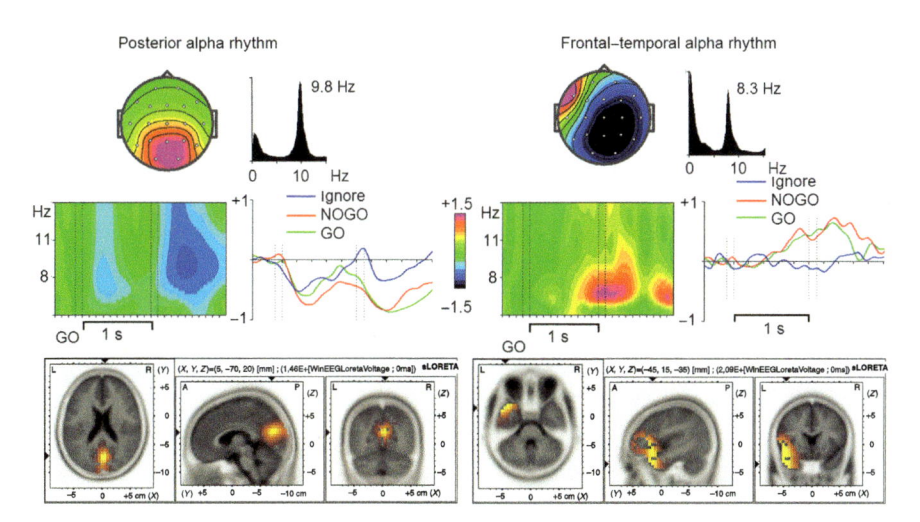

FIGURE 2.2.10 **Alpha rhythm generated in inferior frontal/superior temporal lobe in a subject with a history of depression.** Two independent components with clear rhythmicities in the alpha band are extracted from spontaneous EEG: posterior alpha rhythm (left) and frontal–temporal alpha rhythm (right). They show different synchronization/desynchronization patterns: the posterior alpha rhythm desynchronizes in response to visual stimuli, while the frontal–temporal rhythm synchronizes in GO and NOGO trials.

NO ALPHA RHYTHMS: LOW-VOLTAGE FAST ELECTROENCEPHALOGRAMS

As mentioned above, alpha rhythms are the predominant EEG waveforms in most individuals during states of alert relaxation. Power in the alpha frequency domain is highly heritable. However, in about 10% of healthy individuals **no alpha rhythms** can be found in the resting state. This does not mean that their thalamocortical networks cannot produce alpha rhythms because under some conditions (like after consuming a substantial amount of alcohol) prominent alpha rhythms can be found in EEGs. The reason for having this subtype is probably associated with a deficiency of inhibitory transmission. The EEG subtype is named the "low-voltage alpha" or **low-voltage fast EEG** (Fig. 2.2.11).

In a 1999 study by Mary-Anne Enoch and coworkers from the National Institutes of Health (United States) in the total group of unrelated individuals, alcoholics were significantly (three times) more likely to show the low-voltage fast EEG trait than were nonalcoholics, and individuals with anxiety disorders were significantly (three times) more likely to exhibit the low-voltage trait than were those without anxiety disorders.

Giuseppe Moruzzi and Horace Magoun from Northwestern University in 1949 were the first to show that stimulation of the brain stem reticular formation activates the cortex and transforms a high-voltage, low-frequency EEG into a low-voltage, high-frequency EEG. This effect is mediated,

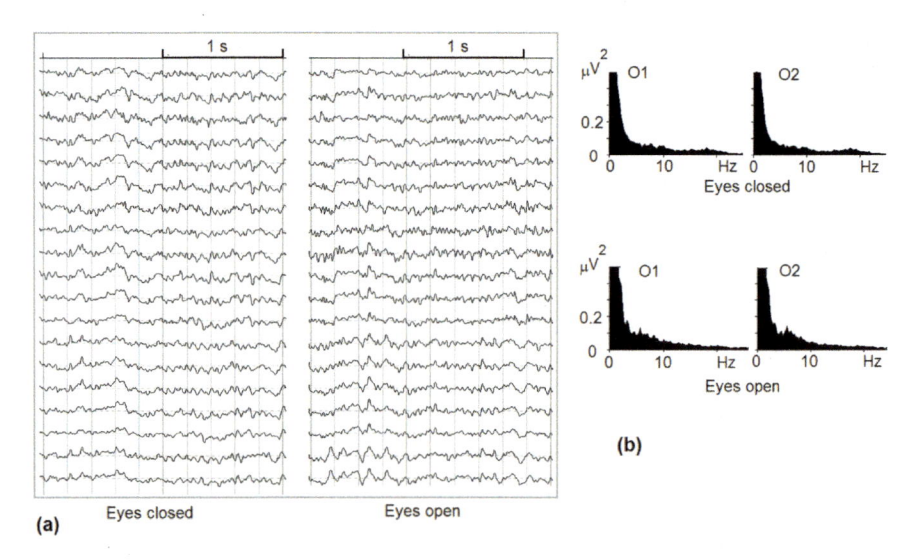

FIGURE 2.2.11 Low-voltage fast EEG phenotype in a 50-year-old healthy subject. (a) Two-second EEG fragments under eyes-closed and eyes-open conditions in the common average montage. Note no prominent alpha rhythms under the eyes closed condition; and (b) EEG spectra for O1 and O2 electrodes under eyes-open and eyes-closed conditions.

in part, by thalamic midline nuclei. Electrical stimulation of intralaminar nuclei of the thalamus alone also induces the same effect. Actually, the reaction of activation consists of suppressing alpha rhythms (reaction of desynchronization) and increasing high-frequency activity. As we show below this activation is maintained by the cholinergic, dopaminergic, and serotoninergic pathways that originate in different brain stem nuclei and that project to the cortex directly, or indirectly, through thalamic nuclei.

HERITABILITY

Twin studies of EEG power spectra, in general, and alpha power, in particular, show a striking similarity (with correlation coefficients > 0.8) of EEG parameters in monozygotic (MZ) twins. Of the different EEG frequency bands, alpha band power (8–13 Hz) shows the highest heritability. Correlations are as strong as the corresponding test–retest correlations within the same individual. Moreover, correlations for a sample of MZ twins reared apart are virtually at the same level as for MZ twins reared together. EEG heritability is illustrated in Fig. 2.2.12 taken from the 2014 review by Andrei Anokhin from Washington University School of Medicine.

NEURONAL MECHANISMS

Views regarding the neuronal mechanisms behind alpha rhythm generation have evolved during the last 50 years. Initially, views were based on an association between sleep spindles (14-Hz rhythms during the second stage

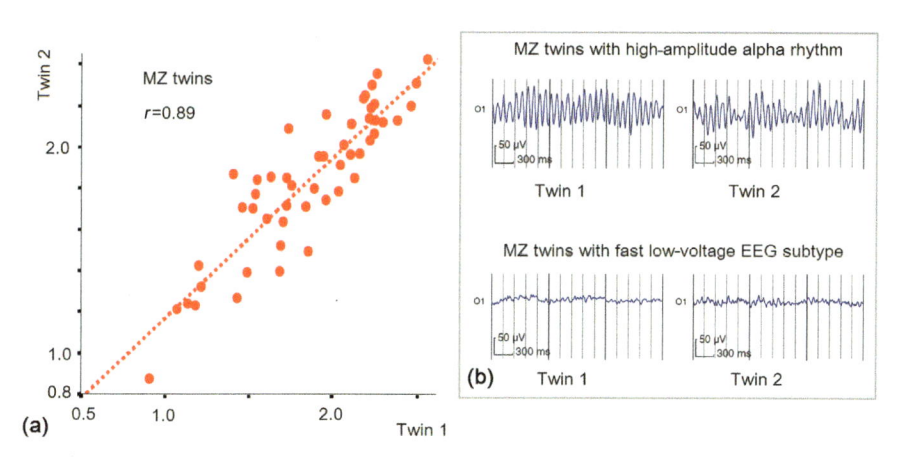

FIGURE 2.2.12 **EEG heritability.** (a) Scattergram of intrapair correlation in monozygotic (MZ) twins; and (b) EEG phenotypes in twins. Upper panel, regular alpha variant; lower panel, low-voltage fast EEG. *Adapted with permission from Anokhin (2014).*

of sleep) and burst discharges of thalamocortical neurons. It was shown that these neurons operate in two modes: (1) **tonic mode** in a depolarized (activated) state, and (2) **burst mode** in a hyperpolarized (inhibited) state. In tonic mode the impulse activity of neurons follow sensory input from the retina, so that they are able to transfer sensory information reliably. In burst mode, thalamocortical neurons fire in bursts that follow a period of hyperpolarization (inhibition). Hyperpolarization that lasts relatively long deactivates calcium channels and produces a low-threshold calcium spike which in turn generates a burst of conventional action potentials. Burst mode leads to unreliable information transmission. Tonic and burst modes were considered to reflect "open" and "closed" thalamic gates, respectively. In line with these facts, alpha oscillations in the awake brain were associated with the periodic burst mode of thalamocortical neurons.

However, three recent discoveries led to a new view of the mechanisms behind alpha rhythms. First, at the cellular level in animal experiments Stuart Hughes and his coworkers from Cardiff University in 2004 discovered a novel form of thalamic burst firing in a depolarized state. They showed that thalamocortical neurons under certain conditions can generate **high-threshold calcium potentials** which in turn produce bursts of spikes with interspike intervals longer than 10 ms. The bursts are synchronized by gap junctions and correlate with scalp EEG oscillations. Depending of the level of cortical feedback the oscillations can be in the alpha frequency band (relatively high level of activation) or in the theta frequency band (relatively low level of activation) (Fig. 2.2.13).

Second, at the behavioral level it was recently shown that alpha oscillations in the primary sensory cortex are not associated with a "closed thalamic gate" but rather produce an optimal level of functioning of the cortex. The evidence comes from near-threshold psychophysical experiments and from combined EEG–TMS experiments. An example of a near-threshold experiment is given by a recent study by Klaus Linkenkaer-Hansen and coworkers at Helsinki University who found that detection of somatosensory stimuli by humans was equally probable for both small and large prestimulus alpha-frequency band oscillations in the primary somatosensory cortex (Fig. 2.2.14).

The relationship between prestimulus alpha oscillations and upcoming perception could be simply correlative or causal (ie, the oscillation in the alpha band is causally shaping perception). The **causal function** of the alpha rhythm was tested in a 2010 study by Vincenzo Romei and coworkers from the University of Glasgow by directly stimulating visual areas via short trains of rhythmic TMS. It was shown that target visibility was significantly modulated by alpha stimulation relative to theta and beta stimulation control conditions. These frequency-specific effects were observed both for stimulation over the occipital and parietal areas of the left and right hemispheres and were short lived: they were observed by the end of the TMS

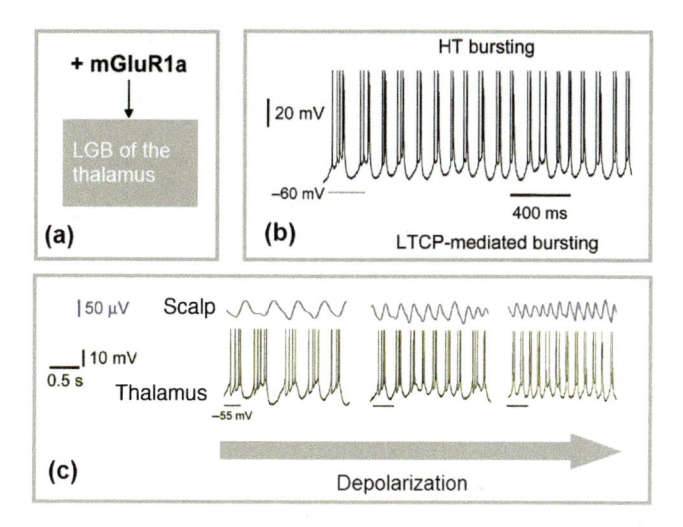

FIGURE 2.2.13 **High-threshold (HT) repetitive busting as a mechanism behind alpha rhythm generation.** (a) In vitro experiments: when corticothalamic feedback is mimicked in a slice of the lateral geniculate body of the thalamus by activating mGluR1a (a subtype of the metabotropic glutamate receptor), thalamocortical neurons became depolarized; (b) A subset of TC neurons exhibit high-threshold bursting in addition to single-spike activity; (c) Decrease in the depolarization level of thalamocortical neurons leads to transfer of alpha oscillations (8–13 Hz) into theta oscillations (4–8 Hz). *Adapted with permission from Hughes & Crunelli (2007).*

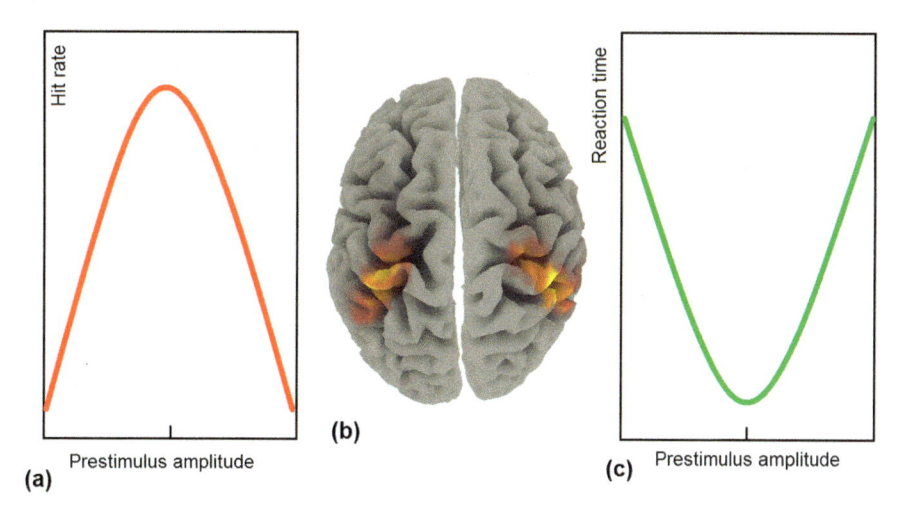

FIGURE 2.2.14 **Alpha rhythms determine optimal performance.** Subjects' finger tips were stimulated electrically at the threshold of sensation. The subjects were instructed to react to perceived stimuli. MEG was recorded. (a) intermediate amplitude of the prestimulus alpha oscillations in the sensorimotor cortex facilitates stimulus detection (Hit rates); (b) schematic representation of areas of recordings; (c) inverted U-law for the reaction time versus amplitude of the prestimulus alpha activity in the somatosensory cortex. *Adapted from Linkenkaer-Hansen et al. (2004).*

train but were absent 3 s later. This shows that the posterior alpha rhythm is actively involved in shaping forthcoming perception and, hence, constitutes a **substrate** rather than a mere correlate of visual input regulation.

MODEL

The thalamus like a gate regulates information flow to the cortex. Thalamic neurons are in position to inhibit irrelevant information and enhance relevant information. In theory, a mechanism of lateral inhibition can implement this function. In experiment, different types of lateral inhibition have been observed in the reticular nucleus of the thalamus—a layer of inhibitory neurons that implement feedback lateral inhibition on thalamocortical neurons.

On the basis of the data available 30 years ago Francis Crick from Salk Institute, USA (1984) proposed a searchlight hypothesis according to which the thalamic reticular nucleus controls the internal attentional searchlight that simultaneously highlights all the neural circuits called on by the object of attention. He speculated that during perception, preparation, and

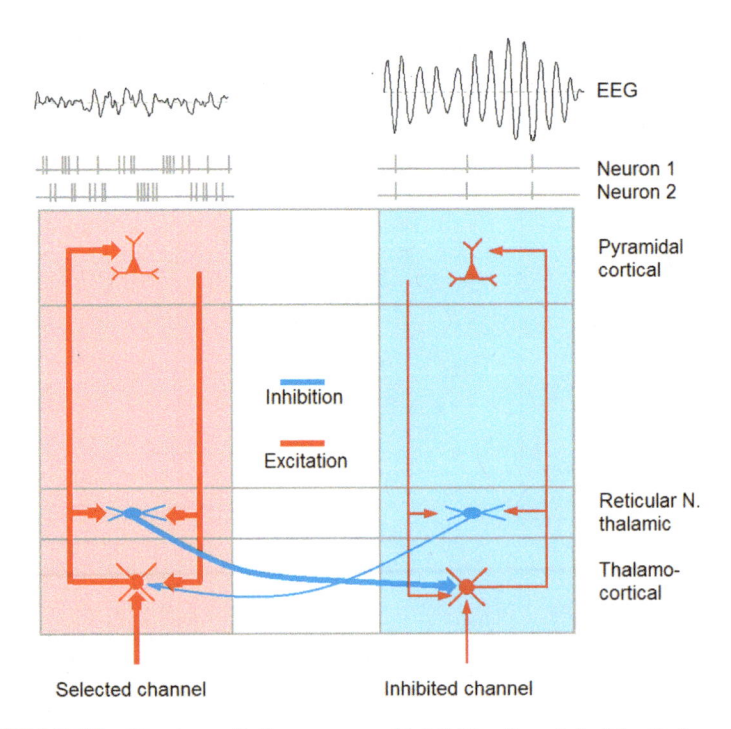

FIGURE 2.2.15 "Center excitation–surround inhibition" model of the thalamocortical networks responsible for alpha rhythm generation.

execution of any cognitive and/or motor task, the reticular nucleus of the thalamus sets all the corresponding thalamocortical circuits in motion. His idea of lateral inhibition was based on the suggestion that neurons in the reticular nucleus have extensive inhibitory axon collaterals, which are supposed to generate large-scale intrareticular nucleus interactions, and successive thalamic burst discharges are what are required of a searchlight.

However, new data show that this is not the case: (1) a weak proportion (about 10%) of neurons display short-range local axon collaterals, (2) rhythmic burst activity in thalamic neurons occurs during states of drowsiness and sleep whereas attentive wakefulness and cognitive tasks are usually associated with gamma oscillations in TC systems. Twenty years after Fransis Crick, Didier Pinault from INSERM, France summarized new knowledge on the structure and function of the reticular nucleus of the thalamus. An attempt to simplify these views is presented in Fig. 2.2.15. This scheme can explain the center desynchronization/surround synchronization general pattern of EEG reactivity reviewed in a 1999 paper by Gert Pfurtscheller from Technical University Graz and Fernando Lopes da Silva from Graduate School of Neurosciences, Amsterdam. This scheme can also explain a similar pattern presented in Fig. 2.2.6 of this book.

CHAPTER 2.3

Beta and Gamma Rhythms

HISTORICAL INTRODUCTION

Beta waves were first described by Hans Berger from Jena University and were associated with focused attention. In 1929 he wrote "… that the beta waves not the alpha waves of the EEG are the concomitant phenomena of **mental activity**." Initially, all waves at frequencies higher than 13 Hz were referred to as beta waves. Later it became clear that the waves in the 13–30-Hz band should be divided at least into two bands—**low beta (13–21 Hz)** and **high beta (21–30 Hz)**—because of the differences in topographies and reactivity to tasks.

Giuseppe Moruzzi and Horace Magoun from Northwestern University in 1949 were the first to show that stimulation of brainstem reticular formation activates the cortex and change the EEG with high-amplitude alpha rhythms into an EEG with low-amplitude beta waves. Electrical stimulation of intralaminar nuclei of the thalamus alone induces the same effect. The reaction to activation consists in suppressing alpha rhythms (reaction to desynchronization) and increasing beta activity.

In the 1980s a higher frequency band (the **gamma band, 30–60 Hz**) became a focus of intensive research within the framework of a **binding problem** in perception. The basic suggestion was that fluctuations in the gamma band were associated with the ability of the brain to bind together different features of the same object into a **single Gestalt**, or, more precisely, to unite different representations of an image activated in a different cortical area into some unitary neurophysiological process that is perceived by the brain as a single percept.

The statement that beta activity is found in almost every healthy subject sounds trivial because EEG spectra, as is the case with any polymorphic pattern, comprise all frequencies including beta. However, in some cases high-frequency activity is quite regular and this regularity is expressed in spectrograms in the form of a distinctive peak. In such cases we are talking about rhythms, in general, and **beta and gamma rhythms**, in particular.

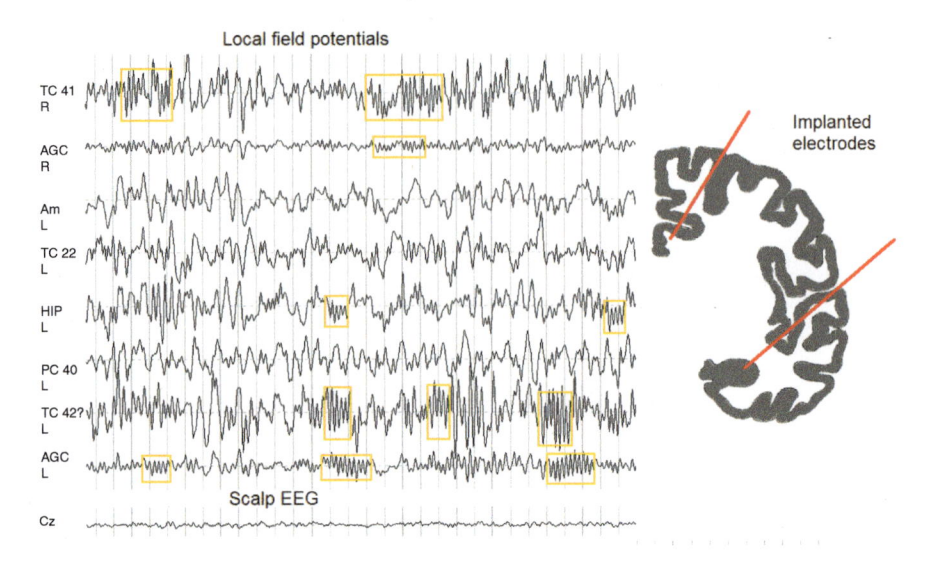

FIGURE 2.3.1 Local field potentials recorded from intracranial electrodes implanted for diagnosis and therapy into an epileptic patient brain simultaneously with the scalp electroencephalogram *(EEG)* at **Cz**. Bursts of local beta rhythms are marked *(boxes)*.

In intracranial recordings (Fig. 2.3.1), beta rhythms are found in many cortical and subcortical anatomical structures including the **limbic system** and **basal ganglia**. These rhythms are **local** and may be quite different at electrodes just a few millimeters apart. These local field potentials when summarized and attenuated at the scalp constitute quite moderate input to raw EEGs.

Because of the nature of local field potentials, there is a big discrepancy between studies of these phenomena and scalp-recorded EEG rhythms. Local field potentials are mostly studied in animal experiments with the ultimate goal of understanding the relationship between local field oscillations (reflective of collective behavior of neuronal networks) and underlying single-neuron processes such as impulse activity and synaptic transmissions. Several ideas have been put forward and got experimental support such as binding-by-synchrony, communication-through-coherence, and phase-coding hypotheses. So, when we are talking about scalp-recorded EEG rhythms, we need to keep in mind that these rhythms represent the collective behavior of large groups of synchronously acting neurons which might be quite different from local operations performed within the brain.

It should be stressed here that just 20 years ago the study of neuronal oscillations was mostly confined to clinical EEG and sleep research in humans and sparse studies on the neuronal origin of delta rhythms and sleep spindles. Today, brain rhythms are a hot topic in many neuroscience laboratories. This paradigm shift is due to the recognition that information is transferred in coordinated packages and that regularities are the best way to implement this role.

The neuronal mechanisms and functional meaning of beta and gamma waves are less understood than those of alpha rhythms. It might be that different types of beta and gamma waves play **quite different roles in cortical self-regulation**. In early years of electroencephalography, beta activities were thought to be more common in psychiatric patients, for example, in patients with anxiety and schizophrenia. However, this subjective experience was only partly supported by the standardized quantitative measures and normative databases that appeared in 1970s.

THE MYSTERY OF MULTIPLE BETA RHYTHMS

Many hypothetical functions have been suggested for beta rhythms. They are discussed in a 2010 review by Andreas Engel and Pascal Fries from the University Medical Center, Hamburg. The most popular of them are: (1) **coordination** among multiple representations in the neocortex; (2) **inhibition of movement**; (3) **motor planning**; (4) **preservation of the status quo**; (5) signaling whether there is enough evidence for **decision making**; and (6) **focusing action–selection** network functions. Beta rhythms come in various overlapping ranges: **beta 1** (13–20 Hz), **beta 2** (21–30 Hz), and **20-Hz beta** associated with motor planning and control. It appears that different beta rhythms might be associated with different neuronal mechanisms.

ROLANDIC BETA RHYTHMS

The best known beta rhythms in brain electrical activity are basal ganglia oscillations in the high beta frequency band synchronous with beta oscillations in motor cortical areas. There is strong experimental evidence that voluntary motor actions are preceded by and correlated with decreased beta oscillations in the **basal ganglia–thalamocortical motor loop**. It is also known that beta oscillations in these areas are **elevated** in basal ganglia dysfunctions such as **Parkinson's disease**. Beta rhythms correlate with the motor symptoms of bradykinesia and rigidity in Parkinson's disease.

In 2011 Ned Jenkinson and Peter Brown from the University of Oxford (United Kingdom) attempted to explain the numerous phenomenologies by the hypothesis that beta rhythms **promote tonic motor activity** at the expense of novel voluntary movement and thus could be considered a measure of the likelihood that a new voluntary action will

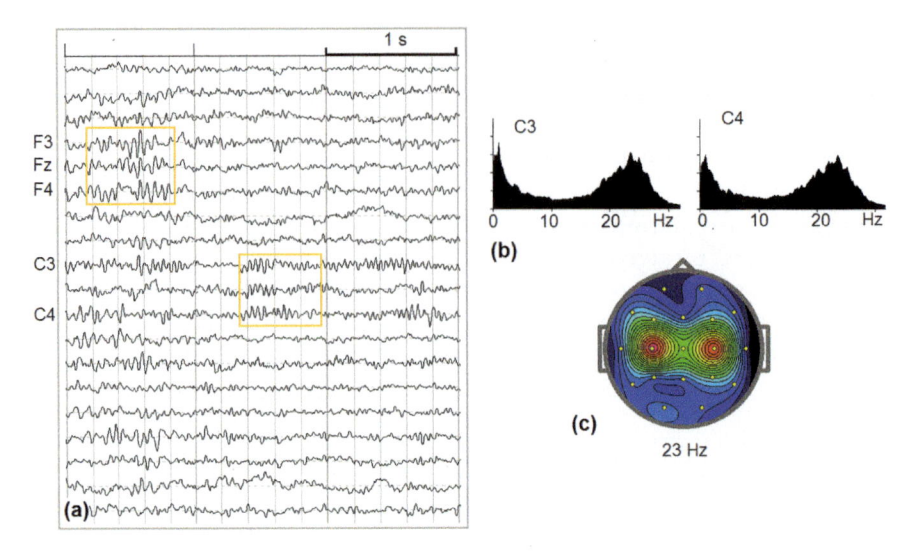

FIGURE 2.3.2 **Rolandic beta rhythms in a healthy 28-year-old subject.** (a) A 3-s fragment of EEG under the eyes-open condition in CSD montage. Note together with beta rhythms over C3 and C4 a burst of the frontal beta rhythm at F3, Fz, F4; (b) EEG spectra at C3 and C4. Note clear peaks at around 23 Hz; and (c) Map of EEG spectra at 23 Hz.

need to be actuated. Oscillatory beta activity is modulated by **dopamine levels in the basal ganglia** which are in turn modulated by the saliency of internal and external cues. The resulting modulation of beta activity enables prospective resourcing and **preparation of potential actions**. Loss of dopamine in basal ganglia, as in Parkinson's disease, destroys this function.

The result of functioning of the motor loop of the basal ganglia–thalamocortical circuit is recorded in EEGs in the form of **Rolandic beta rhythms.** An example of such beta rhythmicity is presented in Fig. 2.3.2. Beta rhythms in spontaneous EEGs are seen as bursts at the C3 C4 electrodes. The average frequency of beta rhythmicity in this case is about **23 Hz**.

Independent components corresponding to these beta rhythms are presented in Fig. 2.3.3. According to sLORETA the Rolandic beta rhythms are generated in the **motor cortical areas** in the left and right hemispheres. In the cued GO/NOGO task, they show slowly evolving desynchronization during **preparation to motor response**. The motor response correlates with strong **desynchronization** in GO trials. In NOGO trials an initial short-lasting desynchronization is followed by **rebound synchronization**. Note that Ignore trials are not associated with changing beta power. The frontal beta rhythm shows absolutely different dynamics in the task and will be discussed in details in later sections.

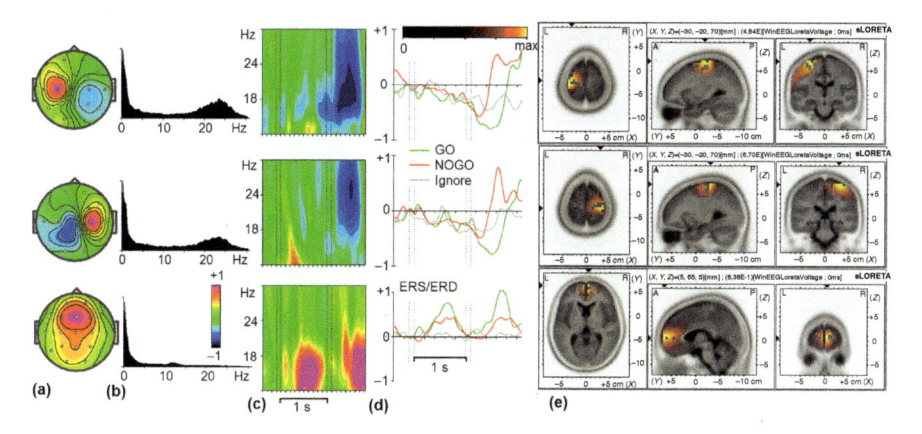

FIGURE 2.3.3 Rolandic beta rhythms (two upper rows) in the healthy subject from Fig. 2.3.2. EEG in CSD montage in the resting state was decomposed into independent components by ICA. Three components corresponding to three beta rhythms are presented. (a) Topography; (b) Spectra; (c) The results of wavelet analysis in the GO trials; (d) ERS/ERD curves for the corresponding frequency bands for GO (*green*), NOGO (*red*), and Ignore (*gray*) conditions. (e) sLORETA images of components.

CORRELATIONS WITH BOLD fMRI

In a study by Petra Ritter and colleagues from Germany and Norway, simultaneous EEG and fMRI were recorded during a bimanual motor task. A blind source separation algorithm was used for the EEG data. For the Rolandic beta the spectral power determined by wavelet analysis **inversely correlated** with the fMRI-BOLD signal in the **precentral cortex** (Fig. 2.3.4).

FRONTAL BETA RHYTHMS

In the example in Figs. 2.3.2 and 2.3.3 the **frontal beta rhythm** can be detected. The rhythm is seen at the F3, Fz, and F4 electrodes in bursts with a frequency around 16 Hz and is frontally located. In contrast to Rolandic beta rhythms, the frontal rhythm **synchronizes** after presentation of a cue in the GO/NOGO task. Sharp synchronization occurs after a delay of around 700 ms in response to cues but does not react to No cue stimuli.

Frontal beta rhythms are usually expressed in spectrograms of frontal leads as wide picks with small amplitude. This appearance on spectrograms reflects the low amplitude of these rhythmicities and their irregular pattern. These waves in the linked ears reference montage seldom—approximately 2% of normal population—exceed 20 μV. This is in contrast to alpha rhythms of central and posterior regions which usually have more regular patterns and are expressed in narrow picks on spectrograms.

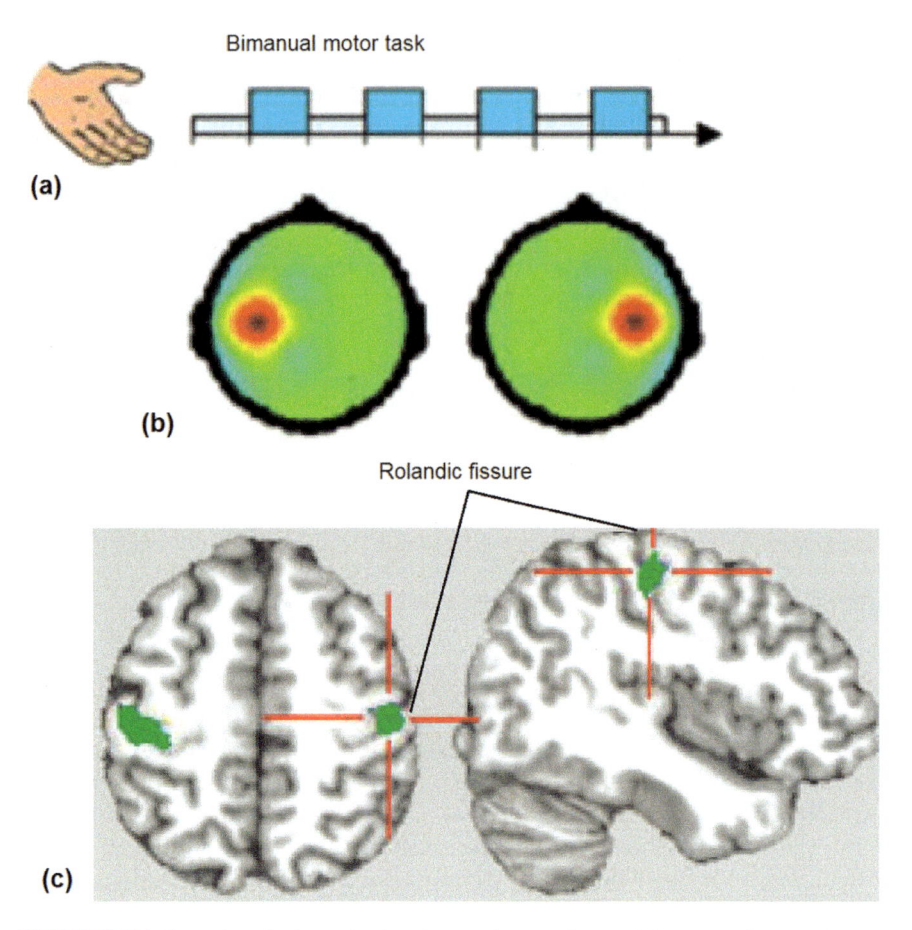

FIGURE 2.3.4 **Rolandic beta rhythm inversely correlates with BOLD fMRI.** (a) EEG and fMRI were acquired simultaneously when subjects performed a bimanual motor task; (b) Blind source separation method was applied to the EEG data. The sources of the components were convolved with the hemodynamic response function. (c) Metabolic activity in the precentral areas negatively correlates with the Rolandic beta rhythm. *Adapted with permission from Ritter et al. (2009).*

The frontal beta rhythm is a consistent pattern in healthy subjects and can be clearly seen at grand-averaged wavelet representations. The activation pattern of the frontal beta rhythm in the cued GO/NOGO task is depicted in Fig. 2.3.5.

In contrast to the suppression of Rolandic beta rhythms associated with preparation to make a movement and movement per se, the frontal beta shows an activation pattern. The rhythm is activated with a 500 ms delay after the cue, GO, NOGO, and Novel stimuli (ie, conditions associated with activation of the frontal lobes). The beta rhythm is not produced as a

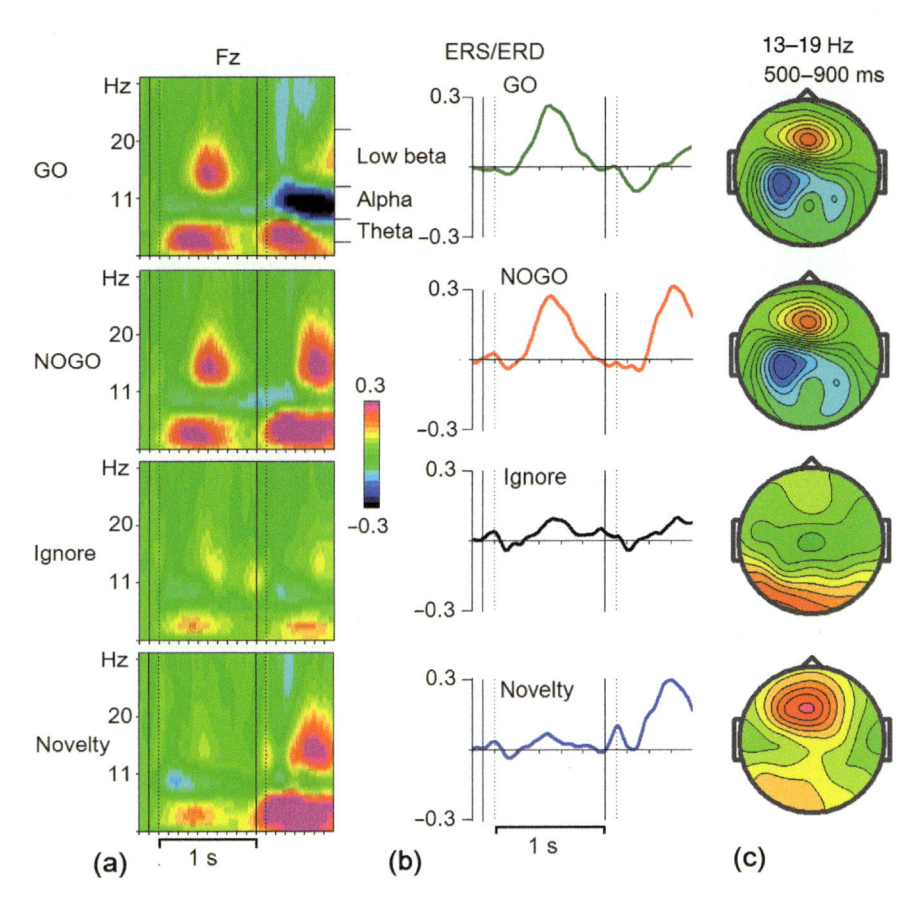

FIGURE 2.3.5 **Frontal beta rhythm in the cued GO/NOGO task.** (a) Results of wavelet analysis for GO, NOGO, Ignore, and Novelty conditions separately—grand average of the data of 312 healthy subjects of age 18–50 years in the CSD montage. Note the appearance of a burst of low beta activity in the 13–19-Hz band during 500–900 ms after stimulus under conditions that presume activation of the frontal lobes; (b) ERS/ERD curves for GO *(green)*, NOGO *(red)*, Ignore *(gray)*, and Novelty *(blue)* trials; and (c) Maps of ERS/ERD in the 13–19-Hz frequency band and time interval 500–900 ms after second stimulus presentation *Data from the HBI database.*

result of presentation of No cue and Ignore stimuli, again demonstrating that frontal lobe activation is needed for its generation.

VERTEX BETA RHYTHMS

A peak at high beta frequency and at the Cz location is sometimes seen in healthy subjects. The rhythm is called the "vertex beta rhythm." An example of such a rhythm is presented in Fig. 2.3.6.

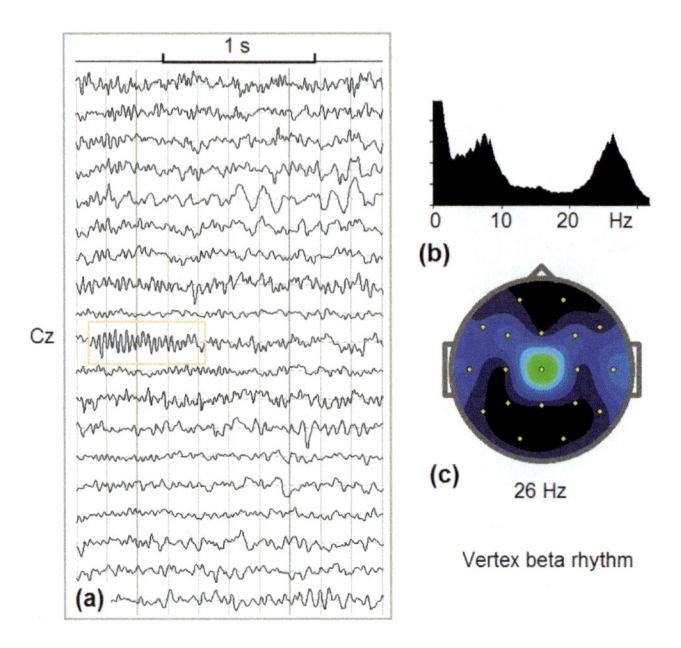

FIGURE 2.3.6 **Vertex beta rhythm in the resting state.** (a) A 2-s resting state EEG fragment in a 27-year-old healthy subject in the CSD montage. A burst of high-frequency beta rhythm is marked at Cz; (b) EEG spectra under the eyes-open condition. A peak at around 26 Hz is seen; and (c) Map of spectra at 26 Hz.

The rhythm is different from Rolandic rhythms in frequency, location and response pattern to tasks. This is illustrated by Fig. 2.3.7, which depicts the responses of the vertex beta in the cued GO/NOGO task for the same subject. Note that the vertex beta rhythm responds to NOGO and Novelty conditions, that is, in unexpected situations, in contrast to the Rolandic beta rhythms that respond to preparation of actions.

OCCIPITAL REBOUND BETA RHYTHMS

Visual stimulus presentation induces suppression of occipital beta rhythms followed by rebound synchronization. This is illustrated by Fig. 2.3.8. Rebound beta synchronization occurs after stimulus presentation with peak latency around 600 ms within the 14–19-Hz frequency band and maps into occipital–temporal areas.

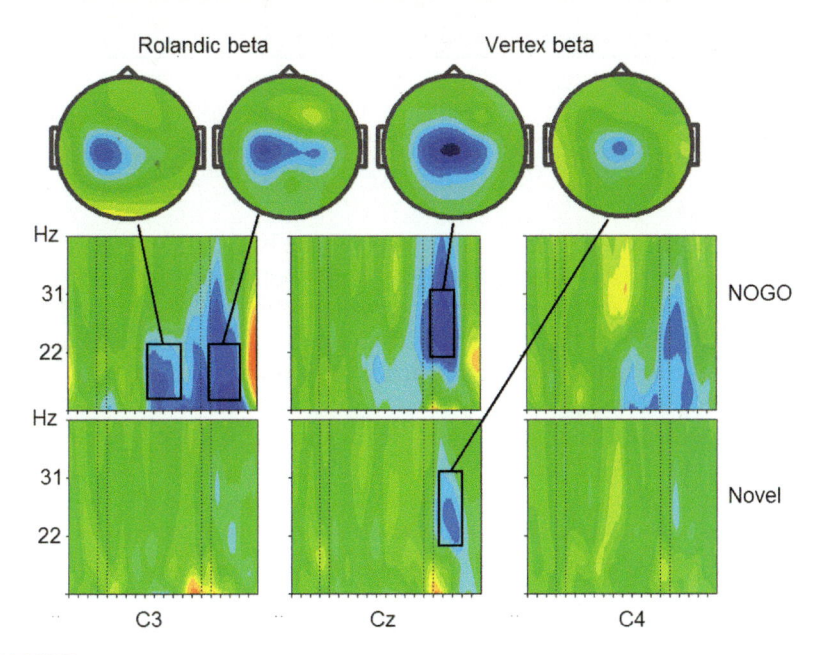

FIGURE 2.3.7 **Responses to the vertex beta rhythm in a cued GO/NOGO task in the healthy subject from Fig. 2.3.6.** The results of wavelet analysis for NOGO and Novel conditions are presented. Note that vertex rhythm desynchronization occurs in a higher frequency band (23–30 Hz) than Rolandic rhythm desynchronization, which occurs at a lower frequency band (13–22 Hz). Desynchronization of the vertex beta rhythm is not shown in the preparation period and is not asymmetrical, in contrast to Rolandic rhythms. Vertex rhythm is desynchronized in response to novel stimuli. Both rhythms did not change under the Ignore condition (not shown).

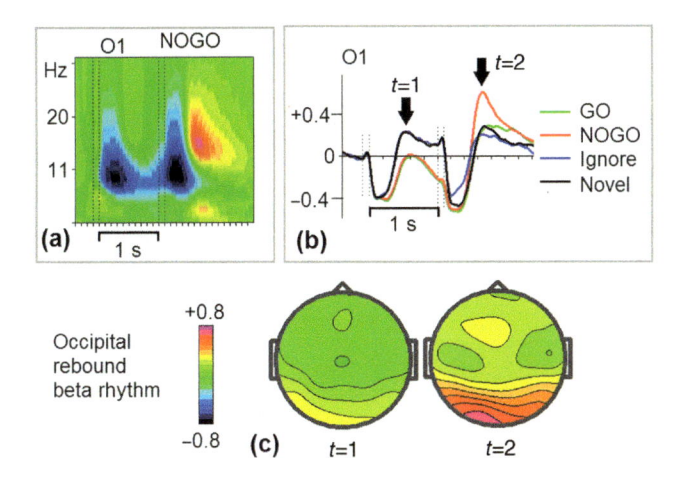

FIGURE 2.3.8 **Occipital rebound beta rhythm in a cued GO/NOGO task.** (a) Results of wavelet analysis for NOGO condition—grand average of the data of 312 healthy subjects of age 18–50 years in the CSD montage. Note the appearance of a burst of low beta activity in the 14–19-Hz band at around 600 ms; and (b) ERS/ERD curves for GO *(green)*, NOGO *(red)*, Ignore *(gray)*, and Novelty *(blue)* trials; (c) Maps of ERS/ERD in the 14–19-Hz frequency band at $t = 1$ and $t = 2$. *Data from HBI database.*

ARRHYTHMIC BETA ACTIVITY AS AN INDEX OF CORTICAL ACTIVATION

In many healthy subjects no clear peaks in the beta band on EEG spectra are found. In these cases the term **arrhythmic beta activity** must be used. In general, there is a close relationship between local indexes of metabolic activity of the human brain and parameters of spontaneous and induced EEG. Arrhythmic beta activity, in particular, appears to be a sensitive index of cortical activation.

One of the most cited attempts in this respect was done in 1998 by Jan Cook and colleagues from UCLA. The authors simultaneously recorded multichannel **EEG and positron emission tomography (PET)** scans in healthy subjects at rest and during a simple motor task. In relation to beta activity the results show that EEG power in the beta frequency range of a local average reference montage positively is correlated with metabolic activity in a corresponding local cortical area (Fig. 2.3.9a). Note that EEG power in other frequency bands also correlates with metabolic activity but in different ways: alpha activity, for example, negatively correlates with perfusion. In a 2002 study by Krish Singh from Aston University, UK and coworkers—magnetoencephalography (MEG) as the counterpart of EEG was studied in connection with fMRI. BOLD changes observed in the covert letter fluency task. The MEG was analyzed using an adaptive beam-former technique which provides continuous 3-D images of cortical power changes. The results (Fig. 2.3.9b) show that frequency-specific, task-related changes in cortical synchronization closely match those areas

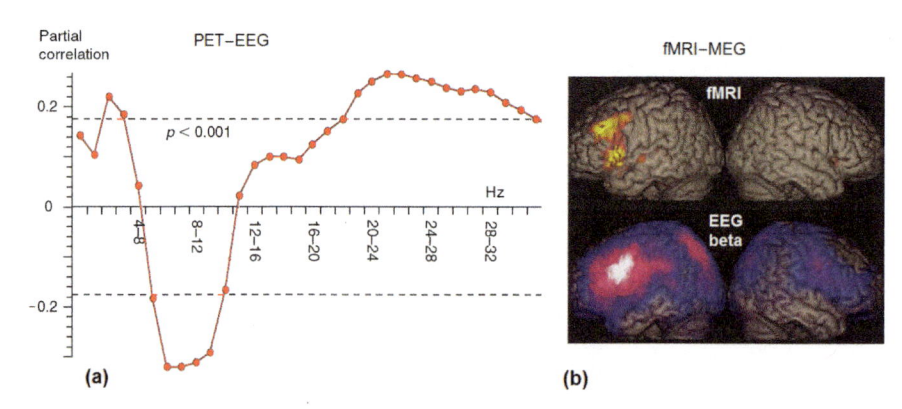

FIGURE 2.3.9 **Beta arrhythmic activity correlates with metabolic activity.** (a) Relationship between local PET perfusion values and EEG for the relative source derivation montage (y-axis, partial correlation coefficient; x-axis, EEG frequency); and (b) Upper row, increased BOLD signal of fMRI in the covert word fluency task; lower row, beta (15–25-Hz) desynchronization of MEG measured in the same task. *Part a: adapted with permission from Cook et al. (1998); Part b: adapted with permission from Singh et al. (2002).*

produced the corresponding evoked cortical hemodynamic response measured by fMRI. The MEG changes include event-related desynchronizations in beta frequency range.

The results in Fig. 2.3.9 must be viewed with caution because (1) beta activities in different parts of the cortex may have different functional meanings, and (2) beta desynchronization is often followed by beta-synchronization so that the averaged effect could be in either direction.

NEURONAL MECHANISMS

The existence of several beta rhythms with different frequencies and with different functional properties presumes that **no single neuronal mechanism** might be suggested to explain all beta rhythms. Indeed, the frontal beta rhythm is synchronized by task demands with about 500 ms delay, Rolandic beta rhythms are desynchronized by motor preparation and motor actions and showed rebound synchronization after motor actions. The vertex beta rhythm is desynchronized in response to novel stimuli. However, reasonable neuronal networks explaining specific behavior of certain beta rhythms under task or neurotransmitter manipulations could be suggested. A good example is provided by a model of **Rolandic beta rhythms**.

It is a well-established fact that this type of beta rhythmicity is sensitive to **GABA agonists** such as benzodiazepine (BNZ). In a 2005 study by Ole Jenson and colleagues from Helsinki University of Technology and colleagues the mechanisms generating Rolandic beta oscillations were studied by administration of benzodiazepines to healthy subjects while recording cortical oscillatory activity by means of magnetoencephalography (MEG) (Fig. 2.3.10). It was shown that BNZ increased the power and decreased the frequency of beta oscillations over Rolandic areas. Observed effects were simulated by a neuronal network comprising excitatory and inhibitory neurons.

In the normal brain, beta activity seems to act as a reset operation that clears all sequences of strong activation in neuronal networks and enables them to process information again and again. In the abnormal brain, when it occurs almost constantly and is reflected in too high level of beta activity (ie, manifested in deviation from normality) it is an index of hyperactivation, irritation.

GAMMA ACTIVITY

Gamma rhythms (>30 Hz) have low energy and are difficult to record. Special measures must be taken to avoid 50- and 60-Hz artifacts as well as muscle artifacts. In animal experiments the synchrony between neuronal

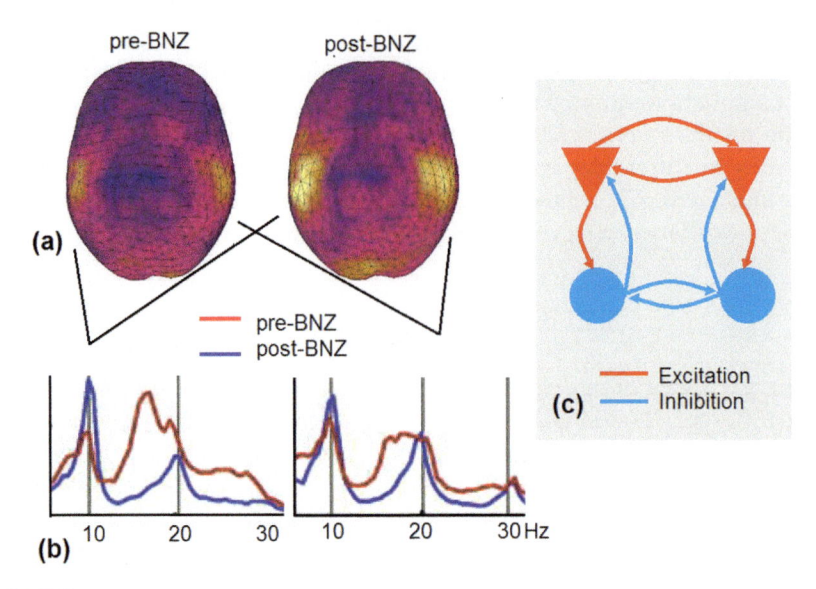

FIGURE 2.3.10 **Mechanism behind generation of Rolandic beta rhythms.** (a) Minimum current estimates of spontaneous beta oscillations in MEG before and after application of benzodiazepine *(BNZ)*; (b) The power spectra of a healthy subject calculated for spontaneous activity measured before (pre-BNZ, *blue lines*) and after (post-BNZ, *red lines*) application of BNZ. Sensors are placed over the Rolandic fissure on the left and right sides; and (c) Schematic diagram of the model constructed of 64 excitatory (e-cells) and 16 inhibitory (i-cells) cells. In simulations the consequences of changing inhibitory connections have been studied and compared with experimental findings. *Adapted with permission from Jensen et al. (2005).*

elements at 40 Hz has been proposed as a special mechanism of neural cooperation, called **temporal binding**. This temporal coordination in spiking spatially distributed neurons is needed to glue together spatially distributed representations of the image into a single percept. Much of the work in this field was done in the Max-Planck-Institute for Brain Research in Frankfurt (Germany) by a group headed by Wolf Singer.

In humans, scalp EEG recordings consistently reveal the existence of synchronized oscillatory activity in the gamma range when subjects experience a coherent visual percept. This was demonstrated by Catherine Tallon-Baudry and coworkers in 1997 from INSERM in France. In this study the authors presented quasirandom dotted pictures to subjects. Naive subjects perceived such stimuli as meaningless blobs. However, when the subjects were trained to detect the Dalmatian dog that was hidden in the pictures those now meaningful dotted pictures **induced gamma activity at 280 ms after stimulus** onset at occipital EEG recordings. A plausible interpretation of these findings is that objects giving rise to a coherent percept recruit visual areas that are synchronized in the gamma range.

ABNORMAL BETA RHYTHMS

To define whether a beta rhythm in a particular person is too large or too small one must rely on comparison with a normative database. When the excess or lack of beta activity is found the interpretation must be done with caution. First, the scalp distribution of beta excess is important. Uncommon locations, such as parietal or temporal, must be paid more attention than sensorimotor or frontal locations. Second, asymmetry of beta activity itself (higher than 50%) must be considered as an additional indication of abnormality. Third, the reactivity of the observed excessive beta rhythm is an important index of its functioning. If excessive beta activity does not change in response to task conditions in the way observed in the normative group, then this observation serves as an additional indicator of abnormality.

It should be stressed here that increased beta activity is seen in areas associated with epileptic focus, for example, during preepileptic auras. Another example can be given by cortical hyperactivity associated with hallucinations. In a 2004 study, Axel Ropohl and colleagues from University of Erlangen-Nuremberg demonstrated that auditory hallucinations in a schizophrenic patient were associated with local increase of beta MEG activity in the left auditory cortex. Sometimes excess beta power in an EEG can be seen in raw EEG traces as spindling beta rhythms. This pattern has been reported to be associated with "cortical irritability," viral or toxic encephalopathies, and epilepsy. This abnormal beta is seen in waxing and waning spindles over the affected cortex. Excess of frontal beta rhythms has been found in fewer than 10% of the Attention Deficit Hyperactivity Disorder (ADHD) population. The abnormal gamma band dynamics was found in schizophrenia. The results were reviewed in a 2015 paper by Peter Uhlhaas from University of Glasgow, UK and Wolf Singer from Max Planck Institute for Brain Research, Germany. The authors associated the observed abnormalities with disturbances of GABAergic mechanisms or/ and alterations of long-range glutamatergic projections.

CHAPTER 2.4

Frontal Midline Theta Rhythm

HISTORICAL INTRODUCTION

In the healthy brain during wakefulness there is only one rhythm in the theta band—the **frontal midline theta rhythm or FM-theta**. Because this rhythm appears in short bursts (of a few seconds) with long and varied interburst intervals in a small (10–40%) group of healthy subjects, and is enhanced by task load, it can be reliably measured by spectral analysis in highly demanding cognitive tasks. That is why the rhythm was observed not early than in 1950s by Alejandro Arellano and Robert Scalp from Massachusetts General Hospital and its name was coined even later (1972) by Tsutomu Ishihara from Osaka University Medical School and Nobuo Yoshi from Toho University Japan who found it in EEG of healthy subjects performing continuous arithmetic addition.

The frontal midline theta rhythm in humans is often associated with the **hippocampal theta rhythms** in animal research. Indeed, oscillations in the range of 3–10 Hz are the most prominent electrical patterns of hippocampal activity in rats. It was hypothesized that hippocampal theta oscillations are involved in memory encoding and retrieval. Similarly, on the basis of the studies of an Austrian group from the University of Salzburg, Wolfgang Klimesch in 1998 hypothesized that encoding new information is reflected by theta oscillations in the human EEG.

Recordings from the **human hippocampus** are available only in rare cases of stereotactic operations in epileptic patients with depth electrodes implanted for diagnostic purposes. In the literature, there are only a few publications dealing with the hippocampal theta. For example, a group of researchers from Harvard Medical School (Cantero et al., 2003) using subdural and depth recordings from epileptic patients was able to show the presence of state-dependent theta waves in the human hippocampus. Unlike the tonic theta in rodents, these oscillations were consistently observed in short (around 1-s) bursts during REM sleep and transitions to wakefulness.

The hippocampal function can be derived from two seemingly contradictory discoveries in the second half of the 20th century: (1) the observation that damage to the hippocampus leads to inability to form new episodic memories (as in the famous patient H.M.) and (2) the observation that spiking activity of hippocampal neurons is associated with the spatial navigation of the rat (space and grid cells in the hippocampus). In 2013 György Buzsáki from New York University and Edvard Moser from the Norwegian University of Science and Technology proposed that the mechanisms of memory and planning have evolved from mechanisms of navigation in the physical world and that the neuronal algorithms underlying navigation in real and mental space are fundamentally the same.

A detailed review of the frontal midline theta rhythm from the **perspective of hippocampal theta** was accomplished by Damon Mitchell, Neil McNaughton, Danny Flanagan, and Ian Kirk from New Zealand and Australia in 2008. They concluded that (1) the frontal midline theta like the hippocampal theta is elicited during processing of memory and emotion, (2) the frontal midline theta is increased by anxiolytic drugs and personality-related reductions in anxiety in contrast to the hippocampal theta which is decreased by anxiolytic drugs, and (3) in rats the frontal cortex and hippocampus often show independent theta activity of different frequencies but, during exploratory activity, can show coherence with frontal areas becoming locked to the hippocampus.

FUNCTIONAL FEATURES

The scalp topography of the rhythm power is frontal with a maximum at Fz, or if the number of electrodes is large enough just anterior to **Fz**. The frequency of frontal midline theta varies from **5 to 7.5 Hz** with an average around 6 Hz. The amplitude is often higher than the average background. The waveform is close to sinusoidal. The rhythm appears in **discrete bursts that last a few seconds** and is **modulated by the task load**: the harder the task the more prevalent the rhythm.

An example of the frontal midline theta rhythm in a healthy 19-year-old subject is presented in Fig. 2.4.1. The subject performed an arithmetic task: adding two numbers presented on the screen for 400 ms and then comparing the result with a number presented 1 s later. As one can see, the effort associated with an arithmetic operation evokes a burst of a few cycles of theta rhythm. The whole task lasts 20 min and the EEG spectra of CSD at Fz with the map at the maximum of EEG power are presented in Fig. 2.4.1c,d.

It should be stressed here that bursts of theta in this particular subject are observed not in every trial but randomly every three or five trials with a time period between separate bursts of around 20 s. These bursts are not time-locked with stimulus presentations and are analyzed by wavelet analysis. The results of wavelet analysis for this subject are presented in Fig. 2.4.2a. Note that the stimulus presentation induces an increase of EEG

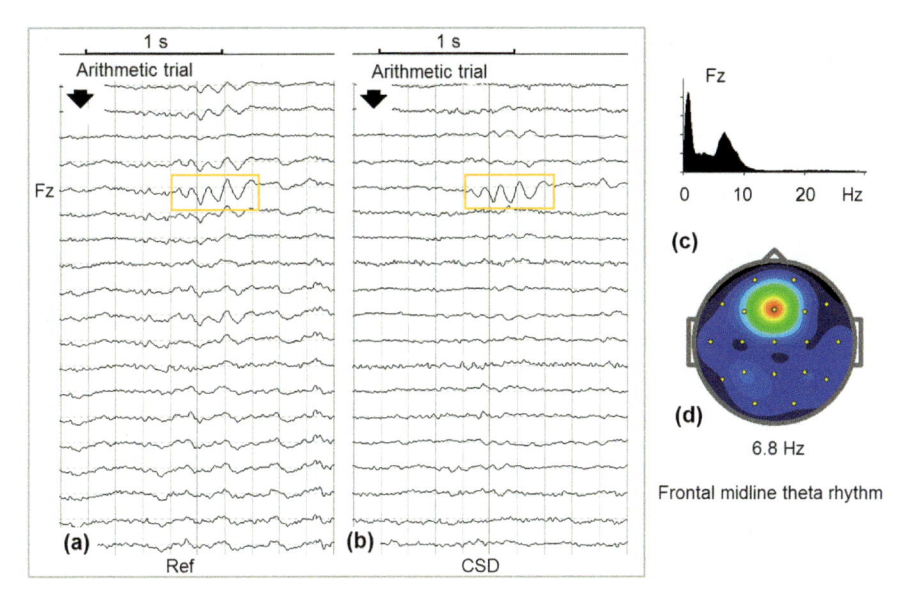

FIGURE 2.4.1 **Frontal midline theta rhythm in a healthy subject performing an arithmetic task.** (a) Referential montage; (b) CSD montage. A healthy subject performs the arithmetic task. The *arrow* indicates onset of stimulus presentation—two numbers that need to be added and compared with a number presented 1 s after. As one can see, around 700 ms after stimulus onset a burst of theta emerged with maximal amplitude at Fz; (c) CSD spectra at Fz for the whole arithmetic task (200 trials) with maximal amplitude at 6.8 Hz; and (d) Map of the spectra at 6.8 Hz.

power in the theta band. The ERS for the 6–8-Hz band is presented in Fig. 2.4.2b while the map at the peak is presented in Fig. 2.4.2c.

LOCALIZATION

Application of independent component analysis for an EEG of the whole task reveals an independent component corresponding to the frontal midline theta (Fig. 2.4.2c). sLORETA images indicate that those bursts in this particular subject are generated in the medial part of the prefrontal cortex (including the anterior gyrus cingulate).

Many attempts to localize the source of the frontal midline theta rhythm by means of EEG, MEG, and different localization methods suggest that the main sources are located in the medial prefrontal and anterior cingulate cortical areas. Intracranial EEG recordings in patients and monkeys confirm these results.

A 2001 study by Sridhar Raghavachari and colleagues from Brandeis University, Massachusetts using implanted depth and cortical surface electrodes in epileptic patients have shown that theta increases during both verbal and

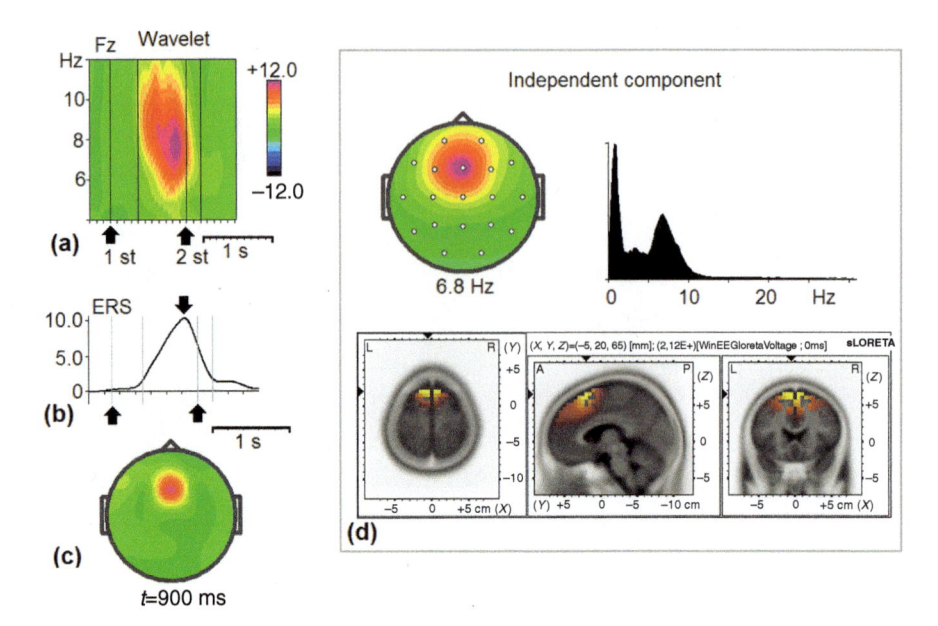

FIGURE 2.4.2 Time dynamics and localization of the frontal midline theta rhythm. (a) Time frequency presentation of induced EEG in the arithmetic task in the subject presented in Fig. 2.4.1. *Arrows* indicate the onset of the first stimulus (two numbers, such as 36 + 53) and the second stimulus (a number, which correspond to the correct result such as 89 or does not correspond to the correct result such as 98); (b) The ERS computed for the 6–8-Hz frequency band; (c) Map of the ERP at its maximum; and (d) independent component corresponding to the frontal midline theta with map, spectra, and sLORETA image.

spatial memory tasks over widespread regions of the neocortex including the temporal lobe, the parietal lobe, the inferior frontal gyrus, and the central sulcus. These neocortical theta oscillations, when synchronized over large regions, appear to produce the frontal midline theta rhythm on the scalp.

PREVALENCE

In the resting state with the eyes open or the eyes closed a few healthy subjects show a reliable frontal midline theta rhythm. The rest of the normal population does not reveal any signs of this rhythmicity. One explanation could be that because of deeply located generators the frontal midline theta becomes visible on the scalp only in rare cases. Indeed, for any cortical rhythm to be recordable from the scalp the following requirements must be fulfilled. First, the sources have to be located near the surface, but not too deep in the brain. Second, the sources must be synchronized over relatively large spatial scales. Third, the sources must be strong enough to be recorded on the skin of the head.

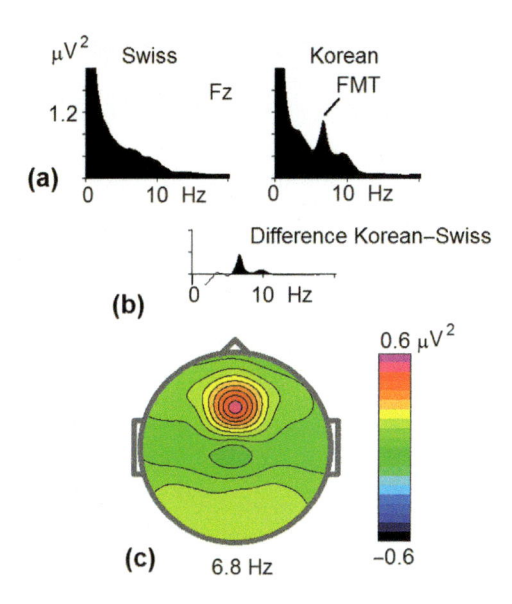

FIGURE 2.4.3 The frontal midline theta is more prevalent in the Asian population. (a) Grand average ($N = 100$) of EEG spectra at Fz in the common average montage under the eyes-open condition in a group of healthy subjects aged 9–12 in Switzerland (left) and in Korea (right). (b) difference Korean – Swiss spectra. (c) map of the difference at 6.8 Hz. The data from the Asian population were collected by Seung Wan Kang and his team from Seoul National University.

GENETIC FACTORS

The other possibility for the rare incidence of the frontal midline theta in healthy populations could be that the rhythm depends on genetic factors and is revealed only in a specific group of the healthy population. This suggestion is supported by comparison of EEG spectra in the European population with those in the Asian population. The results are illustrated by Fig. 2.4.3. As one can see, the average power of the frontal midline theta rhythm in the Korean group of children is much higher than that in the Swiss group of corresponding age and similar task performance.

When EEG is recorded in paradigms known to evoke the rhythm (such as arithmetic or space navigation tasks) the prevalence of the rhythm gets higher.

AGE DYNAMICS

There is reasonable evidence that the rhythm decreases with aging. The incidence of frontal midline theta–emitting subjects reported by Japanese research groups as reviewed by Kazutoyo Inanaga from Chikusuikai Hospital in 1998 is around 40% in young adults (18–28 years of age), and around 10% in older subjects.

PERSONALITY TRAITS

In studies, Kazutoyo Inanaga from Chikusuikai Hospital, Japan in 1998 found that the normal subjects were divided into three groups according to the amount of frontal midline theta generated in an arithmetic task at Fz: low, moderate, and high frontal midline theta groups. The high frontal midline theta group showed the lowest anxiety score, the highest score in the extraversion scale, and the lowest score on the neurotic scale. The low frontal midline theta group, on the other hand, showed the opposite correlation: they showed the highest anxiety score, the lowest extraversion score, and the highest neurotic score. No differences were observed in these three groups regarding the quality of the task. From these results, it is suggested that the appearance of frontal midline theta shows a close relationship to the personality trait and anxiety level of the subject.

Studies by Yasushi Mizuki and colleagues from the Yamaguchi University School of Medicine (Japan) in the 1970s and 1980s showed that the frontal midline theta rhythm is a reliable index of certain personality traits. The authors reported that the appearance of the frontal midline theta rhythm is more likely in extravert, less neurotic, and less anxious subjects. Later Masatomo Suetsugi and colleagues from Yamaguchi University School of Medicine, Japan showed that the appearance of the frontal midline theta rhythm might be related to the relief of anxiety in patients with generalized anxiety disorder.

CORTICAL METABOLISM

Diego Pizzagalli and colleagues from Harvard University and University of Wisconsin in 2003 were the first to study the link between frontal theta midline activity and metabolism in the human brain. Concurrent measurements of brain electrical activity (EEG) and glucose metabolism (positron emission tomography, PET) were performed in healthy subjects at baseline. EEG data were analyzed with a source localization technique that enabled voxel-wise correlations of EEG and PET data. For theta, but not other bands, the **rostral anterior cingulate** cortex was the largest cluster with positive correlations between current density and glucose metabolism.

In the 2007 study by Gebhard Sammer and colleagues from Justus-Liebig University, Giessen (Germany) simultaneous EEG-fMRI recordings were used to investigate the functional correlational topography of theta. EEG-theta was enhanced by a mental arithmetic–induced workload. For the EEG-constrained fMRI analysis, theta reference time series were extracted from the EEG, reflecting the strength of theta occurrence during the timecourse of the experiment. Theta occurrence was mainly associated with activation of the cingulate cortex, insular cortex, hippocampus, superior temporal areas, superior parietal, and frontal areas. The authors argue the involvement of **widely distributed brain regions** implies that scalp-recorded frontal midline theta represents comprehensive functional brain states rather than specific processes in the brain.

WORKING MEMORY

The tasks in which frontal midline theta is most prevalent are those in which the subject is engaged in **focused, but also relaxed, concentration**. An example of such a task is provided by the arithmetic task described above. The grand average spectra in this task compared with the resting-state eyes open condition are presented in Fig. 2.4.4. Note that the task is associated with higher EEG activities in the frequency range from 5.5 to 9.5 Hz and low beta (13–18 Hz). The increase in the theta band corresponds to enhancement of the frontal midline theta rhythm with the task load. The increase in the low beta band corresponds to enhancement of the frontal beta rhythm in cognitive tasks described in chapter: Beta and Gamma Rhythms.

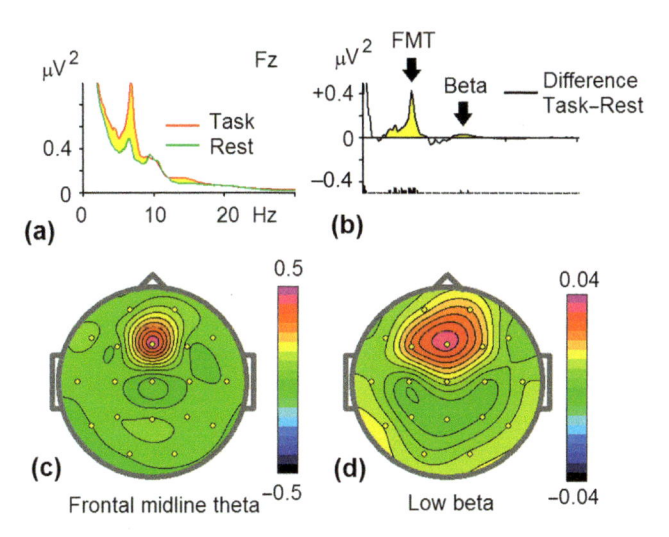

FIGURE 2.4.4 **Increase in the frontal midline theta rhythm with task load.** (a) Grand average EEG spectra in the arithmetic task (Task, *red*) in contrast to the grand average of EEG spectra for the resting state with the eyes open (Rest, *green*) for a group of adolescents aged 13–14; (b) Difference spectra Task − Rest at Fz with p values of the significance at the bottom (the lowest bar corresponds to $p < 0.05$); (c) Map of difference spectra at the frontal midline theta band; (d) Map of difference spectra at the low beta band. *Data from the Human Brain Indices (HBI) database.*

Other examples of tasks associated with induction of the frontal midline theta rhythm are games such as Tetris and driving tasks. In a 1995 study by Seppo Laukka and colleagues from the University of Oulu (Finland) subjects were engaged in a simulated driving task in which they had to learn how to navigate through a series of streets represented as an animation on a computer screen. It was found that the percent of frontal midline theta activity increased as the subjects grew better at the task, making more correct decisions.

Mental arithmetic is a complex action that requires the involvement of multiple processes including number recognition, working memory, mental calculation, decision making, response selection. **Working memory** has been systematically studied in association with the frontal midline theta rhythm. In a 1997 study by Alan Gevins from EEG Systems Laboratory and SAM Technology in San Francisco healthy subjects performed working memory tasks requiring comparison of each stimulus to a preceding one on verbal or spatial attributes. The frontal midline theta rhythm was shown to **increase in amplitude with increased memory load**. Moreover, theta signals increased after practice at the tasks.

Wolfgang Klimesch and colleagues from Salzburg (Austria) gained evidence for the existence of **three types of theta activities** with different associations to memory. The first type of activity constitutes long-lasting, tonic oscillations in the theta frequency band which are related to sustained attention or working memory. Phasic bursts of frontal theta activity are supposedly associated with the retrieval operation while the parietal theta activity is associated with encoding of episodic memory.

In a 2014 paper Liang-Tien Hsieh and Charan Ranganath from the University of California at Davis reviewed evidence demonstrating the relationship between frontal midline theta oscillations and **working memory, episodic encoding and retrieval**. Fig. 2.4.5a demonstrates tonic increase of the frontal midline theta rhythm during a delay period of trials in which the temporal order of visual items or items themselves had to be kept in working memory during 4-s delay periods.

A condition that is also associated with the appearance of frontal midline theta is hypnosis. To be hypnotized, people have to focus their attention on themselves and on the induction. If they are distracted, the hypnosis is broken. In a 1994 study, Helen Crawford from Virginia Technical University found that highly hypnotizable people exhibit more theta before and during hypnosis then lowly hypnotizable people.

Another uniquely human condition linking theta activity to attention is meditation. During Zen meditation, EEGs of experienced practitioners have been recorded (Takahashi et al., 2005). As the Zen practitioners reach deeper and deeper meditative states or trances, the typical patterns of EEG activity begin with alpha activity (8–10 Hz) and gradually move into the theta range as the trance progresses.

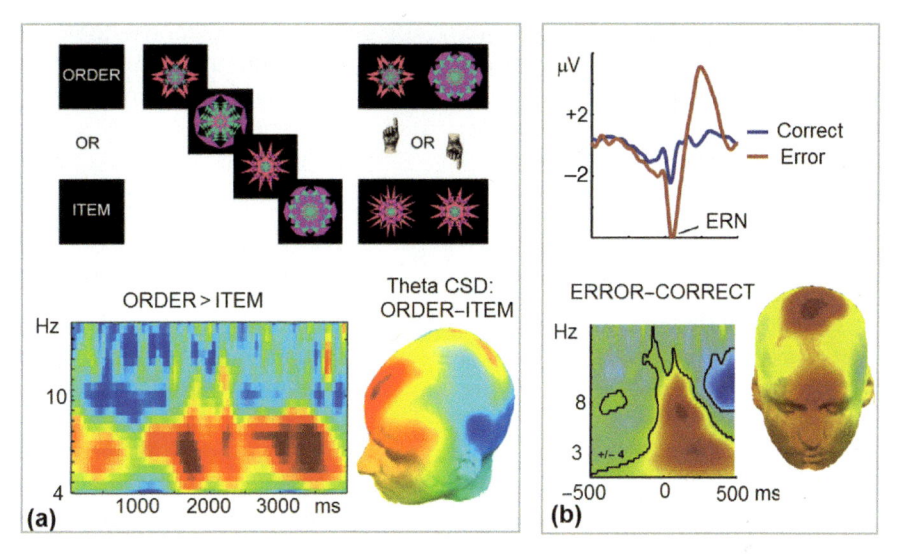

FIGURE 2.4.5 Reflection of working memory and conflict monitoring in frontal theta activity. (a) Tonic power increases in the frontal midline theta rhythm during maintenance of temporal order information. Top left, schematic diagram of working memory tasks; top right, power spectra in a healthy subject during a delay period in temporal order trials; bottom left, results of wavelet analysis illustrate the averaged difference in oscillatory power for frontal electrodes between correct order *(ORDER)* and correct item *(ITEM)* trials during working memory delay (*x*-axis, 4-s delay period); bottom right, topographic map of difference in oscillatory power at CSD montage between *ORDER* and *ITEM* trials in the theta frequency band; and (b) Theta phasic induction in response to erroneous trials. Top, error-related negativity *(ERN)* (a frontally distributed negative wave in response to errors in comparison with correct trials); bottom left, the results of wavelet analysis illustrate a sharp increase in EEG power in the theta band; bottom right, topography of theta increase. *Part a: Adapted from Hsieh and Ranganath (2014). Part b: Adapted with permission from Cavanagh & Shackman (2014).*

CONFLICT MONITORING AND ANXIETY

Accumulated experimental evidence shows that frontal theta activity is induced also in response to situations associated with **conflict detection and conflict monitoring**—operations of cognitive control. These data were reviewed in a 2014 paper by James Cavanagh from the University of New Mexico and Alexander Shackman from the University of Maryland. In a series of metaanalyses, the authors show that frontal midline theta signals reflect activity of the mid-cingulate cortex—the dorsal region of the rostral cingulate—and, **moderated by anxiety, predict adaptive behavioral adjustments**. Fig. 2.4.5b demonstrates phasic changes in frontal theta activity in response to erroneous trials in comparison with correct trials.

It should be stressed here that these phasic signals **are not necessarily related** to the frontal midline theta rhythms described above. These frontal midline theta signals are phasically generated in response to novel or error-related stimuli which evoke N2/P3 waves with a clear theta component of these ERPs. Indeed, when Figs. 2.4.5a and b are compared the differences in scalp distribution and spectral content of tonic and phasic responses are noticed.

MODEL

The frontal midline theta rhythm in human EEG is often associated with **hippocampal theta** in rats that is recorded by implanted electrodes in all parts of the hippocampus and in the majority of interconnected anatomical structures. This type of global, extracellularly recorded phenomenon reflects the cooperative behavior of a large number of hippocampal pyramidal cells which fire periodically in synchrony to produce theta oscillations.

The structures in which theta rhythms have been observed are depicted in Fig. 2.4.6a. They represent a loop which was first described in 1937 by

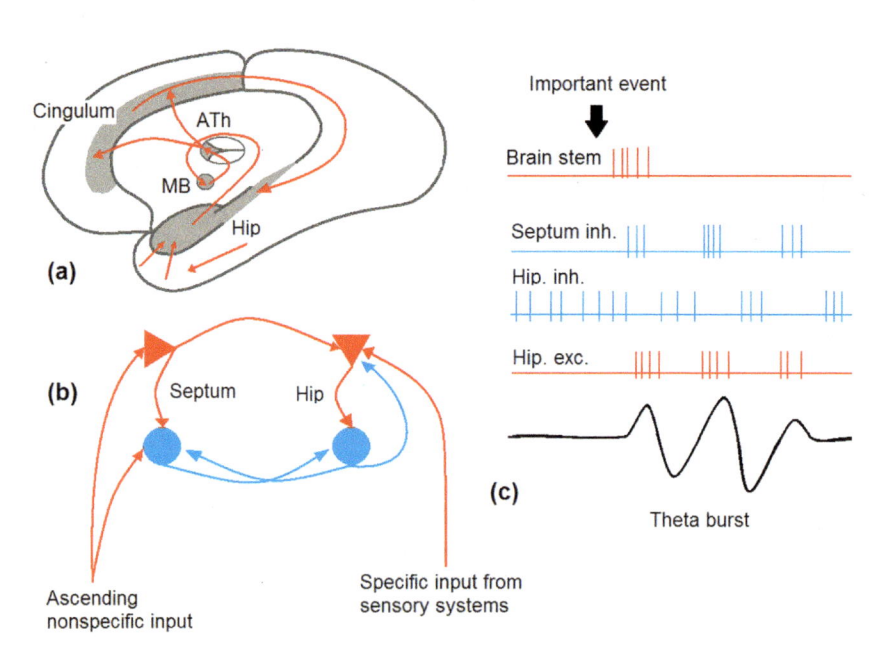

FIGURE 2.4.6 Hypothetical neuronal mechanisms of the hippocampal theta rhythm. (a) Brain structures in which theta activity is recorded form a closed loop and (b) The theta is generated in the septal–hippocampal network with excitatory and inhibitory interconnections. In the resting state, hippocampal pyramidal cells are inhibited by local inhibitory neurons and the gate to the loop is closed. An important event activates septal inhibitory neurons which inhibit inhibitory hippocampal neurons and thus remove inhibition and open the gate. This cycle is repeated by reciprocal excitatory/inhibitory hippocampal–septal connections.

an American neuroanatomist John Papez and was named after him. Papez proposed that this circuit plays a critical role in emotional experience. Unlike its persistence as an anatomical entity the function of the **Papez circle** has been less resilient. The early notion that Papez's circuit is associated with emotions has been abandoned and replaced by the proposal that it is primarily involved in memory functions. This proposal was made on the results of studies showing that lesions in the major components of the circuit disrupt episodic memory but not affective functions.

The information flow in this loop could be simplified as follows: (1) the hippocampus receives **polysensory signal** through the rhinal cortex from the sensory systems and from the executive system of the brain through the cingulate cortex, (2) the **hippocampus integrates** this information into a chunk of episodic memory, and (3) sends the results back to the cingulum via the mamillary bodies of the hypothalamus and anterior nucleus of the thalamus, (4) the cingulum closes the loop through its connection to the hippocampus.

As clinical evidence indicates, the hippocampus is critical for **consolidation of episodic memory**. Despite intensive experimental research the link between memory-related phenomena at the cellular–molecular level, on the one hand, and computations at the network level, on the other hand, remains unknown. One of the key issues of current research is to understand how the theta rhythm is generated by the collective behavior of single neurons and what role this rhythmicity plays in memory formation. With this goal, hippocampal theta rhythms have been extensively studied in animal models, especially in rats. These studies revealed a **model of theta generation** in the hippocampus (Fig. 2.4.6b,c).

According to this model the theta rhythm originates in the **brainstem as nonrhythmic neuronal activation**. The neuronal elements of the brainstem include noradrenergic cells in the locus coeruleus, serotoninergic cells in the raphe nuclei, dopaminergic cells in the ventral tegmental area and the substantia nigra pas compacta. Activation of these cells directly by sensory collaterals or indirectly through feedback projections from the cortex leads to activation of cells in the septal nuclei of the basal forebrain. The activation of septal neurons starts a burst of theta rhythms.

A critical role in theta rhythm generation is played by the **reciprocal connections** of neurons in the septal region with neurons in the hippocampus. In the resting state, background activity of inhibitory neurons in the hippocampus suppresses pyramidal cells so that no sensory information is encoded into a memory trace. If a behaviorally meaningful stimulus (eg, emotionally competent event) is presented, this stimulus through the ascending activating system of the brainstem activates septal excitatory and inhibitory neurons. Septal inhibitory neurons inhibit inhibitory cells in the hippocampus. This inhibition of inhibitory cells removes background inhibition from pyramidal cells and eventually gates the pyramidal cells. This sequence of neuronal events starts a theta cycle. Further on, the activation of pyramidal cells in the hippocampus inhibits inhibitory cells in the septal nuclei, which in turn removes inhibition from hippocampal inhibitory neurons, and they in turn inhibit pyramidal cells. This ends the cycle.

In this scheme the septal region is considered the nodal point where ascending nonrhythmical inputs from the ascending activation system are converted into rhythmical signals of septal neurons that are further transmitted to the hippocampus. The hippocampal theta depends strongly on the strength of the input from the septal region: when a certain critical level of activation of septal neurons is reached, the hippocampal theta rhythm materializes.

Several lines of evidence support the concept that the theta rhythm plays an important role in specific memory operations. First, long-term potentiation as a memory mechanism is sensitive to the phase of the theta rhythm, with potentiation favored at the peak of the cycle and depotentiation favored at its trough. Long-term potentiation was discovered in the mammalian hippocampus by Terje Lømo from the University of Oslo (Norway) in 1966 who observed that high-frequency electric stimuli of a synapse strengthen (potentiate) the synapse for minutes to hours and days. Second, a number of in vitro and in vivo studies have reported that induction of long-term potentiation is optimal when the time interval between stimuli is approximately 200 ms (ie, in theta range). Third, septal lesions not only block theta in the hippocampus but also produce severe impairments in memory function. These findings suggest that that the theta rhythm acts as a windowing mechanism for synaptic plasticity.

The theta cycle may be considered an **information quantum**. This quantum seems to serve as a functional linkage of different limbic structures allowing encoding of a certain episode of our daily life into a **memory chunk**.

ABNORMAL THETA RHYTHMS

There is only one theta rhythm in the healthy brain—the frontal midline theta with certain functional features. If a theta rhythm is observed in a different location or shows different functional features this rhythm must be considered abnormal.

One type of abnormality may be associated with a low level of cortical activation. Recall that Fig. 2.2.13 shows that decreased depolarization of thalamocortical neurons due to the low level of cortical activation leads to transfer of alpha oscillations to theta oscillations. This kind of theta abnormality is often seen in patients with autism in whom theta rhythms are observed over temporal areas. Another example is given by an EEG recorded in a patient after several days of closed brain injury and presented in Fig. 2.4.7. One can see the high-amplitude theta rhythm at frontal locations the power with topography different from that of the frontal midline theta rhythm.

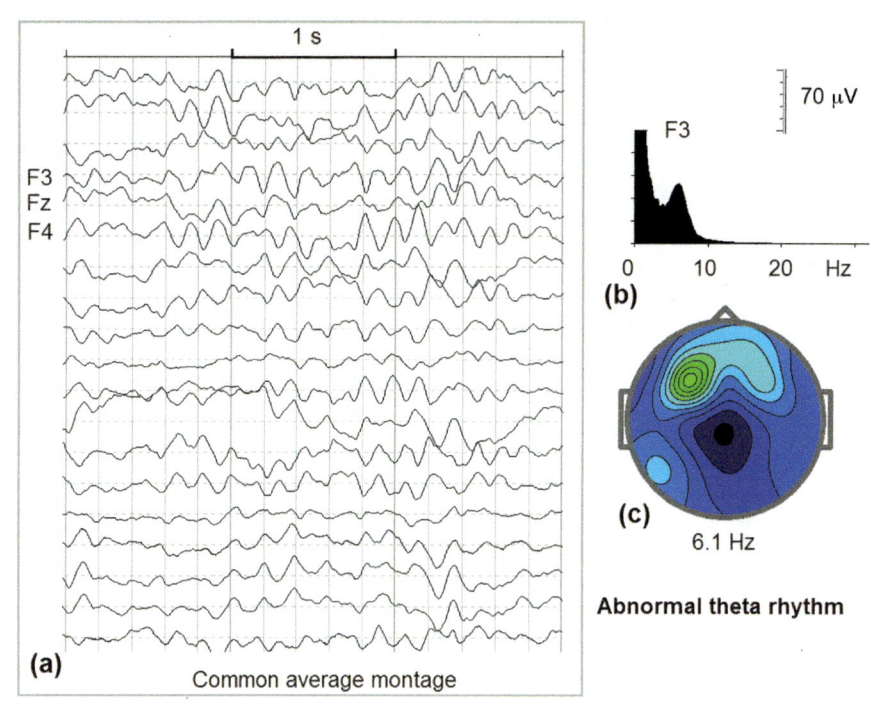

FIGURE 2.4.7 **Abnormal theta rhythm in closed brain injury.** (a) A 3-s fragment of EEG in eyes closed condition recorded in an adult patient 5 days after closed brain injury; (b) EEG spectra at F3 with a clear peak at 6.1 Hz; and (c) Map of EEG spectra at 6.1 Hz.

A second type of theta abnormality is seen in a special group of patients diagnosed with ADHD. The group exhibits poor social relationships and an inability to correct behavior. From the electrophysiological point of view, this ADHD subtype is characterized by very long runs of the frontal midline theta rhythm which is expressed in an extremely high and sharp peak on EEG spectra at Fz in the frequency range 5.5–8 Hz and extremely low theta synchronization in response to meaningful stimuli under the two-stimulus GO/NOGO and mathematical task conditions.

PART 3

INFORMATION FLOW WITHIN THE BRAIN

CHAPTER

3.1

Sensory Systems and Attention Modulation

INTRODUCTION

Information about the external and internal world is perceived by the brain through different **receptors**. A receptor cell transfers a sensory stimulus in a specific modality into impulse activity of output cells in receptor organs. For example, cones in the retina are sensitive to light and transfer their reactions to the light toward ganglion cells of the eye.

Receptors are divided into **extra- and inter**receptors. Extrareceptors tell us about external interactions with the environment, while interreceptors tell us what happens inside the body. The activity of some interreceptors usually does not reach our consciousness. Sensory elements for perception of the external world are highly specialized from species to species. Mammals, for example, cannot see in infrared light or perceive a magnetic field. The architecture and function of vision of hunting animals is different from those that eat grass. Moreover, the ability of perception depends very much upon experience. Eskimos spend their lives among snow and can distinguish many hues of snow.

The brain regions where neurons respond to stimulation of a certain type of receptors are called **sensory systems**. There are different **sensory modalities** that give us sensations of images, sounds, body movements, and produce vision, hearing, pain, taste, touch, smell. The visual modality is the most studied sensory system (Fig. 3.1.1).

For discovery of the columnar organization of the primary visual cortex, **David Hubel and Torsten Wiesel** from Harvard University were awarded the 1981 Nobel Prize. In particular, they were the first to show that V1 is organized into columns of cells that are strongly activated by lines of a given orientation. Charles Gross from Princeton University and Keiji Tanaka from the RIKEN Brain Science Institute in Japan contributed significantly to our understanding of visual processing in the inferior temporal cortex.

Functional Neuromarkers for Psychiatry. http://dx.doi.org/10.1016/B978-0-12-410513-3.00011-5
Copyright © 2016 Elsevier Inc. All rights reserved.

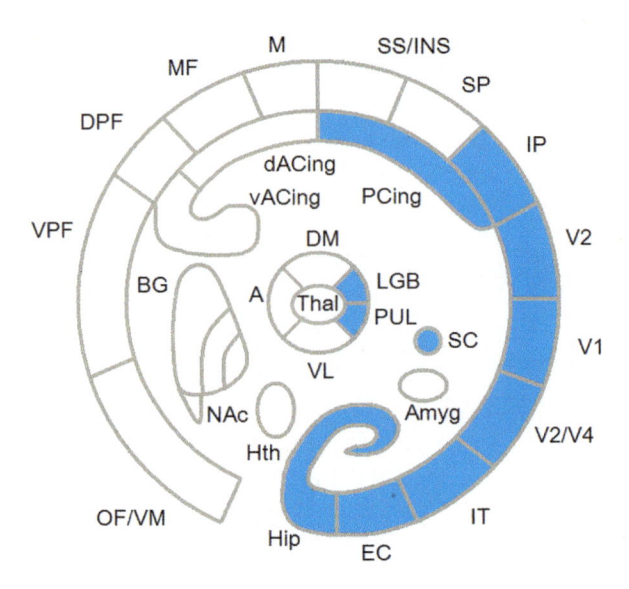

FIGURE 3.1.1 **Visual sensory system.** The visual sensory system includes cortical visual fields of the occipital, inferior temporal, inferior parietal, and hippocampal areas. Subcortical structures include the lateral geniculate body *(LGB)* and pulvinar nuclei of the thalamus as well as the superior colliculus.

One of the basic discoveries in sensory systems, in general, and in the visual system, in particular, is decomposition of the visual (auditory, somatosensory) pathways in several streams dealing with different properties of the sensory world. The ventral "what" and the dorsal "where" visual streams were first described by **Leslie Ungerleider** and **Mortimer Mishkin from the National Institute of Mental Health**.

Information flow in sensory systems is modulated by **attention**. To stress a critical role of attention in human behavior the famous American psychologist **William James** wrote in 1890 "My experience is what I agree to attend to." To get an idea of the operations underlying attention, imagine you are sitting in a dark room with a mechanical clock in it. At first you do not hear the clock, but then you start noticing the ticking sound. Attending to it looks like "turning on" the loudness of the loudspeaker: the ticking is getting louder. Attending to another source of sound such as a whisper of your friend sitting in a corner "turns off" the sound of the clock in your perception. As this simple example shows, attention is associated with enhancement of relevant sensory information and suppression of irrelevant sensory information. These **enhancement/inhibition operations** are collectively called **selection operations**. Attention as a psychological process can be determined as selection operations in sensory modalities with a goal to process relevant sensory information more accurately and to inhibit irrelevant information.

Systematic studies of attention started in the late 1960s when the **information-processing paradigm (George Miller, Eugene Galanter, Karl Pribram)** replaced behaviorism with its reflexes. Attention had become the hot topic of this approach. Using ERP method, many laboratories were able to separate the stages of information processing and show how these stages were modulated by attention. Various degrees of attention from hyperfocus such as in hypnosis and meditation to deep sleep (complete loss of attention) became areas of intensive research.

Later, in the 1970s, recordings of the impulse activity of neurons and local field potentials in animals showed that attention was associated with two joint operations: **facilitation** of responses of neurons to attended stimuli and **suppression** of neuronal responses to unattended stimuli. Lesion studies reveal a specific phenomenon—an impairment of spatial attention—called **neglect**. Neglect was associated with lesions in numerous cortical and subcortical structures including the parietal, frontal, anterior cingulate, basal ganglia, and thalamus. These findings indicate involvement of a brain complex system in the implementation of attention.

Studies of sensory systems in the human brain revealed several ERP waves modulated by attention. In the 1970s, several ERP phenomena such as mismatch negativity, an automatic brain response to regulatory changes in acoustic stimulation, were discovered. The studies of **Steven Hillyard** from UC San Diego and **Risto Näätänen** from the University of Helsinki contributed substantially to our understanding of attention mechanisms in visual and auditory modalities. After a few decades of detailed investigation, some ERP waves, such as mismatch negativity, were introduced into clinical practice, for example, in predicting recovery from coma or conversion to psychosis.

SEPARATION OF VENTRAL AND DORSAL VISUAL STREAMS BY fMRI

As mentioned in chapter: Event-Related Potentials (ERPs), there are two visual streams in the brain: ventral and dorsal. **Local lesions** in the **ventral stream**, such as lateral and inferior temporal areas lead to **visual object agnosia**, prosopagnosia (inability to recognize familiar faces), and achromatopsia (cortical color blindness) whereas elementary visual capacities remain preserved. In contrast, local lesions in the **dorsal stream**, such as occipital–parietal cortical areas produce optic ataxia (misreaching), visuospatial **neglect**, constructional and gaze apraxia, akinetopsia (inability to perceive movement), **simultagnosia** (inability to perceive more than a single object at a time), as well as disorders of spatial cognition. Single-unit recording studies in monkeys show increased visual receptive fields as well as **increased selectivity** of responses to more complex object features while moving in the ventral stream. At the highest levels of this pathway some neurons preferentially respond to faces. Responses of many neurons in the ventral pathway are **invariant over various transformations** such as retinal position, size, distance in depth, and degree of illumination. Neurons in the dorsal visual stream respond selectively to location of stimuli and their spatial relationships as well as to the direction and speed of stimulus motion when the animal visually tracks a moving object.

Good spatial resolution of **fMRI** provides a neuroimaging tool for separating ventral and dorsal streams in humans. In a study by Mark Pennick and Rajesh Kana from the University of Alabama at Birmingham healthy subjects viewed images of common household objects presented in a blocked design and were instructed either to recognize the object or to locate its position. The location detection task in this study elicited greater activation in the dorsal visual stream while the task of object recognition showed greater activation in the middle occipital gyri, left inferior temporal gyrus, and in the left inferior frontal gyrus.

There is another way of separating dorsal and ventral streams: by simply presenting different types of stimuli. This approach is based on the observation that ventral and dorsal visual streams receive distinct inputs from the magnocellular (M) and parvocellular (P) pathways originated in the retina and segregated within different layers of the thalamus and the primary visual cortex. The dorsal pathway is dominated by M input from the striate cortex. The inferior temporal pathway receives both P and M input.

Parvocells and magnocells get their names from the layers of the **lateral geniculate nucleus** to which the majority of them project. Parvocells in comparison to magnocells have smaller receptive fields, slower conductive rate, higher contrast sensitivity, and higher spatial frequency sensitivity. Parvocells process information about shape and color. Magnocells process information about motion and orientation in space.

It should be stressed here that despite this functional segregation the two streams are heavily interconnected and form a joint network in organizing perception and action. This interplay, for example, enables the subject to select between actions (dorsal stream function) on the basis of the visual form (ventral stream function).

ATTENTION MODULATION EFFECTS IN fMRI

The main function of attention is selecting relevant sensory information according to subject goals. Peripheral receptors continuously send a vast amount of information to the brain. The optic nerve alone comprises about one million axons. On the basis of ongoing goals the brain makes decisions about what part of the available information is relevant and should be **selected for detailed processing**, and what part of the information is irrelevant and should be discarded. When viewing a visual scene, the brain serially selects stimuli in some temporal order with the most relevant being processed in more detail. This serial searching is reflected in saccadic eye movements with sequential analysis of the most salient parts of the image.

fMRI activation patterns in the following task conditions are presented in Fig. 3.1.2: (1) peripheral stimulus without attention, (2) peripheral stimulus attracts attention, and (3) subject focus attention without peripheral stimuli. As one can see, peripheral visual stimulation without attention activates only areas of the visual cortices, but when the subject directs attention to a peripheral target location and performs an object discrimination task, a distributed frontoparietal network is activated including the supplementary eye field, frontal eye fields, and the superior parietal lobe. The same network of frontal and parietal areas is activated when the subject directs

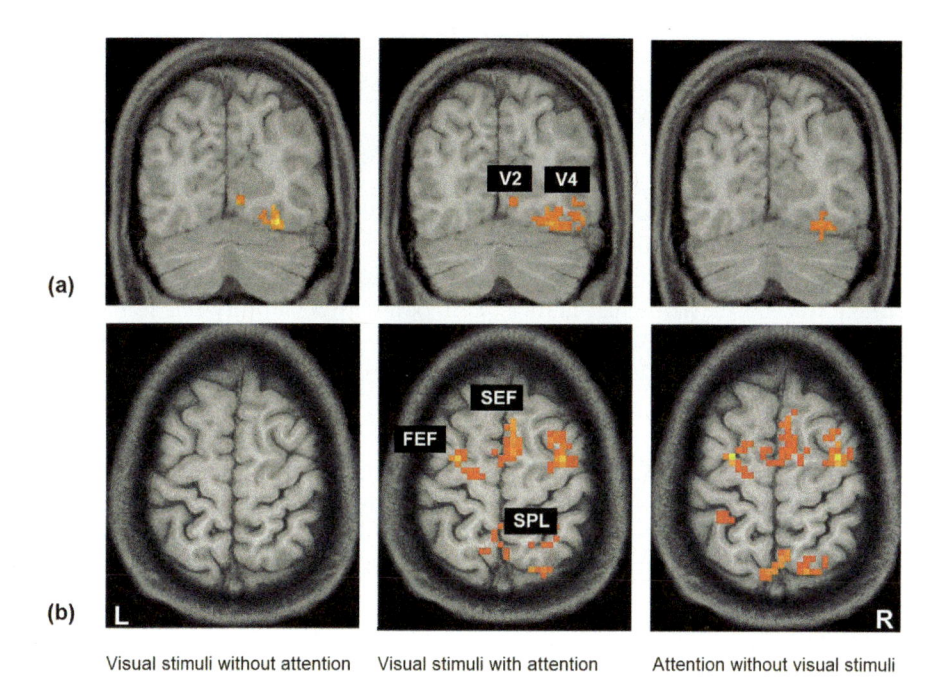

FIGURE 3.1.2 **fMRI activation patterns modulated by attention.** (a) Cross section at the visual cortex during unattended, attended, or expected visual stimuli. The subject was tested under three different conditions in the same scanning session: (left, L) visual stimuli without attention—stimuli were presented to the periphery, while the subject performed a letter-counting task at fixation; (middle) visual stimuli with attention—the same visual stimuli but the subject attends to one of the peripheral stimuli while performing pattern discrimination; (right, R) attention without visual stimuli—attention is directed to a peripheral target location in the expectation of stimulus onset; and (b) axial slice through frontal and parietal cortex. The same subject and experimental conditions as in (a). *Adapted from Kastner & Ungerleider (2001).*

attention to the peripheral target location in the expectation of the stimulus onset, that is, in the absence of any visual input.

This example shows that sensory-related cortical areas (**occipital–parietal**) and action-related parts of the cortex (**frontal**) are simultaneously activated during attention. Visual signals arriving in the cortex are analyzed and processed through multiple stages, objects are recognized and locations identified, a decision of some kind is made, and an action (such as saccadic eye movement to a peripheral location) is selected. Equally important is the reverse process by which information about the motor action is fed back to earlier stages of sensory processing, allowing action to influence perception.

VISION AS AN ACTIVE PROCESS

The idea that action has an impact on perception is an old one. More than a century ago, Hermann von Helmholtz from the University of Heidelberg in 1866 proposed that the reason the world appears to *stay still* when the **eyes move** is because the "effort of will" involved in the generation of a saccade simultaneously adjusts perception to take that eye movement into account.

A simple experiment convinces most observers that Helmholtz's account is correct. When the retina is displaced by pressing on the eye, the world does seem to move. In contrast, we are oblivious to changes in the retinal image that occur with each glance. This perceptual stability has long been understood to reflect the fact that what we see is not a direct impression of the external world but a construction, or internal representation, of it.

It is this **internal representation** that is adjusted, or updated, in conjunction with eye movements. The interaction between the sensory and action operations led Ulric Neisser from Cornell University in 1979 to formulate the idea of **schemata**. According to his view, schemata are **anticipatory cognitive structures** that prepare the perceiver to accept certain kinds of information rather than others. This idea in the explanation paradigm of the 21st century is replaced by concepts of **bottom–up and top–down operations of attention**.

BOTTOM–UP AND TOP–DOWN SELECTION OPERATIONS

The processing of sensory information relies on interacting mechanisms of **sustained attention** and **attentional capture**. Sustained attention is a tonic process while attentional capture is a phasic process. Both of them operate in space and on object features. The ability to sustain attention and to ignore irrelevant stimuli is a fundamental operation of goal-directed behavior. On the other hand, the ability to be captured by sudden changes in the environment which could be potentially dangerous events is critical for surviving and avoiding harm. Here we are talking about two different mechanisms of attention: **bottom–up and top–down processes**.

Bottom–up processes depend primarily on **current sensory input**. An example of bottom–up processes in the visual modality is detecting a yellow flow on a green grass (Fig. 3.1.3a). One of the mechanisms of these processes appears to be associated with lateral inhibition at the initial stages of visual information flow starting in the retina and continuing in the primary visual cortex. For example, lateral inhibition in the primary visual cortex shapes the orientation and spatial frequency selectivity of neuronal responses in this part of the cortex.

Top–down information streams depend on an internal state of the brain called a **preparatory or attention set**. The preparatory set includes subjective goals, plans, and previous experience that supposedly reside

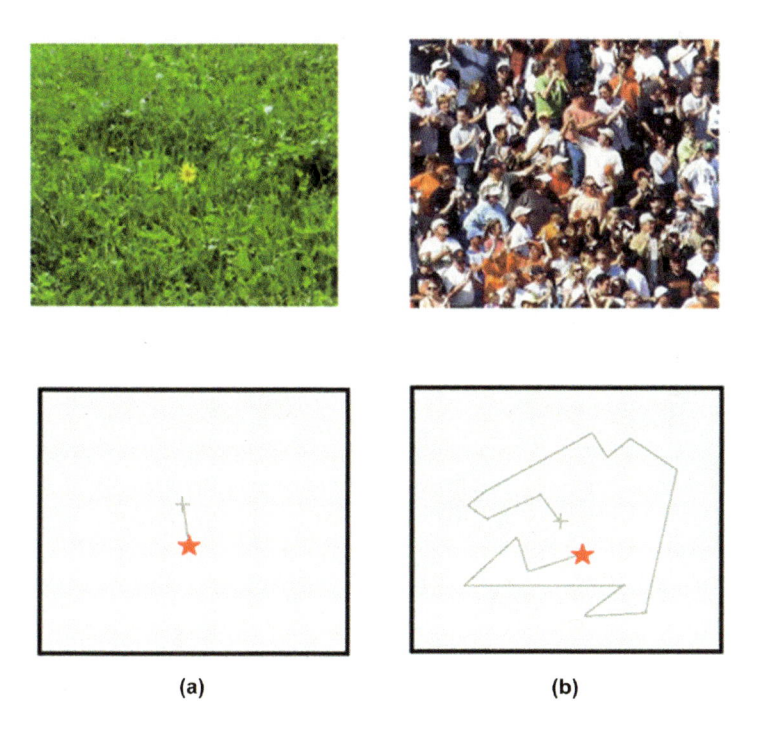

(a)   (b)

FIGURE 3.1.3 **Two mechanisms of attention.** (a) Bottom–up mechanism which relies primarily on sensory input which attracts attention like a yellow flower on green grass, and (b) top–down selective attention which relies primarily on the higher order preparatory set such as looking for a friend in a crowd. Given pictures are schematic representations of saccadic eye movements while viewing the scene: in (a) the stimulus immediately attracts attention, in (b) the focus of attention is shifting from one subject to another until the anticipated friend is found.

in higher order cortical areas. These higher order cortical areas control information flow in the sensory systems by feedback cortico-cortical pathways as well as by corticothalamocortical projections and by corticobasal ganglia–thalamocortical projections. An example of top–down operation is searching for a friend in a crowd of people (Fig. 3.1.3b).

The preparatory set involves activation of the frontal lobe networks which, via **feedback projections** to posterior sensory cortical areas, control information flow in these sensory networks. A hypothetical mechanism of membrane depolarization of the selective sensory neurons via the feedback projections from the frontal areas in the preparatory set is presented in Fig. 3.1.4.

It should be stressed here that the poor time resolution of fMRI cannot provide insights into how visual information is processed in time. In contrast to fMRI, ERPs offer a reliable neuroimaging technique for obtaining a detailed picture of visual processing. The stages of information flow at

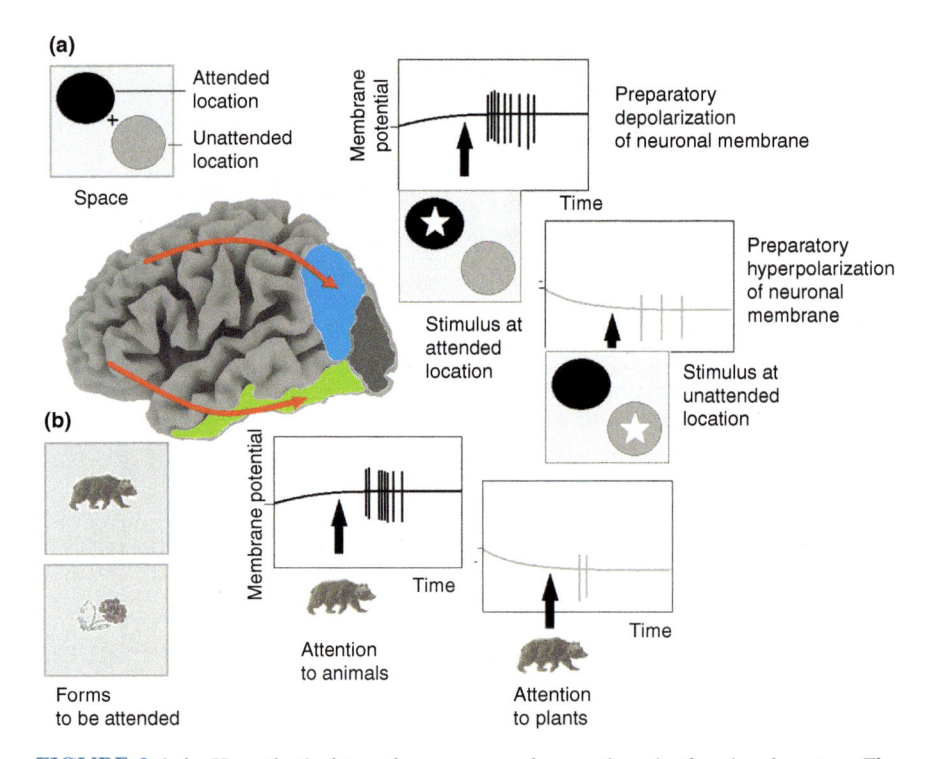

FIGURE 3.1.4 **Hypothetical top–down neuronal operations in the visual system.** The preparatory set is formed in the frontal lobes by instruction: (a) to attend a certain location in space (upper part) which activates the dorsal visual stream; and (b) to attend to a certain category of stimuli, for example to animals (lower part) which activates the ventral visual stream. In both cases, feedback projections hypothetically selectively depolarize membranes of the corresponding sensory neurons to make them more sensitive to expected stimuli.

millisecond time resolution are reflected in the canonical ERP to visual stimulus.

BOTTOM–UP OPERATIONS IN THE C1 WAVE OF EVENT-RELATED POTENTIAL

The first wave of the canonical ERP in response to a brief presentation of visual stimulus (Fig. 1.6.8) is the **C1 wave**. It is generated in the primary visual cortex and codes simple visual features such as orientation and spatial frequency. In an elegant 2012 study by Xilin Zhang, Li Zhaoping, Tiangang Zhou, and Fang Fang from Peking University subjects participated in a modified Posner paradigm in which the cue was a texture stimulus (Fig. 3.1.4a) presented for 50 ms followed by a 100-ms mask (Fig. 3.1.4b) so that the stimulus was not consciously perceived by the subject but subconsciously cued the location. Each texture stimulus contained 15×29 low-luminance bars in a regular Manhattan grid in the lower visual field on a dark screen. All bars were identically oriented except for a foreground region of 2×2 bars of another orientation. The foreground region was at 7.2 degrees eccentricity in either the lower-left or the lower-right quadrant. The orientation of the background bars was randomly chosen from 0 to 180 degrees. There were five possible orientation contrasts between the foreground bars and the background bars: 0, 7.5, 15, 30, and 90 degrees (Fig. 3.1.5).

A nonzero orientation contrast made the foreground region salient enough to attract attention. The high luminance mask was used for isolating the bottom–up saliency signal and minimize top–down influences. As the results show, local texture contrast modulated the C1 wave. The authors suggested that the **bottom–up saliency map** is formed in the primary visual cortex.

N1 WAVE AS INDEX OF VISUAL DISCRIMINATION

The intention to make a visual discrimination can also influence neural responses. For example, Walter Ritter and his coworkers from the Albert Einstein College of Medicine in 1983 conducted several experiments comparing two tasks: a **simple response** task in which subjects made a speedy response as soon as they detected a stimulus regardless of its form and a **discriminative response** task in which subjects made different responses depending on the form of the stimulus. The initial P1 sensory response was the same for these two conditions, but the subsequent N1 wave was larger in the discriminative response condition than in the simple response condition, beginning around 150 ms poststimulus (Fig. 3.1.6).

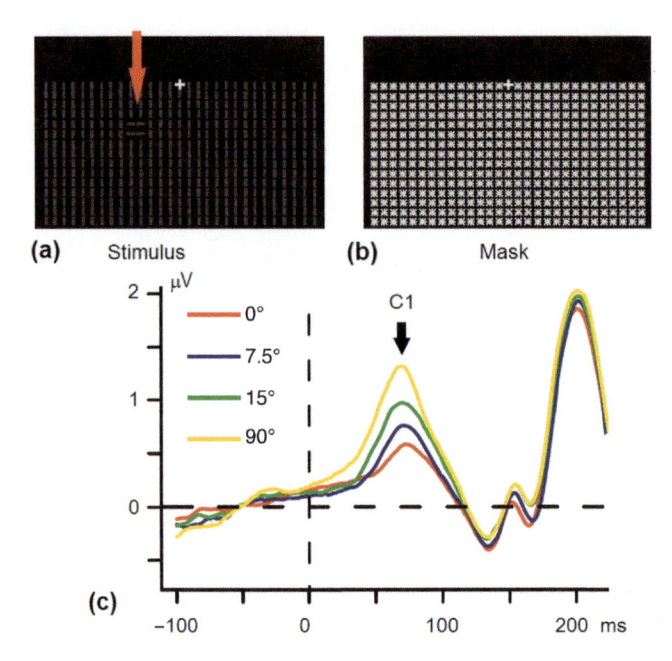

FIGURE 3.1.5 **Automatic orientation detection in the primary visual cortex.** (a) A low-luminance texture stimulus presented in the lower visual field. Orientation contrast between the foreground and the background bars is marked *(red arrow)*; (b) mask represents a high-luminance stimulus. A texture stimulus was presented for 50 ms as a cue, followed by a 100-ms mask and a 50-ms fixation screen. The texture stimulus was invisible to subjects but was able to produce the cuing effect; and (c) ERPs in response to the masked texture stimuli at orientation contrasts 7.5, 15, 90, and 0 degrees. *Adapted with permission from Zhang, Zhaoping, Zhou, & Fang (2012).*

VISUAL MISMATCH NEGATIVITY AS INDEX OF REGULARITY VIOLATION

When stimuli are repeatedly presented in a particular order the brain forms a **neuronal model** of sensory stimulation so that when a new stimulus violates the model hypothetical **change detectors** are activated. In the visual modality, activation of the change detectors is reflected in a **visual mismatch negativity (MMN) wave**. In contrast to the auditory MMN in oddball tasks that was discovered by Risto Näätänen and coworkers from the University of Helsinki in 1978, it took more than 30 years to discover its visual counterpart. For many years the existence of visual MMN was questioned because of the differences between echoic and iconic memories.

Visual MMN has been most frequently observed in the **visual passive oddball task** in response to infrequent deviants (D). In the difference waves obtained by subtracting ERPs in response to standard stimuli from

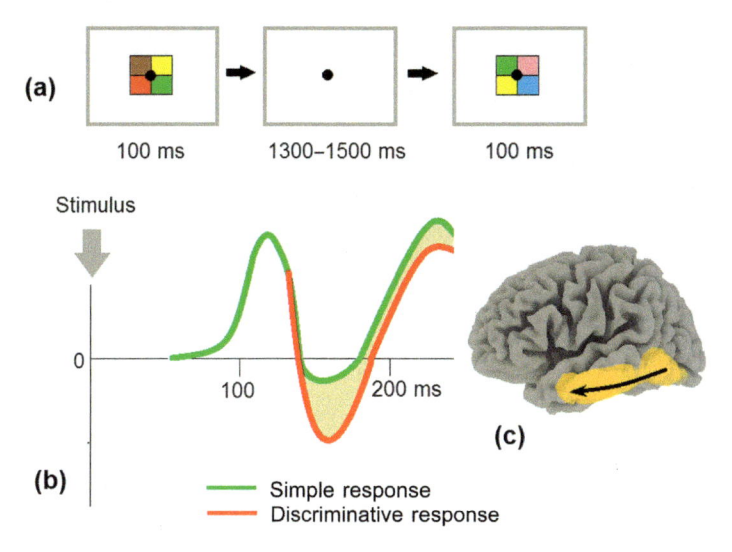

FIGURE 3.1.6 **Visual N1 modulation by discrimination operation.** (a) Example of the task from Hopf et al. (2002) in which subjects were instructed to press a button to every colorful stimulus (simple response task) or were instructed to press with one hand if the red color was present and with the opposite hand if the red color was absent (discriminative response); (b) schematic presentation of enhancement of N1 wave in the discriminative response; and (c) schematic localization of the sources of N1 modulation in the ventral stream.

those elicited by deviant stimuli, visual MMN constitutes only a later part. The early part is simply due to the **adaptation effect** produced by different probabilities of deviant and standard stimuli. To compensate for the adaptation effect a control sequence of stimuli with equivalent probabilities is presented to the subject. The ERP difference **deviant − control** represents the pure visual MMN wave. So far, visual MMN has been observed in response to deviations in several visual features, such as location, direction of motion, orientation, spatial frequency, contrast/luminance, color, shape, duration, conjunction of color and orientation, and facial expression (see the 2012 review by Motohiro Kimura from the National Institute of Advanced Industrial Science and Technology in Japan). Visual MMN is a negative-going ERP component with a posterior (most typically, a **right parietooccipital maximum**) scalp distribution that usually peaks at around **150–400 ms** after the onset of visual sensory events (Fig. 3.1.7).

Source localization analyses further demonstrated that the main generators of visual MMN were located in the right occipital visual extrastriate areas and the right medial prefrontal areas including the orbital gyrus. Numerous experimental findings indicate that, rather than physical stimulus deviations, visual MMN is sensitive to violations of a regular sequential pattern that is repeatedly presented in a stimulus sequence. This leads to the assumption that the memory representation that

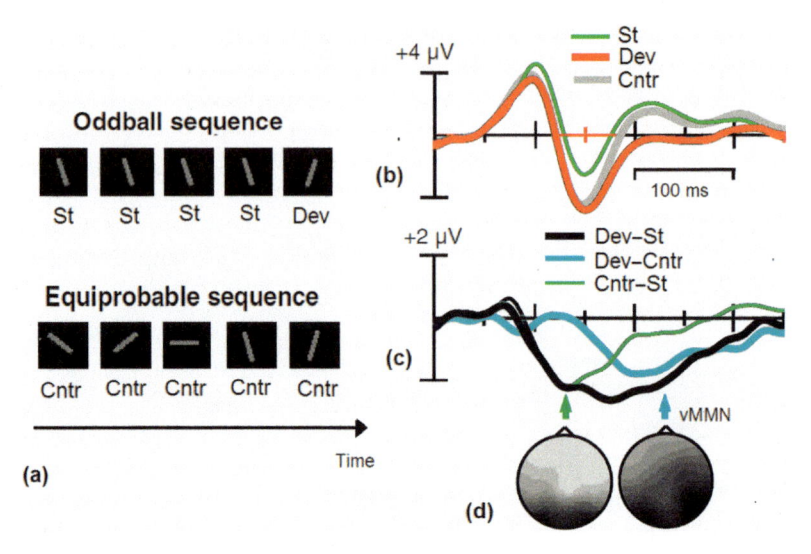

FIGURE 3.1.7 Visual mismatch negativity *(vMMN)*. (a) Experimental design: two sequences are used to obtain the MMN: (1) oddball sequence with 80% of standards *(St)* and 20% of deviants *(Dev)*, (2) equiprobable sequence with each stimulus of 20% *(Cntr)*; (b) ERPs at a right parietooccipital electrode (PO8) for *St, Dev,* and *Cntr* stimuli; and (d) subtraction of ERPs elicited by control stimuli from those elicited by deviant stimuli reveals a genuine visual MMN. *Adapted with permission from Kimura (2012).*

underlies the generation of visual MMN must encode a regular sequential pattern that is repeatedly presented in a stimulus sequence, that is, **regularity representation** as suggested in the 2010 review by Istvan Czigler from the Institute for Psychology of the Hungarian Academy of Sciences. According to this suggestion, visual MMN is elicited when a current visual stimulus violates the regular sequential pattern. Importantly, visual MMN findings were obtained under experimental conditions in which the participant's task was unrelated to the stimulus sequence or the participant's attention was not directed to the stimulus sequence. This indicates that the generation of visual MMN is largely automatic.

VISUAL N170 REFLECTS ACTIVATION OF PERSONAL MEMORY

The **N170 wave** was first reported in 1996 by Shlomo Bentin and coworkers from the Hebrew University in Israel. In his study subjects viewed images of human faces and other objects. Images of faces and face parts (such as eyes) elicited an enhanced N170 wave in comparison to other stimuli, including animal faces, body parts, and cars. This response was maximal over **occipitotemporal electrode** sites. The N170 displayed right-hemisphere lateralization and was linked with the encoding of faces (Ahveninen et al., 1996).

The three visual-related independent components extracted from the HBI database in response to images of plants and faces are presented in Fig. 3.1.8. The stimuli were irrelevant to the task and the subjects simply viewed them (ignore condition).

As one can see, the dynamics of the occipital component in response in plants and faces almost coincide. The topography of the occipital component can be modeled by a single dipole with relative residual energy (RRE) of about 0.01. sLORETA image shows activation of the occipital cortex around the calcarine fissure.

In contrast to the occipital component, the temporally distributed components show considerable difference between time dynamics in response to images of plants and faces. Namely, they show larger negative parts peaked at about 170 ms after the onset of stimulus presentation with the

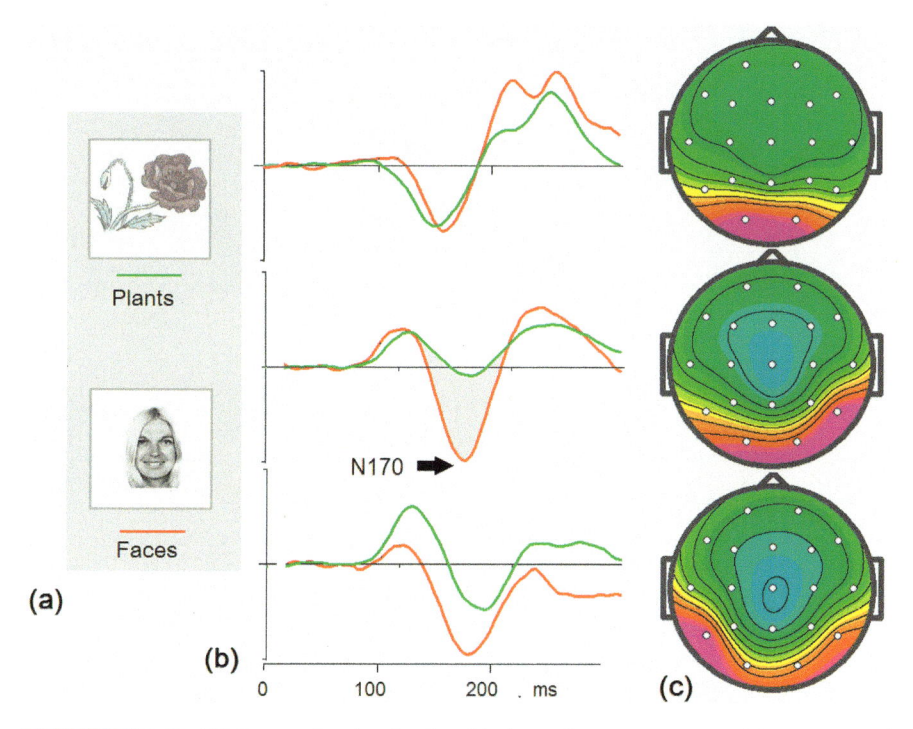

FIGURE 3.1.8 **N170 wave in visual-related independent components.** (a) Examples of visual images presented to the group of healthy adults; (b) timecourses of three independent components extracted by independent component analysis (ICA) from a collection of ERPs computed in response to images of plants *(green lines)* and faces *(red lines)*. No motor responses to stimuli were required; and (c) topographies of the independent components. *Data from HBI database.*

150 3.1. SENSORY SYSTEMS AND ATTENTION MODULATION

most strong effect on the right side. According to sLORETA, these components are generated in widely distributed temporal areas including the fusiform gyrus and lateral surface of the temporal lobe (Fig. 3.1.9). Negative parts of temporally distributed visual-related independent components correspond to the N170 ERP wave.

Recent research shows that the N170 wave is modulated by face inversion and emotional expressions. Inverted faces generate an N170 wave with longer latencies in comparison to upright position faces. This finding

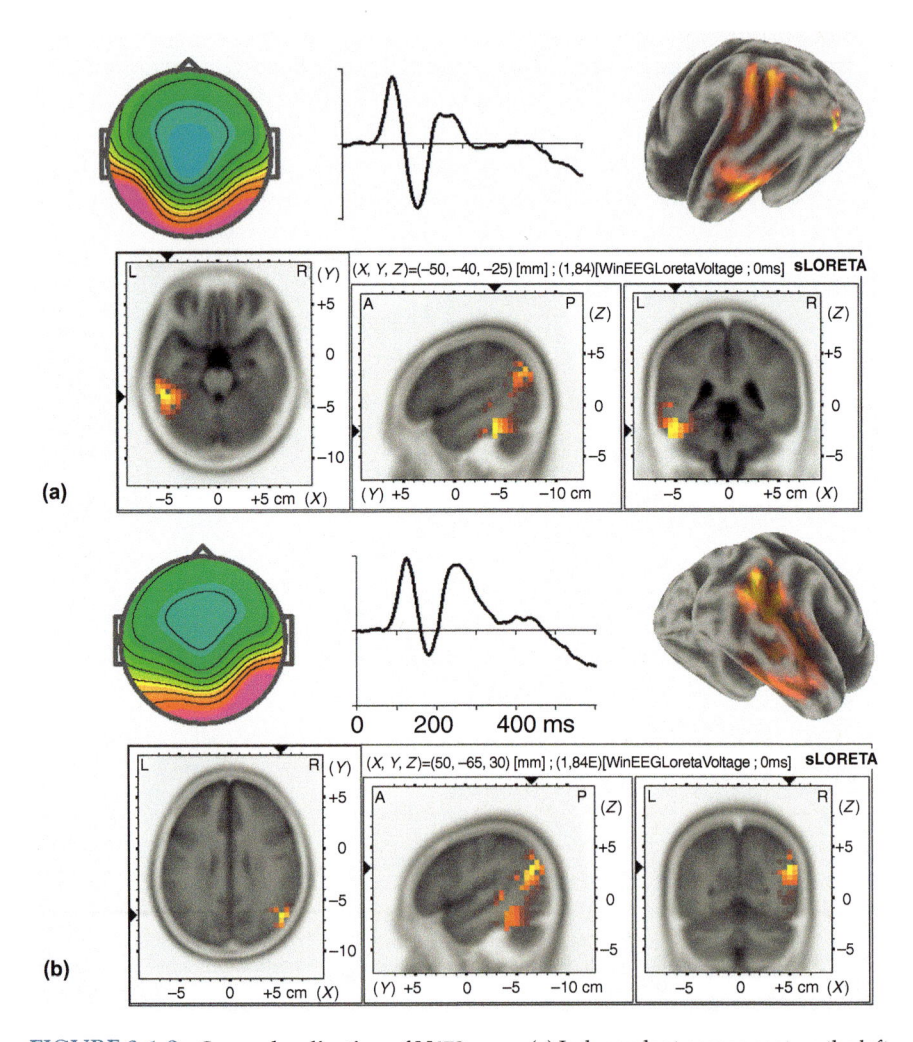

FIGURE 3.1.9 **Source localization of N170 wave.** (a) Independent component on the left side (topographies, timecourses, and sLORETA images); and (b) independent component on the right side.

corresponds to a psychological observation according to which subjects have difficulty recognizing inverted faces. Fearful stimuli generate enhanced potential of N170 in comparison to neutral faces.

The convergence of neuropsychological and fMRI research shows the importance of the **fusiform gyrus** of the inferior temporal cortex in recognition of faces. There is no consensus regarding the source of N170, however; some authors localized the N170 wave in the fusiform gyrus, some others localized the source in the posterior part of the superior temporal sulcus. One explanation for this inconsistency is that the N170 wave is not a homogeneous entity but rather a linear sum of potentials generated in widely distributed occipital–temporal–parietal cortical regions.

VISUAL N250 REPETITION EFFECT

The first report on the visual **N250 wave** repetition effect was published in 2008 by Stefan Schweinberger and coworkers from Friedrich-Schiller University Jena. The authors investigated the immediate repetition effects in the recognition of famous faces by recording ERPs (Fig. 3.1.10). No influence of repetition was observed for the N170 component, but strong modulation was observed in the N250 wave. A face-selective negative response around 250 ms to repetitions emerged over right temporal regions, consistent with a source in the fusiform gyrus. The N250 wave was largest for human faces, clear for ape faces, nonsignificant for inverted faces, and completely absent for cars.

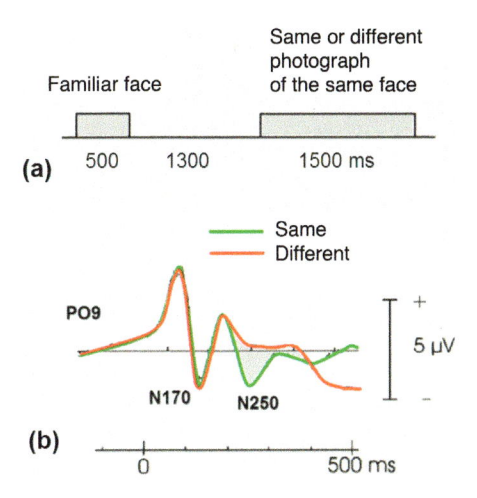

FIGURE 3.1.10 Repetition effects in human N250 ERP wave. (a) Scheme of experiment; and (b) ERPs at PO9 lead in response to presentation of the same familiar face or of a different photograph of the same face. *Adapted from Neumann & Schweinberger (2008).*

In a 2011 study by Lara Pierce form McGill University, Montreal and coworkers the N250 was shown to be unspecific for faces but rather an index of individuated representations in visual working memory.

As previously mentioned, searching for a face in a crowd is an example of top–down attention in which the activated template in the working memory is comparing faces in a crowd. Kartik Sreenivasan and coworkers from the University of Pennsylvania in 2011 examined the neural correlates of this process by instructing subjects to compare the memory trace of a target face with faces that varied in their similarity to the target and to press a button to the target (Fig. 3.1.11). The stimulus presentation evoked N170 and N250 waves that both had a focus in the right parietooccipital electrode PO8, but the peak N250 electrode varied considerably across subjects. Only the N250 wave was modulated by probe–target similarity. The amplitude of N250 was

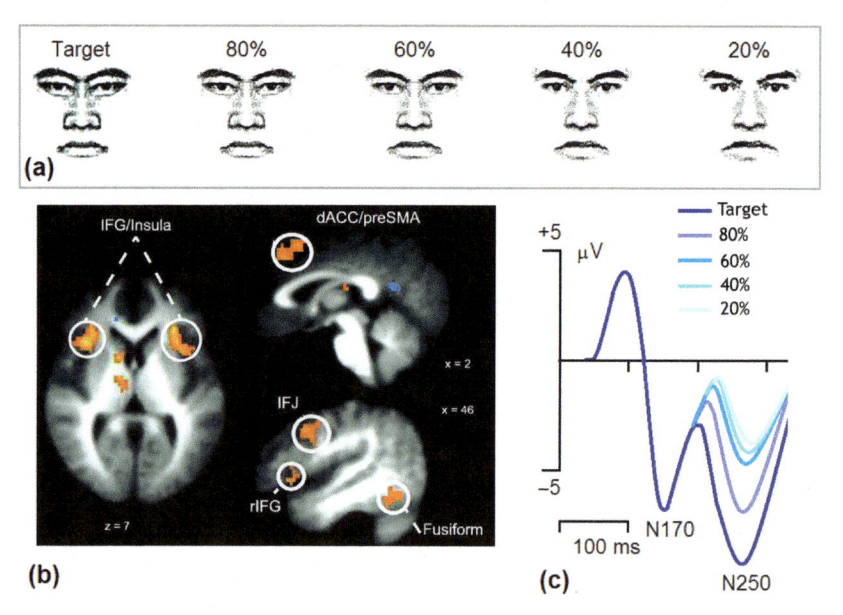

FIGURE 3.1.11 **N250 wave as an index of similarity to the template in working memory and its fMRI correlate.** (a) Face stimuli used in continuous target detection. The target face was maintained in working memory. Nontarget probe faces contained a percentage of target facial features (80, 60, 40, and 20%, respectively). Participants responded to the presentation of the target face; and (b) whole-brain results of fMRI study—regions exhibited a greater linear effect during the Memory task relative to the Control task. The regions include the fusiform gyrus (perceptual region) and presupplementary motor cortex (executive region); (c) grand-averaged ERPs for each stimulus type. N170 does not differ across conditions. The N250 wave's amplitude is largest for targets and decreases progressively for 80, 60, and 40% faces. *Parts a,b: adapted with permission from Sreenivasan et al. (2014); part (c) adapted from Sreenivasan et al. (2011).*

largest for probes that were similar to the target, and decreased monotonically as a function of decreasing probe–target similarity. These data show that the N250 wave is an index of **similarity of a visual stimulus with the template in the working memory**: the closer the stimulus to the template, the higher the N250. The authors suggested the working memory templates are maintained by biasing activity in sensory neurons that code for features of items being held in memory.

In the following 2014 study Kartik Sreenivasan, now from University of California, Berkeley, in similar tasks but using fMRI, demonstrated that activity within face processing regions as well as some frontal regions varied linearly with the degree of similarity between the probe and the target (Fig. 3.1.11b). The authors suggested that the features of the Target face were stored in the perceptual regions. Furthermore, directed connectivity measures in the fMRI study revealed that the optimal direction of information flow was from the perceptual regions to dorsal prefrontal regions, supporting the notion that sensory input is compared to representations stored within the posterior perceptual regions and transferred to the frontal executive regions.

VISUAL P2 DISCREPANCY EFFECT

In many situations the **detection of discrepancy** (but not similarity) is vital for adaptive behavior. Evolution constructed several mechanisms for detecting different types of discrepancy. One such mechanism is associated with the visual MMN described earlier. This mechanism is responsible for automatic detection of discrepancy when the regularity pattern of visual stimulation is violated. The other mechanism requires effortful forming of the model of expected stimulus. This mechanism is expressed in the temporally distributed **P2 wave** (Fig. 3.1.12).

The P2 wave appears in the delayed match-to-sample task when the **sensory model** of expected stimulus is explicitly and actively formed in the working memory after cue presentation while the delayed target does not match this trace; the prepared action must be changed. The P2 appears to reflect the **amount of the discrepancy**. The P2 may appear in conjunction with the N250 but usually is alone.

In the cued GO/NOGO task in which the second image of an animal is the same as the first (the GO condition corresponds to animal–animal "repeat") both the **repetition N250** wave appears for GO trails and the discrepancy **P2 wave appears for NOGO trials** (Fig. 3.1.12b). In the cued GO/NOGO task in which the second image of an animal is different from the first the repletion N250 wave disappears and the discrepancy P2 **stays alone** (Fig. 3.1.12c). In the cued GO/NOGO task when the subject is instructed to press a button to a plant that followed the image of an animal

3.1. SENSORY SYSTEMS AND ATTENTION MODULATION

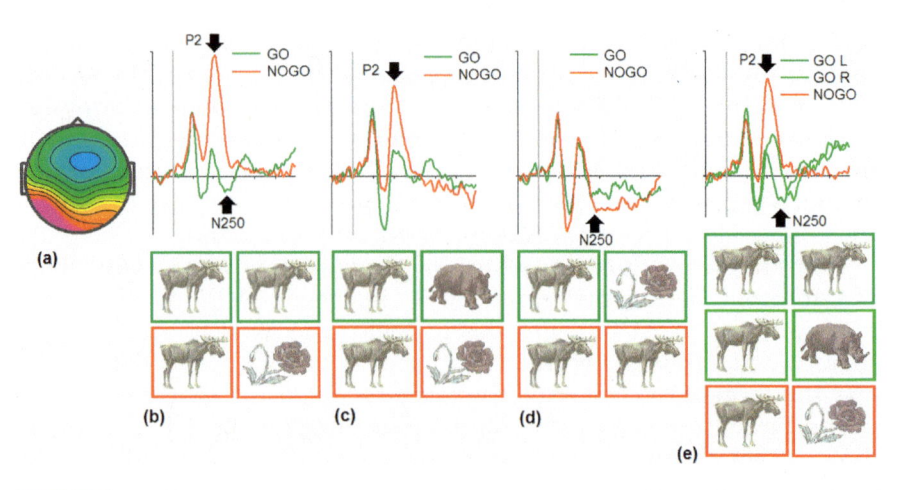

FIGURE 3.1.12 **Visual similarity N250 and discrepancy P2 effects.** (a) Topography of the temporal independent component, (b), (c), (d), and (e) Top—time dynamics of the component in GO and NOGO trials for the four variants of the cued GO/NOGO task in which images of animals (*a*) and plants (*p*) are presented in pairs "*a–a*," "*a–p*," "*p–a*," and "*p–p*." where: (b) the second animal in "a–a" target is always repetition of the first one; (c) the second animal in the "a–a" target always differs from the first one; (d) the second animal in "a–a" nontarget is always repetition of the first one but "a–p" is the target; (e) the second animal in "a–a" target in 50% of cases is the same animal and in 50% of cases is a different animal with the subject task to press the left button to repetition of animals and to press the right button to nonrepetition. Bottom— examples of GO and NOGO pairs are presented (*dark green*, GO pairs with pressing the left button; *light green*, GO pairs with pressing the right button; *red*, NOGO pairs). *Data from HBI database.*

the sensory model of expected GO stimulus cannot be formed and the discrepancy P2 **disappears** while the repetition N250 wave appears in response to NOGO stimulus (Fig. 3.1.12d). Finally, in the cued GO/NOGO task when the second image of an animal repeats the first image in 50% of GO trials and in 50% of GO trials the second image of ananimal is different from the first the repetition N2 appears to GO "repeat" trials and the discrepancy P2 appears to GO "different" trials, while the animal–plant trials produce a stronger P2 discrepancy wave (Fig. 3.1.12e).

LATENT EVENT-RELATED POTENTIAL COMPONENTS OF VISUAL PROCESSING

Observing natural scenes involves a bunch of **hypothetical operations** such as attending to a particular part of the visual field, detecting the object, discriminating its category, comparing the image with the expectation in working memory. As animal experiments show, the neuronal correlates of these operations are widely distributed over the hierarchically organized sensory cortical areas so that the activity of a single neuron in a particular cortical area can be modulated by each of the previously mentioned operations. In other words, the neuronal representations of psychological operations in visual perception are **spatially distributed and overlapped in time**. There is no reason to expect that a separate psychological operation (such as visual recognition) would correspond to a separate wave of canonical ERPs. This fact lies at the heart of the problem of separating **functional components** of ERPs. Functional components are presumed to be hidden in raw ERPs.

One way to solve the problem is to **manipulate a hypothetical operation** in a behavioral task so that under one task condition (but not under the other) the hypothetical operation is selectively involved. For example, one can present stimuli at the threshold of recognition so that in 50% of cases the subject recognizes the visual stimulus but in the other 50% of cases the subject simply detects the appearance of the stimulus but could not recognize it. However, in this case the modulation of response could also be due to fluctuation of attention, and this factor should be taken into account. In another example, visual MMN is measured as the difference wave between ERPs to deviant stimuli in the oddball task and ERPs in the control task with stimulus presentation at equal probabilities. But, even in this case, there is no reason to suggest that the difference wave obtained is a homogenous phenomenon: experience says just the opposite, even the simplest ERP components represent **mixtures of separate subcomponents**.

The other way to solve the problem is to use the blind source separation approach described in chapter: Event-Related Potentials (ERPs). The method does not presume any a priori knowledge about the signals but suggests that they in some sense or another vary independently from each other in the whole population of healthy subjects.

A combination of the two approaches was recently performed in our laboratory and published in *Psychophysiology* in 2015 by Juri Kropotov and Valery Ponomarev. The **category discrimination** and **comparison to working memory** operations were studied in a delayed match-to-sample task and its variants. Our working hypothesis suggested that in the delayed match-to-sample task three distinct neuronal operations are taking place during the first 300 ms after stimulus presentation. They are (1) visual category discrimination, (2) comparison of the sensory input with WM content, and (3) activation of the motor action if needed. The hypothesis implicitly presumes the existence of neuronal representation of a prepotent sensorimotor model of behavior. It is hypothesized that the model is proactively constructed by the brain to insure the most efficient forthcoming behavior.

The concept is not a new one and is similar to the schemata concepts in Neisser's perceptual cycle proposed in 1976 and in the supervisory attentional system by Norman and Shallice proposed in 1986. Briefly, the prepotent sensorimotor model represents a hypothetical neuronal network, mapping stimulus to response. Activation of the network prepares the behavior by proactively presetting neuronal thresholds at the corresponding sensory and motor cortical areas. When the incoming stimulus deviates from the content in WM, it activates the comparison detectors that in turn initiate executive control over a nonroutine situation. A similar hypothesis was suggested by Risto Näätänen from the University of Helsinki in 1978 for the explanation of neuronal processes of the MMN wave elicited in different modifications of the auditory oddball paradigm.

The task consists of sequential presentation of two stimuli s1 (**sample cue**) and s2 (**match or mismatch**). The subject has to discriminate stimuli and compare s2 with s1 to perform different actions (such as to press a button with the left/right hand) depending on the match/mismatch condition. The interstimulus was 1000 ms, thus exceeding the duration of iconic memory so that s2 had to be compared with the contents of working memory.

The three operations were **independently modulated in five experimental tasks**. The stimuli were of two categories presented in four equally probable combinations. In Experiments 1–4, various combinations of images of animals (a) and plants (p) served as stimuli with different instructions: in Experiment 1 to press a button to a–a, in Experiment 2 to press a button to p–p, in Experiment 3 to press a button to a–p, and in Experiment 4 to press a button with the left hand to a–a and with the right hand to a–p. In Experiment 5 the stimuli were physically similar (digit 6 and 9) with the task to press a button to 9–9 trials.

To separate the latent components hidden in ERPs we used a method of blind source separation based on second-order statistics and adapted for ERPs [see chapter: Event-Related Potentials (ERPs)]. Four latent components having the largest power and explaining around 90% of the signal are presented in Fig. 3.1.13. As one can see, according to sLORETA the components are generated in different parts of the ventral and dorsal visual streams.

Three different behavioral patterns of the difference waves across the five experiments were separated. They corresponded to the hypothetical behavioral patterns. The category discrimination operation was attributed to three latent components with peak latencies of 130–170 ms, which were generated in different parts of the prestriate cortex. The comparison to working memory operation was attributed to a latent component that is generated in the temporal cortex and manifested in a positive deflection with a peak latency of 250 ms after s2. Category discrimination and comparison to WM effects were dissociated spatially and temporally from attention and action selection effects.

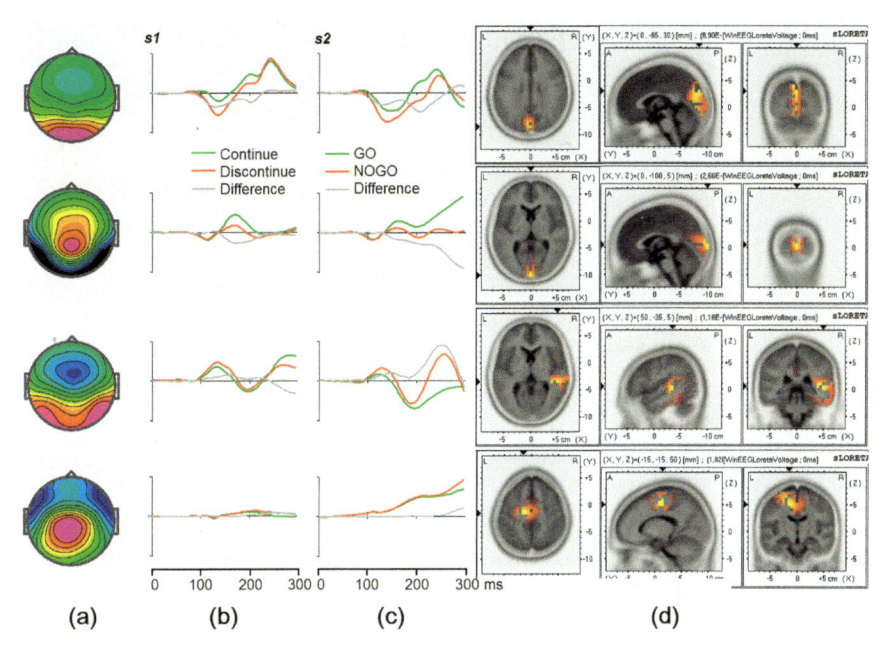

FIGURE 3.1.13 **Visual-related latent components.** The components are extracted from the large collection of ERPs recorded in a group of healthy adults in the cued GO/NOGO task using the blind source separation approach. (a) Topographies of components, (b) timecourse for continue *(green)* trials, discontinue *(red)* trials, and the difference discontinue − continue *(gray)*, (c) Timecourse for match, GO *(green)* and mismatch, NOGO *(red)* trials and the difference mismatch − match *(gray)*, and (d) sLORETA images. *Adapted with permission from Kropotov & Ponomarev (2015).*

A NEURONAL MODEL

A realistic neuronal model of top–down modulation of information processing was suggested by Markus Siegel and coworkers from the University of Zurich in 2000. The neuronal model is based on recently discovered intracellular mechanisms. First, it has been shown that somatic action potentials can actively **back-propagate** into the apical dendrite. Second, voltage-dependent calcium conductances can initiate slow dendritic **calcium spikes** inducing in turn bursting behavior of the cell. Third, these two effects interact if the excitatory input at the apical dendrite is paired with an action potential of the postsynaptic neuron leading to a drastically lowered threshold for generation of a burst. These neuronal processes are schematically presented in Fig. 3.1.14.

The model suggests that pyramidal cells with long apical dendrites could serve as **coincidence detectors**: in the presence of back-propagating action potential, the subthreshold synaptic input at the apical dendrite can trigger a dendritic calcium spike leading to a burst of axonal action potentials.

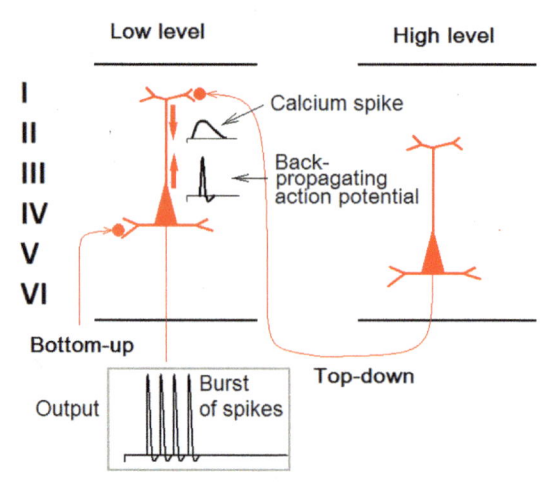

FIGURE 3.1.14 **Pyramidal cell as a coincidence detector.** If the excitatory input of top–down projections is strong enough and bottom–up input initiates an action potential that propagates back into the apical dendrite, a dendritic calcium spike is triggered. This calcium spike in turn initiates a burst of action potentials. *Adapted from Siegel, Kording, & Konig (2000).*

PRINCIPLES OF INFORMATION FLOW IN THE VISUAL SYSTEM

When teaching courses on ERPs the author of the book was often asked about some theoretical principles of information flow in the brain that would guide a newcomer in ERP research. For many students the field appeared as pure descriptive science without any theory behind experimental facts. The facts themselves obtained in different laboratories sometimes turned out to be contradictory and confusing. In the ERP field there are guidelines on how to make proper recordings but there are no theoretical guidelines. Here an attempt of formulating basic theoretical laws of sensory-related ERP is made:

Principle 1. Feedforward and feedback projections within the visual system play critical roles in bottom–up and top–down operations of visual processing.

Principle 2. The interplay between sensory cortical areas is not limited by corticocortical reciprocal connections. **Corticothalamocortical projections and corticobasal ganglia–thalamocortical projections** contribute to the interplay between sensory cortical areas. Corticothalamocortical connections seem to play a critical role in assessment operations. Corticobasal ganglia–thalamocortical projections play a critical role in selection of appropriate sensorimotor actions.

Principle 3. The first 300 ms of sensory information processing is divided into **two stages**: the fast (early) stage based on feedforward bottom–up processing and the slow (late) stage based on feedback top–down processing. In the visual modality, the early stage is associated with P1/N1 waves over occipital/temporal areas whereas the late stage is associated with P2/N2 waves.

Principle 4. A **sensory model** of the environment is formed during repetitive stimulation even in the absence of attention. Comparison of the incoming stimulus with the sensory model is done automatically and seems to rely on bottom–up processes. In the visual modality, this operation is reflected in visual MMN.

Principle 5. When attention is applied, the sensory world is actively transformed into representations maintained in the **working memory** by top–down processes. Comparison of attended stimulus with working memory activates cognitive control operations. In the visual modality the comparison to working memory operations are reflected in N2 and P2 waves.

WHAT AND WHERE STREAMS IN THE AUDITORY MODALITY

The cortical part of the primate auditory system is organized into a **core** of primary auditory cortical areas that project onto a surrounding **belt**, with the belt projecting onto **parabelt** areas. The core or primary auditory cortex is located in the first transverse gyrus of Heschl (BA 41) (Fig. 3.1.15). The primary auditory cortex receives direct input for the medial geniculate body of the thalamus.

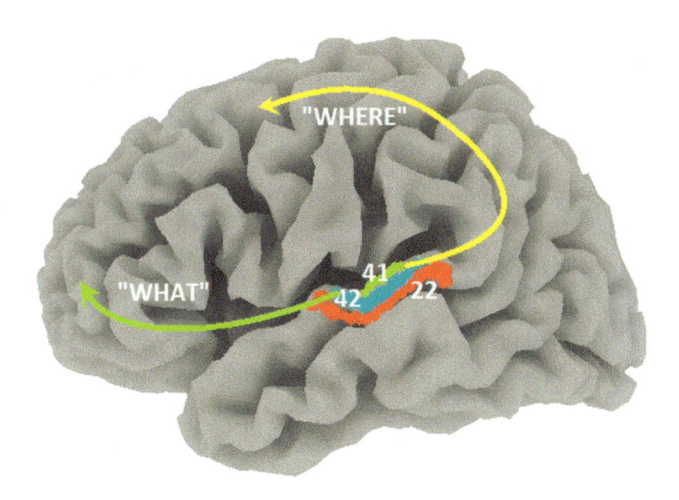

FIGURE 3.1.15 **What and where streams in auditory modality.** Schematic representation of "what" (*green arrow*) and "where" (*yellow arrow*) streams in the auditory system. Brodmann areas (BA) are shown in *white*.

Ordered maps of sound frequency selectiveness are present at all auditory areas with finer **tonotopic organization** in the core regions. Activation specific to speech is observed in the left anterior superior temporal sulcus. This region is multimodal, receiving projections from auditory, visual, and somatosensory cortex in primates, and is important in representing or accessing the meaning content of utterances. In humans, support for the role of this part of the cortex in extracting semantic information comes from patients with semantic dementia, who having gray matter loss in the left anterior temporal lobe exhibit progressive deterioration in the comprehension of single words.

Word production requires activation of cortical areas that control face and articulator movements. These areas are located in the inferior motor cortex and adjacent **inferior prefrontal** areas. The articulations cause sounds that activate neurons in the auditory system, including areas in the superior temporal lobe. Strong fiber bundles between inferior frontal and superior temporal areas provide the substrate for associative learning between neurons controlling specific speech motor programs and neurons in the auditory cortical system stimulated by self-produced language sounds.

The concept of multiple, parallel processing streams similar to those in the visual modality has been established for the auditory modality (Fig. 3.1.15). Electrophysiological studies in primates and fMRI studies in humans indicate a functional dissociation between anterior and posterior streams. The **what stream** is associated with vocalizations and presumably is associated with auditory object identification. It involves the anterior belt and parabelt, that are further projected to anterior temporal and ventrolateral frontal regions. The **where stream** for sound localization involves the posterior belt and parabelt, posterior temporal and parietal regions, and dorsolateral frontal regions. Posterior auditory regions have been shown to respond to spatial auditory cues in a similar way to parietal neurons responding to visual spatial cues supporting the existence of a posterior temporal–parietal stream in the processing of auditory spatial information.

AUDITORY N1/P2 WAVE

Reactivity of the auditory cortex can be assessed using the **N1/P2 wave** of auditory-evoked potentials. A typical ERP to a brief (100 ms in duration) presentation of a pure tone is depicted in Fig. 3.1.16. Note that ascending and descending parts of the N1 wave show different spatial distributions indicating that it is **not a homogeneous** wave.

There is converging evidence from magnetoencephalographic studies, from intracranial recordings, and from lesion studies that the N1/P2 wave is composed of overlapping subcomponents generated by primary as well

INDEPENDENT COMPONENTS

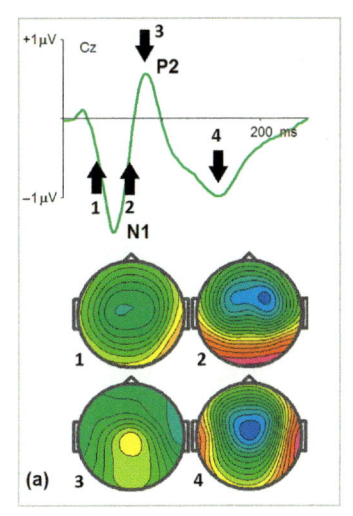

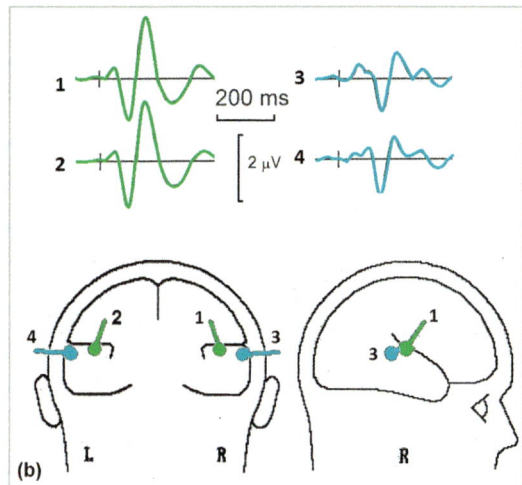

FIGURE 3.1.16 **Auditory ERPs.** (a) An example of grand-averaged ERPs computed for 134 healthy subjects from 17 to 45 years old in response to 1000-Hz tone of 100-ms duration. *Arrows* indicate different latencies at which the maps (below) are made. Note clear differences in topography at different latencies; and (b) radial *(blue)* and tangential *(green)* dipoles approximating the auditory ERPs. Brain Electric Source Analysis is performed on grand mean auditory-evoked potentials from 32 healthy subjects. With 2 equivalent dipoles per hemisphere more than 98% of the variance in scalp potentials (32 channels) in the time range of the N1/P2 component can be explained. Most of the variance (about 80%) is explained by the tangential dipoles (1 and 2) which are supposed to mainly reflect activity of the primary auditory cortex in the superior temporal plane. The N1/P2 dipole activity of the radial dipoles (3 and 4), reflecting activity of secondary auditory areas, occurs about 40 ms later than that of the tangential dipoles (1 and 2). *Part a: data from HBI database. Part b: adapted with permission from Hegerl, Gallinat, & Juckel (2001).*

as secondary auditory cortices. Fig. 3.1.16b demonstrates dipoles that approximate the N1/P2 wave.

INDEPENDENT COMPONENTS

Application of independent component analysis to collection of ERPs recorded in response to 100-ms tones of 1000 Hz enables us to separate functionally meaningful subcomponents of the brain response. Fig. 3.1.17 shows the components found for the right hemisphere. Symmetrical components were found for the left hemisphere (not shown).

The results demonstrate that the N1 wave is decomposed into **five components with different topographies and timecourses**. The first two components are generated in the primary auditory cortex on the left and right hemisphere. The second two components are generated at the lateral

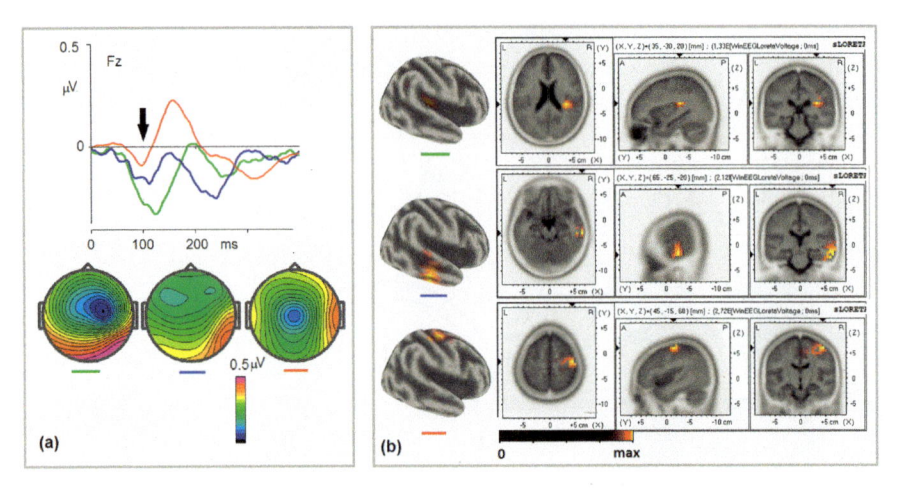

FIGURE 3.1.17 Decomposition of the auditory N1/P2 wave into independent components. (a) Timecourse (top) and maps (bottom) of three independent components (only the components with the right-side distribution are shown); and (b) sLORETA images of the three components. *Data from HBI database.*

temporal surface of the left and right hemisphere—the ventral auditory stream. Finally, the fifth component is generated in the dorsal auditory stream in the parietal cortex. The fifth component makes the largest contribution to the P2 wave.

AUDITORY MISMATCH NEGATIVITY

The auditory world is reflected in continuous activity of neurons of the auditory system. Some part of the world remains constant for a relatively long time and often does not reach consciousness. But if a change occurs unpredictably, such as an engine of a car suddenly breaking its regularity, this change may enter the consciousness. So, the auditory system is constantly **comparing incoming stimuli** with a **sensory model** formed by a previous stimulation.

In the auditory modality the involuntary attention induced by a sudden stimulus change is studied in the **passive oddball paradigm**. In its simple form, the passive auditory oddball task consists of a repetitive sequence of an auditory stimulus—a standard auditory stimulus (eg, 1000-Hz tones of 100-ms duration) which is rarely interrupted by a deviant auditory stimulus (eg, 1100-Hz tones of 100-ms duration). The subject usually performs some other task not involving the auditory stimuli such as reading a book or watching a film. In this task, ERPs computed to the deviant stimulus are contrasted with ERPs to the standard stimulus. If the standard and deviant stimuli are similar physically but still could be discriminated by a subject the ERP difference wave (Deviant − Standard) is of negative value, peaks at about 140–160 ms at frontal electrodes, displays a frontal distribution, and is labeled an **MMN wave** (Fig. 3.1.18a).

The components of MMN have been recorded in **local field potentials** from intracranial electrodes implanted into the brain of neurological patients (Fig. 3.1.18b). In these studies it was shown that responses in the primary auditory area (BA 41) to repetitive auditory stimulation did not habituate and encoded physical properties (such as tone frequency) of the auditory stimulus. Responses in the secondary auditory area (BA 42) strongly habituated (ie, decreased with consecutive repetition of the same stimulus). Synaptic depression in this area might be responsible for this type of habituation. Operation of comparison of the input stimulus with the memory trace of a repetitive signal appears to be carried out in the association auditory area (BA 22). If the stimulus does not match the memory trace developed by the previous repetitive stimulation, this area generates a strong negative component followed by a positive component.

The MMN was discovered in 1978 by **Risto Näätänen** and coworkers from the University of Helsinki. He interpreted MMN as an automatic change detection process in which a mismatch between the deviant stimulus and memory representation (trace) plays a critical role. However, a different view was presented by Patrick May and Hannu Tiitinen from Helsinki University of Technology in a 2010 review. According to their

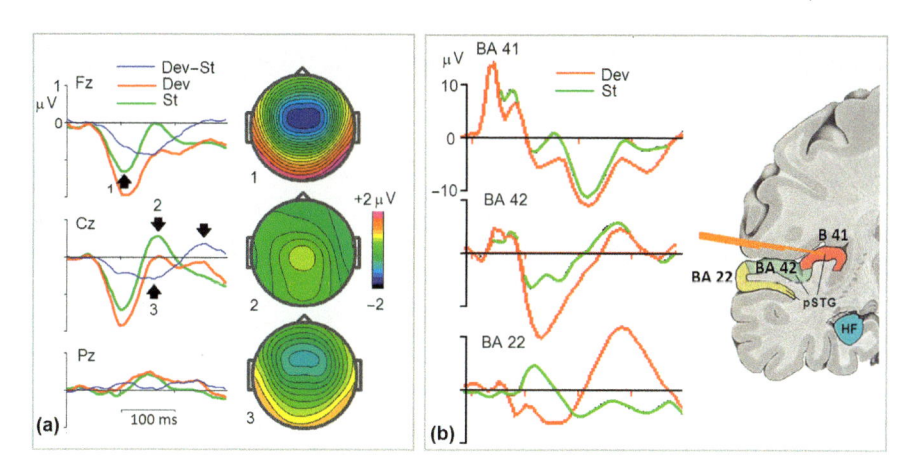

FIGURE 3.1.18 **Mismatch negativity (MMN) in the passive auditory oddball task.** (a) Grand-averaged MMN recorded from the scalp of healthy adults. Subjects are reading the book while passively listening to the standard *(St)* tones of 1000 Hz and duration of 100 ms and deviant *(Dev)* tones of 1300 Hz in the auditory oddball task. Note frontal distribution for the negative wave *Dev–St*; and (b) MMN-like activity in local field potentials recorded by implanted electrodes from BAs 41, 42, and 22. *Part a: data from HBI database. Part b: adapted from Kropotov, Ponomarev, Kropotova, Anichkov, & Nechaev (2000).*

theory the MMN is in essence a latency- and amplitude-modulated expression of the auditory N1 response, generated by fresh afferent activity (disadaptation) of cortical neurons that are under nonuniform level of adaptation.

A vast amount of empirical knowledge has been accumulated since the discovery of the MMN in 1978. One empirical fact concerns the **heterogeneity** of MMN. The existence of several sources of MMN is demonstrated by **independent component analysis** applied to a large collection of individual ERPs recorded in 202 healthy subjects (Fig. 3.1.19).

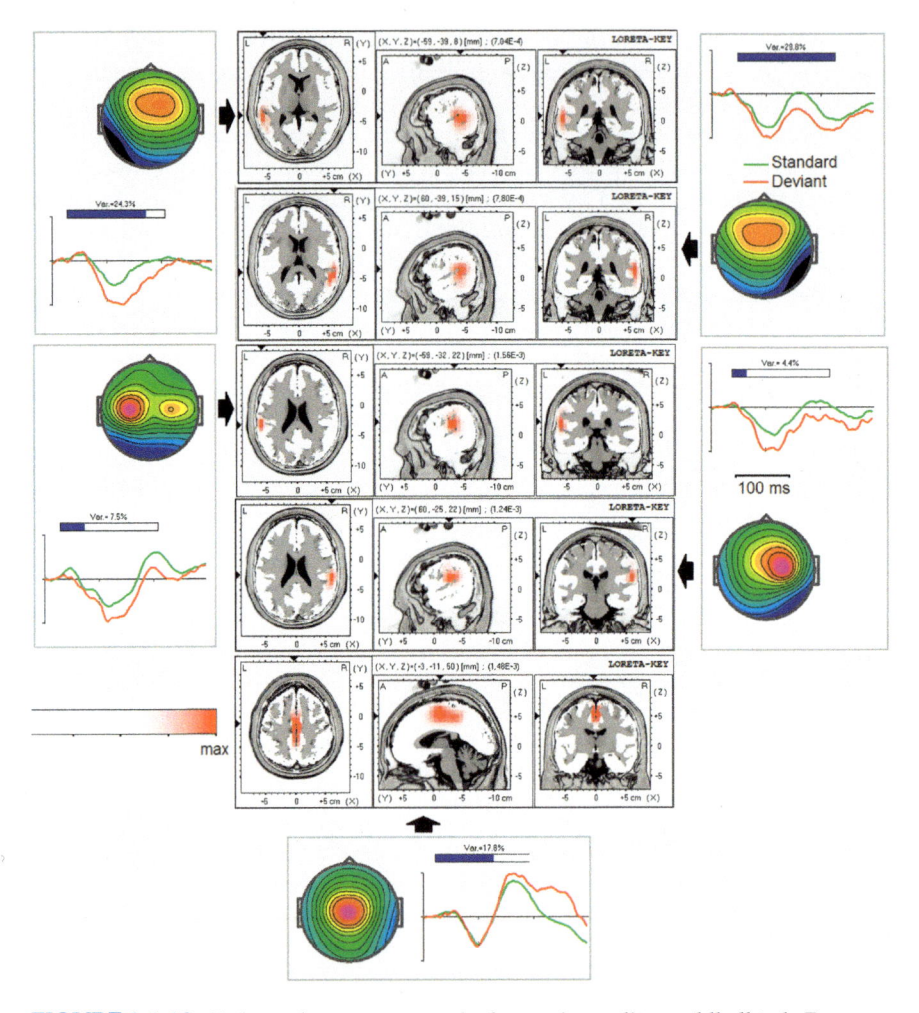

FIGURE 3.1.19 **Independent components in the passive auditory oddball task.** Two symmetrical (left and right temporal components) and one medial (supplementary motor cortex) components are presented. *Green*, activation pattern for standard tone; *red*, activation pattern for deviant tone. *Arrows* indicate the corresponding *LORETA* images. *Data from HBI database.*

ERPs were recorded in the passive oddball task. As one can see, five sources of the N1 are separated. In the temporal frontal area there are two symmetrical (left–right) sources. Both of them are modulated by frequency deviation in the N1 time frame. The fifth source shows only P2 additional fluctuation in response to the deviant stimulus.

ORIENTING RESPONSE

Note that in the previous section we demonstrated MMN in response to deviant stimuli that were physically quite similar to standard stimuli. But what happens if the difference between standard and deviant stimuli gets larger? From our experience we know that such large deviations from the background enter our consciousness. According to Ivan Pavlov, they generate the **orienting response** or "what is it?" reflex. It is an organism's involuntary response to a sudden change in the sensory environment. Pavlov distinguished two phases of the orienting response: (1) inhibition of ongoing activity, (2) shifting attention to the source of stimulation.

The orienting response in the ERP field has been studied in the three-stimulus paradigm. In this paradigm rare novel stimuli (ie, human vocalizations, mechanical noises, or digitally synthesized nonsense sounds) are inserted unexpectedly during the otherwise classical oddball paradigm. The novel stimuli in this paradigm elicit a **Novelty P3 or P3a** wave (Fig. 3.1.20).

According to a 1998 review by Carlos Escera and coworkers from the University of Barcelona the distraction-related operations in the Novelty task include **three sequential stages**. The first stage represents a sensory-adaptive, attention-independent filter which appears to minimize the information load on the limited processing resources of the brain. This stage in turn is divided in two subprocesses: first-order and second-order change detection operations reflected in the **N1 wave and MMN**. According to the hypothesis, when activation of the change detection neurons exceeds some threshold it triggers an involuntary attention-switching process that is reflected in the centrally distributed **Novelty P3 wave**. Finally, if the attention switch is not followed by immediate adaptive changes, the initial attention set is restored. This third hypothetical stage is reflected in the frontal–central reorienting negativity (RON), which peaks at 400–600 ms after novel stimulus.

Independent sources of the human brain response to novel stimuli in the passive three-stimulus task are presented in Fig. 3.1.21. These sources were extracted by independent component analysis applied to a collection of ERPs that were recorded in 175 healthy subjects. As one can see the topographies of temporal components are the same as

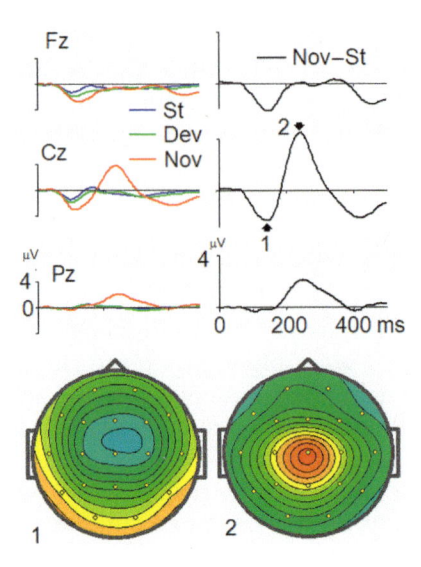

FIGURE 3.1.20 **Novelty P3 wave (P3a).** Grand-averaged ERPs in the common reference montage for healthy adults recorded in the three-stimulus paradigm. Frequent standard *(St)* stimuli, tones of 1000 Hz, 100-ms duration, and probability 80%. *St* are interspersed with rare deviants *(Dev)*, tones of 1300 Hz, 100-ms duration, and probability 10%, and rare novels *(Nov)*, multifrequency artificial sounds of 100-ms duration and probability 10%. Note that a novel stimulus evokes an N2/P3 response in comparison to deviant stimulus. Novelty P3 shows a central distribution. *Data from HBI database.*

were extracted in the active oddball paradigm. The third component is generated in the anterior cingulate cortex and responds to deviation in a strength-dependent manner. But the components generated in the hippocampus and supplementary motor cortices are quite specific for the novelty condition.

On the basis of these data the scheme of brain response to novelty given in Fig. 3.1.22 can be suggested. This scheme is supported by numerous studies of change and novelty detection in fMRI and ERP fields.

The first stage involves different areas of the **superior temporal cortex**. Neurons in the superior temporal cortex adapt to auditory stimuli to different extents. Appearance of a new stimulus activates "fresh" neurons as well as specific change detectors in different cortical areas. If the disadaptation is large enough the signal is registered in the **hippocampal formation** and in the **anterior cingulate cortex**. The hippocampus plays a critical role in acquiring new knowledge while the anterior cingulate is important for action control. Further, registered information is transferred to the **presupplementary motor cortex**. This area is a part of the dorsal attention network and plays an important role in orientation to the source of novelty.

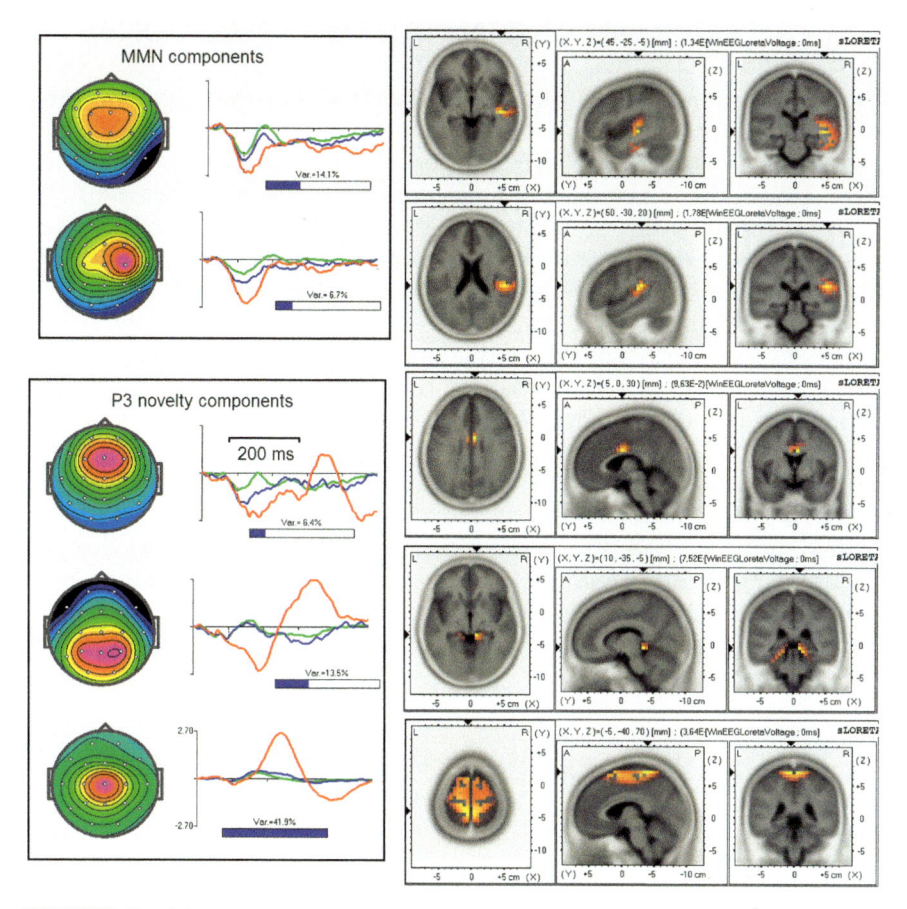

FIGURE 3.1.21 **Independent component of response to novel stimuli.** From left to right, topography, timecourse, and *sLORETA* images of five independent components. Note that symmetrical temporal components for the left hemisphere are not shown, The components are extracted from the collection of individual ERPs in the passive three-stimulus task. *Data from HBI database.*

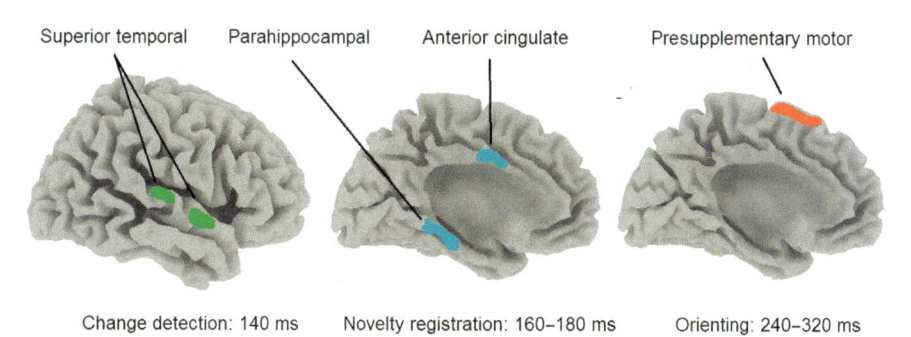

FIGURE 3.1.22 **Stages of information processing during brain response to novelty.** See text for explanation.

ROLE OF DOPAMINE IN ORIENTING RESPONSE

John Polich in his 2007 integrative theory of P3a and P3b associates the phasic response in **dopamine (DA) production** with the Novelty P3 wave. This suggestion fits our data on decomposition of the P3a into three subcomponents localized in the hippocampus, anterior cingulum, and presupplementary motor areas—parts of the cortex **heavily innervated by DA pathways from the ventral tegmental area**. Following the **inverted U-law** for modulatory effects of dopamine, it can be suggested the P3a is maximal for optimal levels of dopamine and is low at low levels of DA (as in Parkinson's disease) and is also low at high levels of DA (as in schizophrenic patients). And again, this suggestion fits experimental data obtained in schizophrenic patients and Parkinsonian patients.

It should be noted that all areas receiving DA innervation from the ventral tegmental area send back glutamatergic connections. So, when a new stimulus is presented to the subject it activates neurons in the hippocampus, anterior cingulum, and presupplementary motor area which in turn phasically activates neurons in the ventral tegmental area. The corresponding burst of DA activation facilitates (increases the gain) of corresponding cortical areas.

LOUDNESS DEPENDENCE OF AUDITORY N1/P2 WAVES

The P1/N1 waves in auditory processing are associated with fast-acting neurotransmitters of the brain such as glutamate as excitatory neurotransmitters and gamma-aminobutyric acid (GABA) as inhibitory neurotransmitters. These waves are tonically modulated by serotonin as a slow-acting neurotransmitter. High concentrations of cortical serotonin have been found in the primary sensory cortices, especially in the primary auditory cortex. Therefore, tonic activity of the serotonergic system is well suited for modulation of primary sensory processing.

The loudness dependence of auditory-evoked potentials (LDAEP) has been proposed as a marker for central serotonergic neurotransmission (Fig. 3.1.23). One possibility is to measure the difference amplitude between N1 and P2 peaks of ERPs in their dependence on the loudness of the auditory stimulus. Subjects with steep increase in amplitudes of the N1/P2 wave to stimuli of increasing intensity are characterized as "augmenters"; those who present with a shallow increase or decrease in amplitudes are called "reducers" (Buchsbaum and Silverman, 1968).

LDAEP is a neuromarker that can be used in clinical practice since subgroups of patients with a serotoninergic dysfunction can be identified and

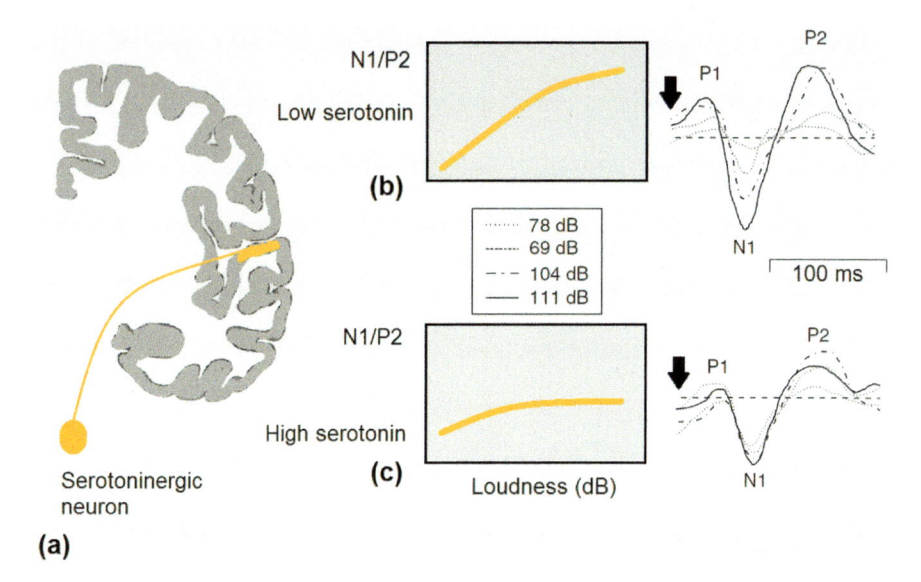

FIGURE 3.1.23 **The LDAEP is defined by intracortical serotonin.** (a) Schematic representation of serotonergic modulation of the primary auditory cortex, (b) the high steepness of the LDAEP corresponds to a low serotonin level in the primary auditory cortex of the rat, and (c) the low steepness of the LDAEP corresponds to a high level of serotonin. *Adapted from Wutzler et al. (2008).*

treated more specifically. For example, in depressed patients, a significant relationship between a strong LDAEP and a favorable response to selective serotonin reuptake inhibitors (SSRI) has been found.

CHAPTER
3.2

Executive System and Cognitive Control

INTRODUCTION

At psychological level, **cognitive control** includes a broad class of processes that provide **goal-directed flexible** behavior. The concept of cognitive control appeared in 1950 with the shift from behaviorism to information processing paradigm. Indeed, the stimulus–reaction contingencies of behaviorism don't need the executive system, while the information flow must be controlled by a higher cognitive control system. Similar shift occurred in clinical practice where the minimal brain dysfunction in children became Attention deficit disorder. The impairments of cognitive control were found in other psychiatric conditions.

Concepts of **attentional control** and **executive control** are often used as synonyms of cognitive control. In 1950–60s executive functions were associated with a unitary **"central executive"** that controls and coordinates task-specific sensory processing modules. Donald Broadbent, the British psychologist, was one of the first who separated controlled processes from automatic responses and introduced concept of selective attention. The term cognitive control was used by the American psychologist Michael Posner. Timothy Shallice, Joaquin Fuster, Robert Knight, Alan Baddeley, and Donald Stuss laid down the foundations of our current understanding of cognitive control.

The executive functions were initially ascribed mostly to the frontal lobes. Later neuropsychological data proposed an alternative view— different executive operations are **segregated**. In 1990s this view was supported by functional MRI studies. These data showed that different executive operations like conflict-monitoring, inhibition of prepared actions, working memory emerge from functioning of separate distributed neuronal systems.

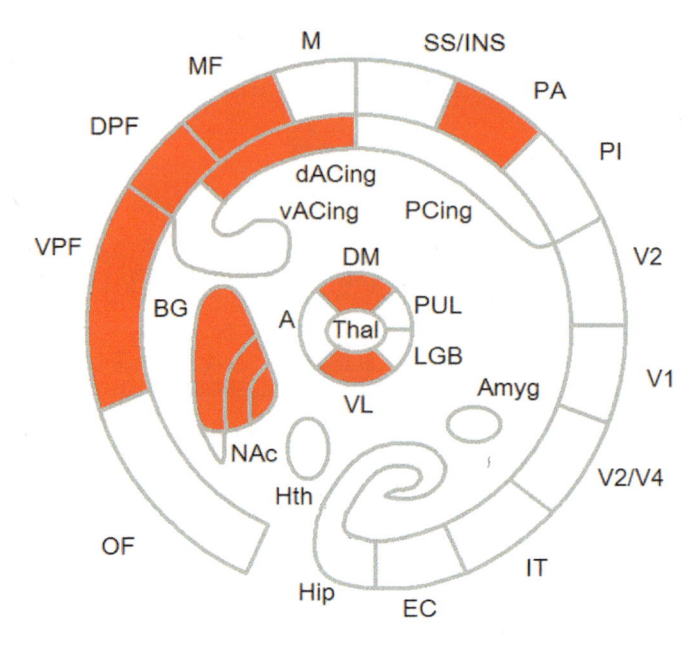

FIGURE 3.2.1 **Executive system.** The executive system includes the prefrontal-parietal network interconnected by the cortico-cortical and cortico-basal ganglia-thalamo-cortical pathways.

The current view on neurophysiology of cognitive control presumes that frontal lobes are involved by dynamic interaction with the temporal-parietal areas both directly **through cortico-cortical** pathways and indirectly through and **cortico-basal ganglia-thalamic** pathways (Fig. 3.2.1). This view is based on the lesions studies showing that lesions in parietal and basal ganglia areas may produce similar executive dysfunctions, on stimulation studies showing that similar affects are produced by stimulation of different nodes in the frontal-parietal circuits, and on the fMRI showing widely distributed brain areas in the tasks on cognitive control.

The cognitive deficits are common for many neurologic disturbances of the frontal lobes and basal ganglia including frontal dementia and Parkinson's disease, as well as for psychiatric conditions including ADHD, schizophrenia, and OCD.

OPERATIONS OF COGNITIVE CONTROL

One of the basic goals in human behavior is **selection** from a large repertoire of possible behavioral options those actions that are more likely to promote human well-being. The action selection process needs to take into account many factors including estimations of the sensory model and probable stimulus-action contingencies, the action-outcome probabilities, accessibility of the program in a given behavioral context, the previous experience. Moreover, because action selection presumes a preparation stage for action implementation during the preparatory set a new situation may appear that does not fit to the planned action so that the initial plan must be **suppressed** and a new plan selected.

The previously mentioned operations are united under a general concept of "cognitive" or "executive control." The hypothetical operations include: (1) **action selection,** (2) **shift from one action to another,** (3) **action preparation,** (4) **action execution,** (5) **working memory,** (6) **suppression of prepared action,** (7) **inhibition of ongoing activity,** (8) **detection of conflict,** that is situations when a new context does not match the planned action, (9) **adjusting future behavior in order to avoid conflicts.**

MODES OF COGNITIVE CONTROL

The executive operations listed in the previous paragraph can be divided into two large categories: the first one is associated with **maintenance of action** (such as keeping task setting, performing the task as fast as required, keeping it as long as needed without slowing), the second one is associated with **monitoring** the sensory-motor results of actions (ie, constantly comparing them with the template in working memory for quality control and **shifting** from the current action to a more appropriate action).

These two categories of operations require opposite properties of the underlying neuronal networks. The first property is associated with **stability** of neuronal activity in order to pursue/maintain a goal and to **inhibit irrelevant sensory information and actions** for relatively long time required for achieving the goal.

The second property is associated with **labile characteristic of neuronal activity** required for continuous monitoring the environment for potentially significant information even if this information is not relevant to the task in order to be able to suppress the ongoing or prepared actions and immediately shift to the most appropriate goal. As we are going to show later in the chapter these two qualities of cognitive control are implemented by different neuronal circuits which in turn are modulated by different dopamine receptors D1 and D2.

The idea of two types of neuronal operations of cognitive control is expressed in **the dual mechanisms of cognitive control** framework proposed by Todd Brevor from Washington University in St. Louis and reviewed by him in 2012. The framework included proactive and reactive modes of cognitive control in order to explain large inter- and intraindividual variability of cognitive control. **Proactive control** reflects the **sustained and anticipatory maintenance** of internal goals to enable optimal performance in future, whereas reactive control **reflects transient, stimulus-driven** mechanism that is mobilized just in time when cognitive control is urgently needed.

PREPOTENT MODEL OF BEHAVIOR

In experimental paradigms on action inhibition concepts of **prepotent response or prepotent model of behavior** are often used. By definition, the prepotent response is automatically (habitually) performed by a subject in a certain situation. This concept may be extended to responses elicited by the brain according to a dominant (most probable) model of behavior. This model appears represents a network "connecting" a certain stimulus to the prepotent response. In experimental conditions the network is formed by conditioning in animals or by verbal instructions in humans. In this context certain cognitive control operations are activated when it is necessary to override the prepotent response.

The prepotent automaticity makes our life **efficient and easy** in many respects—it can free limited mental resources from the numerous routine requirements of life. However, automatic processing is not always a blessing—it is **inflexible and therefore difficult to control**. Moreover, the **automatic prepotency** may interfere with a contrary intention.

There are at least three types the prepotent model: **innate, habitual, and motivational**. The innate model is genetically determined and does not need a prior training. An example is orienting reflex. The habitual model needs considerable training. An example is pressing a break by a driver. The motivational model needs a dominant motivation, for example, a thirsty man in a desert would be motivated to take a glass of water standing in front of him.

BEHAVIORAL PARADIGMS

There are many functions of cognitive control and there are many tasks to study these functions. Moreover, many laboratories are using their own tasks which differ in stimuli, responses, probabilities, interstimulus intervals, task settings etc. In spite of this diversity the whole repertoire of the tasks used to study cognitive control could be separated into few categories. A noncomplete attempt for classification of cognitive control paradigms is presented in Fig. 3.2.2.

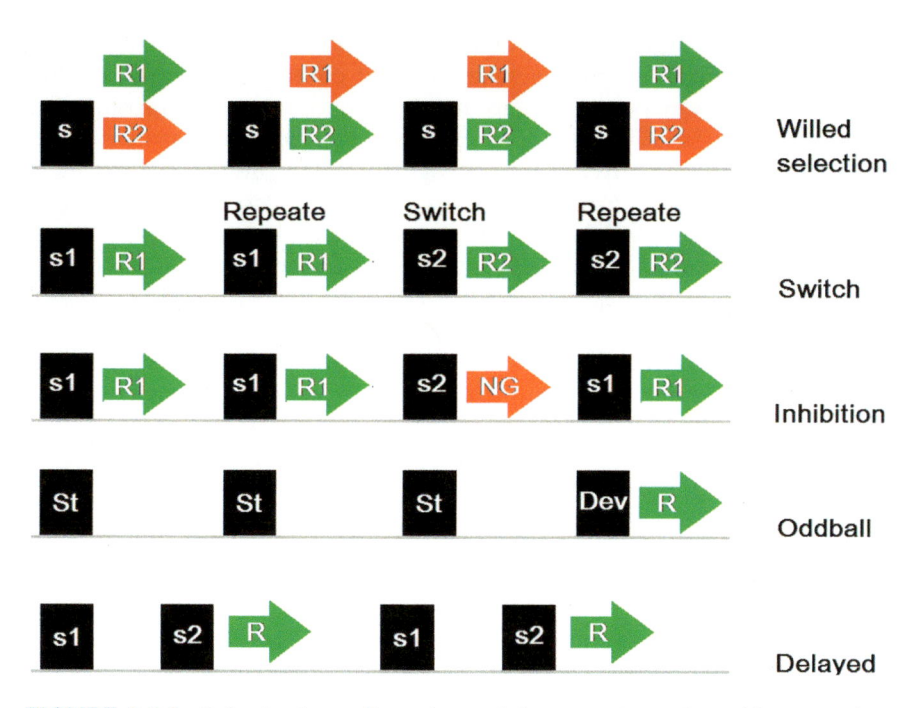

FIGURE 3.2.2 Behavioral paradigms for studying operations of cognitive control. *s*, *s1*, and *s2*—stimuli indicating the required response. *St*, standard frequent stimulus; *Dev*, deviant rare stimulus; *R*, *R1*, and *R2*—actual actions where *R1* is different from *R2*. *NG*—no action is requires (action inhibition). Other explanations are in text.

Willed selection paradigm. The main idea is to study the willed selection of an action from two or many alternatives. In some tasks the selection is prompted by a stimulus, in other cases the subject can decide by himself when to initiate the action.

Task switching paradigm. In this paradigm, participants have to switch between two or more simple cognitive tasks (eg, a letter- and a digit-categorization task). It is found that performance is better on task repetitions (eg, two letter tasks in a row) than on task switches (eg, a letter-categorization task after a digit-categorization task). These **switch costs** (ie, decreased performance resulting from switching between tasks) have long been taken as a direct measure for the amount of cognitive control needed to flexibly switch between cognitive operations.

Inhibition paradigm. In these tasks the prepotent model of behavior (such as stimulus-reaction contingency—GO) is formed by probability of the response or by expectancy of the response. When a NOGO stimulus appears instead of GO or Stop signal is presented after the GO stimulus the subject is instructed to withhold from the prepotent action.

Oddball paradigm. In the oddball paradigm the propotent model is just attending to the stimulus sequence and consequently only sensory model may be formed. The probability of standard stimuli in the sequence is usually high (80–90%). The deviant stimuli are presented randomly and infrequently, and require a reaction (such as button press, counting, etc.). The deviant stimulus mismatches with the sensory model and evokes so called mismatch potentials. The deviant stimulus also activates neuronal networks required for producing the response or for updating the content in the working memory. Some laboratories are using the paradigm a passive construction, that is, no response is required.

Delayed tasks. Delayed tasks are designed to study the proactive cognitive control. The trials usually consist of two stimuli s1 and s2 match. The task could be to compare s2 with s1 and to respond according to the result of comparison (eg, a match to sample delayed task) or simply to wait to s2 and to respond as fast as possible, In this case s1 serves as a warning stimulus and s2 serves as a trigger stimulus.

STROOP TASKS

Separately from the earlier mentioned paradigms stand the **Stroop tasks** which have been designed by John Stroop from David Lipscomb College in 1935 to study the **interference** of the automatic prepotency with the task setting. The task includes a combination of selection, inhibition and switch operations. In the conventional **Stroop task** subjects are presented with a color word written in a conflicting color (eg, "RED" printed in blue) and are asked to name the color of the word (ie, blue). The subjects show difficulty suppressing the automatic prepotency to read the word. In this example, an irrelevant stimulus dimension, that is, the word's meaning, influences subjects' responses despite their efforts to ignore the irrelevance. Note that the degree of prepotency critically depends on the level of familiarity with a language used in the task. A subject who has never been exposed to English should not be subjected to the Stroop interference effect when color words are written in English.

The automatic prepotency seems **enhanced** in some disorders of the brain. For example, people suffering from a neuropsychiatric disorder, such as obsessive–compulsive disorder and drug addiction, show considerable difficulty suppressing their maladaptive, impulsive behaviors—such as recurrent, unwanted thoughts, and repetitive drug-seeking behavior. In another example of frontal lobe damage some patients show utilization behavior, that is, **inability to inhibit** actions afforded by everyday instruments, such as matches, scissors, and combs, even when those actions are contextually inappropriate. The enhancement of automaticity in these disorders may be caused by **dysfunction** of cognitive control.

MODELS OF COGNITIVE CONTROL

There are many models of the cognitive/executive/attentional control operations. They are defined by the author's personal experience and the corresponding perspective from which the operations are viewed. A few models are presented to illustrate this statement.

Baddeley's model has been derived from psychological perspective in 1974 (Baddeley & Hitch, 1974) and revised by Alan Baddeley from University of Cambridge in 2012. According to the revised model a central executive system regulates three other subsystems: the **phonological loop**, which maintains verbal information; the **visuospatial sketchpad**, which maintains visual and spatial information; and **episodic buffer**, which integrates short- and long-term memory, holding and manipulating a limited amount of information from multiple domains in temporal and spatially sequenced episodes.

Barkley's model was suggested by Russel Barkley from Medical University of South Carolina on the basis of his studies of behavioral control impairment in ADHD population and considered the cognitive control from perspective of self-regulation. This model published in 1997 in the book *ADHD and the Nature of Self-control* divides executive functions into three main elements. One element is working memory that resists interfering information. A second component is the management of emotional responses in order to achieve goal-directed behaviors. A third component is internalization of self-directed speech for control rule-governed behavior and for generating plans for problem-solving. The last component includes analysis and synthesis of information into new behavioral responses to meet the goals. Changing one's behavioral response to meet a new goal or modify an objective is a higher level skill that requires a fusion of executive functions including self-regulation, and accessing prior knowledge and experiences.

Miller and Cohen's model was suggested by Earl Miller from Massachusetts Institute of Technology and Jonathan Cohen from Princeton University in 2001. They argue that cognitive control is the primary function of the prefrontal cortex (PFC), and that control is implemented by increasing the **gain of sensory or motor neurons** that are engaged by task- or goal-relevant elements of the external environment. The aggregate effect of these bias signals is to guide the flow of neural activity along the pathways that establish the proper mappings between inputs, internal states, and outputs needed to perform a given task. According to this theory the selective attention mechanism is in fact just a special case of cognitive control–one in which the biasing occurs in the sensory domain.

REPRESENTATIONS IN WORKING MEMORY

Working memory can be defined as a temporary retention of **representations** that just recently experienced (but no longer exist in the environment) or recalled from long-term memory. These internal representations can be kept long periods of time by active maintenance or rehearsal strategies, and can be used for cognitive control. The neuronal mechanisms of working memory have been studied in animals by recording neurons and in humans by means of fMRI.

The animal experiments show that the behaviorally meaningful stimulus activates neurons located in different parts of the **frontal-parietal network** (Fig. 3.2.3) For example, the posterior parietal cortex of rhesus monkeys has been found to encode the behavioral meaning of categories of sensory stimuli initially processed in the **primary sensory areas**. He Cui and Richard Andersen from California Institute of Technology in 2007 demonstrated that when monkeys are instructed with visual cues to make either eye or hand movements to a target, the parietal neurons show specificity depending on which effector (eye or hand) is instructed for the movement. These data show that the posterior parietal cortex serves as a **sensory-motor interface** containing an intentional map. The prefrontal neurons receive sensory-motor information from the posterior areas of the brain and code the plans of voluntary actions. These plans are further transformed into motor commands in the primary motor cortex (Fig. 3.2.3)

Earlier in 2005, Cisek and Kalaska from University of Montreal demonstrated that while a monkey chooses between two actions, its motor system first represents both options and later reflects selection between them. Neurons in the frontal lobe show the response bias to the **action** (Fig. 3.2.3b), while neurons in the parietal-temporal areas show the response bias to **stimulus** (Fig. 3.2.3c). But they both show sustained activity between stimulus and response.

The cortical-basal ganglia-thalamo-cortical pathways are not shown in the scheme for simplicity. Numerous intracranial recordings in delayed tasks in monkey and patients with implanted electrodes demonstrate that the subcortical neurons show activation patterns similar to those recorded from the cortex. Fig. 3.2.4 demonstrates the response patterns of the subcortical neurons in the visual recognition delayed task. Impulse activity of neurons in the basal ganglia and thalamus was recorded in the visual recognition delayed response task in patients to whom electrodes were implanted for diagnosis and therapy. One can see that there are three groups of neurons in these subcortical structures: excitatory sensory-related and motor-related neurons distributed in the striatum and thalamus and inhibitory neurons distributed in the internal part of the globus pallidus.

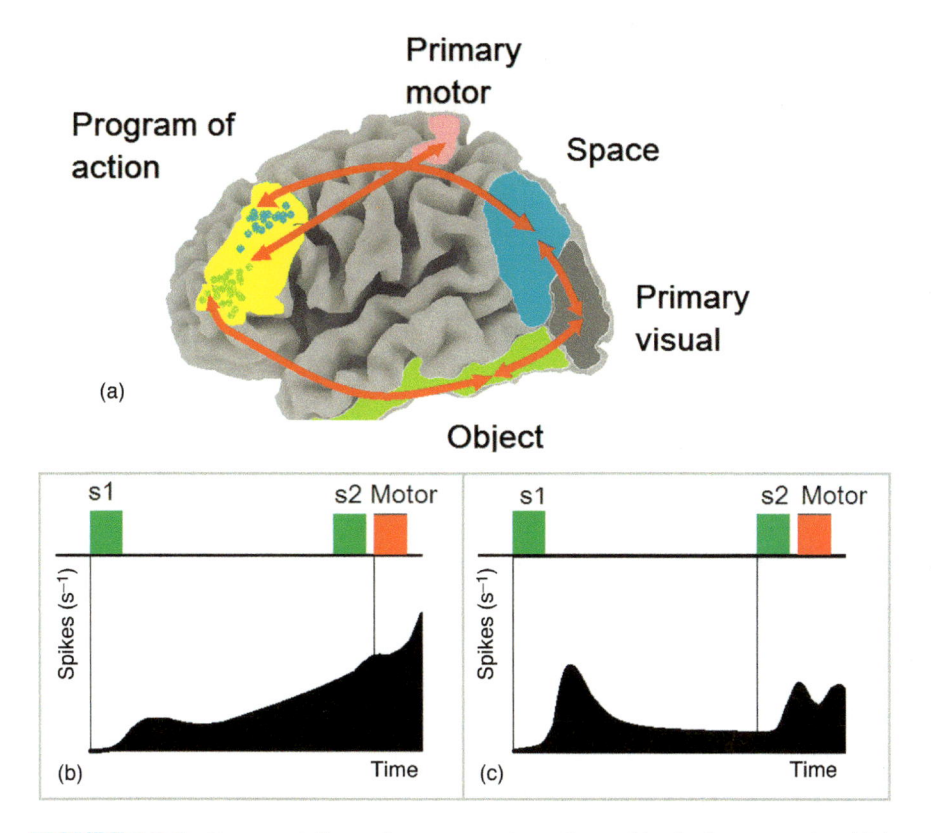

FIGURE 3.2.3 **Representations of sensory-motor actions of in the human cortex.** (a) A scheme is presented for the visual modality. Object-related and spatial representations of the visual image are extracted in the dorsal and ventral posterior streams. The plan of action is formed in the prefrontal cortex. The stimulus–response mapping is reflected in activation pattern of the frontal-posterior networks. Two patterns of neuronal responses in the two stimulus delayed response task are observed in the frontal-posterior networks: (b) the motor-related pattern with maximum of activation just before the motor action and (c) the stimulus-related pattern with maximum shortly after stimulus.

PREPARATORY CORTICAL ACTIVITIES

Proactive cognitive control mode consists of different theoretical operations including preparation to receive a stimulus and preparation to make a movement. These operations may last for several seconds and are associated with **working memory** in general and with **attention** and **motor preparatory sets** in particular.

In ERP correlates of brain preparatory activities are associated with negative slow shifts of electrical potential recorded from the scalp. Historically negative waves were named depending on experimental paradigms in which they were recorded. Three types of negative preparatory waves

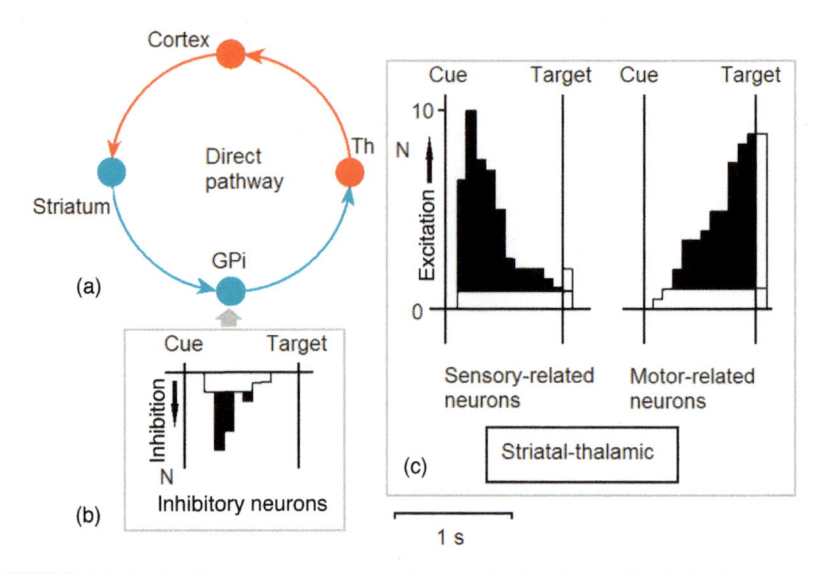

FIGURE 3.2.4 Profiles of neuronal reactions in the basal ganglia-thalamic structures. (a) Schematic presentation of the basal ganglia thalamo-cortical circuit; (b) inhibitory neurons of the internal part of the globus pallidus. X-axis time between the cue and trigger in the stimulus response paradigm. Y-axis, number of neurons reactive with inhibition; and (c) sensory-related and motor-related neurons in the striatum and thalamus. Y-axis, number of neurons that show significant excitation in the corresponding time interval. *Adapted from Bechtereva et al. (1990).*

are distinguished: (1) the Bereitschaftspotential in German transcription or **readiness potential** (RP), (2) the **contingent negative variation (CNV)**, and (3) the **stimulus preceding negativity (SPN)** (Fig. 3.2.5).

The readiness potential was first recorded by Hans Helmut Kornhuber from University of Ulm and Lüder Deecke from University of Vienna in 1964 as a result of search for cortical electrical correlates of **voluntary movement**. An example is presented in Fig. 3.2.5a where the RP precedes volitional rapid flexions of the right index finger. The vertical line indicates the onset of EMG of the agonist muscle. Recording position is Pz. The readiness potential has two components, the early one lasting from about −1.2 to −0.5 and the late component from −0.5 to shortly before muscle activity. The early part is generated by the supplementary motor area (SMA) and the pre-SMA, while the late component is generated by the primary motor area.

The **CNV component** was first described by Grey Walter and his co-workers from Burden Neurological Institute in Bristol, in 1964. It was the first of ERP waves to be discovered. The negative CNV waves develops around 260–470 ms after the warning stimulus, it appears bilaterally symmetrical with the largest values at the vertex (Fig. 3.2.5b). When the interstimulus interval is longer than 3 s two negative waves can be visually separated from the CNV. The first wave follows the warning

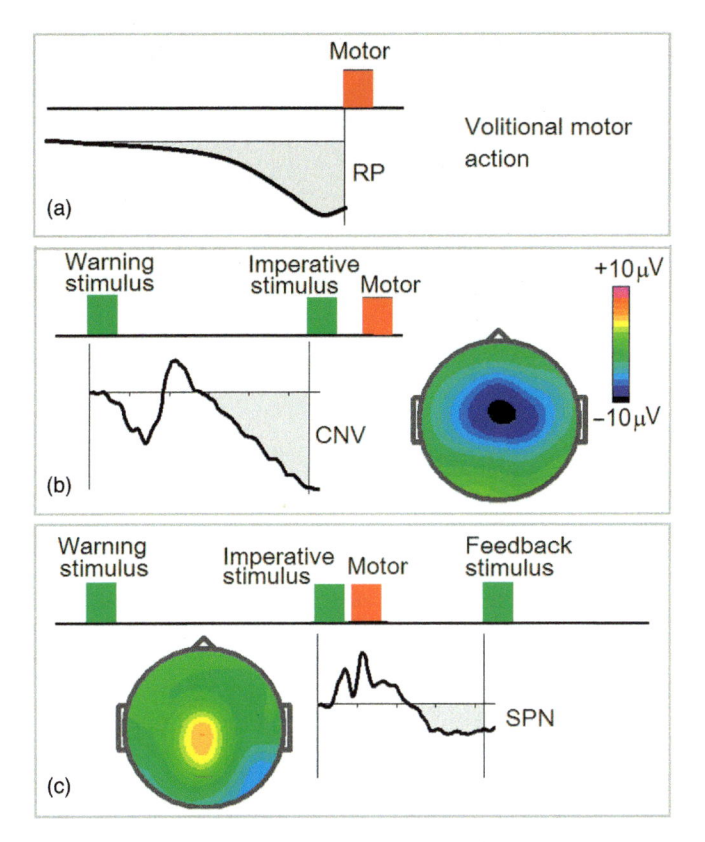

FIGURE 3.2.5 **Cortical preparatory negativities.** (a) RP precedes voluntary flexions of right finger; (b) CNV evolves after the warning stimulus and precedes the trigger stimulus; and (c) stimulus preceding negativity (SPN) is generated prior of the feedback stimulus requiring no response. *Part a: adapted from Deecke, Grözinger, & Kornhuber (1976); Part b: data obtained from a healthy subject from the HBI database performing the cued GO/NOGO task.*

stimulus, peaks at frontal electrodes and is called the **orienting wave**. The second wave precedes the imperative stimulus and is called the **expectancy wave**.

When no response action is required and the subject is just waiting for a visual stimulus a Stimulus preceding negativity is generated over the temporal electrodes (Fig. 3.2.5c).

Latent ERP Components of Preparatory Activity

Let us consider the cued GO/NOGO task in which the subject is instructed to ignore trials started with presentation of a plant and to prepare for the second stimulus in trials started with presentation of an animal. In the cued trial the subject is preparing to receive the same image of an

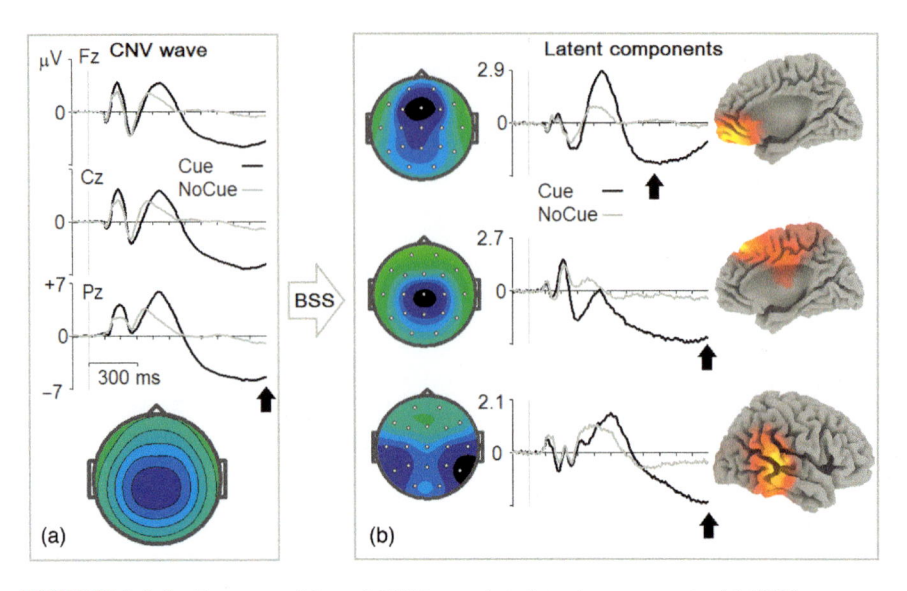

FIGURE 3.2.6 **Decomposition of CNV wave into latent components.** (a) CNV waves at Fz, Cz, and Pz for a group of healthy subjects (34–40-year-old) with the map at 1000 ms after stimulus. Black line, ERPs to cues; grey line, ERPs to No cues. (b) Latent components in standard units. From left to right: topography, time course for Cue and NoCue and sLORETA image. *Data from HBI database.*

animal at the second position and to respond to it as fast and as precise as possible. This is a typical situation for eliciting the CNV wave.

The grand average CNV wave for adult healthy subjects in the cued task is shown at Fig. 3.2.6a. One can see that preparation for the second stimulus generates a strong and widely distributed negative shift of potential. Application of the blind source separation method, described in Part Methods, enables us to decompose the CNV wave into three different components (Fig. 3.2.6b). The components are generated in different frontal-temporal areas and demonstrate quite different dynamics.

The component generated in the supplementary motor and pre-supplementary motor areas (Fig. 3.2.6b middle) is characterized by a negative fluctuation to Cue (animal) stimuli in comparison to NO-cue (plant) stimuli. This N2 is followed by a developing negativity which resembles the expectancy part of CNV. The component generated in the temporal cortex (Fig. 3.2.6b bottom) is characterized by a late positive fluctuation at 440–530 ms interval followed by a negative fluctuation preceding the second stimulus and might be associated with stimulus preceding negativity. Finally, the component generated in the anterior cingulate cortex (Fig. 3.2.6b bottom) is characterized by a negative fluctuation peaked at 700 ms and can be associated with the orienting part of CNV.

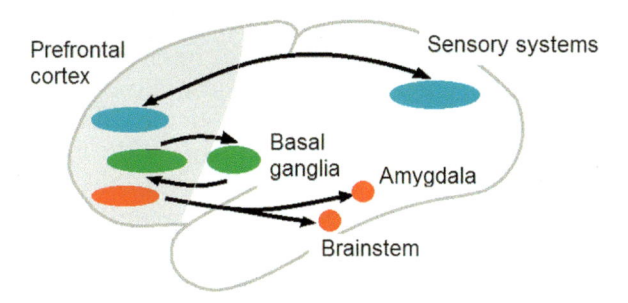

FIGURE 3.2.7 **Frontal lobe interconnections.** The sensory (marked by blue), executive (green) and affective (red) functions are segregated within the frontal lobes. The subregions of the prefrontal cortex mapping these functions are interconnected with the rest of the brain (marked by arrows).

In animal studies the preparatory activity in neuronal discharges is seen not only in the frontal-temporal-parietal areas but also in the basal-ganglia thalamic nuclei. There is uncertainty in what extant the preparatory negativities observed in humans from scalp electrode reflect cortical-cortical and cortical-basal ganglia-thalamo-cortical interactions. However, undoubtedly the frontal lobes play a critical role in these processes.

FRONTAL LOBE FUNCTIONS

The functions of cognitive control are not the only functions involving the frontal lobes. They also modulate sensory, affective, and memory systems of the brain. The basic interactions of the frontal lobes with the rest of the brain are schematically illustrated in Fig. 3.2.7. The prefrontal cortex regulates **sensory information flow** in the posterior sensory systems. The prefrontal cortex originates **planning, initiation, and inhibition of actions** through the basal ganglia and motor cortical/subcortical areas. The prefrontal cortex modulates the **affective system** of the brain though amygdala and brain stem connections.

These sensory, executive, and affective functions are **segregated** within the frontal lobes. For example, dorsal regions regulate "top–down" spatial attention through extensive projections to parietal cortex, whereas ventromedial and orbito-frontal cortex regulates emotion through extensive projections to the amygdala, the nucleus accumbens of the basal ganglia, and brainstem

BASAL GANGLIA-THALAMO-CORTICAL LOOPS

From the theoretical perspective, the frontal-parietal-temporal networks with their reciprocal "positive" excitatory interconnections can potentially implement working memory functions including the sustained activity in the delayed tasks between the first stimulus and the action. In other words, these networks may provide the basis for **stability** of neuronal activity in order to pursue/maintain a goal. However, as mentioned earlier, the **flexibility** of neuronal activity in order to shift between options and to inhibit inappropriate actions needs large scale "negative" inhibitory connections. To provide flexibility of behavior, the evolution created these circuits in a form of the **cortical-basal ganglia-thalamo-cortical loops**.

An example of the importance of the basal ganglia in cognitive control is given by **Parkinson's disease**. This disease is characterized by **depletion of DA** in the basal ganglia due to degeneration of dopaminergic neurons in the substantia nigra. Although hypokinetic movement disorder pops-out when observing these patients, more sophisticated neuropsychological assessment reveals **impairments in cognitive control**. Indeed, Parkinsonian patients used to say that rigidity and tremor are not the core of their illness, but rather inability to initiate and suppress an action creates significant impairment. I recall a patient in our clinic who complained that the most difficult for him was crossing a street. When he sees the green light he has difficulty to initiate the movement and is delayed so that when he starts walking the light turns red, but he continues going because now he has difficulty of stopping the movement.

Historically the basal ganglia were considered as **movement-related noncortical pathways** until the 1970s. The basal ganglia were viewed as a funneling mechanism by which movement initiation could originate from diverse cortical areas. According to this view the basal ganglia integrate projections from diverse parts of the cortex and project this information, via the thalamus, to the motor cortex and supplementary motor area.

Recent neurophysiological and anatomical studies have shown that instead of funneling the cortical-basal ganglia-thalamic pathways are **segregated**. Specifically, it was demonstrated that the inputs to the basal ganglia from a certain cortical area terminates within a specific basal ganglia area, which in turn is connected to similarly specific parts of the thalamus. The thalamic nucleus projects back to the same areas of the cortex from which the circuit originates thus forming a closed loop. These **segregated reentrant loops** are able to influence widespread areas of the frontal lobe and play a role in far more than simple motor functions.

Different parallel circuits were separated on the basis of the functions of the cortical areas from which they originate. They are **motor, oculomotor, prefrontal associative, and limbic** areas. Although some uncertainty remains regarding the degree of segregation and whether the circuits are pure closed or partially open, the basic scheme has been accepted by most researchers.

The **canonical basal-ganglia thalamo-cortical circuit** (Fig. 3.2.8) provides a framework for understanding the diverse symptoms seen in the basal ganglia disorders as well as in some other psychiatric conditions.

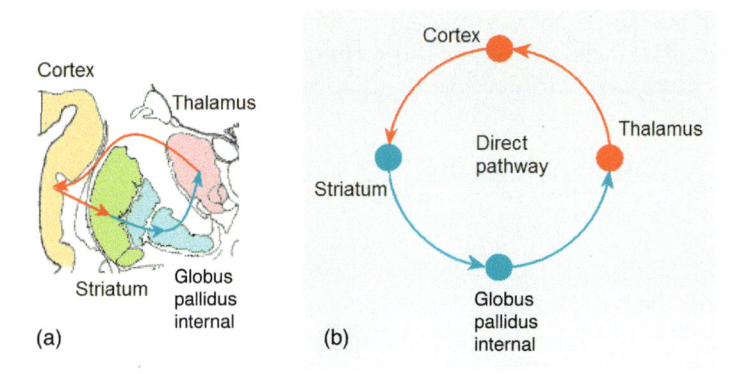

FIGURE 3.2.8 Canonical cortical-basal ganglia-thalamo-cortical circuit. (a) Anatomical localization of thalamus and basal ganglia on the coronal brain section. The basal ganglia include the striatum and the globus pallidus. The thalamus and basal ganglia are located close together; (b) for simplicity, only the "direct" network involved in GO responses is depicted. Excitatory (glutamatergic) connections between cortical areas, the basal ganglia and thalamus are depicted by red arrows. Inhibitory (GABAergic) connections are shown in blue.

The parallel loops appear to perform similar operations but on different pieces of information such as **sensory-motor, spatial, visual/auditory, affective**. (Fig. 3.2.9) These basic operations include: (1) selection of a plan for the relevant action and (2) inhibition of the planned action when the situation become inappropriate. Although the basal ganglia neurons

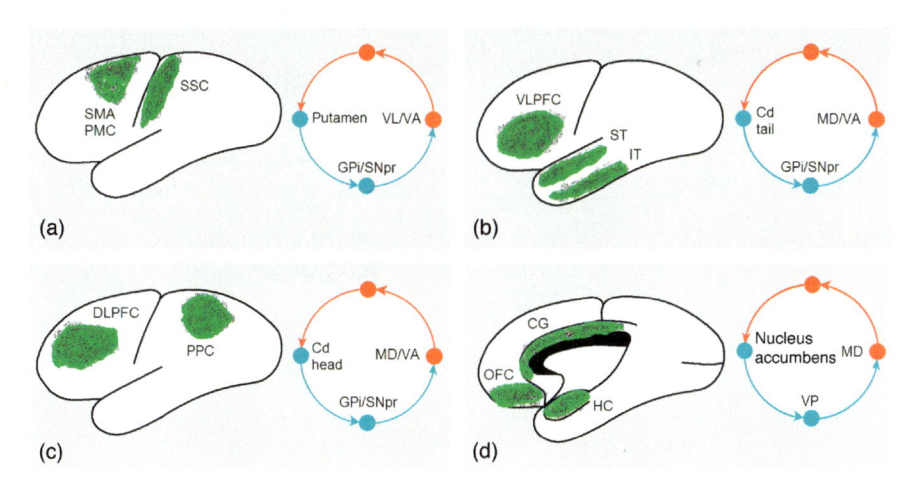

FIGURE 3.2.9 Parallel cortical-basal ganglia-thalamo-cortical circuits. (a) Motor, (b) visual, (c) spatial, and (d) affective. *SMA*, supplementary motor area; *PMC*, premotor cortex; *SSC*, somatosensory cortex; *VLPC*, ventrolateral prefrontal cortex; *ST*, superior temporal; *IT*, inferior temporal; *DLPFC*, dorsolateral prefrontal cortex; *OFC*, orbitofrontal cortex; *PPC*, posterior parietal cortex; *CG*, cingulate gyrus; *HC*, hippocampus; *GPi*, internal sector of the globus pallidus; *SNpr*, substantia nigra pars reticulate; *MD*, mediodorsal; *VA*, ventral anterial; *VL*, ventrolateral nuclei of the thalamus; *VP*, ventral pallidum. *Modified from the original scheme of Alexander & Crutcher (1990).*

show the sustained activity similar to that in the frontal-parietal-temporal circuit, it's unclear whether these circuits are involved in the working memory per se or simply reflect the cortical activity.

NEURONAL CORRELATES OF COGNITIVE CONTROL IN THE BASAL GANGLIA

As numerous animal studies show the **action selection (GO)** and **action inhibition (NOGO)** operations are associated with activation correspondingly of **GO** and **NOGO neurons** in the basal ganglia-thalamo-cortical networks (see for example studies of Okihide Hikosaka from National Institute of Health and National Eye Institute in Japan). Recordings of impulse activity of neurons in GO/NOGO tasks in monkeys demonstrated existence of GO and NOGO neurons (ie, neurons selectively reacted either in GO or NOGO condition) in different parts of the prefrontal cortex, the basal ganglia and thalamus (Fig. 3.2.10). These findings show that the NOGO response is an **active operation** requiring activation of a separate network different from that involved in GO response. They also stress the importance of the basal ganglia-thalamic network in the NOGO process.

Activation of the GO neurons appears to be associated with selection of the **direct cortical-basal-ganglia-thalamo-cortical loop** which disinhibits the thalamic neurons, provides a positive feedback to the cortex and thus

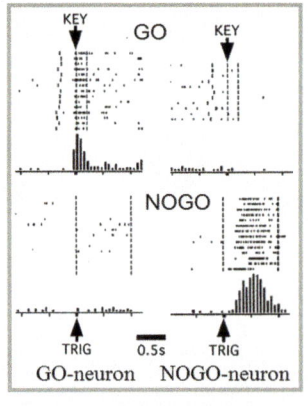

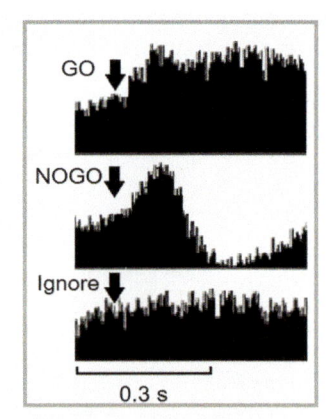

(a) Monkey putamen   (b) Human thalamus

FIGURE 3.2.10 GO and NOGO selective neurons in (a) monkey putamen, and (b) human ventro-lateral thalamus. (a) In raster displays, small dots denote occurrences of action potentials and longer ticks indicate various events during the task. Each horizontal row represents a single trial. In the histograms below each raster, neuronal discharges are summated. (b) Postimulus-time histograms for a neuron in the human thalamus during the cued GO/NOGO task *Part a: adapted with permission from Inase, Li, & Tanji (1997); Part b: adapted from Kropotov (2009).*

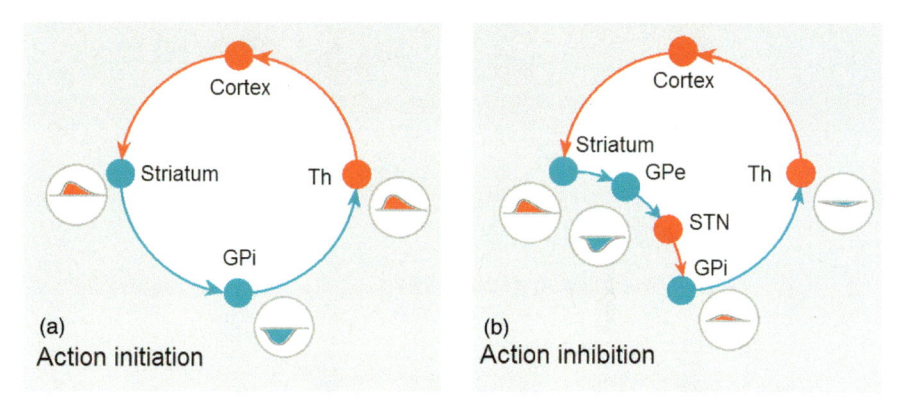

FIGURE 3.2.11 **Hypothetical GO and NOGO circuits associated with direct and indirect basal-ganglia thalamo-cortical pathways.** (a) GO neurons and (b) NOGO neurons. The graphs in gray circles indicate neuronal excitation (up) and inhibitory (down) neuronal responses. The scheme is based on data presented in the 2008 monograph *A Theory of the Basal Ganglia and Their Disorders* by Robert Miller from University of Otago, New Zealand and on the data obtained in the author's laboratory.

facilitates the prepared action (Fig. 3.2.11a). Activation of the NOGO neurons is associated with selection of the **indirect circuit** which inhibits the thalamo-cortical neurons, provides a negative feedback to the cortex and thus leads to suppression of the prepared action (Fig. 3.2.11b).

Numerous animal findings can be summarized as follows: the striatum receives a variety of input from the cortical output, and uses this information either **to release or to suppress specific patterns of behavior**. A combination of striatal input and behavioral output is formed by the previous experience to enable the subject enhance or suppress behavioral activity to obtain the motivationally favorable consequences.

During 1970–80 the author's laboratory had a unique opportunity of recording impulse activity of neurons in the cortical-strio-pallido-thalamic networks in Parkinsonian patients to whom electrodes were implanted for diagnostic purposes. Some of the results relevant for this part of the book are presented in Fig. 3.2.12. One can see that there are **two types of neurons**: those that have a bias of responding to sensory stimuli—sensory-related neurons and those that have a bias of responding to motor actions. All neurons respond to **behaviorally meaning stimuli**—if the stimuli are ignored they don't activate the subcortical neurons. Different nuclei show different patterns: for example, inhibitory responses are found in neurons located in the globus pallidus, the nucleus that gets inhibitory connections from the striatum, whereas the striatum itself shows only excitatory responses with high signal to noise ratio.

The experience of analyzing neuronal reactions in GO/NOGO tasks in patients with implanted electrodes enabled the author formulating a **theory of action programming** first described in 1989 and revised by the author

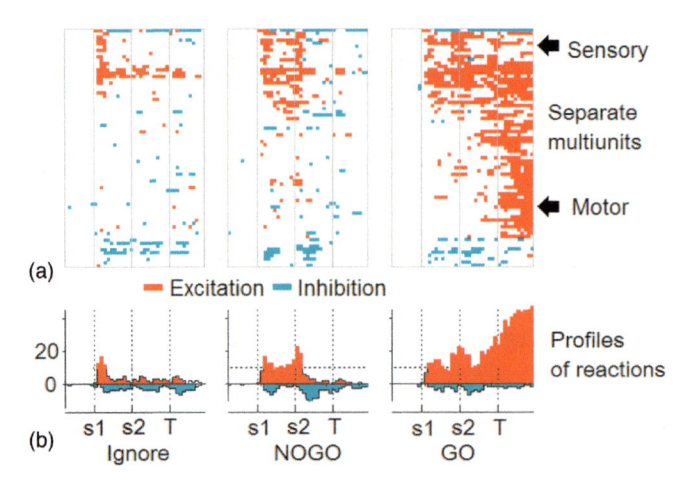

FIGURE 3.2.12 **Responses of neurons in the human basal-ganglia circuits in the cued GO/NOGO task for Ignore, GO and NOGO conditions.** (a) Color-coded reactions of single multiunits with red indicating significant ($p < 0.01$) increase and blue indicating decrease of the discharge rate. Sensory-related neurons and motor-related neurons are marked by arrows; and (b) profiles of reactions with number of excitatory reactions (upward and red) and number of inhibitory reactions (downward and blue). *Adapted with permission from Kropotov et al. (1997).*

and Susan Etlinger from University of Vienna 10 years later. Briefly, according to this theory the programs (representations) of actions are stored in distributed frontal-parietal-temporal cortical networks. The executive functions are defined as operations performed on these representations. They are implemented through the cortical-basal ganglia-thalamo-cortical circuits. The striatum plays a role of action selector through its inhibitory interconnections. The thalamus gates the selected action (Fig. 3.2.13).

It should be stressed here that the activity of the basal ganglia neurons **can't be directly recorded on the scalp** in a form of EEG or ERP for at least two reasons: (1) the nuclei have no laminar structure and can't generate strong external electric fields, and (2) the nuclei are located deep in the brain and can't provide strong potentials at the scalp. So the ERPs measures in the tasks on cognitive control provide **only indirect correlates** of the subcortical processing on the cortical level.

fMRI OF COGNITIVE CONTROL

For studying fMRI correlates of cognitive control different tasks described earlier have been employed. In a simple variant of GO/NOGO paradigm participants are consequently presented with GO and NOGO stimuli on a computer screen. In order to create a strong prepotent response the GO trials are made more frequent than NOGO trials so that

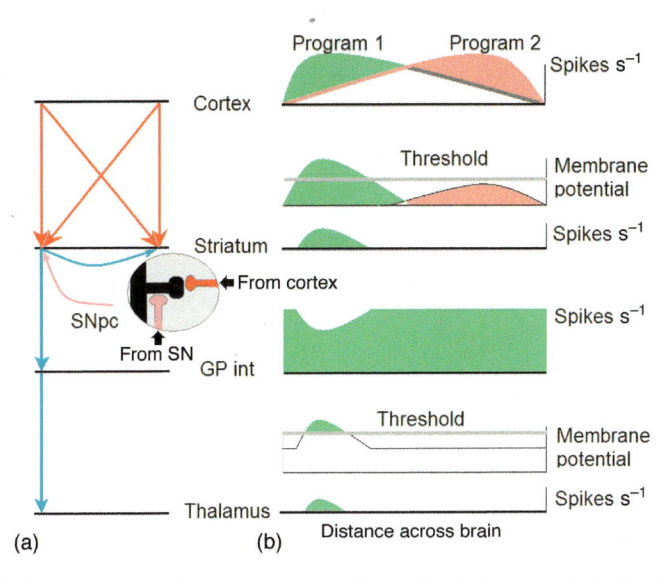

FIGURE 3.2.13 **Action selection in the basal ganglia thalamo-cortical circuit.** (a) Schematic representation of neurons in the cortex, striatum, internal part of globus pallidus (GPi), and thalamus with connections between them: excitatory connections in red, inhibitory connections in blue. In insertion—dendrite of a striatal neuron with a distal synapse from a glutamatergic cortical neuron and proximal synapse from dopaminergic neuron of substantia nigra; and (b) schematic representation of spatial activation patterns at different levels of the circuit. At the cortical level—overlapping representations of two actions (Program 1 and Program 2) are activated. At the striatal level—two overlapping programs are mapped into distinct parts of the striatum. Because of strong lateral inhibition in the striatum, only one of the programs is selected according. At the pallidal level—spontaneously active neurons corresponding to the selected program are inhibited. At the thalamic level the corresponding neurons are disinhibited and gate the thalamocortical pathway for intensifying the selected program of action. X-axis, space. Threshold, a threshold of firing neurons. *Adapted with permission from Kropotov & Etlinger (1999).*

NOGO trials require overriding or inhibition of the prepotent response. In STOP-signal tasks the necessity to override the prepotent action is indicated by presentation of STOP cues. Numerous fMRI studies show that both GO/NOGO and STOP-signal tasks generate similar patterns of BOLD activation.

If the selection of action (GO) and overriding (inhibiting) the prepotent response (NOGO) are opponent operations do they share a common area of activation? This hypothesis was tested in a 2014 study of Charlotte Rae from University of Cambridge, UK and coauthors. The anatomical relationship between selection and stopping of voluntary action was analyzed by two methods: a meta-analysis of previous studies, and by an original fMRI study in a task, in which selection and stopping were combined. The **pre-SMA** was identified by both methods as common area for selection and stopping.

However in majority of fMRI studies the NOGO response is **contrasted** to GO response. In this way a number of brain areas activated during successful response inhibition were identified. The activated brain areas include the lateral and ventrolateral prefrontal cortex, inferior frontal gyrus, inferior parietal lobe, presupplementary motor areas, anterior cingulate cortex, cuneus, and basal ganglia. It should be stressed here that because of the relatively small number of subjects involved in a single study (usually around 20) the results of fMRI studies are often inconsistent. In 2013 a group of researchers from USA reported an fMRI investigation of 102 participants who completed a difficult GO/NOGO task requiring the response inhibition of a prepotent response. The results are presented in Fig. 3.2.14. The whole pattern shows that **widely distributed frontal-posterior-basal ganglia networks** are involved in inhibition of the prepotent response.

ERP CORRELATES OF COGNITIVE CONTROL

Influential discoveries have been made by Japanese scientists Hisae Gemba and Kazuo Sasaki from Kyoto University who in the 1980s in studies with monkeys found that excitation of neurons in the principal sulcus during responses in the GO/NOGO paradigm decreased neuronal activity in the primary motor cortex and eliminated overt responses. ERPs studies in humans commonly find that NOGO stimuli elicit late positive fluctuations peaking around 300 ms with more frontal topographies than P3s elicited by GO stimuli (Fig. 3.2.15). When contrasted with the GO condition the **P3 NOGO wave** is preceded by a frontally distributed negative fluctuation named the **N2d (or N2 NOGO).**

The N2 and P3 NOGO waves were first reported by Simson and coworkers in 1977. These waves are also observed for saccadic responses and for counting. The N2 NOGO wave is increased in subjects who make few inhibition errors, is increased when inhibition is made more difficult, and when participants are required to respond quickly to GO stimuli. On the basis of these facts the early research considered N2 as an index of **action inhibition**. However, later it was shown that the N2 wave could be elicited by infrequent GO stimuli which suggested the **conflict detection** hypothesis of the N2. According to this hypothesis the N2 reflects detection of response conflict elicited between the "press" and "do not press" representations. In line with inconsistency of its functional meaning, the N2 wave has been localized in various cortical areas including the anterior cingulate cortex, the inferior prefrontal and left premotor areas, the medial posterior cortex, and the right lateral orbitofrontal areas.

In a 2011 study by William Randall and Janette Smith from University of Newcastle, Australia subjects were required to inhibit a planned

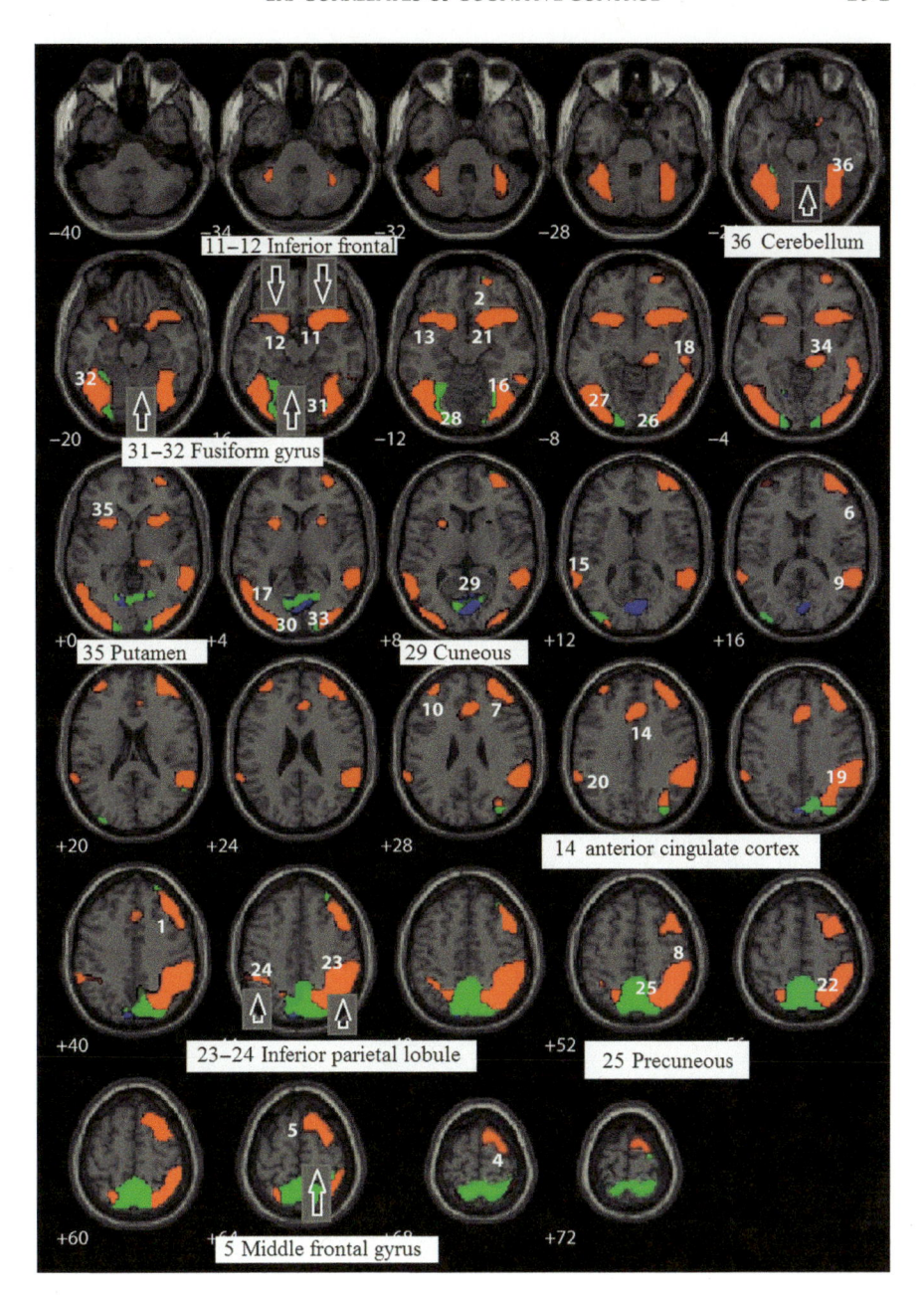

FIGURE 3.2.14 **Activation pattern for NOGO stimuli is contrasted to that for GO stimuli.** Red: NOGO has as a large positive response relative to GO smaller response of the same direction. Green: NOGO has a positive response while GO has a negative response. Blue: NOGO has a small negative response relative to GO large response of the same direction. *Adapted with permission from Steele et al. (2013).*

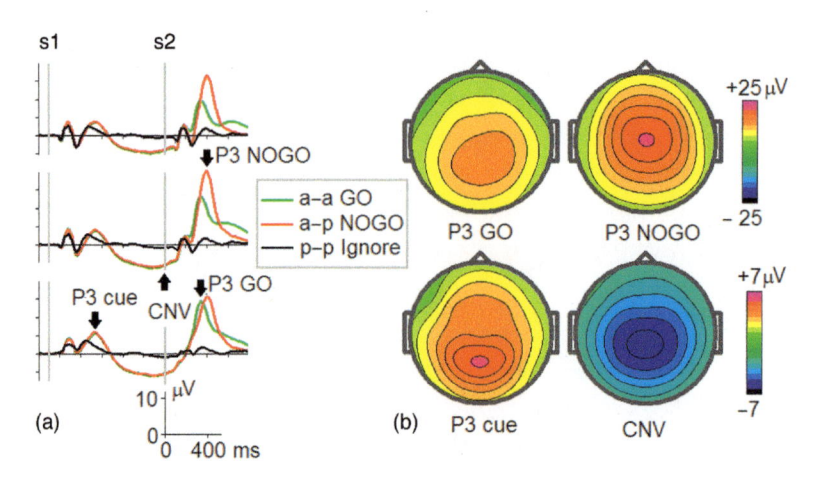

FIGURE 3.2.15 ERPs correlates of cognitive control. (a) Grand average ERPs in referential montage for the group of healthy subjects of age 35–50 in the cued GO/NOGO task. The Cue in contrast to NoCue elicits the P3 cue and CNV waves. The GO and NOGO stimuli in contrast to Ignore condition elicit late positive P300 waves with different amplitudes and latencies; and (b) topographies of P3 cue, CNV, P3 GO, and P3 NOGO waves. Note that P3 NOGO is distributed more frontally in comparison to the parietally distributed P3 GO. *Data from HBI database.*

response (NOGO target after GO cue), change a planned response to a different one (invalid cueing), and activate an unexpected response (GO target after NOGO cue). N2 was more negative whenever the presented target required a different response to what was expected based on the cue. In contrast, P3 was increased when participants had to change or inhibit a planned response, but not when executing a response where none was planned. According to the authors the data support the **conflict hypothesis of the N2 wave** and the **response inhibition hypothesis of the P3 wave**.

INDEPENDENT COMPONENTS OF COGNITIVE CONTROL

As shown in chapter: Event-Related Potentials (ERPs), one way of identifying multiple sources of ERPs is performing group ICA. Recall that group ICA requires a large number of individual ERP to obtain enough training points to compute the robust unmixing matrix. Fig. 3.2.16 shows the results of applying ICA to the NOGO ERP waves recorded in 193 healthy subjects.

One can see that instead of the N2/P3 dichotomy the ICA approach gives another view: the NOGO wave is generated by **two main sources**. One is localized in the supplementary motor area and the other is localized in the anterior cingulate cortex. These components show different latencies and, as we are going to demonstrate later, have different functional meanings.

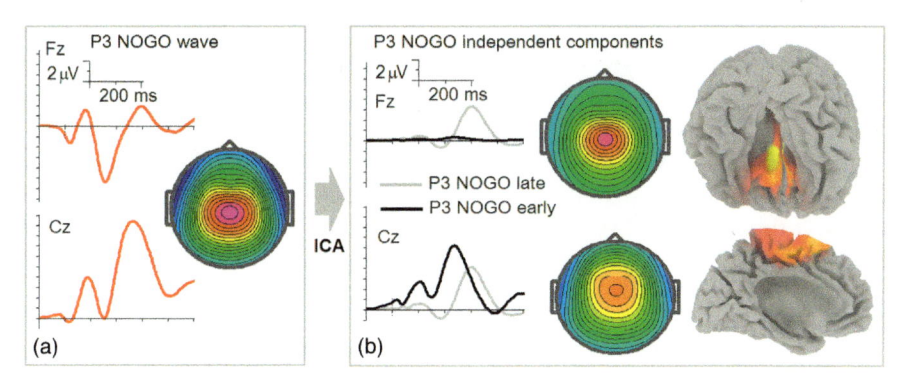

FIGURE 3.2.16 **Decomposition of P3 NOGO wave into two independent components: P3 NOGO late and P3 NOGO early.** Infomax algorithm of independent component analysis (ICA) was performed on 19-channel NOGO ERPs recorded in193 healthy subjects (mean age 24.4 years) during the cued GO/NOGO task. (a) Grand average ERPs at Fz and Cz for NOGO condition; and (b) two independent components (P3 NOGO early and P3 NOGO late) contributing to the P3 NOGO wave. From left to right: time course, topography and sLORETA image. *Adapted with permission from Brunner et al. (2015).*

LESION STUDIES

A unique opportunity to identify the functional meaning of ERP components is given by **stereotactic operations** performed on patients with uncontrollable addiction. A rationale for stereotactic operations is given in the hypothesis of addiction presented in a 2009 book of the author (Kropotov, 2009).

Briefly, three stages of addiction are separated. During the **expectation stage** most prefrontal cortical areas activate the motor cortex to initiate a new and possibly pleasurable action. During the **consolidation stage** the orbito-frontal and anterior cingulate cortical areas receive strong and emotionally meaningful inputs from the limbic system. The representation of the reward is strongly consolidated (imprinted) due to normal mechanisms in the circuit orbito-frontal cortex—nucleus accumbens—globus pallidus—anterior nucleus of the thalamus. After memory of the reward has been imprinted, any cue (either internal, just a subjective recall from memory, or external, associated with the drug) reactivates this critical circuit. At the **habituation/sensitization** state the activity from the orbital-frontal cortex recalled from the memory about the reward (eg, how good it was during the first drug consumption…) is compared with the representation of the real situation (eg, life is boring) in the anterior cingulate cortex. The result of the comparison reflecting the inconsistency activates the basal ganglia thalamo-cortical loop associated with the anterior cingulate cortex. In its turn, activation of the anterior cingulate cortex starts drug-seeking behavior by activation of premotor and motor areas of the cortex. To disrupt this pathological circuit local lesions in the anterior cingulate cortex have been suggested.

At the Institute of the Human Brain in Saint Petersburg a stereotactic **local lesion in the anterior cingulate cortex** for restraining compulsive behavior in heroin drug addicts has been suggested. More than 300 patients with heroin addiction have been treated with a good success. Of these patients, 15 were recorded during the cued GO/NOGO task pre- and postoperation. Fig. 3.2.19 represents the results of the study. One can see that the P3 NOGO late independent component is selectively reduced after lesions in the anterior cingulate cortex. This observation supports the sLORETA localization of the P3 NOGO late component in the anterior cingulate cortex. It also supports the involvement of the anterior cingulate cortex in forming obsessive–compulsive behavior Fig. 3.2.17.

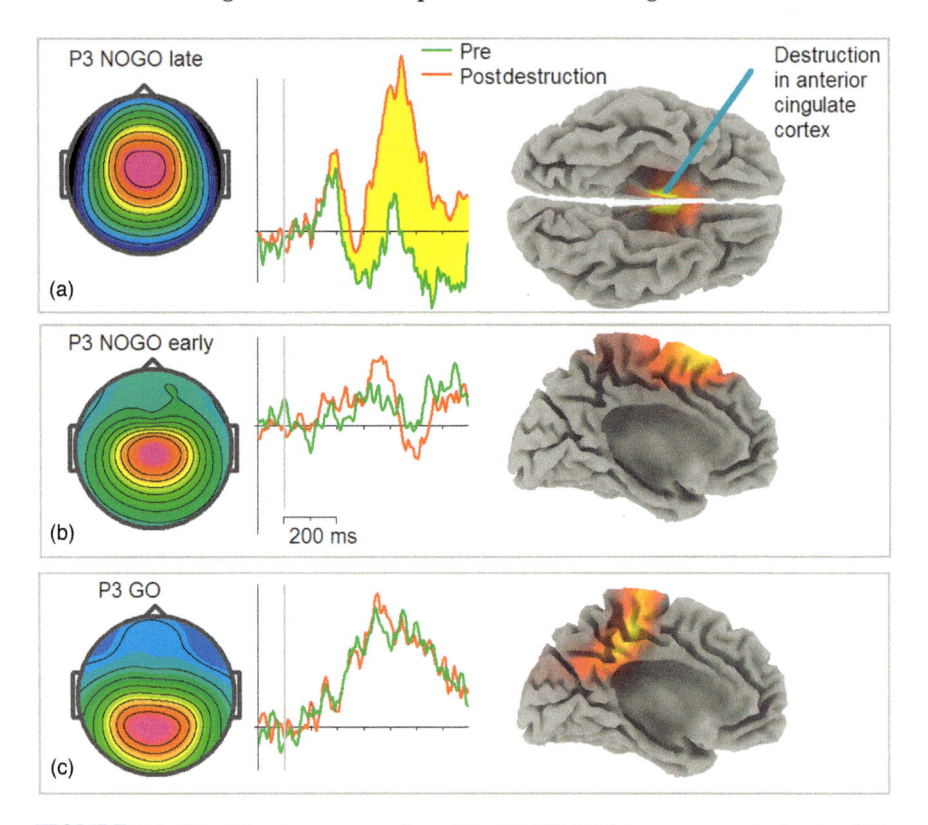

FIGURE 3.2.17 **Selective suppression of the P3 NOGO late component after local bilateral lesions in the anterior cingulate cortex.** Fifteen patents were recorded in the cued GO/NOGO task before (pre) and after (post) stereotactic operations with local destruction of a part of the anterior cingulate cortex. (a) Topography, time course and sLORETA image of the P3 NOGO late independent component. Stereotactively implanted electrode for making destruction is schematically shown; (b) topography, time course and sLORETA image of the P3 NOGO early independent component; and (c) topography, time course and sLORETA image of the P3 GO independent component. *Adapted from Kropotov et al. (2007).*

CORRELATION WITH NEUROPSYCHOLOGICAL PARAMETERS

Neuropsychological studies by Donald Stuss and his coworkers from the Rotman Research Institute (Canada) showed that attentional control can be fractionated into specific processes of energization, monitoring and task setting. **Energization** refers to a process **that facilitates and boosts other processes of cognitive control**, especially those necessary for initiation and maintenance of optimal response patterns. The **monitoring** process is thought to provide quality control of behavior by checking task performance and outcome over time. The **task setting** process refers to the formation of a criterion of how to respond to a defined target, and to organize the schemata to complete a specific task. According to the model by Stuss and coworkers the effect of deficient energization can be observed as **slowing in cued reaction time** tasks and in performance on **fluency tasks**. Monitoring is reflected in *all* types of errors. Recall (chapter: Psychometrics and Neuropsychological Assessment) that the detectability parameter d' is based on both commission and omission errors, indexes how well a subject is able to respond differentially to targets and nontargets, and is sensitive to the monitoring process. The **task setting** process can be assessed behaviorally by measures reflecting poor criterion setting.

According to the model, the behavioral parameters of different neuropsychological tasks can be grouped into the domains of energization, monitoring, and task setting. Jan Brunner from NTNU in Trondheim and his coauthors in a 2015 study correlated the integrated parameters of these three domains with amplitudes of the independent components of cognitive control extracted in the cued GO/NOGO task. The results of the study show that (1) the amplitude of the P3 NOGO early component strongly correlates with the neuropsychological integrated index of energization (Fig. 3.2.18) and does not correlate with the monitoring domain; (2) in contrast the amplitude of the P3 NOGO late component correlates with the neuropsychological integrated index of the monitoring operation and does not correlate with the energization domain (Fig. 3.2.19).

According to the authors, the energization process must be differentiated from the motivation process by describing **motivation** as neural mechanisms **effortlessly** arousing an organism to act toward a desired goal, while energization is the ability to **voluntarily invest** attentional effort to optimize behavior for achieving the goal. The process of action monitoring is needed in all situations where nonroutine actions are performed, and where errors are likely to occur.

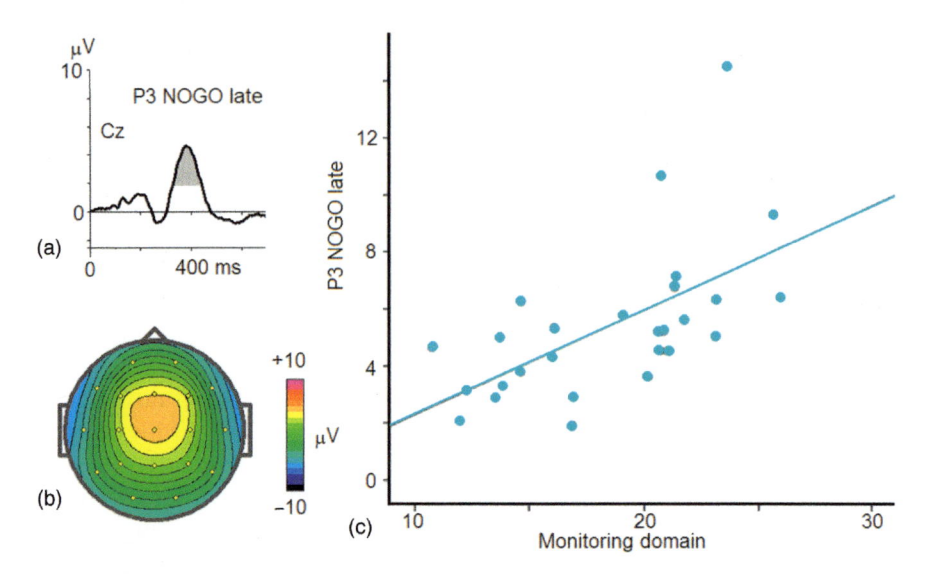

FIGURE 3.2.18 **Component "P3 NOGO early" correlates with neuropsychological domain of energization.** (a) The grand average time course of the component back projected at Cz; (b) topography of the component; and (c) correlationship of the component amplitude with the integrated score of neuropsychological domain of energization. *Adapted with permission from Brunner et al. (2015).*

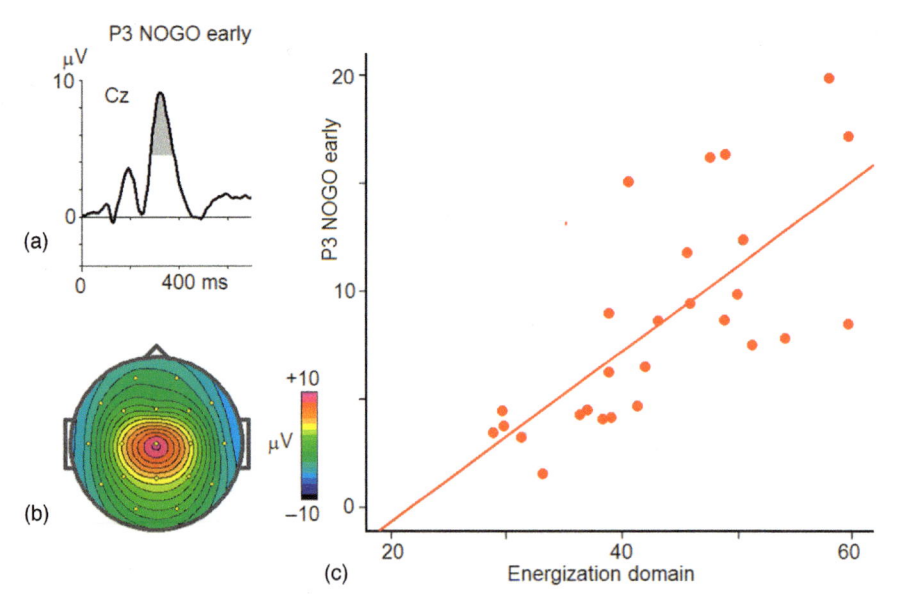

FIGURE 3.2.19 **Component "P3NOGO late" correlates with neuropsychological domain of monitoring.** (a) The grand average time course of the component back projected at Cz; (b) Topography of the component; (c) correlationship of the component amplitude with the integrated score of neuropsychological domain of monitoring. *Adapted with permission from Brunner et al. (2015).*

LATENT ERP COMPONENTS OF REACTIVE COGNITIVE CONTROL

It should be stressed here that the independent component described in the previous section were obtained by ICA applied to a large collection of ERPs in the common reference montage. Similar, but not identical, components are obtained when a different blind source separation approach, such as the joint diagonalization of covariance matrixes (see Chapter 1.6), is applied to a collection of ERPs in the referential montage. The results of decomposition of ERP for GO and NOGO conditions in the cued GO/NOGO task for the group of 454 healthy adult subjects are presented at Fig. 3.2.20. As one can see the ERPs to s2 for GO and NOGO stimuli from Fig. 3.2.15

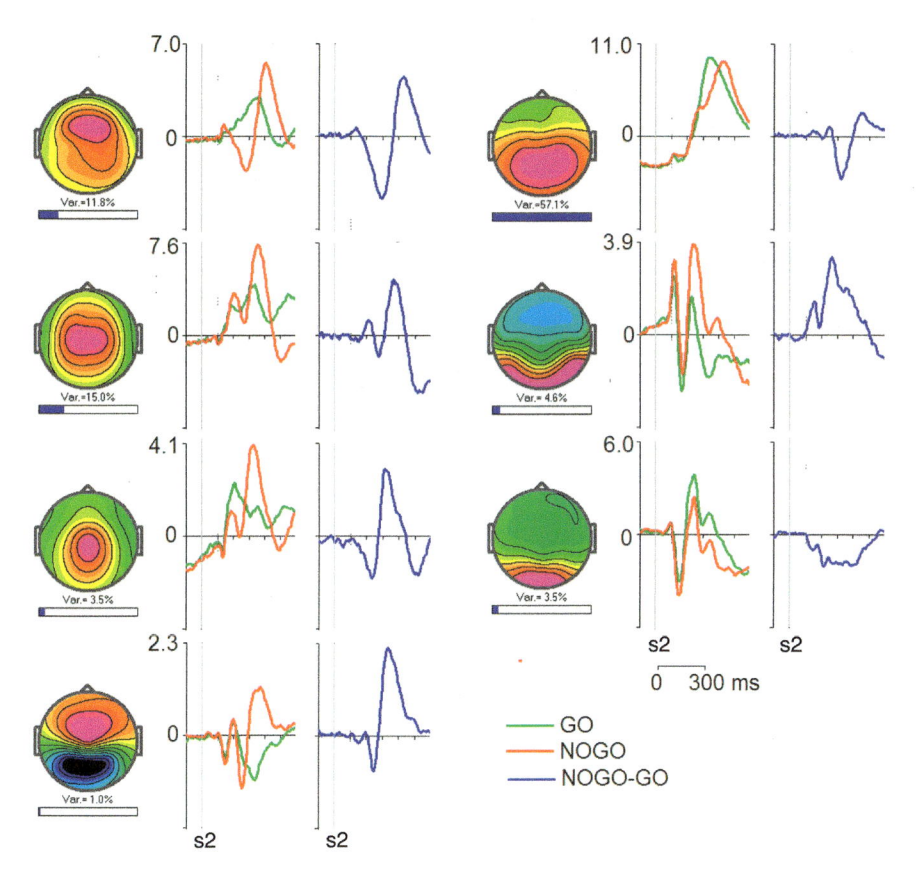

FIGURE 3.2.20 **Latent components for GO and NOGO stimuli.** The joint diagonalization method of blind source separation has been applied to the collection of GO and NOGO ERP in the referential montage in the group of 454 healthy adults. Each component is characterized by (from left to right): (1) topography with the horizontal bar indicating the relative (%) power of the component, (2) time courses for GO and NOGO conditions in standard units, (3) the NOGO-GO difference. The time course is presented for the group of age 24–33. *Data from HBI database.*

in the blind source separation approach are presented as **a sum of seven latent components** with different topographies and time courses.

FUNCTIONAL MEANING OF LATENT COMPONENTS

To define the functional meaning of the extracted components we performed a separate study. Forty healthy subjects aged 24–33 performed 4 tasks that used the same visual images and the same temporal structure as in the conventional cued GO/NOGO task described in Fig. 1.6.10. Recall that in the conventional cued GO/NOGO task two categories of stimuli were used: 20 different images of animals (a) and 20 different images of plants (p). The stimuli were presented in different combinations a–a, a–p, p–p, p–a. The interstimulus interval within pairs was 1 s, and the interval between trials was 3 s. One hundred trials for each category were presented. It should be stressed here that in the conventional cued GO/NOGO task the image of the second animal in a–a pairs was identical to the image of the first animal in the GO trials. In this task the NOGO condition when contrasted to GO condition includes the mismatch operation in the visual domain: the different image in NOGO condition vs. the same image in GO condition. The working hypothesis tested in this study was as follows.

The central concept of cognitive control is the **prepotent model of behavior**. This model is formed by task setting. The model depends on the probabilities of GO and NOGO stimuli and availability of the behavioral program. When the model is formed, two situations are present in cognitive control tasks: the stimulus matches the model or the stimulus mismatches the model. In case of mismatch, there is conflict and the model must be updated. Two hypothetical operations are associated with the mismatch of the prepotent model: **conflict detection and context updating**.

Mismatch can be linked to withholding from response but can also be linked to another nonprepotent response. The mismatch can be linked to withholding from response but can be also linked to another nonprepotant response. In the first situation (withholding from response) a hypothetical operation of **response inhibition** takes place: the operations of response inhibition and conflict detection occur in the same trial. In the second situation (nonprepotent response) the conflict detection occurs without response inhibition.

Together with these two operations of cognitive control, sensory operations take place in the tasks. There are at least two basic sensory-related operations: **discriminating the category** of the stimulus and **comparing the image with the expectation in WM**. The cued GO/NOGO task consists of sequential presentation of two stimuli s1 (cue, sample) and s2 (match or

FUNCTIONAL MEANING OF LATENT COMPONENTS 199

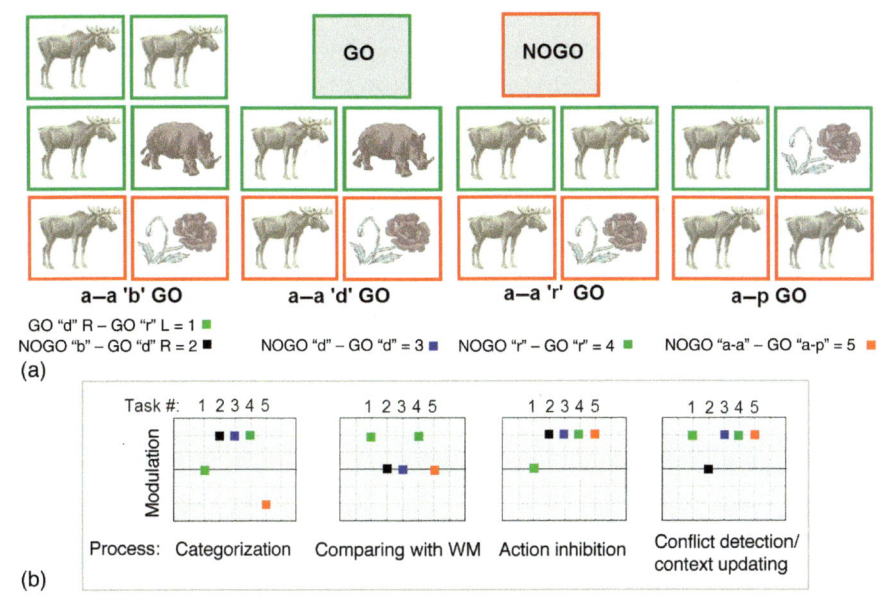

FIGURE 3.2.21 Modulation effects of four hypothetical operations in four variants of the cued GO/NOGO task. (a) Schemes of experiments. Green, GO trials; red, NOGO trials. The task a–a "r" GO—the subject responds to the two identical images of animals ("nonidentical" trials are absent). The task a–a "d" GO—the subject responds to the two nonidentical images of animals ("identical" trials are absent). The task a–a "b" GO—the subject responds to the two identical images of animals with one finger, and to the two nonidentical images of animals with another finger. The task a–p GO—the subject responds to the pair animal-plant. The contrasts used to study modulation effects are shown later. (b) Behavioral pattern of the four operations: (1) categorization, (2) comparing to WM, (3) action inhibition, and (4) conflict detection/context updating. X-axis—the number of task. Y-axis—presence, inversion of sign, absence of the modulation. *To be published from HBI database.*

mismatch with the sample). The subject has to discriminate stimuli and to compare s2 with s1 in order to perform the cued GO/NOGO task.

The **four previously mentioned hypothetical operations** were independently manipulated in the **four tasks** (Fig. 3.2.21). The two categories of images served as stimuli: images of different animals, a; images of different plants, p. The trials consisted of presentations of paired stimuli with interstimulus intervals of 1000 ms and intertrial intervals of 3000 ms. The duration of stimuli was 100 ms. Four categories of trials were used (Fig. 3.2.21): a–a, a–p, p–p, and p–a. The trials starting with the presentation of p at the first position and indicating no further action in the trial will be referred to as discontinue trials. The trials starting with the presentation of "a" at the first position will be referred to as continue trials. The continue trials in turn consisted of two types: GO trials required pressing the button, and NOGO trials required suppressing the prepared action.

3. INFORMATION FLOW WITHIN THE BRAIN

In the task a–a "r" GO the images of animals in a–a pairs were **identical** ("r" stands for repetition) (experiment 4 on Fig. 3.2.21, dark green color). The subject was instructed to press a button to a–a pairs. The NOGO-GO contrast was used. In this contrast all four operations are present.

In the task a–a "d" GO the images of animals in a–a pairs were **different** ("d" stands for different) (experiment 3 on Fig. 3.2.21, blue color). The subject was instructed to press a button to a–a pairs. The NOGO-GO contrast was used. In this contrast, comparison with the WM operation is absent because the sensory-specific trace of the second stimulus could not be formed in the WM.

In the task a–a "b" GO the images of animals in a–a pairs were **both identical** (in 50% of trials) and **nonidentical** ("b" stands for both). The subject was instructed to press a button to a–a identical pairs with one finger, and to a–a nonidentical pairs with another finger. The reaction time to identical a–a pairs was 100 ms faster than to that in response to nonidentical a–a pairs which indicates that the prepotent model was used to respond to identical animals. In the "GO nonidentical–GO identical" contrast the categorization and action inhibition operations are absent while other operations of comparing with WM and conflict detection are present. In the "NOGO–GO nonidentical" contrast the conflict and the comparison with WM operations are absent.

The task a–p GO included the same stimuli as the a–a "r" GO task but the instruction was to **press a button to a–p pairs**. No sensory template could be formed during the preparation and, consequently, no comparison with WM could take place. The categorization effect changes the sign.

Fig. 3.2.22 demonstrates the two components associated with operations of **categorization and comparison with WM**. According to topographies the components have an occipital-temporal distribution. The early time window of both components is associated with the operation of categorization. Categorization-related differences start as early as 80 ms and peak at 110 ms for the occipital component and are delayed for about 20 ms for the temporal component. The differences associated with the comparison with WM peak at around 240 ms.

Fig. 3.2.23 demonstrates the three components associated with operations of **action inhibition** and comparison to WM. According to topographies the components have near-central distribution. The peak differences for action inhibition are around 300 ms. They are preceded by activation patterns associated with comparison to WM.

Finally, Fig. 3.2.24 demonstrates the two components associated with the operations of conflict detection and context updating. According to topographies the components have a parietal-frontal distribution. The frontal component reacts faster than the parietal component with peak latency around 250 ms. The results show that the hypothetical operations are reflected in distinct, but partly overlapping, time windows and in

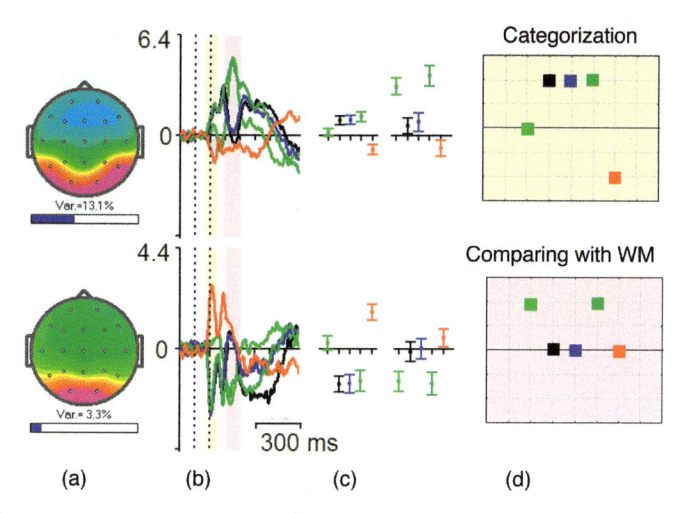

FIGURE 3.2.22 **The sensory related latent components in the GO/NOGO tasks.** (a) Topography of component; (b) the component differences for the contrasts from Fig. 3.2.16; (c) mean values and 95% confidence intervals computed for the two time windows (left, the first time window; right, the second time window) for the contrasts; and (d) behavioral patterns of the two operations: categorization and comparing to WM. The color corresponds to the time interval at (b) in which the behavioral pattern is observed. *To be published from HBI database.*

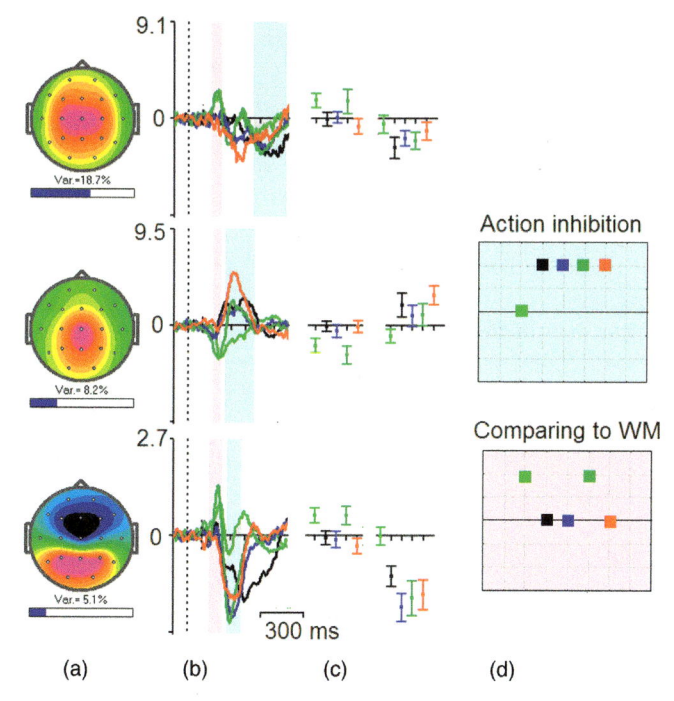

FIGURE 3.2.23 **The action inhibition components.** Other legend as in Fig. 3.2.17. *To be published from HBI database.*

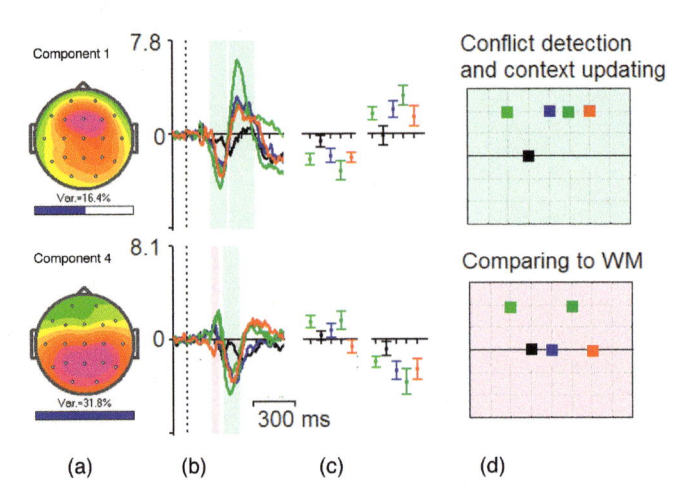

FIGURE 3.2.24 **The conflict detection and context updating components.** Other legend as in Fig. 3.2.17. *To be published from HBI database.*

distinct, but overlapping cortical areas, whereas a separate latent component is associated with at least two hypothetical operations. As will be shown in Part 5 the latent components can serve neuromarkers not only for normal hypothetical psychological operations but also could be used for discriminating different categories of psychiatric patients with impairments in cognitive control.

TARGET P3 (P3b) IN ODDBALL TASKS

Samuel Sutton and coworkers from New York State Psychiatric Institute were the first to discover the P3 (P300) wave in 1964 (published in 1965). They presented subjects with either a cue that indicated whether the following stimulus would be a click or a flash, or a cue which required subjects to guess whether the following stimulus would be a click or a flash. They found that when subjects were required to guess what the following stimulus would be, the amplitude of the "late positive complex" was larger than when they knew what the stimulus would be. In many clinical applications the oddball paradigm is used to elicit P3.

Rare targets in oddball tasks elicit a **late positive potential peaked around 300 ms and are distributed posterior** independently on modality. To discriminate it from other late positive waves this parietally distributed wave was labeled **P3b**. In addition to the P3b wave the targets also elicit sensory-related potentials. For example, in the auditory modality the duration oddball task elicits the sensory related waves such as N1, P2, MMN, and P2 described earlier. The components can be separated by group ICA.

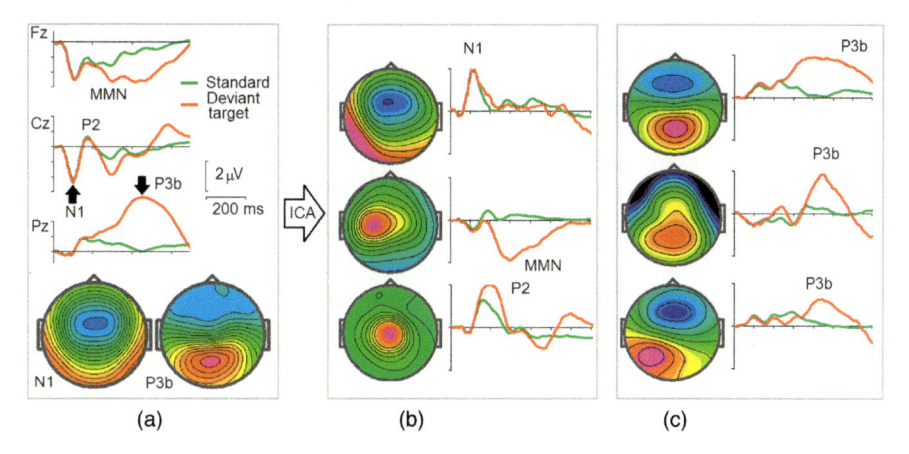

FIGURE 3.2.25 **Target P3 in the auditory oddball task.** (a) Grand average ERP (N = 244, age = 17–45) for standard (P = 0.8, 100 ms) and deviant (P = 0.2, 400 ms) tones in the active oddball task. The group ICA was applied to the collection of ERPs and resulted in independent components presented in (b) and (c); (b) the sensory-related independent components (only left side components are presented for simplicity). The components include the N1, MMN, and P2 components. Note that N1 component does not depend on the task condition, while the MMN component is associated the centrally distributed components; and (c) the P3b related independent components. *Data from HBI database.*

The results of application of group ICA to the large collection of ERPs in the duration oddball task are presented in Fig. 3.2.25. In this task, easily discriminated tones of four different frequencies are used as stimuli. Each tone is presented either with a 100-ms duration (short tones) or with a 400-ms duration (long tones). These eight stimuli are sequentially presented in pseudo-random order so that the proportion of long tones (deviants) is about 20% in comparison with short tones (standards). Stimulus onset synchrony is 1100 ms. The total number of stimuli is 994. One can see that the sensory-related components include N1, MMN, and P2 independent components (only the left-side components are shown in Fig. 3.2.25). The P3b wave is decomposed into three components with different latencies and distributions. The results clearly show the **heterogeneity of the P3b wave**.

The factors affecting the P3b in the oddball paradigm are: subjective stimulus probability, motivational significance, amount of attention paid to the stimulus, the pressure to respond fast, etc. According to Emmanuel Donchin from the University of Illinois the target P300 is a manifestation of activity occurring whenever one's model of the environment must be revised. This **context updating hypothesis** of P3b was reviewed in his 1980 presidential address published in Psychophysiology on 1981. Rolf Verleger from the University of Luebeck in his 1988 critique paper on updating hypothesis suggested that P3b was a correlate of the decision about **what to do with the stimulus**. John Polich from the Scripps Research Institute in California in his 2006 review paper suggests that P3b originates from temporal-parietal activity associated with attention and appears related to **subsequent memory processing**. He further suggests that the P300 and its underlying subprocesses could reflect **rapid neural inhibition of ongoing activity** to facilitate transmission of stimulus/task information from frontal (P3a) to temporal-parietal (P3b). Our data show that P3b is a heterogeneous wave consisting of at least three subcomponents which might be associated with different hypothetical operations.

P3b must be separated from the P3a elicited to novel stimuli in passive three-stimulus tasks and described in chapter: Sensory Systems and Attention Modulation. Note however that the topography of one of the independent components of P3b (middle at Fig. 3.2.25c) coincides with topography of the independent component of P3a localized in the hippocampal area (Fig. 3.1.21 middle of the Novelty P3 components). This fact shows that responses to Novel and Target stimuli share common brain circuits located in the hippocampus of the brain.

P3b AND NORADRENALINE

Recent studies have highlighted the critical role of the noradrenergic (NE) modulatory brain system in regulating operations of engagement in the task. Recall that the cortical noradrenaline is produced by the locus coeruleus (LC) neurons which are located in the dorsal pons of the brain. These neurons innervate widely distributed areas of the cerebral cortex with the highest density over the parietal and motor cortical areas. Noradrenaline neurons in the locus coeruleus are silent during rapid eye movement sleep and have low tonic firing during slow wave sleep, moderate tonic firing and pronounced phasic firing to relevant stimuli during nonstressed waking, and high tonic firing with dysregulated phasic firing during stress.

Based primarily on intracranial recordings from animals, the **adaptive-gain theory of LC–NE function** was introduced in 2004 by Gary Aston–Jones from University of Pennsylvania and Jonathan Cohen from Princeton University. The authors show that relative levels of tonic and phasic LC activity relate to task performance in a manner that reflects the classic inverted U-law: performance and phasic LC responding are optimal at an intermediate level of tonic LC activity, but shifts toward either end of the tonic activity continuum are associated with declining performance and attenuated phasic responses.

The importance of LC in action engagement has been highlighted by pharmacological and genetic studies that corroborate the role of NE as a critical determinant of **engagement** and task performance on tests of attention. In clinical practice, no direct access to LC neurons is available so far. The **P3b** thus could be an important neuromarker of **engagement operation**.

CORTICAL DOPAMINE AND WORKING MEMORY

There is a clear **anterior/posterior gradient** in the cortex for the concentration of DA where it is the highest in the prefrontal cortex. Recall that the ventral tegmental area of the midbrain is the main source of DA in the cortex with meso-cortical projections to the prefrontal cortex/anterior cingulate cortex and with meso-limbic projections to the amygdala and hippocampus. The sustained DA activity in the PFC plays key role in modulating **working memory**—the stable aspect of cognitive control.

In monkey experiments by Tomas Brozoski from Southern Illinois University School of Medicine and coworkers in 1979 the DA depletion in the PFC caused severe impairment of behavior in the **delayed response task**—a test on working memory. D1 receptor agonists in the same monkeys reversed their working memory impairment.

However, recent studies in humans show that the relationship between brain DA and task performance is highly complex. The effects of dopaminergic drugs often seem paradoxical, as both improvements as well as impairments are observed. The paradoxical effects are observed across different individuals who perform the same task or within the same individual across different tasks. These paradoxical effects of DA on behavior are explained by **the inverted U-law of performance dependence on DA level**. The law shows that the effect of DA change on the behavior depends very much on the initial DA level. When the initial level is high the increase of cortical DA leads to decrease in performance. In the same time if the initial level is low the effect would be quite opposite.

STRIATAL DOPAMINE AS REGULATOR OF FLEXIBILITY

Recent studies show that functional role of DA expressed by **D2 receptors in the striatum** might be qualitatively different from that of DA expressed by D1 receptors in the prefrontal cortex. For example, lesions in the prefrontal cortex lead to enhanced distractibility (poor attentional set maintenance), while DA lesions in the striatum reduce distractibility (enhanced attentional set maintenance). It should be stressed that the origin of DA in the striatum is qualitatively different from that in the prefrontal cortex: DA in the striatum is delivered from the substantia nigra and binds to D2 receptors. The DA in the striatum appears to modulate operations of cognitive control such as selection, shifting and inhibition—operations that ensure **the flexibility of the cognitive control**.

CHAPTER

3.3

Affective System, Emotions, and Stress

INTRODUCTION

The concept of "affective neuroscience" was coined by Jaak Panksepp from Bowling Green State University in 1992 to emphasize the role of neuroimaging in studies of emotions. From this perspective an affective system is a network **of cortical and subcortical anatomical structures** that are interconnected with each other to provide a subject with ability to generate and to feel emotional reactions in response to internal or external stimuli.

The system includes (1) structures that **map** sensory stimuli into categories of emotionally meaningful **rewards and punishments** in contrast with **neutral** stimuli, (2) structures that **express emotions** in response to those emotionally meaningful stimuli, and (3) structures that are responsible for **feelings** of those emotional reactions and states. These anatomical structures are schematically presented in Fig. 3.3.1a. Fig. 3.3.1b shows information flow in these structures initiated by presentation of emotionally competent stimulus in the visual modality. The pathway includes (1) the **ultrafast pathway to the amygdala** through the superior collicular (SC) and pulvinar (PUL) nuclei, (2) the **fast pathway to the amygdala, hypothalamus, and orbitofrontal cortex** through the lateral geniculate body, the ventral visual stream, and hippocampus, and (3) the attention-controlled pathways from the amygdala, hypothalamus, and orbitofrontal cortex to the ventral cingulum. Note that although most connections are reciprocal only feedforward connections are depicted for simplicity.

Historically, the scientific approach to the affective system of the brain started in 1937 when **James Papez** from the University of Minnesota College of Medicine and Surgery suggested that the limbic lobe works in cooperation with the hypothalamus which in those days was

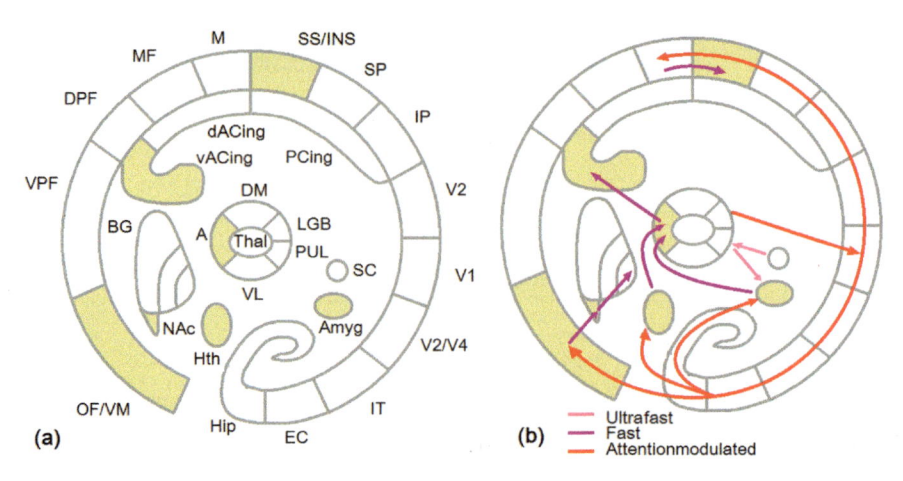

FIGURE 3.3.1 Anatomy and information flow in the affective system. (a) Anatomical nodes of the affective system are the amygdala *(Amyg)*, anterior hypothalamus *(Hth)*, nucleus accumbens *(NAc)* and anterior nucleus of the thalamus, orbitofrontal cortex/ventromedial *(OF/VM)* prefrontal cortex, the ventral part of anterior cingulate cortex *(vACing)*, insular cortex *(INS)*, and somatosensory *(SS)* cortical areas; and (b) information flow within the ultrafast pathway, fast pathway, and attention-controlled pathway.

considered to have a critical role in the expression of emotions. According to Papez the hypothalamus and the cingulate cortex form a reciprocal circuit that, on the one hand, allows emotions triggered by the hypothalamus to reach consciousness at the level of the cingulate cortex and, on the other hand, enables higher cognitive functions from the cingulate cortex to affect emotions. This hypothalamic–cingulum circuit was later coined the **Papez circuit**. Later on, in a series of publications beginning in 1949, Paul MacLean from Harvard Medical School/Massachusetts General Hospital extended the concept of the Papez circuit by adding into it some other anatomically related structures such as the **amygdala**, the septal area, the **nucleus accumbens**, and the **orbitofrontal cortex**. According to this revised scheme: (1) the mammillary bodies of the **hypothalamus** send signals to the **cingulate cortex** through the anterior thalamic nuclei by means of the fiber bundle called the "mammillothalamic tract," (2) in its turn, the hypothalamus receives input from the hippocampus via a bundle of axons called the "fornix," while (3) the **hippocampus** receives projections from the anterior cingulate cortex, thus closing the circle.

Modern anatomical and physiological research shows that from the functional point of view the limbic system **cannot be considered a single entity**, but rather must be separated into different systems playing different functional roles associated not only with emotions/motivations, but with episodic memory and executive functions.

EMOTIONS AS A SEPARATE DIMENSION

Emotions play an important role in our life. The affective system of the brain is designed by nature for survival in the world by using a **dimension different from cognition**. When reasoning is not enough for making a decision, the brain relies on the affective system. For example, we seldom rely on our reasoning when we choose a partner, when we watch a football game played by our favorite team, or when we play with the family dog. Another example of affective system functioning is **releasing fundamental actions** such as seeking for food, water, or a sexual partner.

The affective system adds a new dimension (emotions/motivations) to our perceptions and actions. This new feature regulates the whole behavior by seeking positive emotions and avoiding negative emotions. From the information point of view, the affective system is designed to **enhance the representations of events** that are crucial for human survival in association with the emotional meaning of events.

EMOTIONS AS HABITUAL RESPONSES

The affective system produces certain **habitual responses**, called "emotional reactions," to certain circumstances. These responses are produced when the brain detects an **emotionally competent stimulus**. This concept was introduced by Antonio Damasio from the University of Iowa College of Medicine. A popular version of his work is presented in his 1999 book *The Feeling of What Happens*. These "habits of mind" (routines of the affective system) enable us to associate those emotions (and feelings that follow them) with corresponding experiences. The brain is prepared by evolution to respond to certain emotionally competent stimuli with specific repertoires of actions. However, while some emotionally competent stimuli are evolutionarily determined, some can be learned through experience. So, an emotion is a complex collection of chemical and neuronal responses to certain external and internal events forming a distinctive pattern and reflected in a certain behavior reaction (including facial expressions). These body responses, in general, and facial expressions, in particular, can be perceived by other people and simulate a corresponding reaction in them by a specific brain system consisting of **mirror neurons**. Emotions make us human. All arts (music, painting, literature, etc.) are about emotions.

CLASSIFICATION OF EMOTIONS

As mentioned above, emotionally competent stimuli are divided into two basic classes: **punishers and rewards** (Fig. 3.3.2). They in turn can be divided in smaller entities. Antonio Damasio classifies emotions into

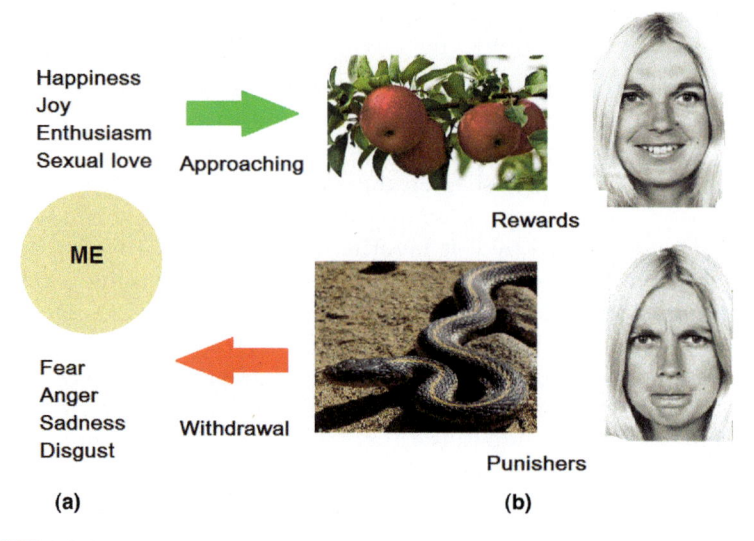

FIGURE 3.3.2 **Emotional stimuli and emotions.** (a) Examples of positive and negative emotions, and (b) examples of rewards and punishers.

three tiers: (1) background emotions, (2) primary emotions, and (3) social emotions.

Background emotions are discouragement and enthusiasm. Primary (basic) emotions include fear, anger, disgust, sadness, surprise, and happiness. These emotions are easily identifiable in human beings across cultures and in nonhuman species. Most of what we know about the neurobiology of emotions comes from studies of **basic emotions**. **Fear** leads the way.

In the functioning of the affective system, emotions and feelings must be separated from motivations and drives. Drives and motivations form **affective states or moods** which must be distinguished from emotional responses. Major examples of drives and motivations are hunger, thirst, curiosity, attraction to the opposite sex. Motivations regulate our behavior by several mechanisms. A major role in motivational behavior is played by the **hypothalamus**.

Neurons in the hypothalamus are very sensitive to different types of **biochemical substances** and reflect changes in body milieu such as blood and cerebrospinal fluid. These hypothalamic neurons in turn regulate the release of various hormones in the pituitary gland (which is attached to the hypothalamus by a stalk) and act on other parts of the affective system by releasing certain types of regulatory behaviors such as drinking, eating. Humans are driven not only by simple motivations to eat and to drink, but also by more complex and sometimes quite abstract drivers. Plans of those drivers reside in the prefrontal areas.

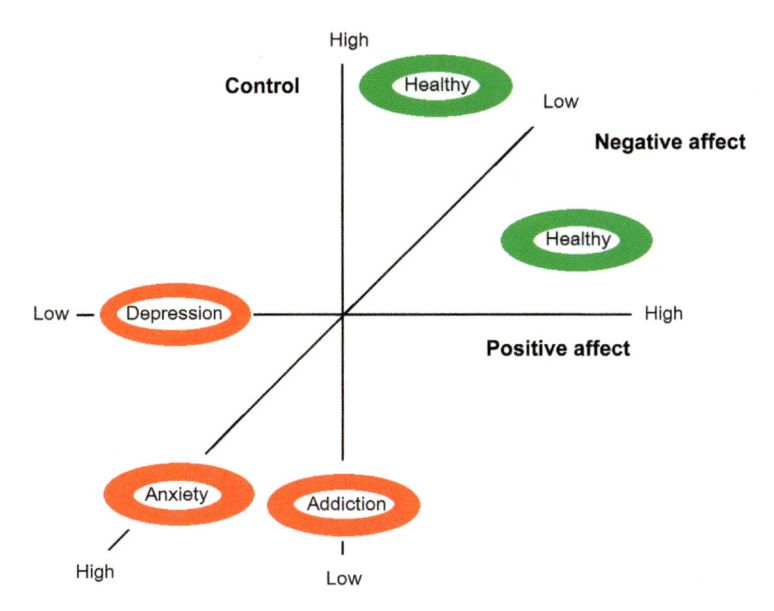

FIGURE 3.3.3 **Three dimensions of temperament.** Three characteristics (positive affect, negative affect, and monitoring) are schematically presented as a 3D space. A neutral state or emotional reaction corresponds to the origin of coordinates. Depression, anxiety, and addiction are represented in this space (*red*).

THREE DIMENSIONS OF TEMPERAMENT

Rewards and reinforcers evoke **positive emotional reactions** which constitute one dimension of the affective system function. Punishers and threats evoke **negative emotional reactions** which constitute another dimension of the affective system function. These two dimensions form the psychological basis of **temperament** (Fig. 3.3.3). In this context the concept of temperament is an attribute of the affective system, while the concept of personality traits is a broader term and is associated with the combined attributes of all brain systems including the affective, executive, attentional, sensory, and memory. However, if we consider temperament in a more broader sense as reflecting not only emotional, motivational tendencies but also the **control/self-regulation** of these affective states, we have to include another dimension: control of emotions (Fig. 3.3.3).

Individuals high on **positive affect** are biased to experience a wide range of positive emotions such as joy, happiness, enthusiasm, and pride and exhibit an active and positive engagement in social and family behavior. Some types of **depression** are associated with low levels in this scale. Individuals high on **negative affect** are biased to experience a wide range of negative moods such as fear, anxiety, sadness, and guilt. They are prone to avoid social communication and inhibit new and unknown endeavors. **Anxiety**, as a psychiatric disorder, is associated with high values in this dimension.

Individuals high on the **control** dimension can effectively control their emotions and affective impulses, they have the ability to wait (and sometimes for a long time) for gratification and reward. Individuals high on this dimension may be described as hard-working, persistent, reliable, and responsible. **Addiction**, as a psychiatric disorder, is associated with a low level in this control dimension.

BRAIN MODEL

On the basis of research on neurophysiological correlates of positive affect, negative affect, and control, Sarah Whittle and her coworkers from the University of Melbourne, Victoria (Australia) in a 2006 study constructed **a brain model of temperament**.

As mentioned above **negative affectivity (NA)** reflects **inhibition, avoidance and punishment** sensitivity. Individuals high on NA have a propensity to experience a wide range of negative moods such as fear, anxiety, sadness, and guilt. Whittle and coworkers showed that the **amygdala and ventral anterior cingulated cortex (ACC)** (involved in the rapid appraisal of affective material) and structures in the **right hemisphere** including the **hippocampus, dorsal ACC, and dorsolateral prefrontal cortex (DLPFC)** (involved in executive processes and the effortful rather than automatic regulation of affective states) are crucial to NA (Fig. 3.3.4a).

Individuals high on positive affectivity (PA) (Fig. 3.3.4b) have an active engagement in the world through **high approach and reward sensitivity** and have a propensity to experience a wide range of positive moods such as joy, happiness, enthusiasm, and pride. The authors of the model demonstrate that a similar, but **left lateralized circuit** linking subcortical limbic structures and dorsal prefrontal structures underlies PA, with reduced involvement of the hippocampus and ventral ACC. The amygdala is suggested to be connected to the NAc in a serial manner, whereby it provides the NAc with information about the associations between discrete stimuli and reinforcements. NAc projections into left prefrontal areas including the dorsal ACC and DLPFC are suggested to promote conscious feelings of pleasure, and influence the drive to engage in pleasurable acts. In return, these prefrontal regions regulate core positive affective processing occurring in the NAc and amygdala via downward projections.

Constraint refers to an individual's degree of control over impulses and emotions, their ability to direct attention and delay gratification. Individuals high on this dimension may be described as **diligent, persistent, reliable, and responsible**. Constraint is a nonaffective factor

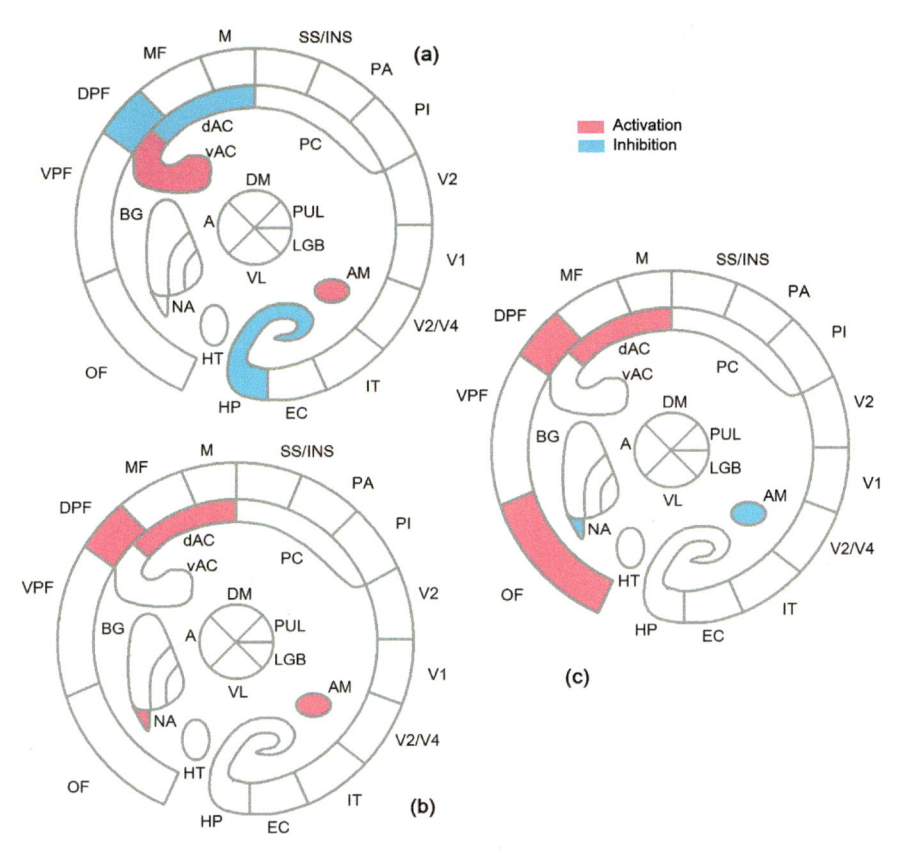

FIGURE 3.3.4 Activation/inhibition brain pattern of three temperamental dimensions. (a) High negative affectivity, (b) high positive affectivity, and (c) high constraint. Note the right DLPFC is particularly involved in NA, the left DLPFC is particularly involved in PA, and right prefrontal structures (DLPFC and dorsal ACC) are particularly involved in constraint. *Adapted with permission from Whittle, Allen, Lubman, & Yucel (2006).*

emerging later in development but interacts with NA and PA by the amount of control one has over approach and withdrawal tendencies. According to the model the amygdala, NAc, dorsal ACC, DLPFC, and orbitofrontal cortex (OFC) are key regions in the constraint circuitry (Fig. 3.3.4c). The dorsal ACC, DLPFC (particularly right), and OFC (lateral sector specifically) have all been implicated in aspects of behavioral inhibition and cognitive control. It is suggested that constraint is positively associated with the degree of inhibitory control that the dorsal ACC, DLPFC, and lateral OFC have over the amygdala and NAc. With a reduction in this inhibitory control, behaviors may be motivated by immediate signals of reward and punishment that are encoded in the amygdala and NA.

MODEL OF LEFT–RIGHT ASYMMETRY IN EMOTIONS

In the Whittle et al. model described above, the left hemisphere is supposed to be involved in positive affect and the right hemisphere in negative affect. The **asymmetric** involvement of prefrontal cortical regions in positive affect (approaching) and negative affect (withdrawal) was suggested as early as the 1930s by observations of patients with brain lesions: patients with left-hemisphere damage showed depressive symptoms, whereas patients with right-hemisphere damage showed manic symptoms. In the 1960s the use of the Wada test supported this suggestion: injecting amytal (Amobarbital) and suppressing activity in the left hemisphere produced depressed affect, whereas injections in the right hemisphere produced euphoria.

In line with the asymmetrical presentation of negative and positive affect, **transcranial magnetic stimulation (TMS)** of the **left prefrontal cortex** for 3–6 weeks has antidepressant effects that are significantly greater than placebo. Remission outcomes are as robust as antidepressant medication. TMS for depression was approved by FDA in 2008.

Left–right frontal asymmetry in human studies can be assessed in EEG spectra by computing an **anterior alpha asymmetry index**. The idea of the index is based on the fact that EEG power in the alpha band is inversely correlated to metabolic activation of the corresponding region of the brain. In a 1993 study of **Richard Davidson** from the University of Wisconsin, USA frontal alpha asymmetry was found to be relatively stable over several weeks and months, enabling him to suggest that this measurement reflects a trait-like tendency to respond differentially to positive (ie, approach-related) and negative (ie, withdrawal-related) stimuli. Davidson and his collaborators conducted several studies for providing support for this suggestion. They recorded baseline EEG from normal subjects, who then viewed film clips designed to elicit either amusement and happiness, or fear and disgust. They found that frontal EEG asymmetry was a stable characteristic of individuals and predicted positive affective responsivity, negative affective responsivity, and affective valence following presentation of the film clips. Asymmetry, however, was **unrelated to baseline mood**.

In a 2009 study by Barbara Schmidt and Simon Hanslmayr from Regensburg University (Germany), **frontal alpha asymmetry** of resting EEG was measured in 16 participants to predict **affective responses** to musical stimuli. Three affective musical stimuli either expressing neutral, positive, or negative mood were evaluated by the subjects concerning "expressed mood" and "level of enjoyment." The results showed that individuals with relatively higher alpha power over right frontal electrode sites rated all stimuli more positive than participants with relatively higher alpha power over left frontal electrode sites (Fig. 3.3.5b).

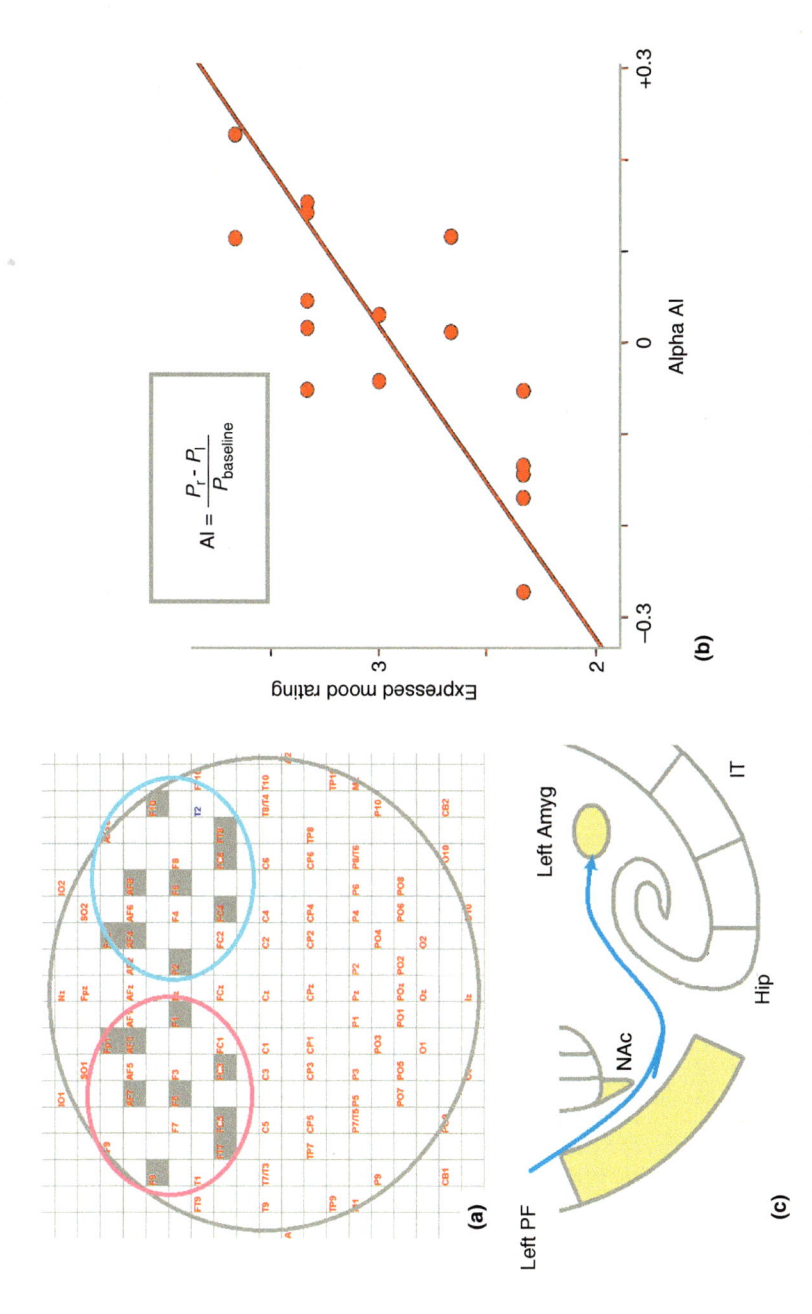

FIGURE 3.3.5 **Left prefrontal cortex (*PF*) modulates emotional response by inhibiting the amygdala (*Amyg*).** (a) Electrode pools for computing the frontal asymmetry index (*AI*), (b) correlation between frontal alpha AI and mean expressed mood rating ($r = 0.80$, $p < 0.001$), and (c) scheme of inhibitory effect of the left prefrontal cortex on the amygdala. *Adapted with permission from Schmidt & Hanslmayr (2009).*

Further, Daren Jackson (a pupil of Davidson) and his coworkers from the University of Wisconsin showed in a 2004 study that subjects with greater activation of the left prefrontal cortex recovered more quickly from the feelings of disgust, horror, anger, and fear evoked by images presented to these subjects. Davidson suggested that activity in the left prefrontal cortex shortens the period of amygdala activation allowing the brain to recover from upsetting experiences—an important property called **resilience** (see inhibitory connections from the left prefrontal cortex to the amygdala at Fig. 3.3.5c).

It should be stressed here that the brain is plastic and is able to "activate" some areas and to "inhibit" others. **Neurofeedback** is one of the neuromodulatory techniques that can be used for this purpose. To test whether individual differences in asymmetric frontal alpha are causally involved in the production of the affective response, John Allen and coworkers from the University of Wisconsin in 2001 used neurofeedback training to manipulate asymmetric frontal cortical activity. Five consecutive days of neurofeedback training provided signals of reward or nonreward depending on whether the difference between right and left frontal alpha power exceeded a criterion value in the specified direction. Subjects were able to voluntarily change the neurofeedback index. Moreover, subsequent self-reported affect of the subjects in response to emotionally evocative film clips was influenced by the direction of neurofeedback training. This kind of neurofeedback was further used in clinical practice for reducing symptoms of depression.

BIG FIVE MODEL

The correlation between psychopathology and personality traits (temperament) is well documented at the behavioral level. At the neurobiological level, research into temperament is still in its **infancy**. Nevertheless, accumulating evidence enables us to start modeling the brain foundations of personality traits.

Neuroscience of personality traits might be key to our understanding of the biological basis of affective disorders. **Personality trait** is a psychological concept that can be defined as a qualitative relatively stable description of a set of unique behavioral patterns of a person in association with his/her internal motivations and emotions, drivers, thoughts, etc. **Heritability** estimates for personality traits are only around **50%**, indicating that personality itself is determined by a complex interaction between the genome and the environment. Personality traits are **invariant** across European, North American, and East Asian samples, suggesting biological universality of these traits.

In the 20th century, psychological studies were mostly concerned with classifying personality traits rather than with explaining their brain basis.

In one of the first studies (1934) by Louis Leon Thurstone from the University of Chicago, 60 adjectives that are in common use for describing human personalities were given to each of 1300 raters. Each rater was asked to describe a familiar person in terms taken from the list. Correlation coefficients for the 60 personality traits were analyzed by multiple factor methods. Thurstone selected five relatively independent factors which could describe the whole complexity of human personality.

Whereas Thurstone found the correct number of personality factors, the honor of the first discovery of what now is known as the **Big Five** (factors) belongs to Donald Fiske from University of Chicago (1949) who analyzed a set of 22 descriptors and found 5 factors replicated across self-ratings, observer ratings, and peer ratings. These factors are now known as **extraversion, neuroticism, agreeableness, conscientiousness, and openness/intellect**.

EYSENCK'S AND GRAY'S MODELS

Although during the last decade neuroimaging and neurochemical correlates of the Big Five were obtained, the Big Five model itself is not based on brain functioning. In contrast, the models of **Hans Eysenck** (1967) and **Jeffrey Gray** (1982) **from the Institute of Psychiatry, Psychology and Neuroscience in London** were derived on the basis of what was known about the brain in those days.

Eysenck assigned personal traits to three factors: **extraversion, neuroticism, and psychoticism** which are almost identical to the corresponding factors of the Big Five with psychoticism roughly reflecting a combination of low conscientiousness and low agreeableness. In his model, Eysenck relied on functions of the **ascending reticular activating system** of the brain by associating extraversion with the reticulocortical circuit and neuroticism with the reticulolimbic circuit. He hypothesized that extraverts in contrast with introverts have lower levels of general cortical activation (arousal) and therefore choose more arousing activities in order to achieve the optimal level of arousal. He also hypothesized that the limbic system of **neurotics** in comparison with emotionally stable people is more easily activated by emotionally competent stimuli. He also hypothesized that psychoticism might be negatively correlated with serotonergic function and positively correlated with dopaminergic function.

Jeffrey Gray, who was a student of Eysenck, focused on functional (defined at the psychological level) systems that could be mapped onto the brain systems (established from animal research) **controlling arousal, reward, behavioral inhibition, and the fight/flight response**. The main components of his model are the behavioral inhibition system, the behavioral approach system, and the fight–flight–freezing system.

The **behavioral inhibition system** is activated by stimuli that one needs or desires to approach but that also contain potential threat thus creating an approach–avoidance conflict. An example of such stimuli is provided by novel, unexpected signals inducing the so-called "orienting response." These stimuli inhibit ongoing activity, orient attention, and elevate the arousal level. According to Gray, anatomically the behavioral inhibition system includes **the septohippocampal circuit** as well as the frontal cortex. Hyperactivation of this system is associated with **anxiety** because anxious people are very sensitive to potentially threatening stimuli.

The **behavioral approach system** is activated by potentially beneficial stimuli. Anatomically, it includes various forebrain structures, such as the striatum, that use dopamine as the modulating neurotransmitter. Hyperactivation of this system is associated with **impulsivity** because the hypersensitivity of impulsive people to reward makes them likely to engage in approach behavior.

The **fight/flight system** is activated by immediately threatening, punishing, or unconditionally aversive stimuli. Anatomically, it includes structures such as the **amygdala** and **medial hypothalamus** that are activated by emotionally negative, fearful stimuli. Activation of this system produces active avoidance (panic and flight) or attempted elimination (anger and attack). The activity of this system relates to psychoticism. In people with high activation of the fight/flight system, rage tends to dominate panic.

BEHAVIORAL PARADIGMS

The affective system in the QEEG/ERP research has been studied to a less extent than the cognitive control system. Most anatomical structures of the affective system are located deep within the brain. This **deep location** of the affective system makes it difficult to record EEG correlates of its functioning from the scalp in the conventional 10–20 system.

There are at least two sets of emotionally competent stimuli often used in studies of the affective system: (1) standardized emotional stimuli from the **international affective picture system (IAPS)** (Lang, Bradley, & Cuthbert, 2005), (2) stimuli that include **facial displays of emotion**, (3) stimuli sets that include **erotic pictures**. Faces are relatively weak emotional stimuli compared with IAPS pictures.

AMYGDALA AS DETECTOR OF FEARFUL STIMULI

The **amygdala** is a nucleus that lies deep inside the anterioinferior region of the medial temporal lobe. The amygdala could be considered an **interface between the sensory world and emotions**. In the visual modality

AMYGDALA AS DETECTOR OF FEARFUL STIMULI 219

the amygdala receives slower inputs from the cortical structure via the **hippocampus** and fast inputs from the **superior colliculus**. The structures in the subcortical route are activated even when subliminal fearful stimuli are presented to healthy subjects and when patients with striate cortex damage discriminate emotional facial expressions.

So, the amygdala receives sensory information, extracts memories stored in the amygdala, and sends the results of the extraction by **feedforward connections to the prefrontal** cortex through the dorsomedial nucleus of the thalamus and by **feedback connection to sensory areas**. For the visual modality, feedback connections are schematically presented in Fig. 3.3.6d.

Feedback projections to visual cortical areas from the amygdala are responsible for enhancement of neuronal reactions in the sensory systems to fearful stimuli in comparison with neutral stimuli (Fig. 3.3.6a). Via

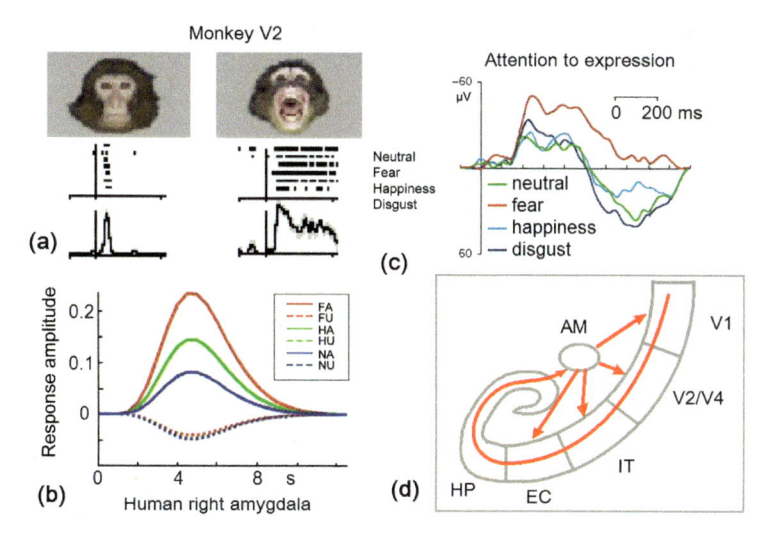

FIGURE 3.3.6 **Sensory and amygdala modulations by emotionally competent stimuli.** (a) Late response of neurons in monkey V2 area is modulated by facial expressions. Two monkey faces are shown, for example, fearful and neutral. Neuronal responses to these faces are shown in the spike density plots below each stimulus during a 500-s period postface onset. There is a delayed and sustained enhancement of neuronal discharges for the fearful expression, following an initial peak of response specific to the stimulus category (faces) but unaffected by expression; (b) attended *(A)* faces compared with unattended *(U)* faces evoke significantly greater fMRI activations in the amygdala for all facial expressions with highest reactions to fearful *(F)* stimuli rather than neutral *(N)* and happy *(H)* faces; (c) local field potentials from the amygdala in patients to each emotional expression show statistically different responses to fear between 200 and 800 ms poststimulus in attention to the expression task; and (d) scheme of input and feedback projections of the amygdala. *Part a: adapted with permission from Sugase, Yamane, Ueno, & Kawano (1999). Part b: adapted with permission from Pessoa, Kastner, & Ungerleider (2002). Part c: adapted with permission from Krolak-Salmon et al. (2004).*

widespread projections, the amygdala may also **facilitate consolidation** of emotionally meaningful events.

Intracranial recordings in the amygdala of animals show that many neurons in it respond to **unpleasant stimuli** and few to **pleasant stimuli**. In a 2004 study by Pierre Krolak-Salmon and coworkers from INSERM, France, local field potentials to facial expressions were recorded in epileptic patients with electrodes implanted for presurgical evaluation. The **fear modulation effect** was found in the amygdala starting at 200 ms after stimulus (Fig. 3.3.6c). The effect was spread at later latencies to the occipitotemporal, anterior temporal, and orbitofrontal cortex.

The amygdala can be described as an anatomical structure detecting threat or potential punishment and thus generating negative emotions such as fear and anxiety; in other words, the amygdala supposedly determines the **negative affective dimension**. In a 2002 fMRI study by Luiz Pessoa and coworkers from the National Institute of Mental Health, fMRI responses evoked by pictures of faces with fearful, happy, or neutral expressions were measured in healthy subjects viewing pictures in attended and unattended conditions. The results, presented in Fig. 3.3.6b, demonstrate that the amygdala has the highest response to fearful stimuli but only when stimuli are attended.

Patients with lesions in the amygdala **do not feel fear and anger**. Antonio Damasio from the University of Iowa College of Medicine in his book *The Feeling of What Happens* describes a patient who lost the amygdala on both sides. Describing this patient he writes "… there was nothing wrong whatsoever with S's ability to learn new facts. … Her social history, on the other hand was exceptional. To put it in the simplest possible terms, I would say the S approached people and situations with a predominately positive attitude." In this patient, and in others with similar damage, the memories for fear and anger seem to be missing, at least for auditory and visual stimuli.

ANXIETY IS A STATE OF PREPARING TO FEAR

Anxiety can be defined as a **mood state** in which the subject is preparing to cope with **future fearful or challenging events**. In this definition, anxiety is distinguished from fear which is an **emotional response** to a **present threat**.

HYPOTHALAMUS IS INVOLVED IN EXPRESSION OF EMOTIONS

The **hypothalamus** is a small (less than 1% of the human brain volume) nucleus that lies very deep in the brain. It regulates many fundamental programs such as keeping the body temperature, eating, drinking, and sexual behavior. The hypothalamus also plays an important role in emotion. Lateral parts of the hypothalamus are involved in emotions such as pleasure and rage, while the median part is associated with aversion, displeasure, and a tendency to uncontrollable and loud laughing. However, in general terms, the hypothalamus has more to do with the expression (symptomatic manifestations) of emotions than with the genesis of the affective states.

ORBITOFRONTAL CORTEX AS A MAP OF REWARDS AND PUNISHERS

The OFC occupies the ventral part of the prefrontal cortex, including BA 11 and 47 (Fig. 3.3.7). The OFC receives polimodal inputs from all sensory systems. The visual, auditory, and somatosensory systems convey information about the external world, about the status of our body, deep tissues, and viscera. Fundamental phylogenetically primitive modalities, such as the chemical senses bring signals to the OFC about smell and taste thus signaling pleasure or danger of food and drink. The amygdala sends emotion-related information mostly associated with fear. The OFC also receives pathways from other areas of the affective system, such as the hypothalamus. In addition, like any prefrontal cortical area, the OFC is self-regulated by means of the so-called "limbic basal ganglia thalamocortical loop" with the nucleus accumbens as the node at the striatal level.

One of the symptoms of OFC selective damage is **lack of affect**. A classic example of the OFC lesion (that also included damage in the anterior

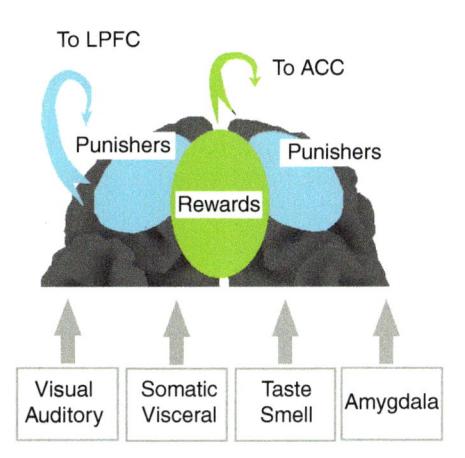

FIGURE 3.3.7 **OFC as a map of punishers and rewards (scheme).** Metaanalysis of the functions of the OFC showed that the medial part of the cortex is related to processing reinforcers while the lateral part is related to processing punishers. *Adapted from Kringelbach & Rolls (2004).*

cingulate cortex) is the case of Phineas Cage, whose frontal lobes were penetrated by a metal rod. After the accident Cage became a different person whose emotional processing and decision making had changed dramatically. In their textbook *An Introduction to Brain and Behavior*, Bryan Kolb and Ian Whishow from the University of Lethbridge, Canada described a patient who underwent **frontal leucotomy**. As the authors report, the first thing that was noticed in this patient was a lack of emotion, and any sign of facial emotional expression. The patient was quite aware of her emotional deficit by saying that she no longer had any feelings about things or other people, she felt empty, much like a zombie. Patients with OFC damage are also impaired in identifying social signals such as emotional expressions of facial reactions and voice intonation (procodity). In addition, they exhibit profound personality changes, problems with self-conduct, social inappropriateness and irresponsibility, and difficulties making decisions in the context of their everyday life.

A metaanalysis of fMRI and positron emission tomography (PET) studies made by Morten Kringelbach and Edmund Rolls from the University of Oxford in 2004 showed that OFC can be considered a **mapping tool** that represents rewards and punishers in a spatial (and, probably, temporal) pattern of activation of the cortex (Fig. 3.3.7). The **medial part** of the OFC represents the **reinforcers (rewards)**. Its close position and strong connection with the ACC indicates that the medial part of the OFC plays a critical role in organizing behavior to approach rewards. The lateral part of the OFC maps **the punishers**. Its close position and strong connection with the lateral prefrontal cortex indicates that it might play an important role in suppressing behavior that is associated with punishers. According to this view, patients with damage in the orbitofrontal cortex lost the ability to effectively map rewards (what is desirable) and punishers (what must be avoided) and consequently lost their ability to make appropriate decisions between selecting rewards and avoiding punishers.

The important functional feature of the orbital cortex is that it is self-regulated by means of the **limbic basal ganglia thalamocortical circuit**. The cortical areas of the OFC are topographically mapped into distinct areas of the ventral striatum (the nucleus accumbens). The long-distance lateral inhibitory connections within the striatum as well as within the thalamus potentially may select an appropriate affective action (eg, positive response) and inhibit an inappropriate action (correspondingly negative response).

VENTRAL ANTERIOR CINGULUM AND ANXIETY

The anterior cingulate cortex is divided into two portions: the affective (ventral) part and executive (dorsal) part. Stimulation of the ventral part can produce intense fear or pleasure, whereas stimulation of the more dorsal part can produce a sense of anticipation of movement. The ventral part of the cingulate cortex participates in the emotional reaction to pain, as well as in the regulation of aggressive behavior: wild animals, submitted to cingulectomy, become totally tamed. The anterior cingulate cortex also receives a strong projection from the amygdala, which probably relays negative, fear-related information.

A large body of EEG data shows that the anterior cingulate is the source of the frontal midline theta rhythm. This rhythm is manifested in short bursts of 5- to 8-Hz EEG signals recorded with maximal amplitude near Fz. This signal appears when the subject is performing a task requiring focused concentration and its amplitude increases with the task load. When the subject is restless and anxious, the signal is reduced or eliminated; when the anxiety is relieved with drugs, the signal is restored. That is why in the literature the frontal midline theta is sometimes associated with relief from anxiety. These data indicate that the anterior cingulate cortex is involved in regulating the emotional state from restless anxiety to focused relaxation. This is also consistent with the common experience that focusing on a cognitive problem relieves anxiety.

CONNECTIONS TO COGNITIVE CONTROL SYSTEM

The affective and cognitive control systems are **mutually interconnected** such that dysfunctioning of the affective system can lead to impairment of functions in the cognitive control system. Even quite mild **acute uncontrollable stress** associated with threat or challenge can cause a rapid and dramatic **loss of cognitive control**, while **chronic stress** may cause permanent **structural changes** in prefrontal neurons.

The first studies on the effects of **stress on cognition** were initiated after the Second World War by the observation that pilots highly skilled in peacetime often crashed their planes in the stress of battle due to errors of cognitive control. Many of the early studies showed that stress exposure impaired the performance of tasks that required cognitive control, but that it could actually improve the performance of automatic tasks. These early studies also showed the essential role of the **subjective sense of self-control**. Subjects, who felt being in control of a situation (even if this was an illusion), were often not impaired by stress exposure, whereas those who felt out of control were impaired.

The reduction in cognitive control during stress is important for understanding human **mental illness**. The loss of self-control during stress exposure can lead to a number of maladaptive behaviors, such as **drug addiction, smoking, drinking alcohol, and overeating**. Prolonged stress is a major risk factor for **depression**, and exposure to traumatic stress can cause **posttraumatic stress disorder (PTSD)**. Stress can also exacerbate the symptoms of schizophrenia and bipolar disorder.

fMRI OF EMOTIONS

In recent years affective neuroscience has begun to elucidate the neural structures involved in emotion regulation. The research is dominated by fMRI studies. For example, hemodynamic neuroimaging indicated that activity in the face-selected fusiform gyrus is modulated by emotions without explicit cognitive control and that modulation is abolished by amygdala damage.

A good example of application of the neuroimaging approach to emotions is given by a 2000 PET study by Antonio Damasio and his coworkers from the University of Iowa College of Medicine. In the experiments subjects recalled and reexperienced personal life episodes marked by sadness, happiness, anger, or fear while were placed in PET scans. The results show that all emotions engaged structures related to the affective system including the insular cortex, secondary somatosensory cortex, cingulate cortex, and nuclei in the brainstem tegmentum and hypothalamus.

For example, **sadness** (Fig. 3.3.8a) induced bilateral, but asymmetric, activations of the insula cortex and the mixed activation–deactivation pattern in the cingulate cortex (activations anterior, deactivations posterior). In contrast, **happiness** (Fig. 3.3.8b) induced activation of the right posterior cingulate cortex and suppression of the anterior third of the left cingulate, activation of the left insula cortex and the right secondary somatosensory cortex.

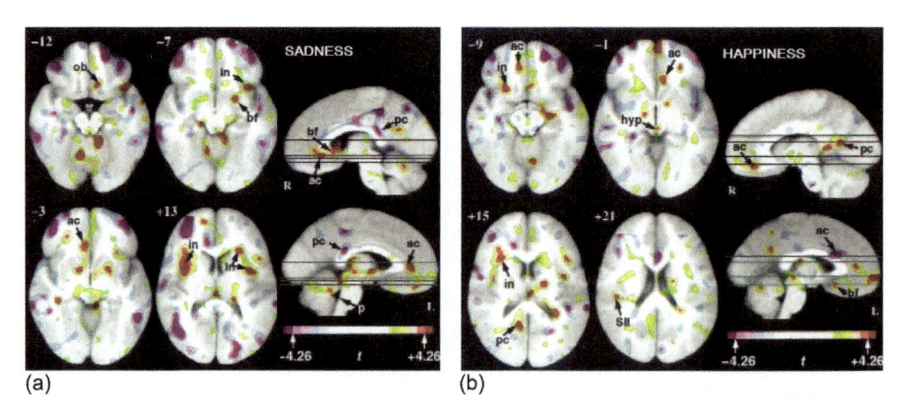

FIGURE 3.3.8 **Neural correlates of feeling sadness and happiness.** Representative axial slices and paramedial brain views of the 3D reconstructed brain of the 39 subjects in the study superimposed on the PET results. *Deep red* areas denote significant activation peaks, *purple* areas denote significant deactivations: (a) sadness and (b) happiness. ob, orbitofrontal; in, insula; bf, basal forebrain; ac, anterior cingulate; p, pons; hyp, hypothalamus; pc, posterior cingulate; SII, secondary somatosensory cortex. *Reprinted by permission from Macmillan Publishers Ltd: Nat Neurosci., Damasio et al. (1999).*

These activation patterns provide distinctive "perceptual landscapes" of an organism's internal state, while differences among those landscapes constitute the critical reason each emotion feels different.

Note that the low time resolution of PET does not allow analyzing temporal patterns of brain activation during separate stages associated with emotional reactions, feelings, and monitoring emotions. EEG and ERPs appear to present a tool that enables us to study the neuronal correlates of distinct stages of information flow in emotions.

STAGES OF REACTIONS OF AFFECTIVE SYSTEM

Damasio's hypothesis suggests that there are consecutive stages of processing emotionally significant stimulus. In the visual modality the critical structures in this processing are: (1) the areas in the temporal and parietal cortex that are responsible for recognition of face expressions and body movements, (2) the amygdala, ventral striatum (nucleus accumbens), and OFC that map the stimulus into 2D space (negative affect and positive affect), (3) motor cortical areas that automatically react with emotional (body and face expression) response, (4) the somatosensory and insular cortex that correspondingly map sensorimotor information of emotional reaction from face/body and visceral organs, and (5) the anterior cingulate cortex that by comparing the expected action and the executed one generates a signal that through the motor part of the cingulate cortex modifies further behavior.

The stages can be classified into (1) the representation (sensation) stage, (2) the emotional reaction stage, (3) the feeling stage, and (4) the monitoring (correcting) stage:

- **Sensation.** Neuronal impulses generated by stimulus reach the amygdala and the primary visual cortex. The short connections between the superior colliculus and the amygdala enable it to react in punishers with short latencies which are important for survival in some dangerous situations.
- **Emotional reaction.** Visual information is processed in temporal (what) and parietal (where and how) areas and reaches the motor cortex. The fusiform gyrus of the temporal lobe, for example, plays a critical role in face discrimination and extracting emotional expressions from face images. This is a stage in which the stimulus is decomposed into spatial patterns in correspondence with 2D affective space: positive affect and negative affect. This is also a stage in which the motor system reacts (partly automatically) to the processed emotional stimulus.
- **Feeling stage.** During this stage information about emotional reaction reaches the somatosensory cortex and insular. These cortical areas

map the external (body/face) and internal (visceral) emotional reactions into distinct feelings such as a feeling of happiness or a feeling of joy. For example, happiness is represented in face expression, in alterations of heart rate, speed of breathing, and other so-called "visceral responses."
- **Monitoring stage.** Finally, the expected emotion and the real emotional reaction and feelings are compared with each other in the anterior cingulate cortex. The results of this comparison operation via the motor part of the anterior cingulate cortex drive the body to avoid this discrepancy or to stay longer in this state.

EVENT-RELATED POTENTIALS TO EMOTIONAL STIMULI

Humans are a social species for whom **emotional facial expressions** bring important information about the emotional state of other people. The encoding of emotional expression of faces seems to be an automatic operation which is hard-wired into the human brain by evolution. This is supported by studies of ERPs in response to faces. Early sensory-related visual waves **P1 and N170** were shown to be enhanced by **negative facial expression** in comparison with neutral and independently on intention. This effect may be due, at least partly, to feedback from the amygdala which rapidly detects the threat carried by the stimulus (Fig. 3.3.6d). It should be noted that not every study observed the early modulation effects indicating that emotional regulation of visual-related early waves occur only under certain conditions. The conditions are not clear though.

When attention or intention is paid to the stimuli, emotionally meaningful stimuli (both negative and positive pictures or complex visual scenes) elicit two additional waves: **early posterior negativity (EPN)** and **late positive complex (LPC)**.

EPN is distributed over **occipitotemporal electrodes** within 150–300 ms after stimulus onset (Fig. 3.3.8). EPN is suggested to reflect enhanced perceptual encoding resulting from automatic visual attention to intrinsically salient stimuli such as images inducing negative and positive emotions. EPN is similar to processing negativity, an ERP wave found in research into selective attention. Processing negativity appears when the subject pays attention to the stimulus (attended condition) rather than the unattended condition.

LPC is distributed over **centroparietal electrodes** around 300 ms poststimulus (Fig. 3.3.9). LPC lasts for several hundred milliseconds and appears to reflect more advanced levels of emotion detection and feelings.

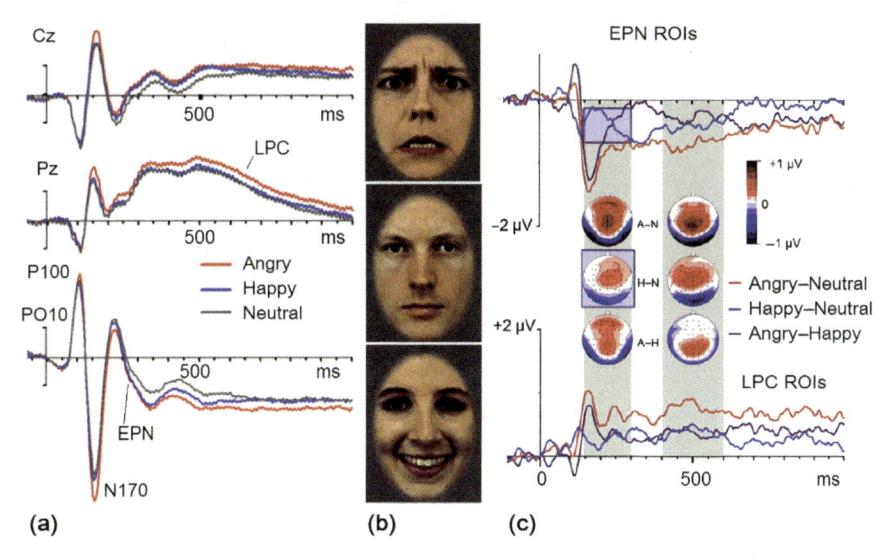

FIGURE 3.3.9 **Emotional modulation of ERPs.** (a) ERPs at selected electrode sites for angry *(red)*, happy *(blue)*, and neutral *(gray)* facial expressions; (b) examples of stimuli; and (c) difference waves for angry relative to neutral *(red lines)*, happy relative to neutral *(blue lines)*, and angry relative to happy facial expressions *(purple lines)*, averaged across typical EPN (EPN ROIs, upper graphs) and LPC electrodes (LPC ROIs, lower graphs). *EPN*, early posterior negativity at 150–300 ms; *LPC*, late positive complex at 400–600 ms. *Adapted with permission from Rellecke, Sommer, & Schacht (2012).*

It was associated, especially in early studies, with late positive waves (including P300) elicited in the oddball paradigm in response to rare targets. However, more recent work on emotion suggests that the source of LPC is different from that of P300.

Increased LPC for emotional rather than neutral pictures has been shown to be larger for more intense stimuli (ie, those rated as more arousing) and for highly arousing pleasant and unpleasant stimuli, such as **erotic** and **threat** scenes. Combined ERP/fMRI studies associated LPC with increased blood flow in the **occipital, parietal, and inferotemporal regions in the brain**.

No studies have explicitly linked LPC to the activity of specific neurotransmitter systems. Hypothetically, LPC might be associated with sustained norepinephrine (NE) response generated by the locus coeruleus. fMRI evidence that subliminal presentation of threatening stimuli activates the locus coeruleus appears to support the hypothesis.

LPC provides a tool to study **individual differences** in responding to emotional stimuli. For example, arachnophobic individuals show an increase in LPC to images of spiders, and abstinent cocaine-dependent participants demonstrate increased LPC in response to cocaine pictures.

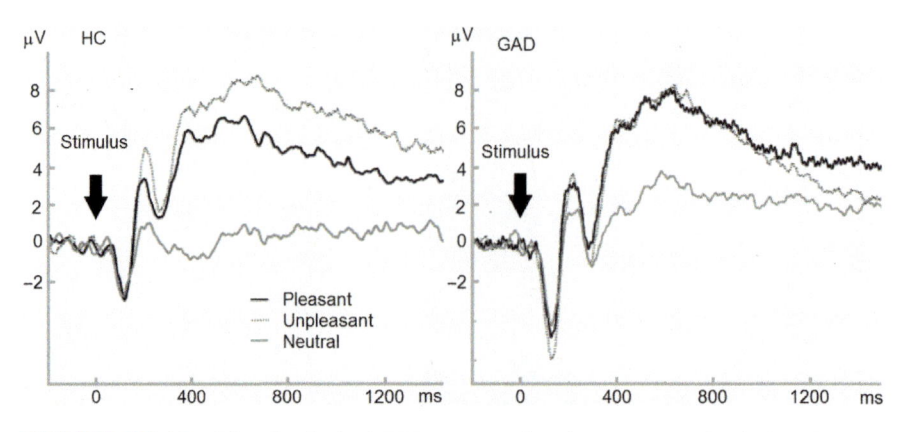

FIGURE 3.3.10 Stimulus-locked ERPs averaged at four centroparietal sites (Pz, Cz, CP1, and CP2) for healthy controls *(HC)* and a group of patients with general anxiety disorder *(GAD)* for pleasant, neutral, and unpleasant images. *Adapted with permission from Weinberg & Hajcakt (2011).*

ANXIETY ENHANCES VISUAL N1 WAVE

Psychological studies show that anxiety facilitates attention to threatening stimuli. This psychological observation is reflected in enhancement of the N1 wave in response to emotionally competent stimuli. In a 2011 study by Anna Weinberg and Gerg Hajcak from Stony Brook University, ERPs were recorded in healthy subjects and patients with general anxiety disorder viewing emotional and neutral images (Fig. 3.8.10). The authors reported an enhanced of early ERP waves for unpleasant rather than neutral images in patients with anxiety. This early **hypervigilance** was followed by a reduction in emotional modulation of LPC in the anxiety group (Fig. 3.3.10).

Note that emotional stimuli selected from the IAPS in this study produced a larger LPC in comparison with those elicited by emotional faces (compare Figs. 3.3.9 and 3.3.10). Note also the enhancement of the N1 wave to neutral stimuli in individuals with anxiety. It appears that enhancement of early ERP waves to visual stimuli in general, and to emotionally competent stimuli in particular, can be considered as a neuromarker of anxiety.

NEUROMODULATORS OF AFFECTIVE SYSTEM

Among all neuromodulators the highest concentration of **serotonin (5-hydroxytryptamine; 5-HT)** is found in the anatomical structures of the affective system. In line with this fact, central serotonergic neurons and receptors are targets for a variety of therapeutic agents used in the treatment of disorders of the affective system.

Serotonergic projections arise primarily from the **dorsal and median raphe nuclei**. They compose two distinct serotonergic systems differing in their electrophysiological characteristics, topographic organization, morphology, as well as sensitivity to neurotoxins and psychoactive agents. The thin varicose axon system arises from serotonergic cell bodies within the dorsal raphe nucleus with fibers that branch profusely in their target areas. They mostly innervate the prefrontal cortex, nucleus accumbens, amygdala, and ventral hippocampus. The thick nonvaricose axon system arises from serotonergic cell bodies within the median raphe nucleus with fibers that mostly innervate the hypothalamus and dorsal hippocampus.

CHAPTER 3.4

Memory Systems

INTRODUCTION

The ancient Greeks, together with other arts, invented an **art of memory** named "mnemotechnics." In the ages before the invention of printing this art was vitally important. Although nowadays we rely on memories stored in books, computers, and the Internet, the importance of our own memory is difficult to overestimate.

A 19th century German philosopher Herman Ebbinghause from the University of Berlin and University of Breslau was the first to study human memory experimentally by asking subjects to remember lists of words and nonwords. This verbal learning approach was complemented by a German Gestalt approach in the 1930s. Frederic Bartlett from the University of Cambridge in England and Alexander Luria from Moscow State University in Russia rejected the learning of meaningless material as an appropriate approach and instead used complex material such as stories or folk tales. After the Second World War, computer metaphors were used for memory models.

A major achievement of recent research on brain mechanisms of learning and memory is the recognition that there are **different types of memory involving different brain systems**. It is not a trivial fact because some readers may recall Lashley's pessimistic conclusion from his series of experiments as stated in his famous 1950 article, "In search of the engram." With all respect to Karl Lashley from the University of Minnesota, the existence of several different forms of memory with differing neuronal substrates was not recognized at that time, nor were modern analytic techniques available then.

TEMPORAL ASPECTS OF MEMORY

As mentioned above, there are several types of memory. One distinction can be made on the basis of **temporal dynamics**. An influential model of memory traces was proposed **by Richard Atkinson and Richard Shiffrin** from Indiana University in 1971. According to this model, in the visual modality three different stages of memory can be separated: (1) **iconic memory** which lasts for a few hundreds of milliseconds (the name was suggested by **Ulric Neisser** from Cornell University in 1967), (2) **short-term memory** which lasts more than a few seconds, and (3) **long-term memory**, which can last for days, months, or years, requires long-lasting structural changes.

George Sperling from the University of California, Irvine in 1963 was one of the first to study iconic memory experimentally. He used the **partial report paradigm** in which subjects were briefly presented with a grid of 12 letters, arranged into three rows of four (Fig. 3.4.1). After some delay, subjects received either a high, medium, or low tone, cuing them which of the rows to report. Based on these experiments, Sperling showed that visual memory degraded within a few hundred milliseconds. This type of memory cannot be prolonged via rehearsal.

Short-term memory allows recall for a period of **several seconds to a minute** without rehearsal. Its capacity is also very limited: George Miller from Harvard University in 1956 wrote an influential paper on the limited capacity of short-term memory reflected in the magic number 7 ± 2. Memory capacity can be increased through the process of chunking—organizing items in groups. For example, remembering a telephone number

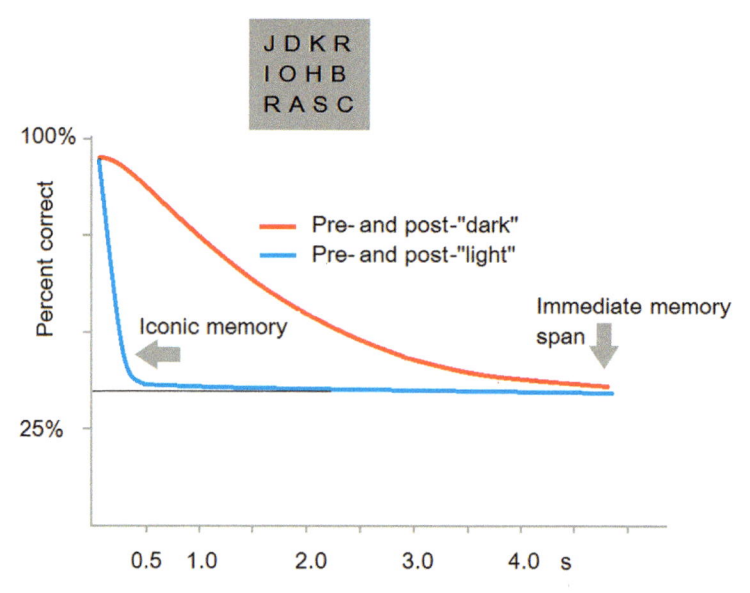

FIGURE 3.4.1 **Decay of information in the visual modality.** Number of letters available using the partial report method as a function of recall delay with *(blue)* and without *(red)* the "light" mask. *Adapted from Sperling (1963).*

of 10 digits can be effectively done by chunking it into three groups (area code, three digit chunk, and four-digit chunk).

Note that in the executive system the concept of **working memory** is used instead of short-term memory to emphasize that this type of memory is responsible not only for temporal storing of sensory information but also for **manipulating** this information according to the task setting.

WORKING MEMORY REPRESENTATIONS

When we are searching for a friend in a crowd, how do we tune our perception to select the friend but not somebody else? As mentioned in Chapter 3.1 a neurophysiological model presumes that the perception is biased by **representations** stored in working memory. Recall Fig. 3.1.4 which demonstrates that working memory representation of the template due to top–down projections decreases the threshold of sensory neurons that code target features thus tuning these neurons to the corresponding bottom-up flow of sensory information.

ERPs have been applied to measure the process of **storing stimulus representations** in working memory. Delayed match-to-sample tasks (in the terminology of the current book) are used for this purpose.

As one can see from Fig. 3.4.2, when subjects are maintaining representations of items in visual working memory initially seen in the right visual field, **sustained negativity** is found over left posterior temporal electrodes. The opposite pattern is found when to-be-remembered stimuli are presented in the left visual field. The corresponding sustained ERP wave was named "contralateral delay activity" (CDA).

In the delayed match-to-sample task of the HBI database, delayed negativity in response to Cue in contrast to NoCue is reflected in the temporally distributed latent component presented in Fig. 3.4.3. The age dynamics of the Cue-NoCue difference of this component is presented in Fig. 3.4.3b, c. Note that the delayed negativity is preceded by a positive P3-like fluctuation which dramatically decays with age. This age-related decay in P3 is contrasted to a relative stability of the delayed negativity in the whole spectrum of age (17–84 years).

TYPES OF LONG-TERM MEMORY

Long-term memory in turn can be divided into two broad categories depending on the type of stored information and neuronal mechanisms. They are explicit and implicit memories.

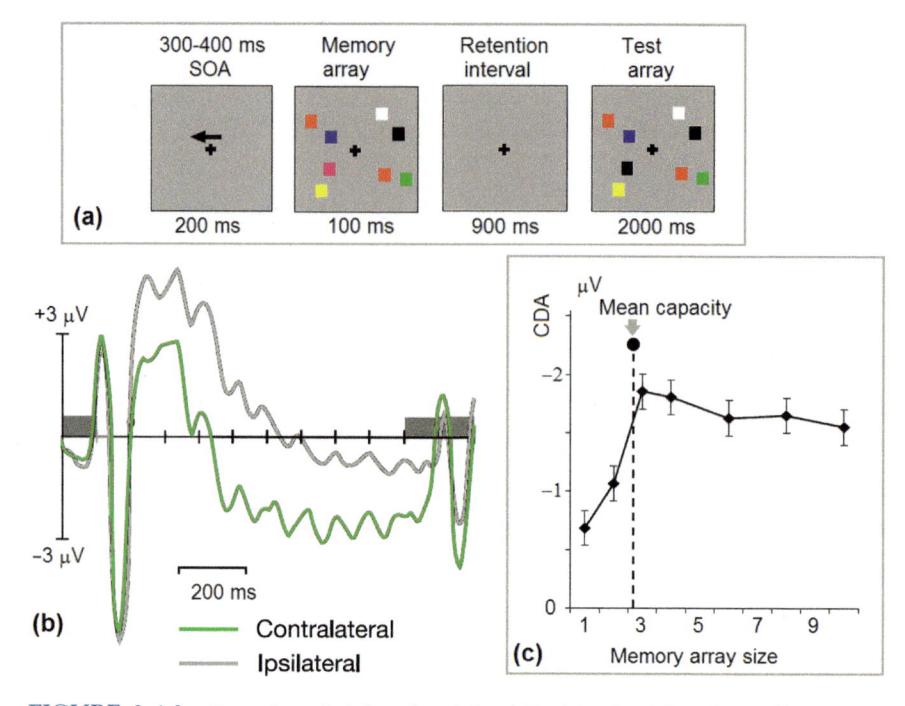

FIGURE 3.4.2 **Contralateral delayed activity (CDA) in the delayed match-to-sample task.** (a) Example of the stimulus sequence, and (b) grand-averaged waveforms from electrodes T5 and T6, contralateral *(green)* and ipsilateral *(gray)* to the location of the cue on each trial. The two gray rectangles reflect the time periods for the memory and test arrays, respectively; (c) mean amplitude of CDA (Y-axis) and visual memory capacity (X-axis) across experiments. Error bars reflect 95% confidence intervals. *Adapted with permission from Vogel & Machizawa (2004).*

Explicit memory involves awareness of the memory whereas **implicit memory** does not necessarily involve being aware of the memory. **Declarative memory** refers to memories that can be consciously recalled as episodes or facts in personal life. Consequently, declarative memory is divided into **episodic** and **semantic** memories.

Procedural memory as a type of implicit memory is based on learning and recalling motor and cognitive skills (eg, learning to ride a bicycle).

The distinction between implicit and explicit learning can be exemplified in language. The lexicon of memorized word-specific knowledge is stored in temporal–parietal–occipital areas and represents an example of declarative memory. This memory is compromised by lesions in the left temporal–parietal–occipital area and is known as **Wernike aphasia**. The grammar that subserves rule-governed combining of lexical items into complex sentences depends on a distinct neural system that includes frontal, basal ganglia, parietal, and cerebellar structures. It represents an

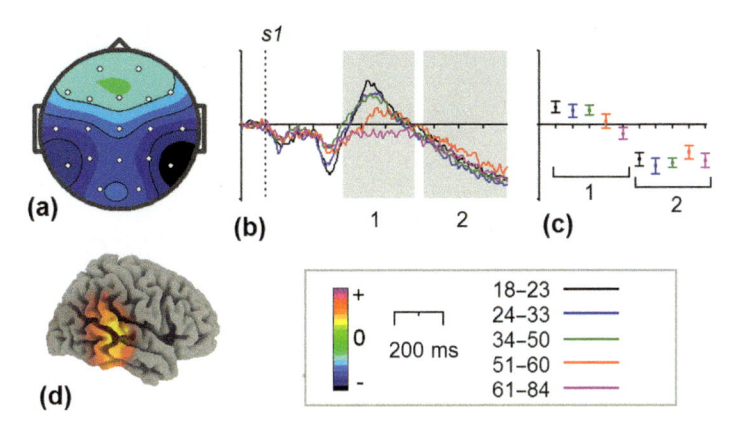

FIGURE 3.4.3 **Delayed negativity in the delayed match-to-sample task.** The joint diagonalization method is applied to the collection of ERPs in the referential montage in response to cue and No cue. The component with delayed negativity at the T5/T6 electrodes is presented. (a) Topography. (b) Cue-NonCue differences for five age groups (18–23, black; 24–33, blue; 34–40, green; 51–60, red; and 61–84, pink colors). Gray areas define time windows (1,2) for which the statistical analysis was performed. Dashed vertical line—s1 presentation. (c) Mean values and 95% confidence intervals computed for the time windows for five age groups. (d) sLORETA image. *Data from the HBI database.*

example of procedural memory. This memory is compromised by lesions in left frontal areas and subcortical structures. This type of aphasia is known as **Broca's aphasia**.

HIPPOCAMPUS AS A REFERENCE TO EPISODIC TRACE

Patient H.M. represents a classic case of severe amnesia. For treatment of medically refractory epilepsy in 1953, he underwent a surgical operation which removed left and right medial temporal areas including the hippocampus. From that day forth he was unable to form a new declarative memory for even the most salient events. Studies of his particular amnesia by more than 100 investigators provided new insights into the neuronal organization of memory.

Medial temporal structures play a critical role in episodic memory (Fig. 3.4.4a). The **hippocampal region** is heavily connected with entorhinal cortex, which is strongly connected with both the perirhinal and parahippocampal cortices, which are in turn connected extensively to **temporal** and **parietal** neocortical regions. The hippocampus projects into the **mammillary bodies of the hypothalamus** and **anterior nucleus of the thalamus**. The last two structures are impaired in another amnesic

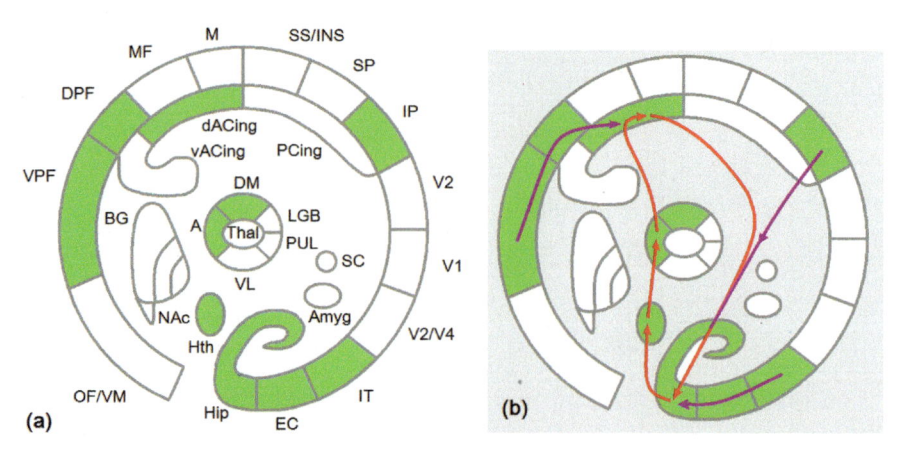

FIGURE 3.4.4 **Episodic memory system.** (a) Brain structures involved in forming episodic memory, and (b) information flow within the episodic memory system; *red*, a closed loop that includes the hippocampus, mammillary bodies of hypothalamus, anterior nucleus of the thalamus, and the dorsal part of the anterior cingulate cortex; *violet*, the input to this loop is from the sensory and executive systems of the brain.

syndrome named **Korsakoff's syndrome** after the Russian clinician Sergei Sergeievich Korsakoff who in 1887 described amnesia (both retrograde and anterograde) that was associated with nutritional (thiamine) deficiency in alcoholics.

The hippocampus also receives strong projections from the anterior cingulate cortex which in turn receives inputs from frontoparietal areas. Thus, the hippocampus is in position to **map the executive and sensory representations of the current episode** (Fig. 3.4.4b). By means of **long-term potentiation** induced by a burst of the **theta rhythm** the hippocampus is able to store a map of the episode in **neuronal form** that is relatively resistant to interference.

It should be stressed that memories themselves are stored in the **temporal–parietal–frontal areas** while the hippocampus serves as a **reference** to memories. Memories eventually become largely independent of medial temporal lobe structures and dependent upon corresponding neocortical regions. In this view the declarative memory system in the visual modality is closely related to the **ventral stream system.** This system is rooted in inferior and lateral temporal lobe structures and underlies the formation of **perceptual representations** of objects. These representations are associated with recognition and identification of objects and the long-term storage of knowledge about objects. The ventral system is thus a memory-based system, feeding representations into long-term (declarative) memory, and comparing these representations with new ones.

FUNCTIONAL NEUROMARKERS OF EPISODIC MEMORY

There are several paradigms to study the neurobiological basis of episodic memory. In all of them, stimuli are presented during the memorization stage and tested during the recalling stage. Different imaging parameters have been used: MRI, PET, scalp EEG in healthy subjects, and intracranial local field potentials recorded from electrodes inserted for diagnostic purposes in epileptic patients.

In **the old–new paradigm**, imaging parameters are measured during the recalling stage in which two types of items are presented: old ones (ie, those that are learned during the memorization stage) and new ones (ie, those that are presented for the first time to subjects). The Old–New effect is reflected in a difference wave that is distributed over the left temporal–parietal area. Those studies are reviewed in a 2003 paper by Michael Rugg from University College London and Andrew Yonelinas from University of California.

In an **encoding paradigm** designed to study neurobiological correlates of episodic memory, imaging parameters are measured during the encoding stage of stimuli that are going to be memorized. EEG and ERP parameters are measured for stimuli that subsequently will be remembered or forgotten. The difference between parameters defines the "remembered–forgotten" effect. Results of the few studies that have analyzed this effect are schematically presented in Fig. 3.4.5.

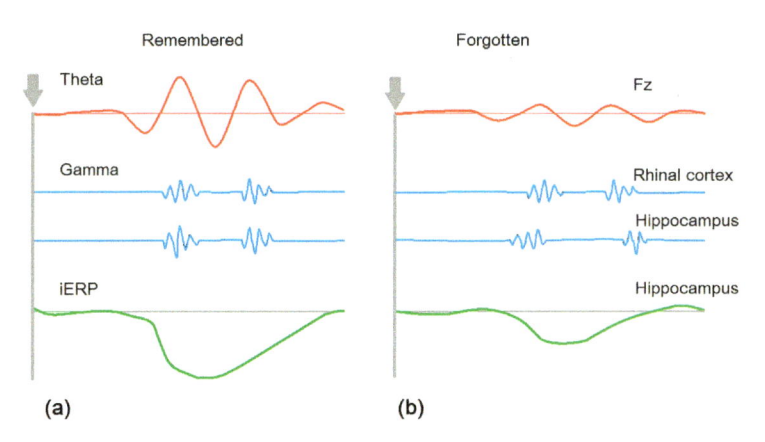

FIGURE 3.4.5 **Remembered–forgotten effect.** EEG responses to stimuli that will subsequently be (a) remembered, or (b) forgotten. *Red*, scalp-recorded theta rhythms in healthy subjects; *blue*, gamma coupling between rhinal cortex and hippocampus in epileptic patients with implanted electrodes; *green*, intracranial ERPs *(iERP)* from the hippocampus in epileptic patients. Schematic representation of results from Klimesch (1999) *(red)*, Fell, Ludowig, Rosburg, Axmacher, & Elger (2008) *(blue)*, and Fernandez et al. (1999) *(green)*.

As one can see, remembered stimuli (1) elicit stronger responses in the hippocampus, (2) higher coherences in the gamma band between the rhinal cortex (the polymodal entrance to the hippocampus) and the hippocampus, and (3) the larger amplitude of scalp-recorded theta oscillations.

NEURONAL MODEL OF EPISODIC MEMORY

A model of the hippocampus that encodes and temporally stores the compressed index of an episode was suggested by Bruce McNaughton from Carleton University in 1989. According to this model the **sensory-related part** of any episode is encoded in the **parietotemporal lobes** while the **action-related part** of an episode is encoded in the **frontal lobe**. Without the hippocampus, the trace in these areas can be kept only temporally in a form of **reverberation of neuronal impulses** in recurrent neuronal networks and is very sensitive to interference. The hippocampus is a place where the two spatially distributed representations of the episode converge into a **single activation pattern**. The hippocampal trace lasts for longer time intervals and is associated with **long-term potentiation** induced by a **burst of theta rhythm**. So, the hippocampus serves as a temporal storage mechanism for the episode. The hippocampal representation later becomes active either in **explicit recall** or in implicit processes such as sleep. This gives rise to reinstatement of the corresponding neocortical memory, resulting in long-term adjustments of neocortical connections: **long-term memory**.

RETRIEVAL OPERATIONS

It is logical to suggest that memory is retrieved in the same areas that are responsible for encoding representations of sensory stimuli. Early evidence that sensory regions are associated with memory retrieval was obtained by **Wilder Penfield** from the Neurological Institute of MacGill University in Montreal in the 1940s and 1950s (Penfield, 1952). Penfield electrically stimulated regions of exposed cortex in awake human patients undergoing surgery for epilepsy and found that stimulation of regions of the occipital and temporal cortex would sometimes elicit memories and that the sensory modality of elicited memories varied depending on the region of cortex stimulated. In these electrical stimulation studies, regions of the superior and middle temporal lobes were associated with auditory memories (I hear singing … Yes, it is *White Christmas*) whereas regions of more posterior temporal and occipital lobes were associated with visual memories (… I saw someone coming toward me as if he were going to hit me).

The following PET and fMRI studies confirmed this inference and gave direct evidence that metabolic activations in **encoding and retrieving stages** of memory task significantly **overlap** with each other. For example, in a 2000 study by Mark Wheeler and colleagues from Georgia Institute of Technology subjects memorized words paired with either sounds or pictures. At test, subjects were asked to recall whether the words had been previously associated with pictures or sounds, thus encouraging retrieval of vivid, modality-specific memories. Encoding resulted in increased activity in the visual cortex (from the calcarine to the fusiform gyrus) for pictures and in auditory cortex (from Heschl's gyrus to the middle temporal gyrus) for sounds. Retrieval of pictures from memory was associated with reactivation of the visual cortex near the fusiform gyrus, whereas retrieval of sounds was associated with the bilateral superior temporal gyrus near secondary auditory regions.

A familiar face or scene might spontaneously trigger a memory trace, but most acts of remembering begin with **a goal-directed attempt to recall**. Areas of the **prefrontoparietal cortices** are involved in this active operation. As we know from Chapter 3.2, these prefrontal areas receive highly processed sensory information from posterior cortical areas, store plans of actions and perform executive functions such as working memory, action selection, action initiation, and action suppression.

Neuropsychological data support this theoretical inference: patients with frontal lesions are significantly impaired at free recall of memorized recently items. Moreover, these patients are not using subjective organizational strategies. That is, whereas healthy control subjects had a tendency to consistently **group** certain words together across retrieval trials (eg, spoon and plate), frontal patients did so to a lesser degree, in essence recalling the words in a more random fashion. Having these difficulties with retrieval effort, some frontal patients **confabulate**. This confabulation behavioral pattern is opposite to that typically observed in memory loss.

ACETYLCHOLINE AS NEUROMODULATOR OF DECLARATIVE MEMORY

The core element of the episodic memory system is the hippocampus. Generation of electrical activity in this anatomical structure, including the hippocampal theta rhythm, is controlled by the neuromodulator **acetylcholine**. This neurotransmitter is produced in the septal nucleus and transported to the hippocampus via the septohippocampal pathway. The intensity of cholinergic input to the hippocampus defines the amplitude of theta oscillations. It is important to stress here that the septum represents an extension of the cholinergic ascending system located in the nuclei of the brainstem and receives strong activation input from the brainstem.

The rest of the cortex including prefrontal, medial temporal, and insular cortical areas receives cholinergic input from the nucleus basalis of Meynert. The basal nucleus in contrast to widespread output receives input mostly from the limbic cortex and, consequently, can be considered an interface between the limbic system and the entire cerebral cortex. The output of this nucleus is also directed to the reticular nucleus of the thalamus thus enabling the basal nucleus to modulate information flow through the thalamus to the cortex.

Theoretically, according to this scheme acetylcholine can modulate the episodic memory trace by two mechanisms: (1) by enhancing the amplitude of the theta rhythm in the hippocampus and, consequently, by enhancing long-term potentiation in this region, and (2) by activating broad areas of the cortex in response to behaviorally meaningful stimuli through activation of the basal nucleus.

PROCEDURAL MEMORY SYSTEM

As we learned from the above, the brain can be roughly divided into two interconnected but separate parts: sensory-related and action-related parts. The sensory-related part (sensory systems, attentional networks, and, partly, the affective system) deals primarily with sensory information and stores this information in distinct elements labeled "chunks" by Wayne Wickelgren from the University of Oregon in 1979. A chunk corresponds to representation of a distinct episode that is maintained in the hippocampal system for a certain period of time before it is consolidated in the frontal–temporal–parietal areas of the cortex. Encoding the chunk into the hippocampal system needs a relatively short time that corresponds to a few oscillations of the hippocampal theta rhythm.

In contrast with the episodic memory system, the procedural memory system is implicated in the **learning of new actions**. Examples of actions are driving a car, playing tennis, playing a piano, writing and typing, articulating words, singing, etc. Such actions are called skills, habits, routines. The number of routines humans learn during life is enormous.

Representations of actions are stored in the **frontoparietal networks**. They are **mapped** into the striatum. As we learned in Chapter 3.2 the striatum plays a critical role in cognitive control including action selection and action suppression. Besides cognitive control functions, the basal ganglia are involved in learning sensorimotor contingencies, in organization sequential movements, reinforcement-based learning (including emotionally competent award-based learning), motor planning particularly if it involves precise timing and multiple motor programs. During the learning procedure, **mapping of the cortical representations onto the striatum** is rearranged. The mechanisms of this rearrangement represent the basis of procedural memory. So, procedural memory critically depends on the basal ganglia and related structures.

Procedural memory, unlike episodic memory, does not need a separate system for encoding and consolidating events. Memories reside in the executive system itself. An engram of procedural memory represents slowly changing synaptic connections in the executive system.

As we know from Chapter 3.2 the basal ganglia receive mapping projections from frontal–parietal–temporal areas which are parallelly processed in segregated circuits. These parallel circuits have the same network structure and perform the same general functions, such as action selection, action initiation, and action suppression.

For example, **Broca's area** projects into the striatum in a similar way to other cortical areas. This indicates that basal ganglia participate in all functions ascribed to Broca's area including procedural learning of grammatical rules. Children learn the grammar of their mother language from the environment gradually and without efforts. However, when we learn a new language it takes tremendous efforts to acquire a new vocabulary, but the most difficult is to use grammatical rules in fluent speech. It seems logical that damage to Broca's area produces a common difficulty known as agrammatism. For those patients, speech is difficult to initiate, it is nonfluent, labored, and halting while language is reduced to disjointed words with great difficulty to construct an appropriate sentence. Moreover, similar speech disturbances are observed in damage or electrical stimulation of all parts of basal ganglia thalamic circuits. For example, George Ojemann from the University of Washington reported that electrical stimulation of the dominant ventrolateral thalamus can produce deficits in language processing that are not seen after similar stimulation of the nondominant ventrolateral thalamus. The nature of language deficit varies, depending upon the location of the stimulation site, as described by Mark Johnson and George Ojemann in 2000 and by Vladimir Smirnov from the Institute of Experimental Medicine, Saint Petersburg in 1976.

Unlike fast encoding subserved by the declarative memory system, learning in the procedural memory system is **gradual** and needs many associations of contextual information with the acquiring action. Recall how long it took you to learn to play tennis or drive a car and compare it with an event of a few seconds, such as a car accident, that was imprinted in your episodic memory for ever. In the same way, recalling from episodic memory is associated with conscious recollection and takes at least a few hundred milliseconds depending on the type of sensory information. Recalling from procedural memory is faster and could be done without any conscious recollection automatically.

NEUROMODULATORS OF PROCEDURAL MEMORY

There are two main neuro mediators of the procedural memory system: (1) dopamine that is transported to the striatum from the substantia nigra and (2) acetylcholine that is produced by specific cholinergic cells within the striatum itself. Recall that dopamine has been viewed in this book as the main mediator of the executive system while acetylcholine has been viewed as the main mediator of the episodic memory system. Taking into account that the striatum plays a critical role in the executive system as well in the procedural memory system we can conclude that both mediators must be involved in modulating procedural memory.

PART 4

METHODS OF NEURO-MODULATION

CHAPTER 4.1

Pharmacological Approach

HISTORICAL INTRODUCTION

I am not an expert in psychopharmacology but when I was introduced to the field by reading textbooks I got an impression of the solid theory that underlies it: all brain dysfunctions were explained as **disturbances in synaptic transmission.** The basic idea was that the **core** of brain functioning was sending **chemical messengers** across synapses so that mental illness was the result of imbalances among these neurotransmitters. It was presumed that imbalances can be treated in the same way as insulin treats diabetes: for example, a lack of serotonin in depression can be treated by selective serotonin reuptake inhibitors (SSRIs). Consequently, the main problem for pharmacology is to find the exact neurotransmitter system involved in a given psychopathology. Taking into account the high heterogeneity of functional proteins (receptors, transporters, ion channels, enzymes) in the brain, the number of selective pharmaceutical agents could be quite big. The search for such highly selective drugs seemed to be the mainstream of pharmacology.

But when I started learning the history of psychopharmacology, I realized that the field was not driven by the theory at all. Its history was rather a remarkable **series of accidental discoveries** made after the Second World War. Among them were (1) the discovery of lithium for treating depression by John Cade in 1949 in Bundoora Repatriation Hospital, Melbourne (Australia); (2) the discovery of chlorpromazine for the treatment of schizophrenia in the beginning of the 1950s; (3) the discovery of thorazine for the treatment of depression by Roland Kuhn in 1956 at the cantonal mental hospital of Münsterlingen: (4) the discovery of benzodiazepines by Leo Sternbach for the pharmaceutical company Hoffmann-La Roche, Nutley, New Jersey in 1957 (Sternbach, 1979).

What is more, some of the theories which were invented to support these accidental discoveries **turned out to be wrong**. For example, in 1965 Joseph Schildkraut from the National Institute of Mental Health offered a theory that depression was the result of imbalances in **dopamine and**

norepinephrine—neuromodulators that were thought in those days to be affected by drugs.

But, a few years later it became clear that antidepressants act mostly by increasing the availability of the neurotransmitter **serotonin**—rather than dopamine and norepinephrine. A new generation of antidepressants—**SSRIs**, including Prozac, Zoloft, and Paxil—was developed to target it. Prozac was approved by the FDA in 1987.

The serotonin imbalance theory, however, has turned out to be just as inaccurate as Schildkraut's theory. While SSRIs indeed alter serotonin metabolism, those changes do not explain why the drugs work in some cases or why they are not more effective than placebo. In the following decade, scientists had concluded that serotonin was only one of **many factors** that modulate the mood and that the causes of depression were far more complex than the chemical imbalance theory implied.

CURRENT CRISIS OF PSYCHOPHARMACOLOGY

Nowadays routine mental health care prescribes multiple medications that are poorly understood and whose long-term effects are unknown. These drugs have become a commodity people learned to consume. Nevertheless, psychopharmacology faces a crisis. In the past few years, one pharmaceutical giant after another (Merck, Novartis, Pfizer, Sanofi) have shrunk or even stopped their neuroscience research projects.

As **Steven Hyman**, former head of the National Institute of Mental Health, wrote recently the notion that "disease mechanisms could ... be inferred from drug action" has succeeded mostly in "capturing the imagination of researchers" and has become "something of a scientific curse. As the result no novel psychotropic drug emerged in the last thirty years."

The Brain Research through Advancing Innovative Neurotechnologies (BRAIN) initiative recently announced by the Obama Administration and the European Human Brain Project with their focus on **neuronal circuits** in the pathogenesis of mental illness may open **new horizons** in psychopharmachology.

CHAPTER 4.2

Neurofeedback

DEFINITION

By definition, neurofeedback (NF) is a **technique of self-regulation** by means of **EEG-based biofeedback**. In this technique, some **current parameters of EEG** recorded from the subject's scalp (such as EEG power in a given frequency band) are presented to the subject through **visual, auditory, or tactile modality**. The subject is supposed **voluntarily** or **involuntarily** (depending on protocol) to **alter** these parameters in a direction that leads to a more efficient mode of brain functioning.

The NF technique in visual modality is schematically illustrated at Fig. 4.2.1. The voltage difference between two places on the head is recorded by two electrodes, amplified and processed by a computer to extract the required NF parameter from the current EEG. The NF parameter is presented on a computer screen located in front of the subject.

The information flow is schematically presented by arrows. One can see that the flow is organized in the closed loop with the brain as a part of it. In simple terms the brain looks at itself as in a mirror and is in a position to adapt accordingly. The main questions that have to be answered by a NF practitioner are: (1) what is the NF parameter to select; (2) where should the electrodes be placed; and (3) which direction the brain is supposed to adapt? The answers to these questions define a **protocol of NF**. Sometimes other names are used for this technology such as EEG biofeedback, neuroregulation, neuronal regulation, EEG-based self-regulation, neurotherapy.

Historical Introduction

The roots of the modern NF approach lie in the Pavlovian method of **conditioned reflexes**. The Russian psychologist **Ivan Pavlov** was the first to study physiological mechanisms of psychological functions and did so at the start of the 20th century using this method. In the Pavlovian method a conditional stimulus, initially having little behavioral significance for an animal, is associated in time with some reinforcement or reward

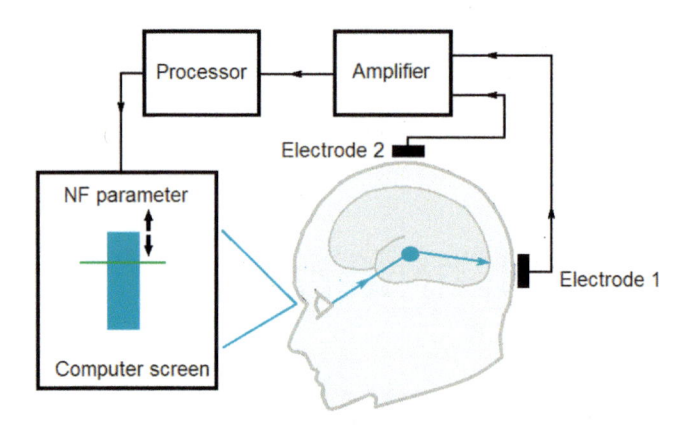

FIGURE 4.2.1 **NF technique.** The voltage difference between two electrodes is amplified and fed into the processor of a computer. The processor online extracts the NF parameter which is reflected on the screen by the amplitude of the *blue bar*. The subject is instructed to keep the parameter (the *blue bar*) above the threshold *(green line)*. The visual information goes to the retina of the eyes and through the thalamus reaches the visual cortex. The visual cortex itself sends information to the affective and executive systems of the brain so that the whole brain can be involved in NF.

(unconditional stimulus in Pavlovian terminology). Conditional stimulus could be a visual or an auditory stimulus. Unconditional stimulus could be a piece of meat for dogs or a small amount of juice for monkeys. The reward—having vital significance for the animal—induced essential changes in its behavior and physiological reactions (salivation, for example). Now, if feeding the dog has been accompanied many times by the sound of a metronome the dog will salivate in response to the sound itself—a conditioned reflex has been established. The experiments were carried out by Ivan Pavlov in the Institute of Experimental Medicine in Saint Petersburg, which was founded in 1890. This is the institute where the author of this book started his scientific carrier in 1972. Our laboratory is located just 200 m from Pavlov's laboratory. In a museum named for him there is a harmony—an old musical instrument similar to an organ—which Pavlov used for his experiments to generate conditional stimuli.

At the end of the 1940s a student of Pavlov's Petr Kupalov invented a methodology called "**situational conditioned reflexes.**" In the West this method was coined **operant conditioning**. In the method **animal behavioral reactions** but not external stimuli served as conditional stimuli.

In 1930, American mathematician Norbert Wiener in collaboration with a Mexican psychologist Arturo Rosenblueth introduced the **concept of feedback** in relation to biological systems. This concept was further evolved into a new science coined by Wiener **Cybernetics** in his book published in 1948. In 1935, a Russian scientist **Petr K. Anokhin**, a student of Pavlov and Bechterev, developed the theory of functional systems. The key element of this theory was neuronal feedback—the interaction between a so-called "acceptor of actions" and behavioral adjustment of the animal.

Following these traditions of the Russian school of physiological science, **Nikolay Vasilevsky**, Professor of the Institute of Experimental Medicine, began his studies of cellular mechanisms of NF regulation in the late 1960s and early 1970s. Vasilevsky was the first to design a biofeedback protocol on the basis of **heart rate variability** and used this methodology to train Soviet cosmonauts. **Natalia Chernigovskaya** from the same institute started using the method of biological feedback to treat some neurological and psychiatric diseases. EEG patterns, muscle electric activity, and slow metabolic processes were used for biofeedback.

Approximately at the same time, **Joe Kamiya**, an American researcher at the University of Chicago observed that subjects could learn to voluntarily control their alpha waves. Thus, though Norbert Wiener and Petr Anokhin formulated the idea of feedback in the 1930s and 1940s, only in the 1960s was it shown that EEG parameters can serve as feedback for self-regulating cortical rhythms.

Barry Sterman from UCLA in his studies with cats introduced a rhythm associated with the sensorimotor system and named it "**sensorimotor rhythm**" (SMR). Using operant conditioning, cats were trained to produce this rhythm for food reward. It was also discovered that overtraining protected these animals from experimentally induced seizers. Shortly after that, training of the SMR was successfully applied to epileptic patients to reduce seizures.

In **1969** the method of brain self-regulation by means of EEG and other physiological parameters was officially named "biological feedback" (**biofeedback**). At the same time, while EEG biofeedback was implemented to treat neuroses and epilepsy, it also started to be used as an antistress rehabilitation procedure for Vietnam veterans. Theta training was added to experimental protocols of EEG biofeedback with the aim of bringing a subject into a changed state of consciousness. At the end of the 1970s, **Niels Birbaumer** and his coworkers from the University of Tübingen (Germany) started using infraslow electrical cortical potentials for biofeedback treatment of epilepsy.

That was a period of euphoria when biofeedback seemed to be a possible panacea for all brain diseases. However, in 1974 an article by James Lynch and coworkers from University of Maryland, Baltimore, showed that subjects who learned to control their own alpha rhythm with their eyes open were not able to increase it more that they could do with their eyes closed. The article showed limitations of human abilities in EEG biofeedback, but the constraints were treated too literally and NF became unpopular.

There seemed to be many reasons NF got a bad reputation in those years. One of them was **misusing** this approach in clinical applications.

Note that it is easy to perform NF in the wrong way because of lack of experience, lack of objective criteria to select an appropriate protocol—recall that in those years QEEG was seldom used before NF treatment—and lack of solid experimental support and validation of the selected protocol. So, the transition from laboratory experiments to clinical practice was too fast and the obvious limitations of NF were used against this approach by its opponents.

However, a few enthusiasts continued to work with EEG biofeedback. Studies by **Joel Lubar** from University of Tennessee played an important role in the field during the 1970–80s. He and his coworkers proved that sessions uptraining the beta rhythm and simultaneously downtraining it significantly reduced hyperactivity and improved attention in **ADHD patients**.

In the early 1970s **Roy John** from New York University developed a new approach in the EEG field coined **neurometrics**. The idea behind neurometrics was to quantitatively compare parameters of an individual EEG with those computed for a healthy control group. It was a revolutionary idea because until then only visual inspection of raw EEG signals had been considered a gold standard in electroencephalography.

In the mid-1980s, two approaches—EEG biofeedback and neurometrics—merged forming a new direction that is now named **neurotherapy**. The American company Lexicor was the first to produce neurotherapeutic devices. At present, the new approach is actively developing and extending. New views for the genesis of EEG rhythms made it possible to form a theoretical basis for the neurotherapeutic approach. Several companies are now specialized in EEG analysis and development of individual neurotherapeutic protocols.

At the beginning of the 21st century a NF protocol with infraslow electrical potentials regained popularity after clinical studies by **Sue and Siegfried Othmar** from the EEG Institute in Woodland Hills, California. In 2014, **Martijn Arns** from Utrecht University, **Hartmut Heinrich** from the University Hospital of Erlangen and **Ute Strehl** from the University of Tübingen reviewed studies of the application of NF for ADHD children focusing on semiactive, active, and placebo control groups. The authors concluded that standard protocols such as theta/beta, sensorimotor, and slow cortical potential NF were well investigated and have demonstrated specificity.

Conventional Electroencephalogram and Infraslow Fluctuation-Based Neurofeedback

As presented in chapter: Infraslow Electrical Oscillations there are two types of phenomena in broadband EEG: (1) spontaneous fluctuations in the 0.5- to 30-Hz band (the **conventional EEG)**; and (2) **infraslow**

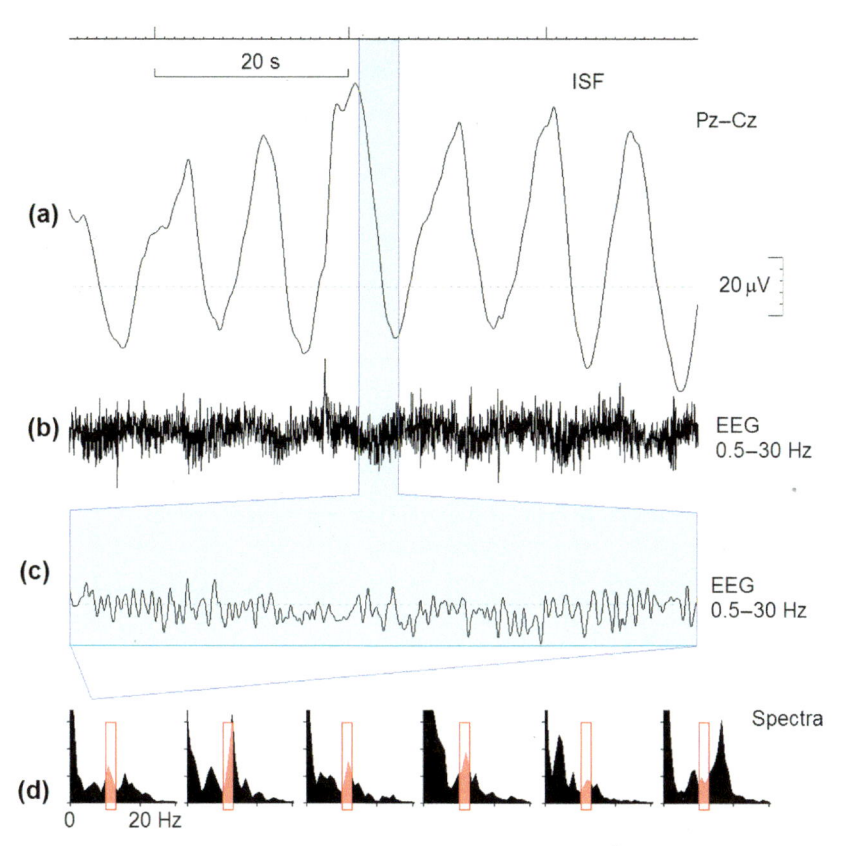

FIGURE 4.2.2 Two variants of NF parameters are extracted from a broadband EEG. (a) A 1-min fragment of resting-state spontaneous infraslow fluctuations in the referential Pz–Cz montage in a healthy subject; (b) a corresponding fragment of a conventional EEG at the time–amplitude scale; (c) a 4-s fragment is expanded to show fluctuations in the alpha frequency band; and (d) EEG spectra for consecutive 4-s fragments are presented with the individual alpha band marked by the *red bars*.

fluctuations (ISFs) with frequencies below 0.1 Hz. Fig. 4.2.2 illustrates these phenomena in a healthy subject in the referential montage. For illustration purposes the Pz–Cz montage is selected. Note that ISFs demonstrate large-amplitude (around 100 μV) fluctuations with a period c. 10 s whereas the conventional EEG is characterized by low-amplitude (<10 μV) oscillations in the alpha frequency band.

As was shown in chapter: Infraslow Electrical Oscillations, the functionally meaningful parameter of ISF could be a phase of fluctuation (increase or decrease) or simply the amplitude itself. For the conventional EEG the functionally meaningful parameter could be the power in a certain frequency band (such as the alpha band), the ratio of spectral power in different bands (such as the theta/beta ratio), or coherence (such as

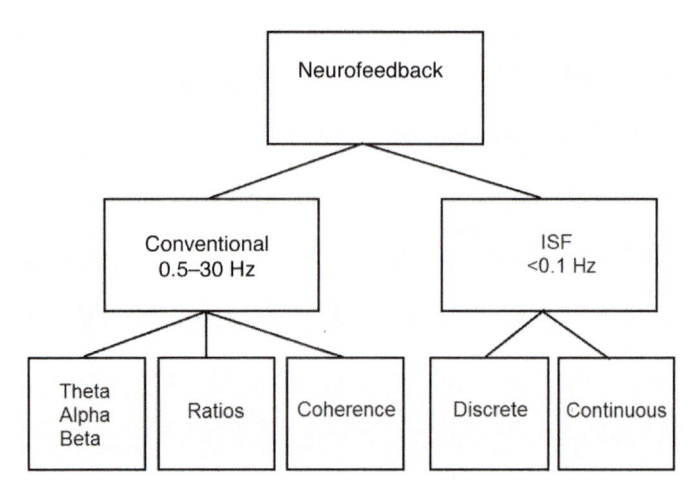

FIGURE 4.2.3 Classification of NF protocols (see text for explanation).

coherence between the fronto- and parietocortical areas). Fig. 4.2.2d illustrates the dynamics of EEG spectra computed for 4-s EEG fragments. The individual alpha band is marked by the *red* bar.

According to the two bioelectrical phenomena in the brain, two variants of NF have been suggested: **conventional EEG biofeedback** and **ISF NF** (Fig. 4.2.3).

Conventional NF uses spectral EEG characteristics as a NF parameter. The characteristics include (1) **EEG power** in fixed or individually tailored frequency bands such as the theta, alpha, and beta bands; (2) **ratios** of the power in different frequency bands; and (3) **coherence**.

ISF NF uses either the amplitude itself or phase (increase vs. decrease) of the voltage fluctuations. NFeurofeedback is done in two forms: (1) **discrete form** when the subject is instructed either to voluntarily increase or decrease recorded potentials in separate trials; (2) **continuous form** when the subject is simply watching a computer game controlled by the NF parameter. The brain is supposed to recognize its activity with respect to the NF parameter, and then it naturally incorporates the signal into its feedback loop (as in Fig. 4.2.1).

Relaxation and Activation Electroencephalogram-based Protocols

There are too many NF protocols in the field and there is no gold standard. Research and commercial groups have been experimenting with various protocols—not bothering much about providing proofs of their efficacy. For didactic purposes we divide the protocols of EEG-based NF into two big categories: **activation and relaxation protocols**. This division is made on the basis of data regarding the correlation between EEG activity in different frequency bands and the metabolic activity measured by PET or fMRI in the corresponding cortical location—see Fig. 2.3.9 and insert (b) in Fig. 4.2.4. Activating protocols include **enhancement of EEG power at beta frequencies** or suppression of EEG power at alpha frequencies. Relaxing protocols include **enhancement of EEG power at alpha frequencies** or inhibition of EEG power at beta frequencies. Although the importance of electrode locations cannot be overstated, below for didactic purposes we consider the protocols without any reference to electrode location.

Beta-activating training is a common method used for a variety of purposes. The protocol is supposed to **activate** the corresponding cortical area. In our experience with **relative frontal beta** training in ADHD children we sometimes encounter an "aha" phenomenon when the trainee discovers the exact type of mental state that is associated with the production of higher amplitude beta activity. In such cases the patients used to say: "Aha, now I know what it means to be attentive." Possible negative

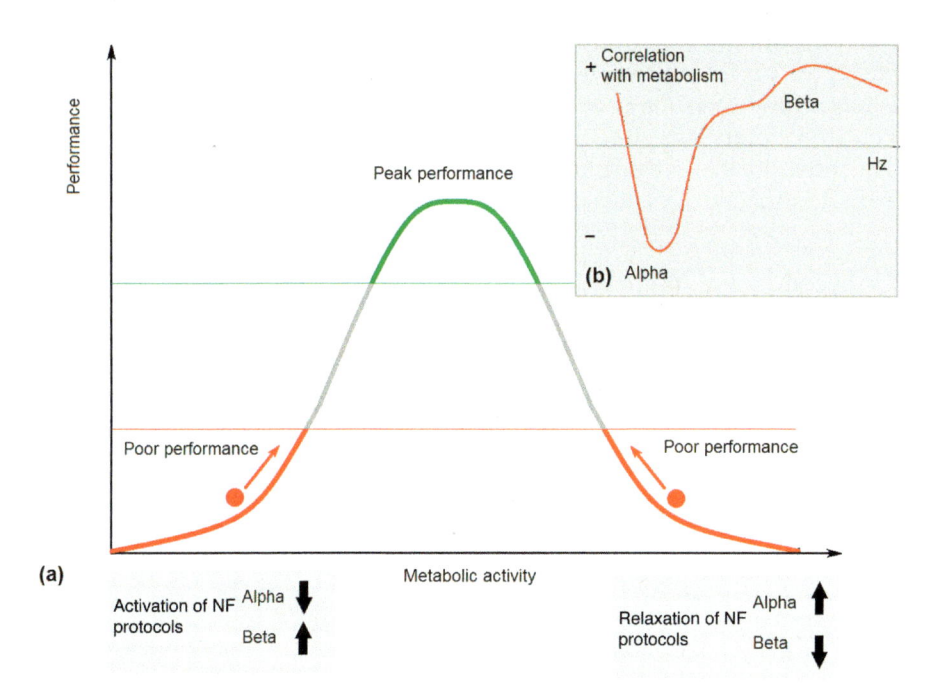

FIGURE 4.2.4 **Activation and relaxation NF protocols (scheme).** (a) The inverted U-law defines the relation between overall (metabolic) activity of a neuronal system and its performance. Too low and too high levels of neuronal activation are associated with poor performance, while intermediate activation is associated with peak performance. Taking into account the opposite correlation of alpha and beta power spectra with metabolic activity, (b) the protocols at the extreme sides of the inverted U curve are different: activation at the low metabolic level and relaxation at the high metabolic level.

side-effects may include agitation, irritability, or a sense of being "hyper." Therefore, sessions may be as short as 10 min, but usually last for 20–30 min. Often, beta training is used at the end of an EEG session to bring the trainee into a state of energy and alertness.

Alpha (8–12 Hz) relaxation training is a common method used for many purposes. It usually deactivates, **inhibiting** the corresponding cortical areas. Relaxation training is often accompanied by presenting auditory stimuli because visual stimuli activate and destroy the brain to a larger extent than auditory stimuli. In the experience of my laboratory, **enhancement of posterior alpha or central mu-rhythms** appears to provide the best results in relaxation training.

The **Peniston–Kulkosky protocol** is a variant of relaxation protocols. Subjects are in an eyes closed and relaxed condition while receiving auditory signals presented on the basis of EEG recorded at the O1 electrode. A standard induction script that uses suggestions to relax and "sink down" into reverie is read. When alpha (8–12 Hz) brain waves exceed a preset threshold, a pleasant tone is heard. By learning to voluntarily produce this tone the subject becomes progressively relaxed. When theta brain waves (4–8 Hz) are produced at sufficiently high amplitude, a second tone is heard, and the subject becomes more relaxed and, according to Eugene Peniston, enters a hypnagogic state of free reverie and high suggestibility. The transition from the state with dominance of alpha activity to the state with dominance of theta activity is called "**crossover**."

Coherence and Virtual Intracranial Electroencephalogram Protocols

It should be stressed here that the methods of construction of NF protocols are not confined by fast Fourier transform or digital filtration of the current EEG. Other NF parameters have been suggested. One of these parameters is a measure of correlation between EEG in different cortical locations, such as **coherence or comodulation**—suggested by Barry Sterman from the University of California at Los Angeles and David Kaiser from Sterman–Kaiser Imaging Lab, Bel Air, in their normative database.

The other approach is to use EEG from **virtual intracortical areas**. The rationale for this approach is the fact that any scalp electrodes pick up activity from widely distributed cortical areas. So, if we want to use NF from a local cortical area, we need to apply tomographic methods that enable us to compute a virtual local EEG on the basis of multiple scalp-recorded EEGs. The NF method based on **Low Resolution Tomography (LORETA)** was first applied in 2004 by Marco Congedo in the laboratory of Joel Lubar at the University of Tennessee.

Low-resolution electromagnetic tomography provides a tool to compute the current density of generators located on the surface of the cortex including gyri and sulci. The solution is **a matrix** which, multiplied by multichannel EEG data, gives the result—the current density of cortical microdipoles. The location of the virtual source within the cortex unequivocally gives the appropriate matrix.

Marco Congedo and coworkers used LORETA-based NF to enhance low-beta (16–20 Hz) and suppress low-alpha (8–10 Hz) current density amplitude (vector length) in a region corresponding approximately to the anterior cingulate cognitive division. The region of interest includes 38 voxels from the total number of 2394 voxels. The authors coined this method as **LORETA NF**. In 2006, the sLORETA based neurofeedback protocol (that is supposed to provide no localization error) was developed in the author's laboratory, and was demonstrated for the first time at the workshop on neurofeedback organized by Lutz Janke from University of Zurich.

Slow Cortical Potential Neurofeedback

This protocol is based on the observation that preparatory psychological operations are accompanied by slow negative fluctuations recorded from the cortex with the largest amplitude at the vertex. These fluctuations are united by a single concept—contingent negative variations (CNVs) (see chapter: Executive System and Cognitive Control). Some studies show that ADHD children evoke smaller amplitude CNV waves in delayed response tasks than healthy controls. From this perspective a NF protocol in ADHD would be to **train the slow activity at Cz**. The rationale behind the protocol is that training the **Cz** slow potential toward negativity would be associated with **activation of the cortex** (increased firing rate of cortical neurons due to long-lasting depolarization of apical dendrites), while training the Cz potential toward positivity would **inhibit corresponding cortical areas**.

In this protocol, subjects learn to **voluntarily generate** negative and positive shifts of potential recorded from Cz. To correct for eye movement artifacts, two or four electrodes record potentials generated by vertical and horizontal eye movements in addition to reference and ground electrodes. Although the authors of this protocol call it the slow cortical potential (SCP) protocol, by definition training potentials belong to ISFs.

The protocol was designed at the University of Tübingen by a team of researchers headed by Niels Birbaumer. SCP feedback is discontinuous and trial based, and consists typically of three phases: a baseline phase (2 s), an active phase (5–8 s), and a reinforcement phase (2 s). During the active phase the participant has to change potential in the cued direction: up for activation (**negativation**) and down for deactivation (**positivation**). A visual stimulus, given as a "reward" during the reinforcement phase if the cued brain state is reached, informs participants whether they were successful. Positivation and negativation trials are presented randomly. About 40 trials constitute a "run." A session may contain two to four runs.

Although the CNV wave is separated into several components with different localization and functional meaning (see chapter: Executive System and Cognitive Control) this is not taken into account, and the protocol simply presumes measuring the wave at the Cz electrode. This one-size-fits-all protocol provides a certain advantage at the current stage of research because it presumes consistency and compatibility among different laboratories.

In a 2003 study by Thilo Hinterberger and coworkers from the University of Tübingen the relationship between negative and positive slow **cortical potentials and changes in the BOLD signal** of the fMRI were examined in subjects who were trained to successfully self-regulate their slow potentials. fMRI revealed that the generation of negativity was accompanied by widespread activation in **central, prefrontal, and parietal brain regions as well as the basal ganglia**. Positivity was associated with widespread **deactivation** in several cortical sites as well as some activation, primarily in frontal and parietal structures as well as insula and putamen.

Slow cortical potentials have been used by the Tübingen group in a thought **translation device**. The method was designed to reestablish communication in severely paralyzed patients. The device relies on the self-regulation of slow cortical potentials, that is, the voluntary production of negative and positive potential shifts. After a patient has achieved reliable control over his or her slow cortical potentials the responses can be used to select items presented on a computer screen. A spelling program allows patients to select single letters by sequential selection of blocks of letters presented in a dichotomic structure with five levels. Several completely paralyzed patients diagnosed with amyotrophic lateral sclerosis were able to write messages of considerable length using their brain potentials.

Infralow-Frequency Neurofeedback

In contrast to the one-size-fits-all protocol of the Tübingen group the infralow-frequency NF protocols designed by **Sue and Siegfried Othmer** are based on clinical reports of clients and are **different for different symptoms**. According to Siegfried Othmer their approach was an evolution of the original Sterman protocol of training the SMR. In the early years of this century Sue Othmer started to optimize the training frequencies of each client for improved training efficiency. This led to a progression to lower frequencies to meet the needs of the most challenging clients. Eventually, this progression led to the adoption of training in the SCP domain, ie, below 0.1 Hz. The authors named this protocol as infralow-frequency (ILF) training. They used bipolar montage in which the measurement is sensitive to differential cortical activation at the two sites.

ILF training is a **continuous training protocol** without any **preset threshold**. It involves signal following in which the brain recognizes its agency with respect to the signal, and then it naturally incorporates the signal into its regulatory schema. The only relevant information the brain has to work with is the real-time dynamics of the signal. The immediate consequence is typically felt by the trainee as a shift toward calmer states, which can take place within minutes. The first publication on the new method was an observational study on the effect of ILF training on chronic pain (Othmer and Othmer, 2006).

Medical versus Nonmedical Application

Biofeedback is used for both medical and nonmedical applications. Until recently, the US Food and Drug Administration (US FDA) recognized only **relaxation training** as an accepted use of EEG biofeedback. Medical applications, for example, protocols for treatment of depression, seizures, headaches, autism, were considered either experimental or unproven. However, nonmedical applications such as protocols for general improvement in concentration and attention and for peak performance did not meet objections. Any EEG system designed to enable the user to train for recreational, educational, or entertainment purposes was not a medical instrument. The FDA or the equivalent agency in other countries allowed such devices to be produced and marketed freely, as long as they posed no undue health risk to the user.

Any EEG-based biofeedback equipment may be provided in two versions, addressing both the medical and nonmedical communities. When marketed for clinical purposes, specific claims are made, and these claims must be reviewed and cleared by the US FDA or its equivalents in other countries. In the nonclinical embodiment the equipment is regarded as an educational and recreational device.

So far, published clinical studies deal with ADD/ADHD, conduct disorders, learning disabilities, anxiety, depression, chronic fatigue syndrome, epilepsy, and autistic spectrum disorders. Neurotherapy has also been applied to help drug addicts relax; rehabilitation after stroke and traumatic brain injury; and enhancing cognitive functions in aging. For nonmedical application, NF is used to improve attention and concentration; for peak performance; assistance in meditation and relaxation techniques; improvement of personal awareness; improvement of mental fitness.

Neurophysiological Basis: General Facts

EEG-based NF as a neuromodulation procedure is based on three **basic scientific facts**:

1. **Brain state** and **brain operations** are **objectively reflected in parameters of EEG** recorded from the scalp. Numerous data show that EEG patterns dramatically change in the sleep–wakefulness cycle

and that alpha, theta, and beta waves respond differently in sensory, cognitive, and affective tasks.
2. In case of active (requiring active involvement of subjects to achieve the goal of NF) protocols the subject can **voluntarily and selectively change the EEG parameters selected**. In case of passive protocols the "brain" can recognize feedback as an index of its own activity and can be involved in the feedback circuit.
3. The human brain has **plasticity** to memorize the desired (and thereby, rewarded) state of the brain. Training during NF sessions can be compared with learning to ride a bicycle: at the beginning it is hard with a lot of trial and error, but when the skill is achieved it is never forgotten.

Neurophysiological Basis: Relative Frontal Beta Training

In this section we show how the general ideas described in the previous section can be specifically applied in ADHD research.

The first basic idea of NF concerns how certain psychological operations are reflected in event-related desynchronization or synchronization of certain EEG rhythms. Fig. 2.3.5 illustrates the **reactivity of beta rhythms** in the cued GO/NOGO task for healthy control subjects. One can see that the focused attention and preparatory set induced by the cue are associated with activation of the frontal beta rhythm. As shown in chapter: Beta and Gamma Rhythms the frontal beta rhythm appears to reflect the activation level of the frontal cortex.

In children with attention problems, reactivity of the beta rhythm is smaller than in healthy controls (Fig. 5.1.8). This observation enabled us to suggest a relative beta-training protocol with Fz–Cz placement of electrodes and the NF parameter as the ratio of EEG power in the 13- to 21-Hz band and EEG power in the 4- to 12-Hz frequency band. When the subject is using the frontal lobe, such as in focusing attention, the beta rhythm is activated. Using this NF parameter the subject can associate the state of the focusing attention with increased frontal relative beta rhythm. This NF protocol was used for training attention in a 2005 study by our group at the Institute of the Human Brain of the Russian Academy of Sciences (Fig. 4.2.5).

The second basic idea concerns the ability of people to voluntarily and selectively change the EEG parameters selected. This idea is illustrated in Fig. 4.2.5a by the training curve. The training curve objectively demonstrates that an ADHD boy is able to learn voluntarily enhancing his relative beta rhythm frontally. Note that although the map of beta increase is widely distributed the maximum is located at frontal areas (Fig. 4.2.5b).

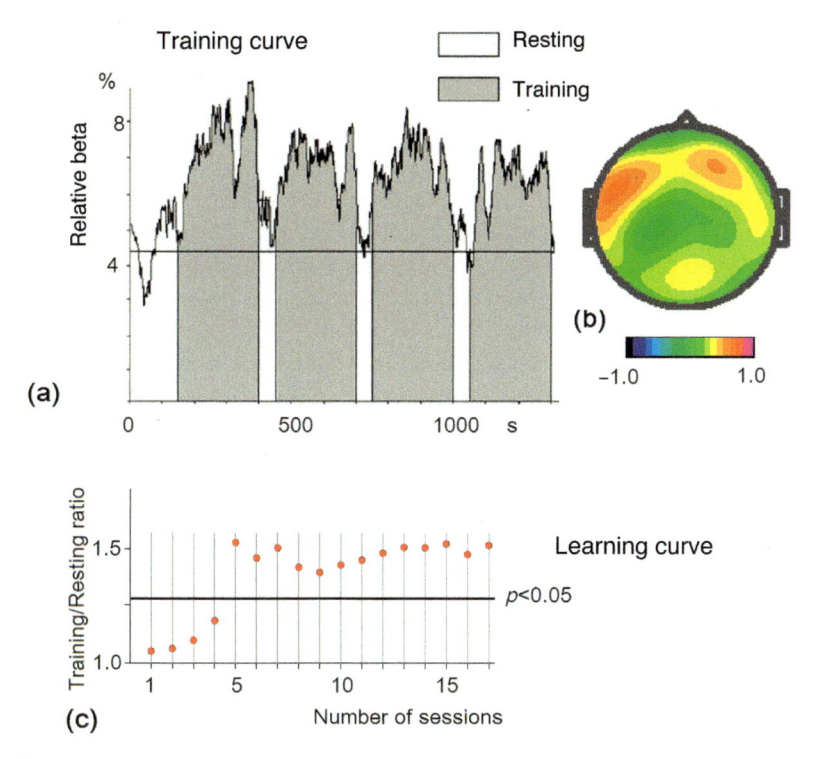

FIGURE 4.2.5 **An ADHD patient is able to learn voluntarily enhancing the relative beta activity.** (a) Training curve—dynamics of relative beta activity during the fifth NF session in an ADHD patient; electrode position Fz–Cz; *white* areas, resting state; *gray* areas, training state; (b) map of changes of beta activity during NF sessions; and (c) learning curve—dynamics of the training/resting ratio for the NF parameter over 17 sessions. *Adapted from Kropotov et al. (2005).*

Third, as far as human brain **plasticity** is concerned, some studies showed that the progress made in NF sessions in the laboratory was transferred into real life in school and family.

A Need for Theory

Compared with the pharmacological approach the treatment of mental illness by NF is a small field. Although the number of NF practitioners are constantly growing, they are still much **smaller** than the number of medical doctors who use the pharmacological approach in their practice.

The history of NF from the theoretical perspective is different from that of the pharmacological approach. At the beginning, NF—such as training the SMR in cats in Sterman's lab—was driven by the **theory of operant conditioning**. To prove the specificity of the effect, early studies used the **"veritable–sham–veritable" approach** with the idea that training NF parameters in the opposite direction (sham situation) must change behavioral parameters in the opposite direction.

However, as in the pharmacological approach, **clinical application** of operant conditioning was discovered **accidently**. In a serendipitous twist, in the 1960s Sterman's laboratory was asked to establish dose–response functions of a highly epileptogenic fuel compound. The cats that had previously taken part in operant conditioning of the SMR were found to display significantly elevated epileptic seizure thresholds compared with untrained animals! This discovery was successfully extrapolated to **epileptic patients**. In clinical research it was shown that seizure incidence could be lowered significantly by feedback training of the human SMR.

Based on the research associating increased theta amplitude with attention deficits, SMR training was later developed into **beta/theta training**. Beta/Theta training rewarded patients for increasing the 13- to 15-Hz band or another subset of the beta bandwidth (13–21 Hz), while simultaneously decreasing theta activity (4–8 Hz).

NF researchers originally studied the effect of NF on **stress, epilepsy, and ADHD**. NF practitioners claim that the method can be applied to a wider range of disorders including **developmental disabilities, traumatic brain injury, stroke, alcoholism, autism spectrum disorder, depression, insomnia, migraines, and chronic pain**. Despite these positive claims the question of the specificity of NF effects still remains to be answered.

One of the problems in NF application seems to be the absence of a neurophysiological theory. An attempt to find such a theory was formulated in the **bulldozer principle of NF**.

Bulldozer Principle

The goal of a NF practitioner is to define the NF protocol for a particular subject on the basis of (1) the client's complaints, (2) spontaneous EEG responses and EEG responses in functional tasks, and (3) on the basis of what is known from current research in terms of the functional neuromarkers of the disease, and (4) on the basis of world-wide experience of using NF protocols for a particular dysfunction.

Those practitioners who apply QEEG assessment as the basis for constructing NF protocols implicitly or explicitly use the **bulldozer principle of NF** (Fig. 4.2.6). According to this principle the aim of NF is to **normalize a pathologically abnormal EEG pattern**. So, if there is an excess in an EEG parameter in a particular patient and in a particular location in the cortex the corresponding parameter must be downtrained. If there is a shortage in an EEG parameter this parameter must be uptrained. In other words the principle works like a bulldozer filling in cavities and excavating

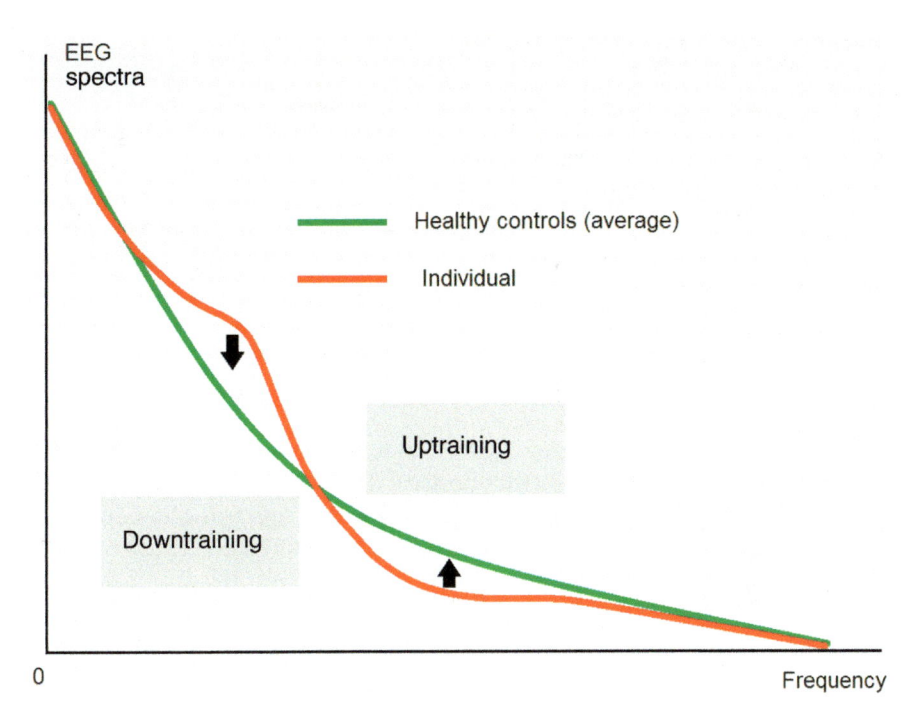

FIGURE 4.2.6 **Bulldozer principle of NF.** The goal is to "normalize" the EEG spectra, that is, for a given individual *(red)* to enhance activity in the frequency band with lower EEG spectra power compared with healthy controls (training UP) and to inhibit activity in the frequency band with higher EEG spectra compared with healthy controls.

bumps. Using this idea some companies offer so-called **z-score training** which train down or up the deviations in the EEG spectra from the reference computed for healthy controls.

The bulldozer principle has not been tested systematically, and there are at least two reasons to be very cautious with this approach. First, the deviation of individual EEG spectra from the average of a healthy control group might reflect individual **uniqueness** of the brain. I saw many EEG spectra of highly performing healthy subjects whose EEG spectra deviated from the mean. Second, the observed deviation from the reference might reflect a **compensatory** mechanism, for example, deactivation of posterior regions of the brain might be compensated by activation of the frontal regions of the brain—a compensatory process seen in elderly subjects. In all these situations it would be dangerous to use the bulldozer principle to suppress deviant activity.

Personalized Approach

As will be shown later (part: Neuromarkers in Psychiatry) a certain brain disorder is associated with a specific pattern of functional neuromarkers in the self-regulation domain (QEEG) and in the domain of information flow (ERPs). There might be several profiles of functional neuromarkers for a single disease. For example, there are several endophenotypes of ADHD in the QEEG domain including excessive central theta and excessive frontal beta endophenotypes. In the first case the QEEG of an ADHD patient is characterized by an excessive theta/beta ratio; in the second case the ratio would be rather small. There is no reason to believe that a single protocol of conventional EEG NF (like relative beta training) would fit every subtype of ADHD.

To define the QEEG endophenotype of a particular patient we need to record **spontaneous EEGs** in this patient. This is done to compute multichannel spectra, map them, and compare the spectral characteristics with data obtained from healthy controls of the same age. Based on large normative and patient databases, there are several studies that have suggested QEEG endophenotypes. Jack Johnston from UCLA, Jay Gunkelman from Q-Metrix, Inc. and J. Lunt from Brain Potential, Inc. by reviewing a large number of clinical EEG and qEEG studies suggested a limited set of individual QEEG profiles (named by the authors EEG phenotypes) that characterize the majority of the population and that require specific neurofeedback protocols for optimization of behavior. In 2008, Martijn Arns from Research Institute Brainclinics, Netherlands and his colleagues confirmed that these EEG phenotypes are identifiable EEG patterns with good interrater reliability. Furthermore, they demonstrated that the same EEG phenotypes occurred in ADHD patients and could be used for predicting treatment response. Andreas Mueller and Gian Candrian from Brain and Trauma Foundation, Switzerland, and the author of the present book in the 2011 book suggested a similar classification of QEEG abnormalities in ADHD.

However, not every patient shows deviations from the mean reference. According to the HBI database a considerable portion of patients in a particular diagnostic category (from 10 to 70% depending on category and age) may show QEEG patterns which do not deviate significantly from the mean spectra of healthy controls. This observation means that in a considerable fraction of the population of mentally sick patients the cortex is **optimally self-regulated**. Usually in such cases strong deviations are found in ERPs—the indices of information flow within the cortex. These cases may need different kinds of neuromodulation such as **TMS, tDCS, or even DBS**.

Specific and Nonspecific Effects

Many studies report improvements after NF training. However, there might be **specific and nonspecific effects** of such interactive intervention. Since NF requires many treatment sessions (20–40 sessions) and thus implicates a substantial amount of client–therapist interaction, **nonspecific effects** occur much as is the case in any other therapeutic relationship, and planned control groups should control for such nonspecific factors.

Nonspecific effects could consist of (1) **cognitive training** by having a child focus on a computer screen for 30–40 sessions, or (2) **positive reinforcement** in the form of feedback shown on the computer screen and of verbal appraisals given to a child by the practitioner during or after session. Early papers on NF clinical outcome were often criticized for not considering nonspecific effects.

Recent studies take nonspecific effects into account by applying different control conditions. These control conditions require similar nonspecific factors such as cognitive training tasks, muscle-training biofeedback, or a different NF protocol. However, the gold standard in any intervention research is the **double-blind placebo control protocol**. This protocol consists of a control condition where everything is identical, except that in this case feedback is not related to the brain activity of the subject. This protocol has some methodological problems in the application of NF which were discussed in a 2014 review by **Martijn Arns** from Utrecht University, **Hartmut Heinrich** from the University Hospital of Erlangen, and **Ute Strehl** from the University of Tübingen.

Advantages and Limitations

Like any other method, NF should not be used without relevant **diagnostic procedures**. Though there are quite a few reports about the negative consequences of biofeedback usage, one should always remember that a patient's brain could be dysregulated and symptoms could worsen if the organization of EEG rhythms is changed in the wrong direction. NF must be performed only by a certified specialist. More and more universities worldwide have started providing courses on biofeedback, in general, and NF, in particular.

There is a parallel between NF and **Eastern self-regulation techniques**. Ancient Eastern arts of self-regulation such as yoga and chi gung are in fact based on the ability of humans to voluntarily control their own physiological processes. However, it takes a long time to master those arts. Moreover, direct tutor–pupil interaction is often needed. In NF the learning process takes essentially **less time** and is **not confined by relaxation protocols**.

Special attention should be paid to **placebo effects** in NF. The placebo effect is based on certain neuronal mechanisms. It can amount to 30% of the "pure" effect of a medical drug, and it would make little sense to negate its presence in NF.

An advantage of NF is that it **minimizes side-effects**. It is not a secret that many pharmacological drugs affect not only the relevant neurotransmission but also other biochemical processes in the brain and in the body. This is why any description of a psychopharmacological drug includes a long list of side-effects. In contrast to pharmacological drugs, NF is supposed to affect brain processes selectively.

Another advantage of NF is the **stability of the effect**. The procedure is often likened to the process of acquiring a specific skill, for example, riding a bicycle. One can hardly balance a bike when getting on it for the very first time, but after several attempts one learns to ride holding the handle bars; some of the more daring, after long training, manage even to ride without holding the handle bars. The same happens in NF: at the beginning, it is difficult for a subject to get the association between his/her own feelings and indexes of bioelectric activity displayed on the screen. However, as the association becomes gradually understandable the penny drops: "Aha, I know how to do it!" Furthermore, skills are improved and one learns to hold the needed condition for a longer time—not only in the laboratory but also at home or in school. After 20–40 sessions, NF can be ceased but the skill will remain for life! Joel Lubar tracked the histories of more than 50 of his patients for 10 years using objective measurement of behavioral parameters. Positive treatment results appeared to sustain for the whole period in almost all the patients; however, some needed additional series of procedures to refresh their skills, and after one or two procedures the skills returned. Many people said that NF had changed their lives.

Does all of this mean that NF has no disadvantages at all? One of the factors restricting usage of NF is that the sessions NF **consume about 20–40 h of labor** from the clinician and the same amount of intense training from the patient. To reduce traveling costs, modern centers for NF offer home trainers. In these cases, several initial sessions of neurotherapy are performed together with a specialist and, once the patient has developed primary skills, training sessions are continued at home. The patient can buy or rent a neurotherapeutic device that can easily be installed at home and connected to a computer or to a TV set.

Another factor to be considered in NF is **the motivation of a patient**. NF like any learning procedure is based on mood and motivation. The higher the motivation level, the better the acquired skills. Some NF centers stimulate children's motivation by rewards in the form of money, toys, or tokens that can be later exchanged for toys or money. **Assistance from the family** also plays an important role in successful NF. Indeed, during NF the brain gradually changes, but to transfer these small changes into behavioral patterns additional efforts are needed. For example, attention is a parameter we are supposed to improve using the relative beta protocol, but attention is not the only parameter that defines behavior. To have good attention is not enough, and children must be able to use this new brain skill to obtain knowledge at school, as well as to adapt their behavior when socializing with parents or other children. Parents and teachers may need special training to be able to help children get over socioenvironmental difficulties.

Nonelectrical Neurofeedback

Real-time fMRI (rtfMRI) of a subject during an MRI scan provides a tool to voluntarily change a local BOLD signal in any region in the brain. One of the first studies in this direction was done in 2005 by R. Christopher deCharms and coworkers from Stanford University when they found that—using rtfMRI as biofeedback—participants were able to learn to control activation in the rostral anterior cingulate cortex, a region putatively involved in pain perception and regulation. When subjects deliberately induced increases or decreases in the BOLD signal in the rostral part of the anterior cingulate cortex, there was a corresponding change in the perception of **pain** caused by a noxious thermal stimulus that was applied. Control experiments demonstrated that this effect was not observed after similar training conducted without rtfMRI information, using rtfMRI information derived from a different brain region, or sham rtfMRI information derived previously from a different subject.

Chronic pain patients were also trained to control activation in rACC and reported decreases in the ongoing level of chronic pain after training. Seven years later in 2012 the first international conference on rtfMRI NF was held at the Swiss Federal Institute of Technology in Zurich. Following this meeting, perspectives on rtfMRI feedback for basic and clinical applications were presented in a 2013 review by James Sulzer and coworkers from the Swiss Federal Institute of Technology.

Functional near-infrared spectroscopy (fNIRS) provides a different measure of local hemodynamics. Literature on the application of fNIRS for diagnosis of different brain disorders is sparse. Consequently, systematic research of clinical application of fNIRS is relatively small. However, numerous papers propose fNIRS as a promising tool for the brain–computer interface.

Brain–Computer Interface

The term NF in a strict sense presumes using EEG parameters such as spectral indices, coherence measures, and intracranial current source density for purposes of self-regulation. Theoretically, any physiological parameter measured within the brain can be fed back to the subject and used for modifying this parameter which in turn can be used not only for self-regulation but for broader purposes such as controlling external devices. This approach is coined the **"brain–computer interface" (BCI)**.

An example of such approach can be given by **an ERP-based BCI**. ERP components as indices of information flow might a priori be applied for BCI. However, technical hurdles such as a small signal-to-noise ratio and the absence of normative databases for ERP components have precluded application of the ERP-based BCI for medical purposes. In 2007 in our laboratory we developed the first version of ERP-based BCI software. The starting point for this approach involved decomposition of ERPs into independent components. Components are characterized by topography and time dynamics. Topographies serve to construct spatial filters to extract components from individual ERPs. The next step of the approach is to compare the amplitude and latency of a patient's individual components with the normative data. Statistically significant deviation from normality in a certain component is used as an indication of impairment of information flow in the corresponding cortical location defined by sLORETA. The component can be further extracted from the multichannel raw EEG in each separate trial and can be fed back to the subject. Other physiological parameters such as slow cortical potentials and fMRI BOLD signals have also been used in BCI.

CHAPTER 4.3

Electroconvulsive Therapy

HISTORICAL INTRODUCTION

Electroconvulsive therapy (ECT) is the oldest and most effective non-pharmacological therapy currently available for psychiatric disease. However, ECT is not effective for all patients and is associated with high relapse rates and adverse effects. Social stigma has limited its utility in certain populations.

Pharmacologically induced convulsive therapy was introduced in **1934 by Hungarian neuropsychiatrist Ladislas Meduna** in the psychiatric hospital at Lipotmezö. He believed that schizophrenia and epilepsy were antagonistic disorders. To treat schizophrenics he applied camphor and cardiazol to induce epileptic-like seizures. In 1937, the first international meeting on convulsive therapy was held in Switzerland and 3 years later cardiazol convulsive therapy was used worldwide.

Italian professor of neuropsychiatry **Ugo Cerletti** from the University of Rome La Sapienza in the 1930s used electric shocks to produce seizures in animal experiments (Cerletti, 1940). He and his colleague **Lucio Bini** suggested application of electricity for convulsive therapy. In animal experiments they found the optimal parameters of electrical shock and tested these parameters on patients. Patients improved significantly after 10–20 sessions of **ECT**.

Retrograde amnesia was a clear side-effect of ECT: patients could not remember the treatments and had no bad feelings about it. Being simpler, cheaper, and less frightening, ECT soon replaced cardiazol convulsive therapy. Cerletti and Bini were nominated for a Nobel Prize but did not receive one. ECT spread all over the world in the following two decades. At the beginning ECT was given without muscle relaxants and resulted in full-scale convulsions.

In order to modify convulsions, psychiatrists began experimenting with curare, the muscle-paralyzing South American poison. As a result a safer **synthetic alternative** to curare, succinylcholine, was introduced.

To reduce retrograde amnesia the use of **unilateral electrode placement and replacement of a sinusoidal current with a brief pulse** were suggested.

In 1976, Blatchley demonstrated the effectiveness of his device that used constant current and brief pulse ECT. At this time a report from the American Psychiatric Association (APA) endorsed the use of ECT in the treatment of depression.

However, the introduction of antidepressants and the negative depictions of ECT in the mass media led to a decline in ECT during the 1960s through to the 1980s. Recall Ken Kesey's novel *One Flew Over the Cuckoo's Nest* in which ECT was depicted as dangerous, inhumane, and misused.

In **1985**, the National Institute of Mental Health and National Institutes of Health organized a **consensus conference on ECT** which concluded that ECT, despite being the most controversial treatment in psychiatry, was still effective in severe psychiatric disorders. Nowadays rates of ECT use vary considerably between countries reflecting social and political decisions concerning its use. In most cases, ECT is indicated only after pharmacological medications have failed. ECT is primarily indicated as a treatment for **major depression, mania, and some cases of schizophrenia.**

An estimate of ECT-related mortality is less than 1 in 20,000 treated patients and similar to that of minor surgical procedures involving general anesthesia. The mortality rate of ECT is lower than that associated with undertreated depression.

Most biologically oriented psychiatrists consider ECT a highly effective treatment that is seriously underused. Prominent side-effects are most often transitory and of minor importance comparing to severity of patients' conditions.

PARAMETERS OF ELECTROCONVULSIVE THERAPY

ECT is nowadays performed under **general anesthesia**, including sedation and muscle relaxation. Electrodes are usually placed either unilaterally (temporal location over the nondominant hemisphere and at the vertex) or bilaterally (**bifrontal or bitemporal**). The electric current parameters are: a pulse width of **0.3–2.0 ms** at 20–120 Hz for 0.5–8 s with a current of **500–800 mA**. Muscle movement and EEG activity are monitored. A course of ECT includes about 10 sessions, given 2–3 times per week. ECT can be easily administered in an outpatient setting provided the patient has appropriate support in the home environment.

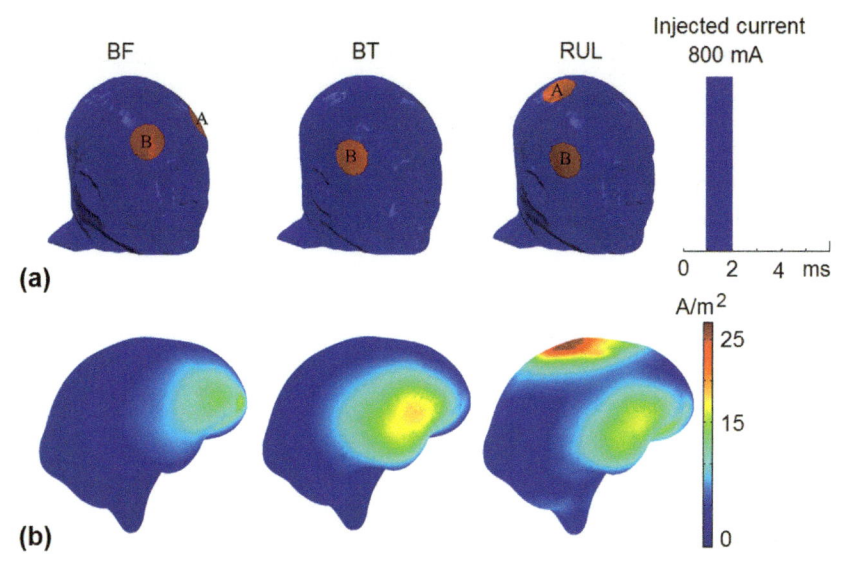

FIGURE 4.3.1 **Computational model of ECT.** (a) Typical electrode placements used in clinical ECT application: bifrontal placement (BF), bitemporal placement (BT), and right unilateral placement (RUL). "A" and "B" are labels for the separate electrodes in each electrode placement; and (b) extracellular current density magnitude in the brain at 0.4 millisecond after ECT stimulus onset in simulations of three conventional electrode placements. *Adapted with permission from Bai et al. (2012).*

NEURONAL MODEL

A **computational neuronal model** of ECT was analyzed in 2012 by Siwei Bai from University of New South Wales in Australia and coworkers using a finite element model of the human head (Bai, Loo, Al Abed, & Dokos, 2012). Maximal current densities were found on the surface of the cortex just beneath the electrodes (Fig. 4.3.1). When the current was set to 800 mA action potentials were initiated in the cortex beneath the electrode at approximately 3 ms after the current injection.

MECHANISMS

Mechanisms of the therapeutic effect of ECT are **not known**. There are several types of mechanisms that must be considered: biochemical, metabolic, and neuronal networks.

Biochemical mechanisms rely on the fact that ECT induces releases of numerous chemicals including neurotransmitters, neuropeptides, endogenic opioids, neurotrophic factors, and hormones. In addition, ECT might modulate gene expression.

Metabolic mechanisms rely on the fact that ECT affects brain metabolism in regions implicated in the pathophysiology of neuropsychiatric illness. Indeed, studies using single-photon emission computed tomography (SPECT) found that bitemporal ECT caused increases in local blood flow in the temporal and laterofrontal cortex bilaterally.

Neuronal network mechanisms presume that ECT promotes the rewiring of pathological networks implicated in psychiatric disturbances.

EFFICACY

As far as ECT efficacy is concerned, two metaanalyses must be mentioned. The **UK ECT Review Group in 2003** completed a metaanalysis of data on short-term efficacy of ECT from randomized controlled trials on patients suffering from depression. The metaanalysis concluded that: (1) real ECT was significantly more effective than sham ECT; (2) ECT was more effective than pharmacotherapy; (3) bilateral ECT was more effective than unilateral ECT.

In a 2005 **Cochrane review by** Tharyan and Adams from Christian Medical College, IndiaECT efficacy on schizophrenic patients was assessed. The conclusion was: (1) in the short term, real ECT was more efficient than sham ECT; (2) no evidence of long-term effects was obtained; (3) a combination of ECT with antipsychotic medication might be considered an option for people with schizophrenia, particularly when rapid global improvement and reduction of symptoms are desired and also in patients who show limited response to medication alone.

RELAPSE

Relapse after a successful ECT course is a major limitation of the therapy. In depressed patients the relapse rate is estimated to be about 50%, with the majority of relapses occurring within 6 months of treatment. Therefore, maintenance therapy after a course of ECT is indicated and includes either drug treatment, another ECT, or combined therapy.

CONTRAINDICATIONS

Contraindications to ECT include severe cardiovascular conditions and conditions affecting the central nervous system such as increased intracranial pressure, recent cerebral infarction or bleeding, cerebral angioma, or aneurysm.

SIDE-EFFECTS

ECT is frequently associated with mild to moderate **retrograde and anterograde amnesia**. Usually, the extent of retrograde amnesia, which is normally specific to autobiographical episodic memories, reduces with increasing time after ECT, but in some patients the recovery will be incomplete and there are reports that ECT can even result in permanent memory loss. At least one-third of patients treated with ECT reported **clinically relevant memory loss** at 6 months after a course of treatment. A growing database suggests that high-dose right unilateral and bifrontal ECT may have similar efficacy to bitemporal ECT with fewer memory side-effects.

Challenges facing the field of convulsive therapy include maintaining response following an effective course of ECT and establishing the most efficient treatment conditions while limiting side-effects.

CHAPTER
4.4

Transcranial Direct Current Stimulation

HISTORICAL INTRODUCTION

The idea to use electricity for the treatment of brain disturbances is so straightforward that its first application is dated **AD 43–48** when Scribonius Largus observed that placing a live torpedo fish—delivering a strong direct electric current—over the scalp of a patient with headache elicited a sudden, transient stupor with pain relief. In the 18th century, Italian scientists Luigi Galvani and Alessandro Volta started their work on the effects of electricity on living organs. In **1804**, Galvani's nephew Giovanni Aldini reported the successful treatment of patients suffering from melancholia by applying galvanic currents over the head. In the middle of the 20th century systematic research into the effect of direct current (DC) on brain tissue was initiated. The discovery that scalp DC can induce **prolonged changes in brain excitability** opened up new approaches to the management of neurological and psychiatric conditions.

The application of direct electric current in Russian physiological research has a long history, which started at the end of the 19th century. In the early 1970s the author of this book worked at the **Institute of Experimental Medicine of the Academy of Medical Sciences of the USSR** together with a group who conducted research on DCs in human physiology. Transcranial DC stimulation (tDCS) in those days was called **micropolarization**. Intracranial DC was applied in patients with implanted electrodes for temporally switching off and on neuronal populations located near the deep electrodes, in parallel with experiments using polarizing electric currents performed on cats and dogs. In the 1980s, several clinics in Leningrad started using micropolarization to treat **neurological and psychiatric patients**. Unfortunately, tragic events (the death of two principal scientists in this field, collapse of the Soviet Union, and cancellation of funding of basic research following the collapse of the USSR) led to interruption of these promising studies. Only at the beginning of the 21st century did research recontinue. In our laboratory, tDCS is now used for correcting the behavior of children with **Attention Deficit Hyperactivity Disorder (ADHD), autistic spectrum disorder, and speech problems.**

In the West, experimental and clinical research on tDCS was renewed in the 21st century in Germany, Italy, and the United States. In 2008 Michael Nitsche and colleagues from the University of Göttingen (Germany) published a review covering the years from 1998 to 2008. The number of papers on tDCS has accelerated almost exponentially since 2008. About 25% of these studies are done in clinical settings.

tDCS is still considered **investigational by the FDA** and officially limited to research use. However, tDCS has been shown to be effective in treating a wide variety of brain dysfunctions with many metareviews to back it up.

PROCEDURE

In a conventional procedure, a small amount of direct **electric current (1– 2 mA)** is applied to the skin of the head by two relatively large electrodes: one is a **stimulating** electrode, localized above the area that is supposed to be stimulated, and the other is a **reference electrode**, localized on some "silent" part of the head or the body such as the neck, shoulder, or mastoid (Fig. 4.4.1a). The electric current flows according to the Ohm's

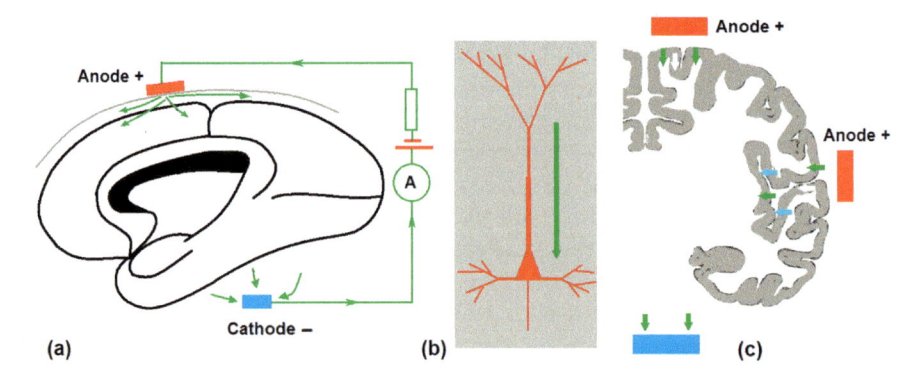

FIGURE 4.4.1 **Transcranial direct current stimulation.** (a) Scheme of tDCS: two electrodes—anode (+) and cathode (–)—are attached to the head, electric current is provided by a battery-driven device which also includes the tool for measuring the current (A). The current usually does not exceed 2 mA while only a small part of it (around 10%) goes through the cortical gray matter; (b) in the cortical layers that are oriented perpendicular to the current, the "anodal" current depolarizes pyramidal cells at their basal membrane (see Fig. 4.4.2 for details); and (c) Depending of the anode location and cortical folding, the anodal current can depolarize the pyramidal cell at the basal membrane (*green arrows*), hyperpolarize the pyramidal (*blue arrows*), or does not change the average potential (*white arrows*).

law. Recall from physics that the law states: the electric current through a conductor between two points is proportional to the voltage across these two points. If the electric field **E** is defined as a vector-force that would be applied on a unit electric charge within the electric field, then the Ohm's law can be rewritten in the vector form in which the electric field **E** is proportional to the current density vector **J** where the coefficient of proportionality is a scalar named conductivity:

$$E = \rho J$$

If the cortical layers are oriented perpendicular to the current, the "anodal" current flows along the apical dendrites and depolarizes pyramidal cells at their basal membrane, while the cathodal current produces an opposite effect—it hyperpolarizes the pyramidal neurons (Fig. 4.4.1b). One can see from Fig. 4.4.1c, the net effect on the cortex depends on the cortical folding so the anodal current can depolarize the pyramidal cell at the basal membrane *(green arrows)*, hyperpolarize the pyramidal *(blue arrows)* or does not change the basal potential *(white arrows)*.

DIFFERENCE FROM ELECTROCONVULSIVE THERAPY

It should be stressed here that tDCS is **fundamentally different from ECT**. Whereas ECT injects pulses of strong currents inducing convulsive activity in the brain, tDCS induces much smaller currents **slightly modulating** brain function by changing spontaneous neuronal activity without inducing seizures. The DC is so weak that it cannot produce a membrane potential shift above the threshold but it slightly changes the membrane potential of neuronal cells which due to collective and cumulative effects are capable of modulating spontaneous cell firing. Available evidence indicates that unlike ECT, tDCS causes no memory disturbance or loss of consciousness; neither does it require the patient to be sedated or receive muscle relaxants. tDCS is so mild that patients cannot subjectively discriminate the feeling produced by tDCS when compared with **sham**.

NEUROPHYSIOLOGICAL BASIS

From basic neurophysiology, we know that a constant electrical field, (which induces the current flow in the electrically conductive tissue), shifts the membrane potential of neurons toward either hypo- or hyperpolarization depending on the direction of the field and localization of the spot on the neuron. This happens because an external electric field according to the laws of physics **sums up** with **the resting state membrane potential** and produces membrane depo- or hyperpolarization depending on the side of the cell.

The pyramidal cell in the gyrus is oriented perpendicular to the surface of the cortex. Consequently, neurons are depolarized at the hillock if the positive pole is applied to the surface of the head (anodal stimulation of the given cortical area) (Fig. 4.4.2b), and neurons are hyperpolarized at the hillock if the negative pole is applied to the surface of the head (cathodal stimulation) (Fig. 4.4.2a).

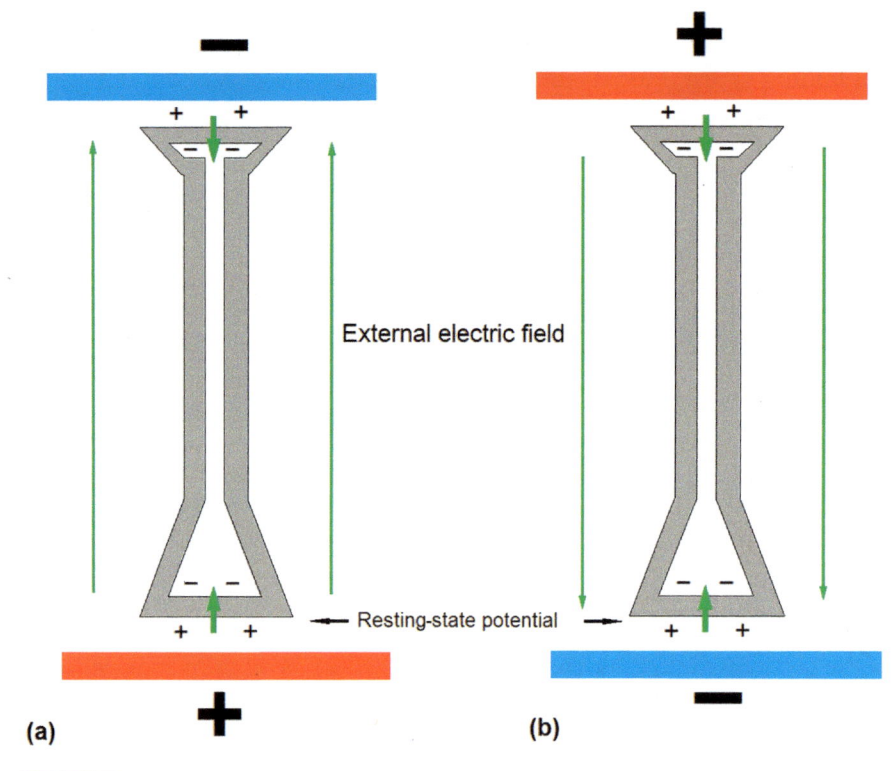

FIGURE 4.4.2 **Schematic presentation of effect of external electric field on membrane of the pyramidal neuron.** The membrane is negatively charged inside the cell which produces resting state trance membrane potential. In this didactic example, a uniform external electric filed is generated by application of a stimulation electrode to the surface of the head and a reference electrode beneath the gray matter. (a) The external and internal fields are of the same direction near the positive pole which in "sum" leads to membrane hyperpolarization of the corresponding membrane; and (b) the external and internal trance membrane fields are of opposite direction near the negative pole which in "difference" leads to depolarization of the corresponding part of membrane.

NONLINEAR COLLECTIVE SHORT-TERM EFFECTS OF tDCS

Numerous experimental findings show that a weak **1-mV/mm** uniform field produces a transmembrane potential change of approximately **0.1 mV**. For spike generation the change must be 150 times higher (ie, approximately 15 mV) as in ECT. However, there is increasing evidence that weak electric fields can acutely modulate coherent network activity in animal experiments and produce long-lasting therapeutical effects in patients. To explain these effects, nonlinear collective phenomena must be taken into account.

We know at least one amplification factor in the nervous system: the collective effect due to recurrent excitatory connections in cortical areas when a small change in membrane potential can lead to large changes in neuronal spiking. From this point of view, tDCS can be considered a subthreshold stimulation technique modulating spontaneous cortical activity and thereby inducing transient functional changes in the human brain.

The **hippocampal and thalamo-cingulate slice** preparations are powerful tools to demonstrate the effects of weak external fields on the collective behavior of these neuronal networks. The effect of weak external fields on collective gamma oscillations in hippocampal slice preparations was demonstrated in a 2010 study by Davide Reato from the City College of the City University of New York. In this study spatially uniform electric fields were oriented along the pyramidal cells while recording of local field potentials was performed in the CA3 region of hippocampus. The negative fields were shown to produce a decrease of gamma oscillations while the positive fields in contrast produced an increase of gamma oscillations in the hippocampal slice preparations.

The attenuation effect of cathodal DCS on spontaneous excitatory postsynaptic currents as well on epileptic excitatory postsynaptic currents caused by the 4-aminopyridine in the thalamo-cingulate slice preparation was observed in a 2015 study by Wei-Pang Chang from National Defense Medical Center in Taiwan (Fig. 4.4.3). It should be stressed here that the effect was observed when the orientation of the electric field was parallel to the direction of axo-dendritic fibers (Fig. 4.4.3a, d), but absent when the field was set perpendicular to the axo-dendritic fibers (Fig. 4.4.3c, e). This study offers a neurophysiological mechanism for the application of cathodal tDCS in drug-resistant seizures.

LONG-TERM POST-tDCS EFFECTS

From the clinical point of view the most important features of tDCS are long-term after-effects expressed in changes in neuronal networks that last from several minutes to hours after switching off the injecting current.

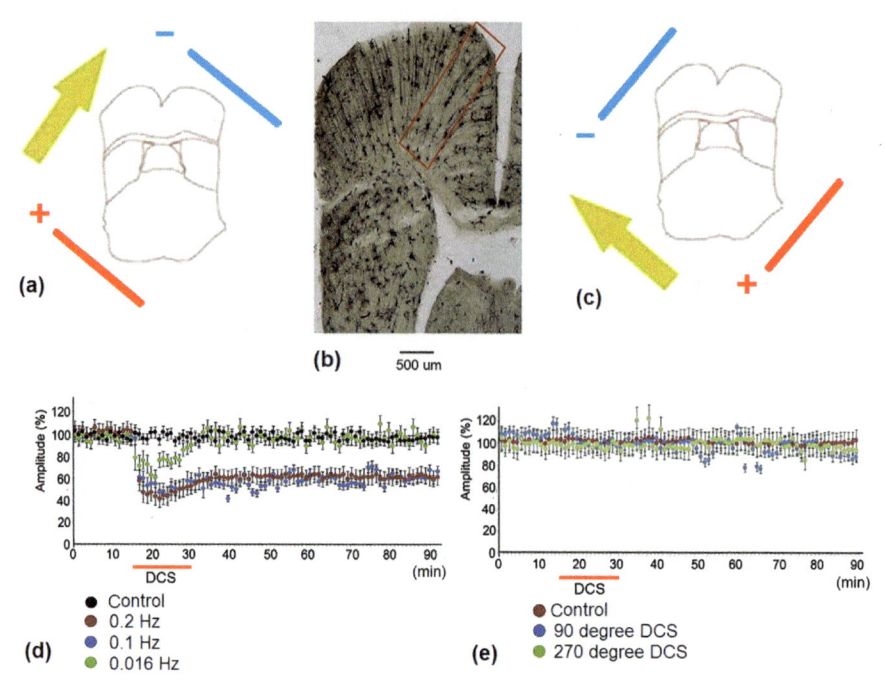

FIGURE 4.4.3 **Attenuation effect of cathodal DCS on thalamocortical postsynaptic currents in the thalamocingulate slice preparation.** (a) Parallel orientation of the electric field and axo-dendritic fibers. (b) Golgi staining of the slice preparation. Note the orientation of axo-dendritic fibers in the red box. (c) Orthogonal orientation of the electric field and axo-dendritic fibers. (d) In the parallel field orientation the cathodal DCS combined with 0.1 and 0.2 Hz synaptic activation produces short- and long-term decrease of amplitude of evoked activity (*y*-axis) while 0.016 Hz synaptic activation produces only short-term effects. (e) No synaptic depression is observed in the case of the orthogonal field orientation. *Adapted with permission from Chang et al. (2015).*

One way of measuring neuronal excitability is the application of transcranial magnetic stimulation (TMS) pulses. For example, a TMS pulse to the motor cortex induces a muscle twitch that can be measured by **motor-evoked potentials (MEPs)**. Fig. 4.4.4 shows that changes in excitability of the motor cortex induced by 5 min of tDCS occur in opposite directions depending on tDCS polarity—increased excitability for anodal tDCS and decreased excitability for cathodal tDCS. The changes last for about 5 min after switching off the injecting current.

Further research demonstrated that longer periods of tDCS induced prolongation (up to an hour) of the effect. For example, a 2001 study by Michael Nitsche and Walter Paulus from the University of Göttingen showed that tDCS after-effects are a function of the **intensity and duration** of tDCS application. Additional reports suggest that weekly repeated tDCS sessions might further increase the duration of its effects on behavioral outcomes.

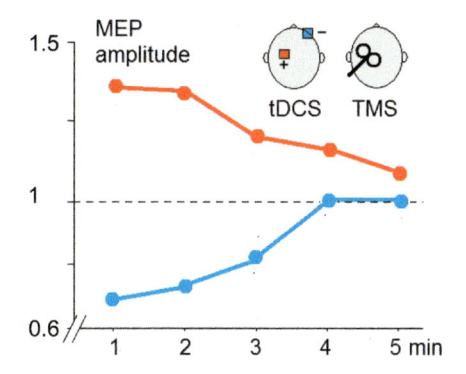

FIGURE 4.4.4 **Polarity-specific effects of tDCS on excitability of the motor cortex.** The y-axis shows the relative amplitude of the motor potential evoked by TMS after tDCS applied to the motor cortex during 5 min. The x-axis shows the time elapsed after the end of tDCS. Heads at the top right schematically show the procedures of application tDCS and TMS. *Adapted with permission from Nitsche & Paulus (2001).*

In a 2006 study by Neri Accornero and colleagues from the University of Rome (Italy), **visual ERPs** were recorded during and after anodal and cathodal tDCS in healthy subjects. The current was 1 mA of two polarities with two durations of 3 and 10 min. A special active electrode of 5×8 cm^2 with an ERP-recording electrode inside were placed on the occipital scalp and the reference electrode was placed over the anterior or posterior base of the neck (Fig. 4.4.5b). ERPs were recorded in response to pattern reversal checkerboard stimuli before, during, and after polarization. Anodal polarization reduced the amplitude of the occipital P1 wave whereas cathodal polarization significantly increased amplitude but both polarities left latency statistically unchanged. These changes persisted for some minutes after polarization ended depending on the duration of tDCS: the longer the duration of tDCS the longer the after-effect lasts (compare Fig. 4.4.5a and c).

NMDA INVOLVEMENT IN LONG-LASTING AFTER-EFFECTS

Experiments with pharmacological agents affecting the membrane and processes in neuronal networks show that short-term effects and long-term after-effects rely on **different neuronal mechanisms**. Short-term effects are defined by nonsynaptic mechanisms associated with **depolarization or hyperpolarization of resting membrane potentials** which corresponds to the model in Fig. 4.4.2. Long-term after-effects occur through NMDA-dependent mechanisms, similar to **long-term synaptic potentiation and depression**. This fact implies that long-term after-effects are the result of maintaining the trace of the electrical disturbance in the collective behavior of neuronal networks heavily interconnected by excitatory glutamatergic synapses.

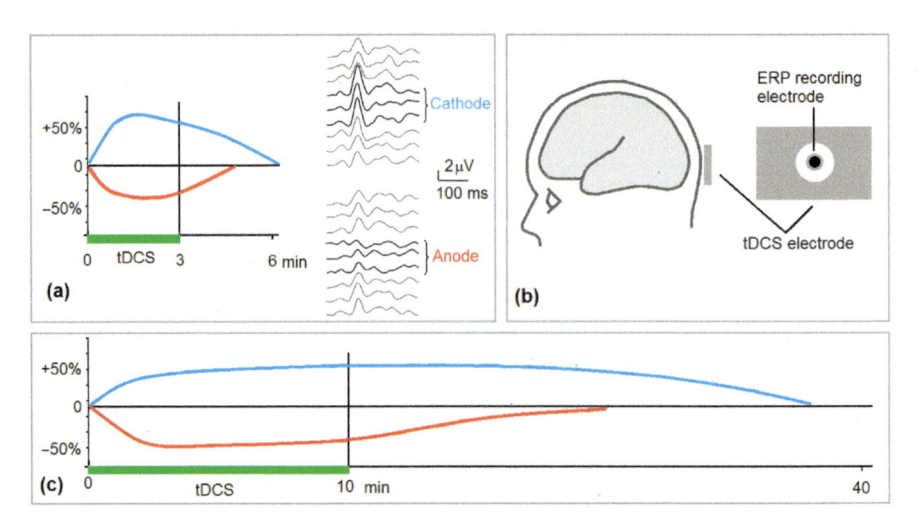

FIGURE 4.4.5 **The tDCS after-affects are measured by ERPs.** (a) Visual P100 relative amplitude modulation induced by a 3 min tDCS. P100 is measured in reference to the pre-tDCS period; (b) position of active electrode on the scalp (left) and construction of the electrode. *Blue line*, cathodal tDCS; *red line*, anodal tDCS; *Green line*, tDCS; and (c) Visual P100 relative amplitude modulation induced by a 10 min tDCS. *Adapted with permission from Accornero et al. (2006).*

SAFETY AND SIDE-EFFECTS

When applied according to standard requirements, tDCS is a **safe procedure.** The reader is referred here to a 2003 paper by Michael Nitsche from University of Goetengen and coworkers. However, in rare cases adverse effects are reported including fatigue (35%), mild headache (11.8%), nausea (2.9%), and a transient tingling, itching, and/or redness in the region of stimulation.

LIMITATIONS

One disadvantage of the conventional tDCS method is its low spatial resolution caused by the large stimulation electrode. Spatial resolution can be improved by several options. One is **decreased size** of the stimulating electrode and increased size of the reference electrode. The other option is using the Laplacian montage of stimulationg electrodes to localize the area of the largest current source density. There have been several methodological studies investigating these manipulations and proving their feasibility.

Another limitation of tDCS is the **difficulty in defining** the treatment protocol (ie, the localization and direction of the current density for a given subject and a selected electrode position). Theoretically, this limitation can be overcome by modeling current density on individual MRI scans.

CHAPTER 4.5

Transcranial Magnetic Stimulation

INTRODUCTION

Compared with ECT and transcranial direct current stimulation (tDCS), transcranial magnetic stimulation (TMS) was introduced in 1985. This was initiated by Anthony Barker and coworkers from the University of Sheffield, UK as a noninvasive method of **activating the human motor cortex** and assessing the integrity of central motor pathways. Since its introduction, the use of TMS in clinical neurophysiology, neurology, and psychiatry has spread widely, mostly in research applications.

PHYSICAL PRINCIPLES

TMS is based on the law of **electromagnetic induction** discovered by Michael Faraday in 1838. According to the law if a pulse of electric current passing through a coil placed over a subject's head has sufficient strength and short duration it generates a rapidly changing magnetic pulse that penetrates the scalp and skull to reach the cortex with negligible attenuation. Pulses of a magnetic field, according to the same law, induce a **secondary ionic current in the brain** (Fig. 4.5.1). Induced brief electric current pulses in the brain can trigger **action potentials** in cortical neurons, especially in superficial parts of the cerebral cortex.

PHYSIOLOGICAL EFFECT

TMS delivered to different levels of the **motor system** provides information about the excitability of the motor cortex, the functional integrity of intracortical neuronal structures, conduction along corticospinal, corticonuclear, and callosal fibers, as well as the function of nerve roots

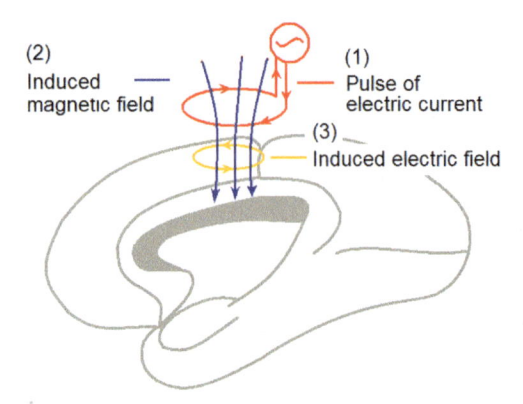

FIGURE 4.5.1 **Scheme of Transcranial Magnetic Stimulation (TMS).** An electric pulse *(red color)* is applied to a coil placed above the head. The current induces the magnetic field *(blue)* that penetrates the skin, scalp, and gray/white matter. The pulse of the magnetic field induces the electric current *(yellow)*. If the initial electric current is strong enough, the induced electrical current can activate cortical neurons.

and peripheral motor pathways to the muscles. When TMS is applied to the motor cortex at the appropriate stimulation intensity, **motor-evoked potentials (MEPs)** can be recorded from contralateral extremity muscles. The **motor threshold** refers to the lowest TMS intensity necessary to evoke MEPs in the target muscle when single-pulse stimuli are applied to the motor cortex. The motor threshold is believed to reflect membrane excitability of corticospinal neurons and interneurons projecting onto these neurons in the motor cortex, as well as the excitability of motor neurons in the spinal cord, neuromuscular junctions, and muscles. The patterns of findings in these studies can help to localize the level of a lesion within the nervous system, distinguish between a predominantly demyelinating or axonal lesion in motor tracts, or predict functional motor outcome after an injury.

A brief magnetic pulse produces **synchronized activation of cortical neurons.** This synchronized activation is followed by **inhibition.** Postimpulse inhibition explains why single pulses or short bursts of TMS can effectively perturb ongoing neuronal processing in the stimulated cortex. This transient disruptive effect of TMS, often referred to as a **virtual lesion**, has been extensively used by cognitive neuroscientists to examine the functional relevance of the stimulated area for behavior.

When long continuous trains or short intermittent bursts are repeatedly applied to a cortical area at a frequency higher than 1 Hz, they are named **repetitive TMS (rTMS).** rTMS can induce changes in neuronal excitability that **persist beyond the time of stimulation**. These neuromodulatory effects of TMS have been exploited in many in vivo studies on cortical plasticity and may be of some use in patients with neurologic and psychiatric diseases to maintain or restore brain function.

Dual-site TMS can induce coordinated stimulation of two interconnected cortical areas. Using a conditioning test approach, dual-site TMS has been applied to assess the corticocortical connectivity of pathways projecting onto the primary motor hand area.

rTMS AT LOW AND HIGH FREQUENCY

The physiological effect of rTMS depends on frequency. If rTMS is pulsed at a **low frequency** (about 1 Hz) cortical excitability generally **decreases**, while **higher frequency rTMS** (usually between 5 and 20 Hz) **increases cortical excitability**. This ability to up- or downregulate cortical excitability, along with its high temporal resolution, suggests that TMS might be a useful tool to manipulate cortical networks in ways that could alter cognitive performance.

The third form of rTMS, known as **theta burst stimulation**, was designed to mimic traditional paradigms of long-term potentiation and depression. The theta burst stimulation consists of 3 pulses at 50 Hz repeated at 200-ms intervals. The after-effects of this type of rTMS are more prominent and longer lasting than those induced by conventional rTMS.

The potential applications of rTMS in psychiatry are exciting, but they presently remain at the **stage of holding promise**. rTMS is a relatively new technology with many options to optimize its use. For example, TMS targeting can be improved by application of individual MRI scans. The parameters of rTMS such as frequency, intensity, stimulus, waveform, and polarity can be optimized.

MODEL

Neurophysiological mechanisms of TMS and rTMS are not fully understood and need further investigations.

SAFETY

rTMS is **reasonably safe** with mild side-effects when performed in compliance with recommended safety guidelines. The most frequent side-effects include mild headache responsive to common analgesics, local pain or paresthesias in the stimulated region, neck pain, toothache, transient changes in audition, and syncope. Induction of a seizure is a possible serious adverse effect, but is a very rare phenomenon found when investigators strictly adhere to recommended guidelines.

CHAPTER 4.6

Deep Brain Stimulation

INTRODUCTION

When psychosurgery as a surgical procedure for the treatment of severe psychiatric conditions emerged in the 20th century, the only method available to neurosurgeons for modulating the brain was **destruction of targeted neural tissue**. In 1935 Egas Moniz and Almeida Lima in Portugal performed the first prefrontal **leukotomies**. In 1949, Moniz was awarded the Nobel Prize for the "discovery of the therapeutic value of prefrontal leukotomy in certain psychoses." However, indiscriminate use of large surgical ablations often led to dreadful consequences such as radical personality changes and cognitive decline. The procedure was shortly regarded as unethical and discarded.

In 1949, the French neurosurgeon Talairach presented the use of a **stereotactic frame** to **selectively coagulate the frontothalamic fibers** in the anterior limb of the internal capsule. Selectively performed **stereotactic psychosurgery** replaced globally performed prefrontal lobotomy. It was applied for various psychiatric disorders. For example, (1) **anterior capsulotomy** was applied for general anxiety disorder and obsessive–compulsive disorder (OCD), (2) **cingulotomy** for addiction, bipolar disorder, depression, OCD, schizoaffective disorder, and schizophrenia, (3) **subcaudate tractotomy** for depression, OCD, and schizophrenia, (4) **anterior callosotomy** for schizoaffective disorder and schizophrenia, (5) **thalamotomy** for Tourette syndrome, (6) **hypothalamotomy** for addiction, aggressiveness, and sexual disorders, and (7) **amygdalotomy** for aggressive behavior associated with mental impairment.

Although stereotactic psychosurgery in the early years almost exclusively used ablative lesions, **experimental deep-brain stimulation (DBS)** in psychiatric patients was performed by several groups in the 1960 and 1970s. For example, at the Institute of Experimental Medicine in Leningrad DBS and intracerebral DC stimulation were performed to induce reversible ablations near implanted electrodes. Before starting these stimulation techniques the neurophysiologists (the author of the book among them) recorded different neurophysiological parameters from implanted electrodes in different functional tasks to map the function into the brain structure.

Since the introduction of **psychoactive drugs** in the 1950s and 1960s, the number of patients requiring stereotactic psychosurgery decreased enormously. At present, it is only applied in the treatment of refractory psychiatric disorders. The stereotactic lesion, while effective, was permanent in nature and therefore associated with the potential for permanent side-effects. There was little room for error in terms of the placement and, once done, there was no means of adjusting the effect except by creating another or larger lesion. Since the 1987 publication of Alim-Louis Benabid and colleagues from Grenoble University (France) on thalamic DBS in parkinsonian patients with tremor, DBS has virtually replaced ablative lesions in stereotactic neurosurgery for both movement and psychiatric disorders.

PROCEDURE

The conventional procedure of electrode implantation consists of the following steps: (1) a **stereotactic frame** is attached to the patient's skull, (2) the patient's brain is imaged with the frame on to localize targets on **MRI or CT scans**, (3) the stereotactic frame is used to place the **electrode** into the targeted area through a small hole in the scull (Fig. 4.6.1b), (4) the electrode is connected to a cable and tunneled under the scalp and skin of the neck to a subcutaneous pocket, (5) where the **internal pulse generator** is placed.

NEURONAL MECHANISM

The exact neuronal mechanisms of DBS are still a matter of debate. One possible mechanism of DBS is the inhibitory effect demonstrated in Fig. 4.6.1. This effect was observed by Jonatan Dostrovsky from University of Toronto and coworkers in a 2000 study. They recorded impulse activity of neurons in the internal part of the globus pallidus (GPi) in Parkinsonian patients undergoing stereotactic exploration to localize the optimal site for placement of a lesion or DBS electrode. The authors suggested that stimulation within GPi activates the GABAergic axon terminals of striatal and/or external pallidal neurons, thereby causing the release of GABA and postsynaptic inhibition of GPi neurons. Thus stimulation-evoked release of GABA within GPi may be one of the mechanisms involved in the therapeutic effects of pallidal DBS, which would explain why the effects are similar to pallidotomy.

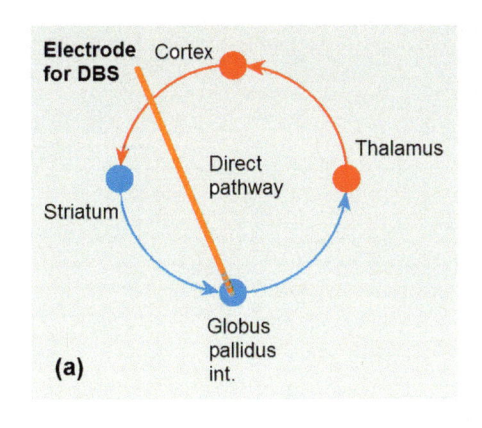

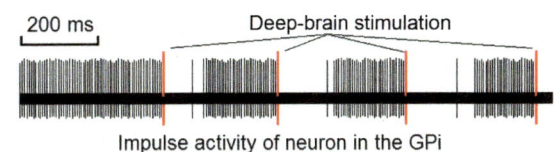

FIGURE 4.6.1 **Inhibitory effect of DBS.** (a) In Parkinson's disease the high frequency activity of neurons in the internal segment of the globus pallidus excessively inhibits neurons in the ventrolateral nucleus of the thalamus. (b) To remove this excessive inhibition the deep brain stimulation induces inhibitory effects on the pallidal neurons.

ADVANTAGES AND LIMITATIONS

DBS has advantages over destruction procedures of neurosurgery. First, the effects produced by DBS are fully **reversible**. Second, stimulation parameters can be **adjusted** according to a patient's changing symptoms and disease progression. Third, stimulation can generally be turned on or off without the **patient's awareness**, which provides a unique opportunity for double-blind studies. Along with the advantages, DBS has some disadvantages, primarily associated with the need to keep the implant within the brain permanently thus producing a **danger for infection**.

PART 5

NEUROMARKERS IN PSYCHIATRY

CHAPTER 5.1

Attention Deficit Hyperactivity Disorder

HISTORICAL INTRODUCTION

Attention Deficit Hyperactivity Disorder (ADHD) is a highly prevalent (**around 5%** of school-aged children) neuropsychiatric condition with onset in childhood. According to the current criteria, ADHD is defined by a persistent and age-inappropriate pattern of **inattention, hyperactivity, and impulsivity**. ADHD is frequently comorbid with other neuropsychiatric and neurodevelopmental disorders and is present in oppositional defiant and conduct disorders, anxiety and depressive disorders, sleep disorders, learning difficulties, and substance abuse disorders.

ADHD is definitely not a disorder of the 20th century as some believe. It has always been present in the population. For example, in **1798** Alexander Crichton in his book *An Inquiry into the Nature and Origin of Mental Derangement* described a condition similar to the inattentive subtype of ADHD. After the First World War and the great encephalitis epidemics of 1917–18, hyperactivity in children was associated with brain damage, and the dysfunction was labeled **minimal brain damage**. Later, when acute brain damage was not confirmed, the name of the dysfunction became **minimal brain dysfunction**.

In the 1970s when the computer metaphor of information processing became a new scientific paradigm, **inattention** as the disability of **filtering information** was postulated to be the core problem in these children. At the end of the 20th century the dysfunction was often associated with impairment of the **executive brain system** with the frontal lobes and basal ganglia as the core nodes of the system. During the last years scientific fashion has shifted toward **cognitive control** mode in which reactive and proactive operations are distinguished and dysfunctions of which are associated with ADHD. These changes in names reflect in part shifts in scientific paradigms and in part the subjective nature of the disorder.

The use of **stimulants** for improving attention in ADHD started in 1937 and regulatory approval of **stimulant treatment for children began in the 1960s**.

Accumulating evidence from twin and adoption studies supports the **genetic contribution** to ADHD and shows the interaction of environmental and genetic factors for manifestation of symptoms.

Despite the long research history and a vast amount of accumulated knowledge, there is **no consensus** regarding the pathophysiology of ADHD which is reflected in quite different theories of ADHD emphasizing different aspects of the experimental data obtained.

SYMPTOMS

The diagnostic criteria for ADHD in DSM-5 are similar to those in DSM-IV. The same **18 symptoms**, divided into **two symptom domains (inattention and hyperactivity/impulsivity)** are used. At **least six symptoms** in one domain are required for diagnosis.

The symptoms are **descriptive**. For **inattention** the patient's symptoms include: (a) often has difficulty sustaining attention in tasks or play activities, (b) often has difficulty organizing tasks and activities, (c) often easily distracted by external stimuli, etc. For **hyperactivity/impulsivity** the patient's symptoms include: (a) often leaves seat in classroom or in other situations in which remaining seated is expected, (b) often has difficulty playing or engaging in leisure activities quietly, (c) often has difficulty awaiting turn, etc. However, several changes have been made in DSM-5 including positioning of ADHD in chapter: **Neurodevelopmental Disorders** to reflect brain developmental correlates with ADHD.

As one can see from the DSM definition, ADHD is a collection of different symptoms. One can arrive at the diagnosis by different combinations of symptoms. Moreover, the behavioral method of diagnosing ADHD and its DSM subtypes is highly subjective and therefore unstable. It depends on assessment made by arbitrary chosen informants, on the types of applied rating scales, and on the way of integrating information from multiple sources (eg, parent and teacher).

The controversy of ADHD diagnoses is demonstrated by the differences between US and European diagnostic criteria. In the **International Classification of Disorder (ICD)** used in Europe, ADHD is defined as hyperkinetic disorder (HKD). Both classifications include children displaying developmentally inappropriate levels of inattention, hyperactivity, and impulsivity that begin in childhood and cause impairment to school performance, intellectual functioning, social skills, and occupational functioning. But ICD criteria are **more restrictive** than the DSM because they need a greater degree of symptom expression. DSM allows for three symptom-based subtypes of

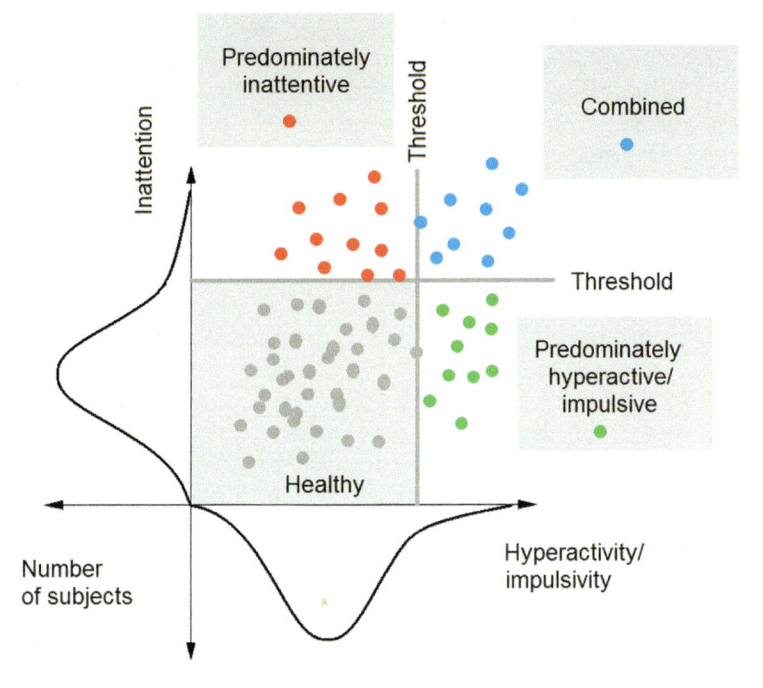

FIGURE 5.1.1 **Multidimensional space of ADHD symptoms.** For simplicity, nine symptoms of inattention in children are presented in the *y*-axis, and six symptoms of hyperactivity with three symptoms of impulsivity are depicted in the *x*-axis. The thresholds (as arbitrarily defined in DSM) separate "healthy" subjects from ADHD patients. Three subtypes of ADHD in the symptom space can be separated: predominately inattentive, predominately hyperactive/impulsive, and combined subtypes.

ADHD: predominately **inattentive** (which is absent in ICD), predominately **hyperactive–impulsive**, or **combined** (Fig. 5.1.1).

LATENT CLASSES IN ADHD SYMPTOMS

Many people (children, in particular) are inattentive, impulsive, and hyperactive; hence the symptoms of ADHD are distributed in the human population and can be found to a lesser degree in normally developing subjects (Fig. 5.1.1).

The question arises: when dealing with the 18-dimensional space of ADHD symptoms can the conventional ADHD subtypes be separated **naturally** without setting a cutoff for the number of positive symptoms required for ADHD diagnosis?

Maria Acosta and coworkers from the National Institute of Mental Health in a 2008 study applied latent class analysis (LCA) to answer this question.

LCA, based on the statistical concept of likelihood, was introduced in 1950. In the 2008 study it was applied to parent reports of ADHD symptoms of a sample of 1010 individuals from a nationwide recruitment of families with at least one child with ADHD and another child either affected or clearly unaffected. The method repeatedly yielded **six to eight clusters**. The clusters included three particularly clinically relevant groups: hyperactive, inattentive, and combined (Fig. 5.1.2). The results demonstrate that the three ADHD subtypes can be naturally selected in the symptom space and that these subtypes might reflect the basic properties of children behavior.

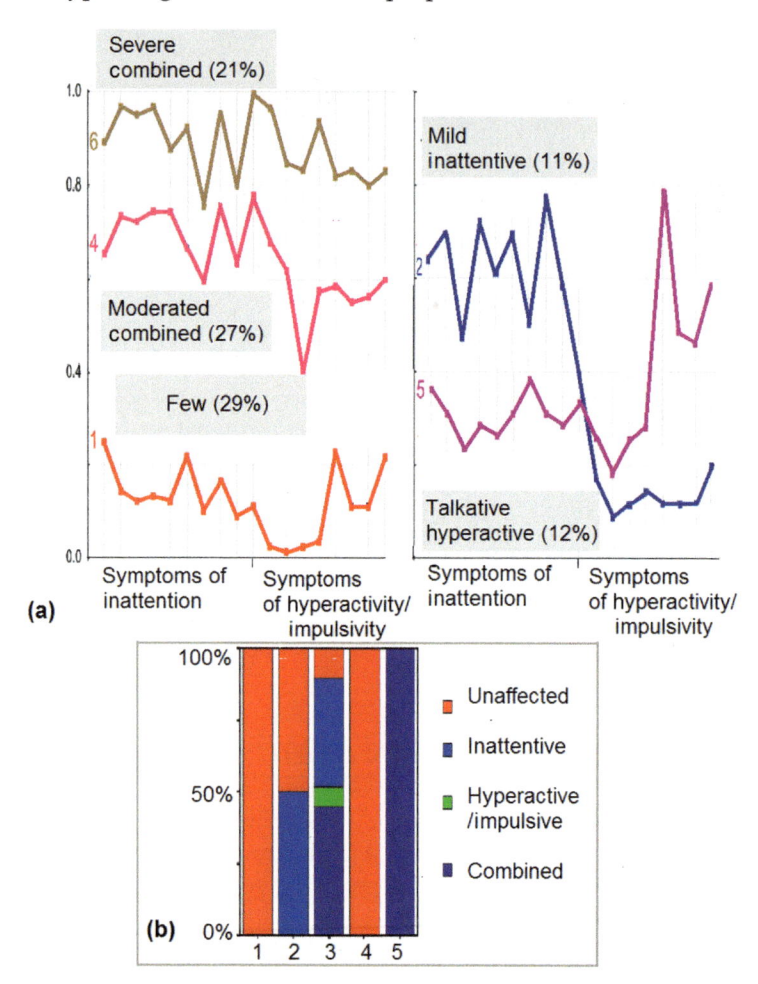

FIGURE 5.1.2 **Latent class analysis in 18-dimensional symptom space.** Children 4–11 years of age have been assessed by the Vanderbilt Assessment Scale for Parents (VAS-P). (a) Each figure shows latent class endorsement probabilities (y-axis) for every VAS-P item (symptoms); and (b) comparison of ADHD status as defined by DSM-IV and cluster membership (each cluster equals 100%). *Adapted with permission from Acosta et al. (2008).*

PREVALENCE

According to a 2003 review by Stephen Faraon and coworkers from Queen's University, Kingston (Canada), ADHD affects **8–12%** of children worldwide. However, some authors speculate that when carefully diagnosed the prevalence of ADHD in childhood can be as small as 1–4%.

The prevalence of ADHD in males could be due to the increased exposure of males to environmental sources of cause, such as head injury. The male-to-female sex ratio is greater in clinical studies than in community studies, which indicates that female individuals are less likely to be referred for services than male individuals, which in turn might indicate that ADHD is less disruptive in females.

AGE ONSET

Most ADHD patients are diagnosed in early age. For example, in a 2014 study by Peyre and coworkers from 807 participants with a lifetime DSM-IV diagnosis of ADHD, 45% had an early age of onset (≤ 7 years), 31% had a late age of onset (>7 and ≤ 12 years), and 18% had very late age of onset (>12 and ≤ 18 years).

PERSISTENCE IN ADULTHOOD

For several decades, the idea that ADHD persisted into adulthood was met with skepticism. However, follow-up studies of ADHD have shown that despite a clear age-dependent decline in symptoms they persist into adulthood and are frequently associated with clinically significant impairments. A 2008 metaanalysis by Victoria Simon from Semmelweis University Budapest, Hungary and coworkers estimated the prevalence of adult ADHD as about 2.5%.

OUTCOME

ADHD puts children at risk of other **psychiatric** and **substance abuse** disorders. ADHD is associated with functional impairments at school and work, family conflict and peer problems, antisocial behavior, injuries, traffic violations, and accidents. Individuals with ADHD are at increased risk of poor educational achievement, low income, underemployment, legal difficulties, and impaired social relationships.

COMORBIDITY

The symptoms of ADHD are not unique to that diagnosis. In children they overlap to a large a degree with symptoms of oppositional defiant disorder (ODD), conduct disorder, learning disability, mood disorders, and anxiety. In adults, ADHD symptoms overlap with antisocial disorders, mood and anxiety disorders, and alcohol or drug addiction.

ENVIRONMENTAL FACTORS

Birth-related risk factors include elevated bilirubin levels, preterm birth, low birth weight, perinatal hypoxia and ischemic events, maternal metabolic disorders such as diabetes and phenylketonuria, as well as maternal alcohol use, smoking, and the use of certain medications during pregnancy. Note that the basal ganglia, which are implicated in ADHD, are the most sensitive to hypoxic insults being one of the most metabolically active structures in the brain. The idea that particular **foods** might cause ADHD received much attention in the media but was not supported by systematic studies. However, lead exposure has been implicated in the pathophysiology of ADHD. **Psychosocial factors** such as marital distress, family dysfunction, large family size, paternal criminality, and low social class are risk factors for ADHD.

GENETIC FACTORS

For many decades, studies have shown that ADHD runs in families. Twin studies estimate the **heritability of ADHD to be 0.75**. Genome studies of individuals with ADHD using **linkage analysis** have so far not produced reliable results suggesting that genes with large effects for the discrimination of ADHD from healthy controls are unlikely to exist.

On the other hand, many candidate gene studies used association methods to determine whether biologically relevant gene variants affect susceptibility to ADHD. In the 2005 review by Faraone and coworkers, mentioned earlier several candidate genes were implicated in susceptibility to ADHD including the family of **dopamine receptor and dopamine transporter genes**.

It should be noted here that genetic variants associated with ADHD are quite frequent and fixed in some populations. This is in contrast to schizophrenia and mood disorders showing lower fertility. According to a 2007 review by Mauricio Arcos-Burgos and Maria Teresa Acosta from the National Institutes of Health, this pattern indicates that ADHD as a behavioral trait had provided **selective advantage in evolution**.

ROLANDIC FOCUS

Having an epileptiform **focus over the Rolandic fissure** is a risk factor of ADHD. As we learned from chapter: Spontaneous Electroencephalogram, having a focus in the cortex does not necessarily mean having epilepsy: the focus can be found in epilepsy-free subjects. The focus in the Rolandic fissure is expressed in central–temporal (Rolandic) spikes. There are subjects without any diagnosis of epilepsy who for some reason show Rolandic spikes in their electroencephalograms (EEGs). Martin Holtman and coworkers from the Central Institute of Mental Health in Mannheim (Germany) in their 2003 study found Rolandic spikes in 6% of children with ADHD without epilepsy which is higher than that reported in healthy children. The authors also found that children with Rolandic spikes tended to exhibit more hyperactive–impulsive symptoms.

On the other hand, children with epilepsy are at higher risk of having ADHD than the general pediatric population. David Dunn and coworkers from Indiana University School of Medicine in their 2003 study found that **38%** of their sample of children with epilepsy meet the criteria of ADHD (mostly the inattentive subtype).

EXECUTIVE FUNCTIONS

Neuropsychological assessment represents an inherently different process than generating a diagnosis on the basis of behaviorally defined criteria. There is no way to suggest any neuropsychological test that would specify a DSM symptom of ADHD such as "often has difficulty organizing tasks and activities." However, the specific nature of neuropsychological assessments does not exclude the existence of a number of neuropsychological tasks in which ADHD patients behave worse than healthy controls.

As mentioned in the historical introduction to the chapter, ADHD is often considered executive dysfunction. To examine the validity of the **executive dysfunction theory** of ADHD a metaanalysis was performed by Eric Willcutt and coworkers from the University of Colorado in 2005. **Eighty-three studies that measured executive functions** in groups with ADHD (total $N = 3734$) and without ADHD ($N = 2969$), were analyzed.

Groups with ADHD exhibited significant impairment on all neuropsychological tasks with medium effect sizes varying from 0.43 to 0.69. Recall that a biomarker can be applied in clinical practice in case of large affect size. Consequently, none of the measured neuropsychological parameters alone can be used as a unitary index of ADHD. In 2002, Josef Sergeant from Vrije Universiteit, Amsterdam on the basis of analysis of executive dysfunctions in ADHD and other psychiatric conditions concluded that these dysfunctions are not specific to ADHD.

Together with executive dysfunctions, **delay aversion and variability of performance** are often considered candidates for neuropsychological markers of ADHD.

HETEROGENEITY OF NEUROPSYCHOLOGICAL PROFILE

In a 2000 study by Alysa Doyle and coworkers from Harvard Medical School, 102 children with ADHD and a corresponding sample of healthy controls were measured in tasks designed for assessment of the three previously mentioned neuropsychological domains: (1) executive functions, (2) delay aversion, and (3) reaction time (RT) variability (Fig. 5.1.3). Children with ADHD differed significantly from controls on all measures, except for delay aversion. **Seventy-one percent** of the children with ADHD had **at**

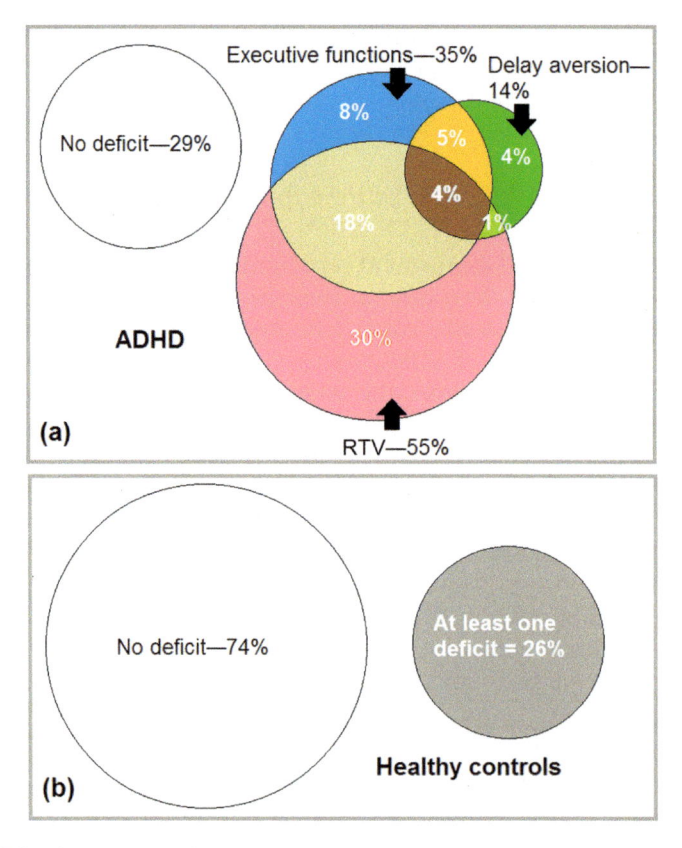

FIGURE 5.1.3 **Heterogeneity of neuropsychological profile of *ADHD*.** (a) Proportion of 7–13-year-old ADHD subjects with impairments in *RTV*, executive functions, and delay aversion. (b) Proportion of healthy controls with neuropsychological impairment. *Adapted with permission from Doyle et al. (2000).*

least one type of **neuropsychological impairment**: executive functioning (35%), RT variability (54%), and delay aversion (14%). **Twenty-nine percent did not show any deficit** in the measurements. At the same time, 26% of the controls had at least one neuropsychological deficit.

This study shows the heterogeneity of both ADHD and healthy control groups. This begs two questions: (1) Could neuropsychological heterogeneity be taken into account by decomposing the whole group of ADHD patients and healthy controls into subgroups with distinct neuropsychological profiles?; and (2) Could this decomposition improve the diagnostic procedure within the selected subgroups?

A 2012 study by Damien Fair and coworkers from Oregon Health and Science University made an attempt to answer these questions. A neuropsychological battery included tasks to measure (1) working memory, (2) response inhibition, (3) response variability, (4) temporal information processing, (5) arousal and activation, (6) interference control, and (7) response speed. The ADHD population ($N = 285$) performed worse on all the tasks in comparison with healthy controls ($N = 213$).

To assess the heterogeneity within ADHD and healthy controls, graph theory was applied in a 2012 study by Damien Fair from Oregon Health and Science University and coworkers. Healthy controls were separated into distinct subgroups: (1) a response variability group (43% of sample), (2) a low-executive group with reduced working memory, memory span, inhibition, and output speed (20% of sample), (3) a low-temporal information-processing group (18% of sample), and (4) a low-arousal group with weak signal detection (18% of sample). The heterogeneity of individuals with ADHD was "nested" in this normal variation so that each subgroup could be identified in both ADHD and healthy controls. Each subgroup was characterized by a **unique pattern of deficits**.

Comparing the neuropsychological measures of healthy children with those of ADHD children within each of these distinct subgroups **increased diagnostic accuracy**. The idea of making discrimination procedures within the subgroups of healthy controls and ADHD takes into account the heterogeneity of the disease, and from the theoretical point of view is a step forward, but practically it's hardly implemented.

INHIBITION DEFICIT

Russel Barkley (1997a,b) suggested that the main deficit in ADHD is **difficulty with inhibition**. Recall his model from chapter: Executive System and Cognitive Control which postulated decomposing inhibition into three interrelated processes: (1) inhibition of a prepotent response; (2) inhibition of an ongoing response; and (c) interference inhibition.

This idea got the support from a 2010 study by Marije Boonstra and coworkers from PsyQ and Erasmus University and VU University

(Amsterdam). In their study 49 carefully diagnosed adults with persistent ADHD, who had never been medicated for their ADHD, were compared with 49 healthy control adults matched for age and gender on a large battery of neuropsychological tests in 5 domains of executive functioning: (1) inhibition, (2) fluency, (3) planning, (4) working memory, and (5) set shifting, as well as several other neuropsychological functions to control for nonexecutive test demands. After control for nonexecutive function demands and IQ, adults with ADHD showed problems in **inhibition and set shifting** but not in any of the other executive functioning domains tested.

DELAY AVERSION

As we know from our own experience, waiting for a long time in a queue evokes negative emotions and restlessness. Children with ADHD particularly dislike such a delay. In a 2009 study by Paraskevi Bitsakou and coworkers from the University of Southampton (United Kingdom) three indices of **delay aversion** were measured in 77 ADHD patients, 65 of their siblings unaffected by ADHD, and 50 non-ADHD controls. The indexes were measured in three delay tasks and included: (1) choice of immediate rather than delayed reward; (2) slower reaction times following delay; and (3) increased delay-related frustration. Children with ADHD differed from controls on all tasks with medium-effect size (0.4–0.7) for different delays but with large-effect size for the **overall index (0.9)**. Highlighting the significant, but limited, role of delayed aversion in ADHD, these results are consistent with recent accounts that emphasize the neuropsychological heterogeneity of ADHD.

It should be stressed here that the neuropsychological concept of delay aversion must be separated from concepts of working memory, set-shifting, planning, and action inhibition implicated in the pathogenesis of ADHD. For example, Mary Solanto and coworkers from Long Island Jewish Medical Center in 2001 compared executive and motivational functions by using measures from a Stop task and a Choice Delay task and showed that these two measures are **not correlated**. They also showed that while performance on either task was only moderately associated with ADHD, applying both measures correctly classified nearly 90% of children with ADHD.

REACTION TIME VARIABILITY

As shown in chapter: Psychometrics and Neuropsychological Assessment, reaction time variability (RTV) is a reliable measure of intraindividual inconsistency in performance. Recall that several parameters of RTV were suggested including the standard deviation (SD) of RT and the variability coefficient calculated as the ratio of SD and mean RT. In ADHD research, most studies demonstrated increased RT variability in the ADHD population using SD of RT. There is consistent evidence that RTV distinguishes individuals with ADHD from typically developing populations with medium- to large-effect sizes for children/adolescents and small- to medium-effect sizes for adults. Most studies of RTV implicitly assume that heightened RTV reflects occasional **lapses in attention**. In addition, these instances of long RTs in children with ADHD may be predictably periodic at a frequency of about 0.5 Hz. These oscillations in RT were unaffected by double-blind placebo and were **suppressed** by double-blind methylphenidate.

An example of excessive RTV in a 14-year-old ADHD patient is shown in Fig. 5.1.4. In this patient the number of omission and commission errors and RT do not significantly differ from mean reference values. RTV is only larger in comparison with healthy controls at $p < 0.02$.

RTV has been suggested as a potentially important index of the stability/instability of a subject's nervous system. However, it **lacks specificity** as increased RTV characterizes populations ranging from ADHD, schizophrenia to dementias and traumatic brain injury.

INTERFERENCE WITH DEFAULT MODE

Edmund Sonuga-Barke and Xavier Castellanos in 2007 proposed that the skewed RT distributions of individuals with ADHD may be a consequence of **interference by the default mode network** of the brain. This

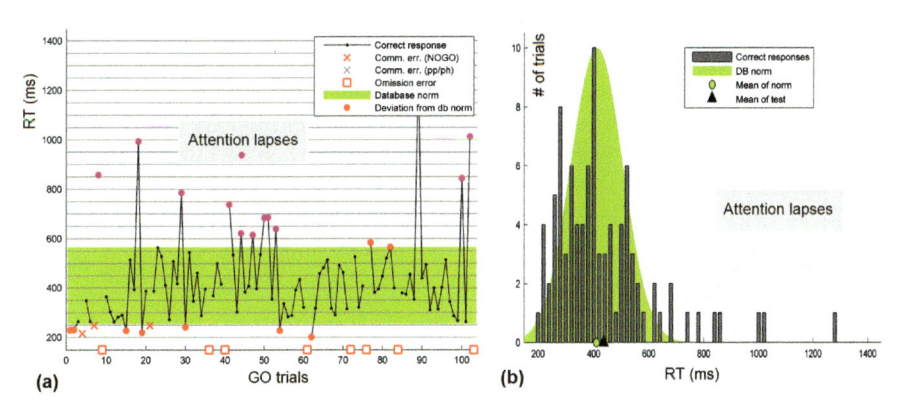

FIGURE 5.1.4 **Attention lapses in an ADHD patient.** (a) Dynamics of *RT* over consecutive GO trials in the cued GO/NOGO task. Confidence intervals of *RTV* in healthy populations are depicted in *green*; and (b) RT distribution of the patient *(black)* in comparison with the distribution averaged across healthy subjects of the same age *(green)*.

theory proposes a recurrent switching between a state of self-reflection (an "Introspective" state) and a state of attentive readiness (an "Extrospective" state). When a task is initiated the magnitude of fluctuations in introspective state is decreased but, over time, this magnitude increases, crossing a threshold. When it occurs an **attention lapse results in a long RT**. In individuals with ADHD, recovery and "default mode interference" is faster and the lapses of attention are more frequent than in non-ADHD individuals performing a task (such as GO/NOGO, Stop, or Stroop), resulting in an increase in intraindividual variance of RTs that creates a skewed distribution.

STATE REGULATION AND ENERGIZATION FUNCTION

According to the cognitive energetic model of ADHD suggested by Joseph Sergeant from the Vrije Universiteit, Amsterdam in 2000, impairments on tasks requiring effortful control of attention and executive processes could be due to deficiencies in three energetic pools—**effort, arousal, and activation**—that control the allocation of cognitive resources rather than impaired cognitive resources per se. **Effort** is conceived as the necessary energy to meet the demands of a task. **Arousal** is defined as phasic responding which is time-locked to stimulus processing. The **activation** pool is defined as tonic changes in physiological activity.

Performance inconsistency, in general, and RTV, in particular, could be considered indexes of poor regulations in effort, arousal, and activation. Studies on patients with stroke show that increased variability or inconsistency in performance correlates with **superior and dorsolateral prefrontal brain lesions** but not with inferior medial prefrontal lesions.

As shown in chapter: Executive System and Cognitive Control the purely cognitive aspects of executive functions ("**cool**") are associated with dorsolateral prefrontal cortex whereas the motivation/affective aspects of behavior ("**hot**") are associated with orbitofrontal and cingulate cortical areas. Xavier Castellanos and coworkers in 2005 suggested that **inattention symptoms** are associated with deficits in cool executive functions, whereas **hyperactivity/impulsivity** symptoms are associated with hot executive deficits.

HYPOAROUSAL HYPOTHESIS

As we learned from chapter: Sensory Systems and Attention Modulation, cortical activation by **noradrenalinergic pathways from the locus coeruleus** plays a central role in optimizing arousal. Recall the inverted U-law which in case of noradrenaline as a neuromodulator presumes that human optimal performance occurs at a relatively narrow rate of tonic activity of locus coeruleus neurons. This optimal tonic rate in turn facilitates phasic responses of locus coeruleus neurons. Recall also the hypothesis that associated the **P3b wave** of ERPs with phasic response of these neurons. Howells, Stein and Russell (2012) from the University of Cape Town suggested two models of pathological locus coeruleus activity: (1) **hypoactivation (hypoarousal) in ADHD**, and (2) hyperactivation (**hyperarousal**) in **anxiety disorders**. The decreased P3b wave in the ADHD population appears to be in agreement with this hypothesis.

A similar hypothesis but from a different perspective was suggested in 2014 by Ulrich Hegerl and Tilman Hensch from the University of Leipzig. They used the concept of vigilance as a synonym for brain arousal and an EEG-based algorithm—Vigilance Algorithm Leipzig (VIGALL)—for objectively measuring the regulation of vigilance during the transition period between wakefulness and sleep onset. According to the hypothesis, **unstable vigilance** is a core pathogenetic factor of ADHD and causes sleep dysregulation in the disorder.

MATURATION DELAY IN NEURODEVELOPMENT

The behavioral pattern of ADHD children is not appropriate for children of a corresponding age but can be found in normally developing children of a younger age. In other words, children with ADHD behave like younger children who are naturally more active, more impulsive, and have a shorter attention span than older children. This simple observation is further supported by the cognitive profile of ADHD children: they show deficits in late-developing higher cognitive functions of inhibitory self-control and attention. The facts that ADHD symptoms tend to improve with age and up to 80% of children (depending on follow-up length and definition of persistence) grow out of ADHD in adulthood further supports the hypothesis of a **maturational lag**.

Indirect neurobiological support for a maturation lag in ADHD comes from cross-sectional structural imaging studies. These MRI studies found corticostriatal brain regions of **reduced size** that are known to develop late in adolescence.

A longitudinal study by Philip Shaw and coworkers from National Institute of Mental Health, Bethesda, and the Montreal Neurological Institute, McGill University on a large population of ADHD and healthy controls provided neurobiological evidence for the maturational delay hypothesis of ADHD. Cortical thickness acquired prospectively on 223 children with ADHD and 223 healthy controls was estimated by means of computational neuroanatomic techniques. The growth trajectory of each cortical point was defined. The trajectory included a phase of **increase**

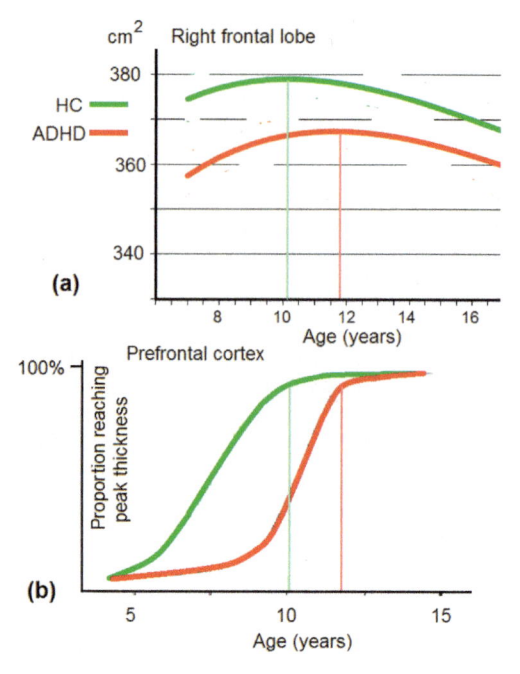

FIGURE 5.1.5 **Maturation lag in *ADHD*.** (a) Developmental trajectories of right frontal lobe surface areas in ADHD and HC. The cortical areas in ADHD reach their peak values at different ages for the frontal cortex: 12 years for ADHD in contrast with 10 years in HC; and (b) Kaplan–Meier curves illustrating the proportion of cortical points that had attained peak thickness at each age for the prefrontal cortex. *Part a: adapted with permission from Shaw et al. (2012). Part b: adapted from Shaw et al. (2007).*

in cortical thickness during childhood that was followed by **adolescent decrease** (Fig. 5.1.5a). From these trajectories, the age of attaining peak cortical thickness was derived and used as an index of cortical maturation. According to this index, primary sensory areas matured before higher order association areas. There was also a marked delay in ADHD in attaining peak thickness throughout the whole brain with the most prominent delay in prefrontal regions.

THETA/BETA RATIO

The absolute characteristics of EEGs vary substantially in healthy and ADHD populations, hence the need for relative characteristics with lower variability. One such characteristic is the **theta/beta ratio (TBR)**. This index shows maximal amplitude over **frontocentral** electrodes depending on age and **reduces with age** (Fig. 5.1.6). Taking into account the maturation lag of

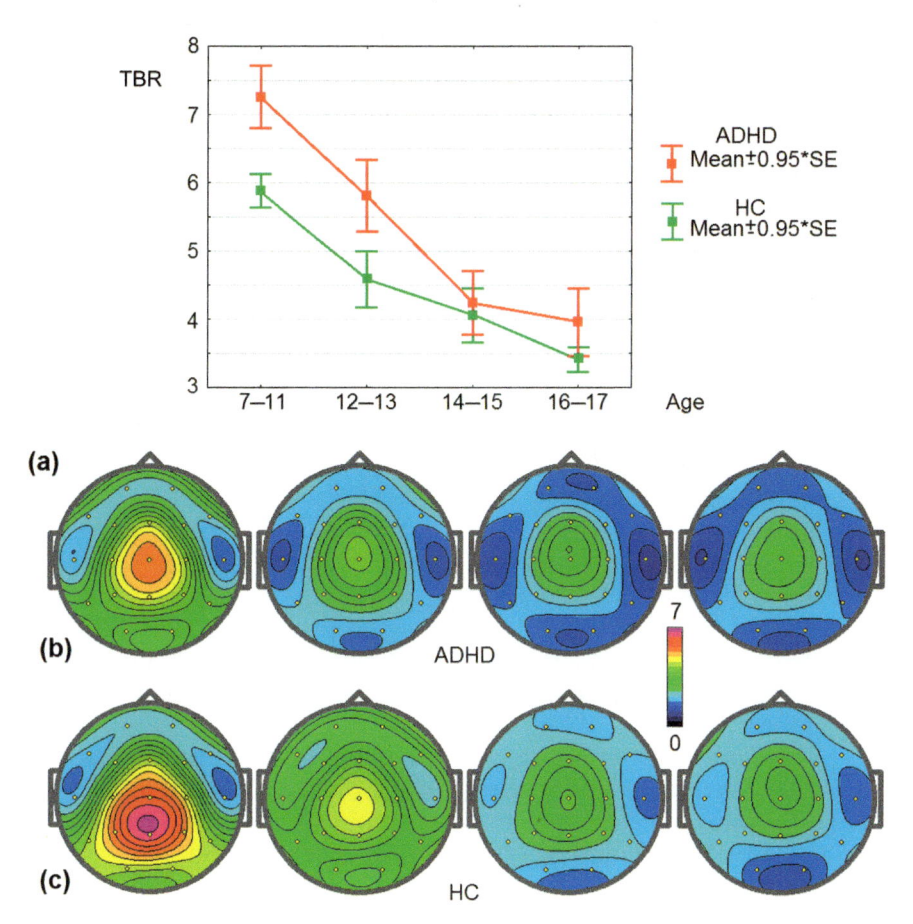

FIGURE 5.1.6 **TBR in ADHD.** (a) Mean value of TBR computed in referential montage at Cz (theta = 4–8 Hz, beta = 13–21 Hz) for different age groups of ADHD (*red*) and HC (*green*); (b) maps of TBR for the same groups in ADHD; and (c) maps of TBR for the same groups in healthy subjects. *Data from the HBI database.*

at least some ADHD patients, it would be logical to suggest that this group of patients shows elevated values of TBR.

This indeed was observed in some ADHD subjects. The first observation was reported by Joel Lubar from the University of Tennessee in 1991. Two 1999 studies of different research groups supported this observation. For example, in a 1999 study by Susan Bresnahan and coworkers from the University of Wollongong (Australia), 50 children (6–11 years), 50 adolescents (13–17 years), and 50 adults (20–42 years)—half of whom were diagnosed with ADHD—were recorded for 2 min under the eyes-open condition. TBR was shown to decline with age and was elevated in the ADHD group in comparison with healthy controls.

In 2012 Martijn Arns and coworkers conducted a **metaanalysis** on TBR in ADHD. TBRs under the eyes-open condition from Cz were analyzed in 9 studies of children/adolescents 6–18 years of age with ADHD (1273 patients) and without ADHD (517 healthy controls). Grand mean effect size for the 6–13-year-olds was **0.75** and for the 6–18-year-olds it was **0.62**. However, the test for heterogeneity was significant; therefore, these effect sizes were considered by the authors as a **misleading overestimation**. The authors concluded that excessive TBR **cannot be considered a reliable diagnostic measure** of ADHD. However, a certain subgroup of ADHD patients do deviate on this measure and TBR has prognostic value in this subgroup, warranting its use as a prognostic measure rather than a diagnostic measure. This conclusion was supported by a 2013 study at the University of Zurich by Liechti et al. (2013) in which **no consistent theta or theta/beta increases** were found in ADHD. In a 2012 study by Geir Ogrim from Østfold Hospital Trust in Norway that used the HBImed normative data an elevated theta at $p < 0.01$ was found in 26% of ADHD patients. All studies stress the heterogeneity of neurophysiological parameters across both ADHD and healthy populations. Fig. 5.1.6 shows the results of comparing TBR between the groups of ADHD and healthy controls from the HBI database. Note that the statistically significant differences are present for the younger group (7–13 years of age) but disappear at the older age.

The previously mentioned studies do not exclude the existence of a **separate ADHD endophenotype characterized by elevated TBR**. As numerous studies show, this endophenotype responds to stimulant medication and to relative beta neurofeedback treatment which confirms the prognostic (not diagnostic) value of the TBR.

However, the TBR could be used in determining **whether symptoms are due to ADHD or another condition** (recall criterion E in the *ADHD Diagnostic Manual* that symptoms are not better explained by another condition). A 2015 study (partly supported by the NEBA company) by Steven Snyder and coworkers from Georgia Regents University showed that the integration method (an addition to the TBR in the assessment procedure) may help improve diagnostic accuracy from 61 to 88%. In this study of 209 patients meeting ADHD criteria per a site clinician's judgment, 93 were separately found by the multidisciplinary team to be less likely to meet criterion E, implying possible **overdiagnosis by clinicians in 34%** of the total clinical sample. Of those, 94% were also identified by EEG, showing a relatively lower TBR. Further, the integration method was in 97% agreement with the multidisciplinary team in the resolution uncertain cases according to a site clinician. Patients with relatively lower TBR were more likely to have other conditions that could affect criterion E certainty.

QEEG ENDOPHENOTYPES IN ADHD

All studies of quantitative EEG (QEEG) in ADHD consistently report **several subtypes of QEEG** patterns in ADHD. For example, Robert Chabot from New York University Medical Center and Gordon Serfontein from the Serfontein Clinic of Developmental Paediatrics and Learning Disorders in Sydney, Australia in 1996 analyzed EEG in 407 children with a DSM-III diagnosis of ADD and found three subtypes: 38% of the sample had **excess theta activity**, 28% had **excess alpha activity**, and 13% had **excess beta activity**.

In a 2001 study by Adam Clarke from University of Wollongong, Australia and coworkers (Clarke et al., 2001c), cluster analysis identified three distinct **EEG-defined clusters—endophenotypes**. The first cluster had increased total power, a relative theta, and a **TBR** which the authors considered an index of cortical hypoarousal. The second cluster was characterized by increased slow-wave activity and deficiencies in **fast-wave activity**, which indicated the presence of a maturational lag in CNS development. The third cluster had **excess beta activity** and was tentatively labeled an overaroused group. However, the overarousal state of the excessive beta cluster was not supported by the latest data of this group showing that the cluster was characterized by a low level of skin–conductance response—an indicator of hypoarousal.

Several other research groups have made attempts to separate subtypes of ADHD on the basis of QEEG. One attempt was described in a 2011 book by Andreas Mueller and coworkers. Another was presented in a 2005 paper by Jack Johnstone and coworkers. The authors suggested 11 general endophenotypes including diffuse slow activity, excess temporal lobe alpha, spindling beta, persistent alpha with the eyes open. As shown in a 2009 paper by Martijn Arns and coworkers from Utrecht University QEEG endophenotypes predict treatment outcomes in ADHD. In spite of methodological differences in the attempts, several clear subtypes of QEEG in ADHD can be separated. Fig. 5.1.7 shows five different QEEG endophenotypes in ADHD.

Cases a, b, c in Fig. 5.1.7 represent 14-year-old ADHD children while cases d and e are taken from adults with ADHD. Case "a" does not show significant deviations from healthy controls in any frequency but reveals a significant increase in **TBR** measured at Cz. Case "b" is characterized by excessive slow activity at Cz which is also reflected in elevated TBR. Cases "a" and "b" are hypothetically associated with **maturation lag** in corresponding brains. With development the frontal-basal ganglia–thalamocortical pathways mature, which is reflected in progressive decrease of TBR with age.

Case "c" exhibits persistent runs of **frontal midline theta rhythm**. By recalling that frontal midline theta rhythm reflects limbic system functioning, we can speculate that this subtype is associated with dysfunction of the

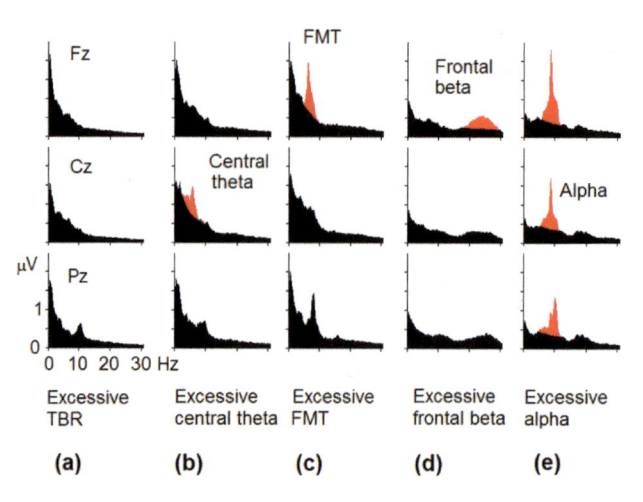

FIGURE 5.1.7 QEEG endophenotypes of ADHD (eyes-open condition, common reference montage). (a) The EEG spectra in an ADHD patient (14-years-old) do not show any statistically significant deviation from the reference, but the TBR is twice higher than in the healthy control group; (b) the EEG spectra in an ADHD patient (14-years-old) show statistically significant excess of central theta activity; (c) the EEG spectra in an ADHD patient (14-years-old) show statistically significant excess of the frontal midline theta rhythm; (d) the EEG spectra in an ADHD patient (38-years-old) show statistically significant excess of the frontal beta rhythm (spindling beta endophenotype); and (e) the EEG spectra in an ADHD patient (39-years-old) show statistically significant excess of the alpha rhythm with maximum at posterior electrodes (not shown). *Data from the HBI database.*

limbic system. This inference is supported by clinical observations of **emotional instability** in patients of this subtype.

Case "d" shows elevated frontal beta rhythm which on raw EEG fragments appears as **beta spindles** at frontal electrodes. Taking into account the positive correlation of the frontal beta with methabolic activity, we can speculate that this subtype is associated with **overactive frontal lobes**. Recall the inverted U-law according to which both underactivation and overactivation of the frontal lobes lead to poor performance on cognitive tasks.

Finally, case "e" is characterized by **excess alpha activity** under the eyes-open condition. Although alpha rhythms in these cases are generated in posterior cortical areas, the excessive alpha in the common reference montage is observed over midline electrodes Fz, Cz, and Pz because of volume conductance. This subtype might be associated with a low level of posterior cortex activation—**hypoarouasal**.

It should be noted that the cases in Fig. 5.1.7 represent extreme deviations from the average spectra of large groups of healthy subjects of corresponding age with p values below 0.001. Consequently, the number of subjects in separated groups should depend on the significance level of the criteria. Experience in our laboratory indicates that approximately 50% of

ADHD patients (especially of adult age) do not show statistically significant deviations from the reference in EEG spectra and cannot be referred to a particular QEEG endophenotype. For example, in a 2012 study Geir Ogrim and coworkers chose a significance level of **0.01** and found only **26% of ADHD patients with elevated TBR** in contrast with 3% of controls.

FRONTAL BETA SYNCHRONIZATION IN CHILDHOOD ADHD

Many ADHD children show not only decreased beta activity in the resting state, they also are **unable to synchronize frontal beta activity** in response to the task load. This can be demonstrated by wavelet analysis of EEG responses in the cued GO/NOGO task (Fig. 5.1.8). Recall that healthy subjects synchronize frontal beta activity in response to Cue, GO, NOGO, and Novel stimuli, that is, to all stimuli associated with activation of the frontal lobes)—but not children with ADHD! This observation served as the basis for the relative beta training (with electrodes placed on Fz–Cz) suggested in the study by Kropotov et al. (2005).

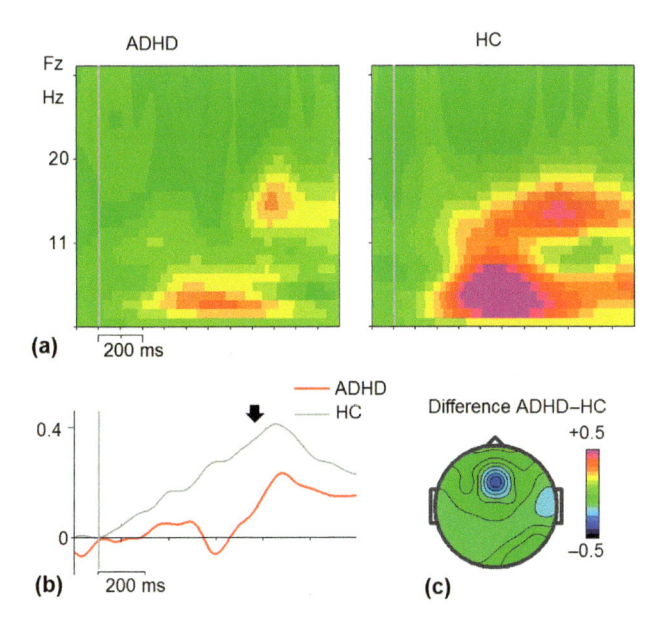

FIGURE 5.1.8 **Decrease of frontal beta synchronization in ADHD.** (a) Grand-average of the power–time frequency representations (wavelet decomposition) of EEG responses to Cue in the cued GO/NOGO task in the group of ADHD patients ($N = 23$) and healthy controls ($N = 39$); and (b) grand average of Event-Related Synchronization (ERS) in the 13–15 Hz band for ADHD and HC. (c) Topographic map of ERS differences between ADHD and HC at time interval indicated by *arrow* at (b). *Data from the HBI database.*

MAGNETIC RESONANCE IMAGING CORRELATES

Both cortical and subcortical structures are associated with ADHD. In subcortical structures the **striatum** has been of particular interest because: (1) it is rich in **dopaminergic synapses**, (2) is vulnerable to **perinatal hypoxic** complications implicated in the disorder, (3) damage due to TBI produces secondary ADHD (Fig. 5.1.9b), (4) **lesions in animals** produce hyperactivity and poor inhibitory control, and (5) in vivo neuroimaging studies in people showed that methylphenidate, which is used to treat ADHD, exerts its effects by binding to **dopamine transporters**, most of which are located in the striatum. Frodl and Skokauskas in their 2012 metaanalysis paper detected reduced right globus pallidus and putamen volumes in MRI voxel-based morphometry studies as well as decreased caudate volumes in children with ADHD (Fig. 5.1.9a).

fMRI CORRELATES

Functional magnetic resonance imaging (fMRI) studies in ADHD revealed **fronto–striato–parietal** dysfunctions during tasks of inhibition and attention. The brain areas associated with cognitive control dysfunctions in ADHD have been reviewed in 2014 by Alexandra Sebastian from Johannes Gutenberg-University, Germany and coworkers (Fig. 5.1.9).

As one can see patients with ADHD display disturbed hypo-activity mainly in ventrolateral and medial prefrontal regions. However the specific patterns depend on the component of the cognitive control. Similar results were reported by Heledd Hart from King's College London and coworkers in a 2013 metaanalysis of fMRI studies in ADHD during inhibition and attention tasks. According to this study, patients with ADHD relative to controls showed reduced activation for inhibition in the right inferior frontal cortex, supplementary motor area, and anterior cingulate cortex, as well as striato-thalamic areas, and showed reduced activation for attention in the right dorsolateral prefrontal cortex, posterior basal ganglia, and thalamic and parietal regions.

DECREASED P3B WAVE

Different ERP waves were tested as candidates for neuromarkers that differentiate ADHD patients from healthy controls. Of those the P3b wave in auditory oddball tasks was studied the most. Decreased P3b in ADHD is a robust observation that has an effect of moderate size.

Fig. 5.1.10 illustrates a decreased P3b wave in ADHD. These maps illustrate the age-specific group differences suggested by the literature, such as

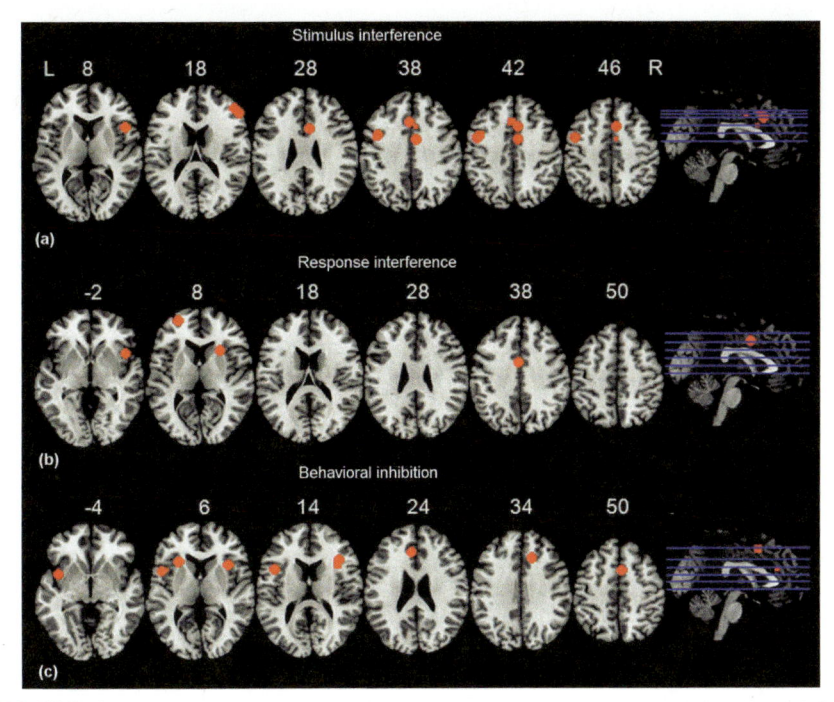

FIGURE 5.1.9 **Prefrontal dysfunctions in ADHD.** Maxima of clusters of prefrontal dysfunctions during (a) stimulus interference, (b) response interference, or (c) behavioral inhibition. L-left. R-right. *Adapted with permission from Sebastian et al. (2014).*

the reduced posterior P3b in *ADHD com* relative to control subjects clearly seen in the child but not the adolescent group.

ERP CORRELATES OF COGNITIVE CONTROL IN CHILDREN

N2 and P3 NOGO waves in GO/NOGO and Stop signal paradigms have been used as candidates for neuromarkers of cognitive control in ADHD. In a 2000 study by Steven Pliszka and coworkers from the University of Texas Health Science Center, San Antonio ERPs in a Stop signal task were measured for a group of children with ADHD compared with healthy controls. The N200 wave in response to Stop signals was markedly reduced in ADHD and significantly correlated with **response inhibition performance** across subjects (Fig. 5.1.11).

In a 2005 review, Leon Kenemans and coworkers from Utrecht University questioned the action inhibition interpretation of the data. They showed that results from stop tasks are consistent with impairments in stopping

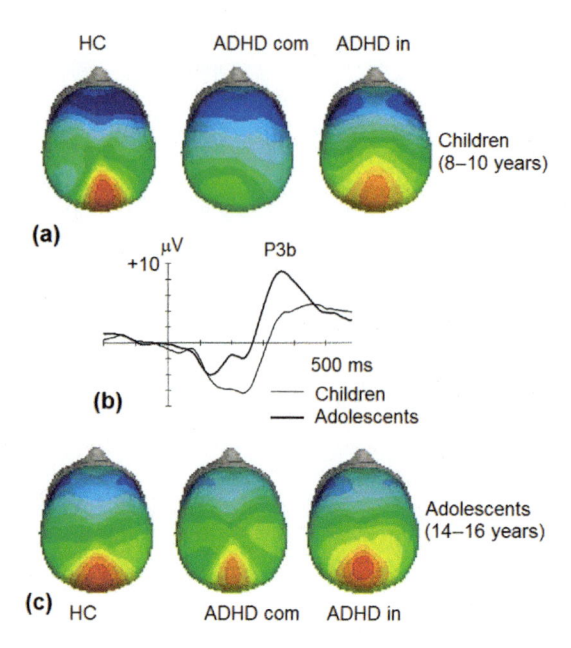

FIGURE 5.1.10 **P3b in ADHD.** (a and c) Maps of the P3b wave in healthy controls *(HC)*, ADHD combined type *(ADHD com)* and ADHD inattentive type *(ADHD in)* for (a) children and (c) adolescents; and (b) representative ERPs to target stimuli in an auditory oddball task averaged across sites and groups and derived from data reported in Johnstone et al. (2001). *Adapted with permission from Barry et al. (2003).*

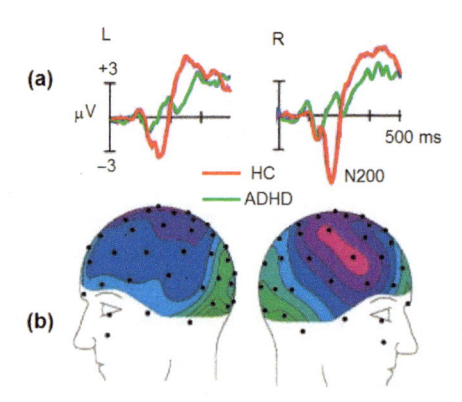

FIGURE 5.1.11 **N200 in response to Stop signals in *ADHD*.** (a) Grand-averaged ERPs for control (*red*) and ADHD (*green*) for successful inhibition trials (zero point, stop signal onset); and (b) topographic maps of controls minus ADHD ERP difference wave for successful inhibition. *Adapted from Pliszka et al. (2000).*

performance in ADHD, but in children these effects cannot easily be dissociated from more general impairments in attention to the task, and therefore an interpretation in terms of inhibitory control is not straightforward. They argue that N2 NOGO waves may rather reflect **conflict monitoring operation** than action inhibition.

In our own study on ADHD children and adolescents with a large sample size of 104 patients, ERPs in the cued GO/NOGO task were compared with ERPs of 172 healthy controls. As we know from chapter: Executive System and Cognitive Control, two late positive waves P3d Cue and P3 NOGO can be obtained in this task. The parietally distributed P3d cue is defined as the difference wave between P3 Cue and P3 NonCue. The frontally distributed P3d NOGO is defined as the difference wave between P3 NOGO and P3 GO.

The two waves in turn can be separated into independent components, but in the current context it is crucial to keep in mind that they reflect activities of distinct neuronal circuits—the parietal circuit of the proactive cognitive control and the frontal circuit of reactive cognitive control. As Fig. 5.1.12 clearly demonstrates, both circuits are suppressed in child and adolescent ADHD. As will be shown below, dysfunction in the frontal circuit in an individual with ADHD predicts a positive response to psychostimulants while selective dysfunction in the parietal circuit predicts a high probability of being a nonresponder to psychostimulants.

EVENT-RELATED POTENTIAL CORRELATES OF COGNITIVE CONTROL IN ADULT ADHD

As was shown in chapter: Event-Related Potentials (ERPs), decomposition of group ERPs into latent components needs substantial sample sizes of the groups. In our study of adult ADHD, we compared the ERPs of 94 ADHD subjects with the ERPs of 352 subjects. Raw ERPs showed a similar pattern to that obtained in children and adolescents (as described earlier). Decomposition of ERPs into latent components with known functional meanings (see section "Functional meaning of latent components" in chapter: Executive System and Cognitive Control) enables us to look at information processing in the ADHD brain from the functional point of view. Extracted latent components for adult ADHD and healthy control groups are compared in Fig. 5.1.13.

Of the 18 components computed for Cue, GO, and NOGO conditions only 4 differ between the ADHD and healthy controls (HC) groups with moderate to large effect sizes (*gray boxes* in Fig. 5.1.13). The **temporally distributed component** to s1 appears to be associated with delayed negativity in delayed match-to-sample tasks (see section "Working

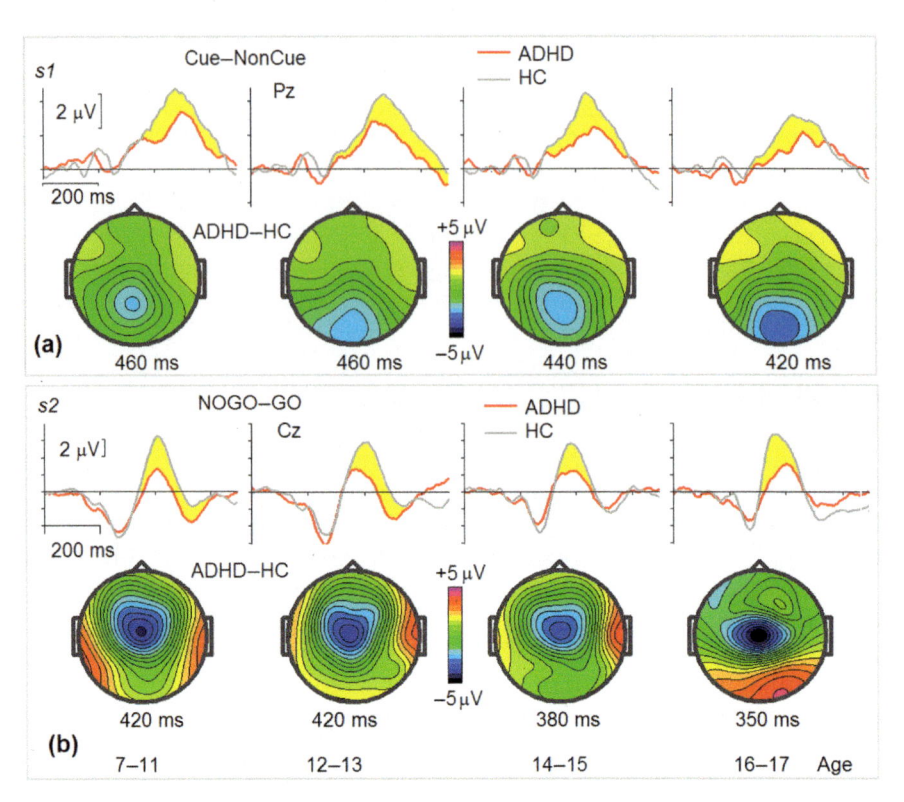

FIGURE 5.1.12 **Impaired cognitive control in ADHD children.** ERPs were recorded during the cued GO/NOGO task in four age groups of ADHD children and adolescents as well as healthy controls (HC) of the corresponding age: 7–11 years (total $N = 105$), 12–13 ($N = 75$), 14–15 ($N = 56$), and 16–17 ($N = 40$). (a) Suppression of P3d cue wave (defined in the difference Cue − NonCue) in ADHD. Upper row, difference waves (Cue − NonCue) in the four age groups for ADHD (*red*) and healthy controls (*HC, gray*). Bottom row, maps of the differences between ADHD and HC made at maximum latencies depicted below. (b) Suppression of P3 NOGO wave (defined in the difference NOGO − GO) in ADHD. Upper row, difference waves (NOGO − GO) in the four age grops for ADHD (*red*) and healthy controls (*HC, gray*). Bottom row, maps of differences between ADHD and HC made at maximum latencies depicted here. *Data from the HBI database.*

memory representations" in chapter: Memory Systems) and reflects a process of storing task-related representations in working memory. The **frontally distributed component** seems to reflect the operation of action inhibition whereas the **parietally distributed component** reflects the operation of conflict detection or context updating. As will be shown later, suppression of these components in ADHD is not a specific phenomenon because it is present in schizophrenia and OCD. It is also worth stressing here that all sensory-related components are intact in adult ADHD in contrast with schizophrenia.

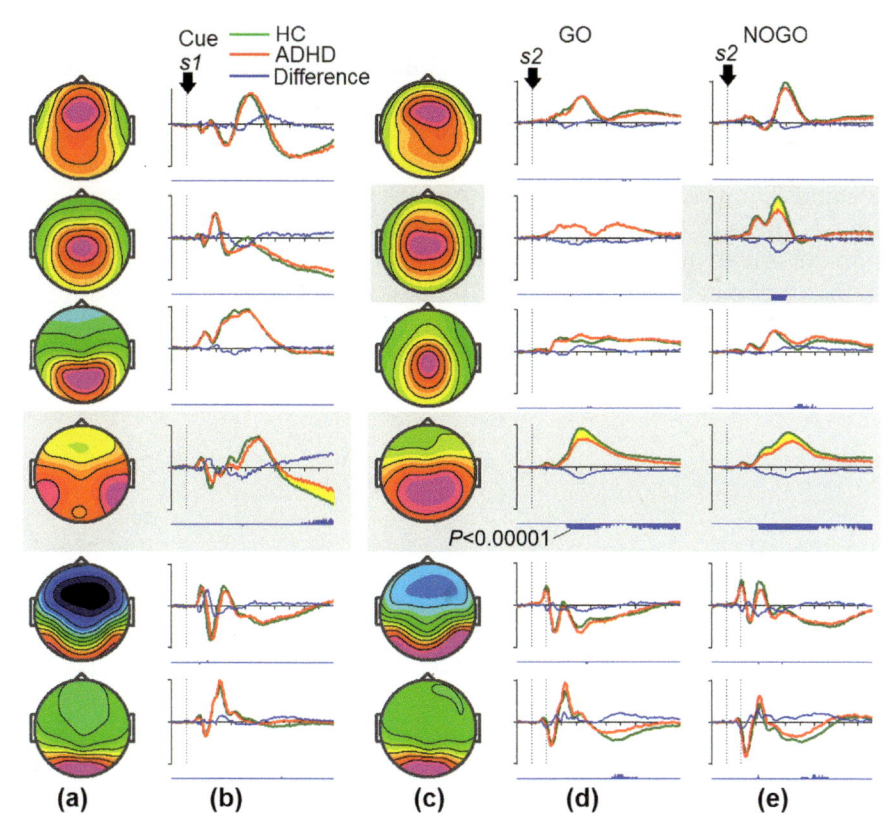

FIGURE 5.1.13 Latent components of cognitive control in adult *ADHD*. ERPs were recorded during the cued GO/NOGO task in adults with ADHD ($N = 94$) and HC ($N = 258$). Latent components were extracted using the method of joint diagonalization explained in chapter: Event-Related Potentials (ERPs). (a) Topographies of latent components extracted from ERPs between s1 and s2; (b) timecourse for ADHD (*red*) and HC groups and the difference ADHD – HC (*blue*) with p-values of statistical significance below; (c) topographies of latent components extracted from ERPs in response to s2; (d) timecourse for ADHD (*red*) and HC groups and the difference ADHD – HC (*blue*) for GO condition with p-values of statistical significance below; and (e) as in (d) but for NOGO condition. The components with the largest differences between groups are shown in *gray* and *yellow*. *Data from the HBI database.*

PHARMACOLOGICAL TREATMENT

The mechanisms underlying the action of drugs licensed to treat ADHD are well established: (1) methylphenidate **blocks DA and norepinephrine (NE) presynaptic reuptake**; (2) amphetamine **blocks DA and NE presynaptic reuptake** as well as release of DA and NE in the synapse; and (3) atomoxetine **blocks presynaptic reuptake of NE** and has an effect on serotonin. The average effect size for stimulants (0.91 for immediate release and 0.95 for long-acting versions) is greater than the average effect for non-stimulants (0.62), although much variability exists within classes.

In a 2000 study, Krause and coworkers from Ludwig-Maximilians-University, Munich demonstrated that methylphenidate lowers increased striatal DAT availability in adults suffering from ADHD (Fig. 5.1.14). Ten previously untreated adults with ADHD were investigated before and after 4 weeks of treatment with methylphenidate by single-photon emission computed tomography (SPECT) with a ligand specifically binding to the dopamine transporter (DAT). Patients with ADHD demonstrated **increased specific binding** of the ligand to the DAT compared with age- and sex-matched controls. After **treatment** with methylphenidate, specific binding **decreased** in all patients.

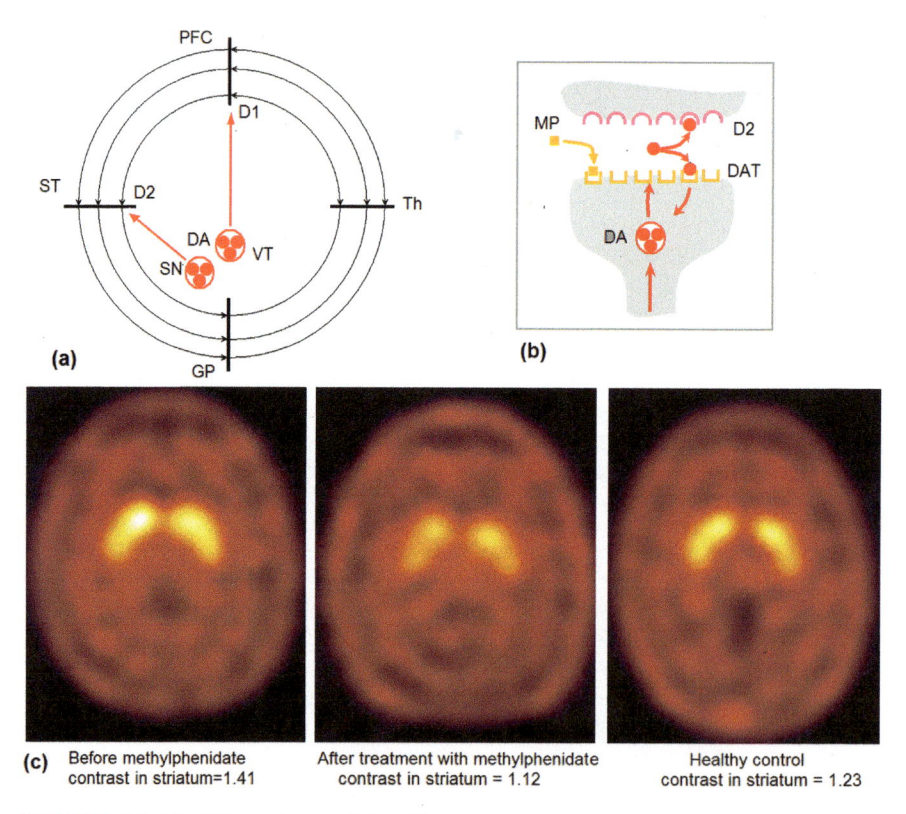

FIGURE 5.1.14 **Effect of methylphenidate on DAT in striatum in ADHD.** (a) Two pathways of *dopamine (DA)* in the brain: (1) from substantia nigra *(NG)* to striatum *(ST)* with D2 receptors, (2) from ventral tegmentum *area (VT)* to the prefrontal cortex *(PFC)* with D1 receptors; (b) circulation of DA in the synapse: (1) binding to D2 receptors, and (2) washing out by binding to dopamine transporter *(DAT)*. Methylphenidate blocks DAT; and (c) representative SPECT scans for an ADHD patients before (left) and after (central) commence of treatment with methylphenidate as compared to an age-matched control (right). *Adapted from Krause et al. (2000).*

In a 2011 review into the effects of stimulant medications on cognitive functions in ADHD, James Swanson from the University of Irvine, Ruben Baler and Nora Volkow from NIH summarize that the optimal dose varies across individuals and depends on the domain of function while stimulant-related enhancements are more prominent on tasks **without an executive function component** (complex reaction time, spatial recognition memory reaction time, and delayed matching-to-sample reaction time) than on tasks with an executive function component (inhibition, working memory, strategy formation, planning, and set-shifting). The authors also suggest that stimulant medication makes brain functioning more efficient by enhancing the signal-to-noise ratio in neuronal networks.

EVENT-RELATED POTENTIAL PREDICTORS OF RESPONSE TO PSYCHOSTIMULANTS

Methylphenidate has shown a therapeutic response in approximately 65–70% of patients. Currently, there is **no reliable method to predict** how patients will respond, other than by **exposure to a medication trial**. The trial sometimes requires several weeks before a conclusion about the effectiveness of a drug is made. When informants evaluate the effects, the presence of side-effects may influence their opinion.

Geir Ogrim from the Norwegian University of Science and Technology used our cued GO/NOGO paradigm for assessing executive functions in ADHD children. His idea was to find an ERP **neuromarker that could predict the therapeutic response to psychostimulants**. He recorded 19-channel EEGs in 82 medication-naive ADHD patients aged 8–17 years during performance of the task two times. The first recording (*Pre*, the baseline) was done before a systematic trial on stimulant medication. The trial lasted for 4 weeks after which patients were categorized as Responders (71%) or non-Responders (29%) on the basis of daily ratings from parents and teachers. The second recording was done some day within the trial around one hour after a single dose of stimulant medication (*Post*, single-dose effect). The main results are presented in Fig. 5.1.15.

The ERPs of Responders and non-Responders deviate from the ERP of healthy controls at the baseline in different ways: non-Responders do not show statistically significant deviations from the reference in the P3 NOGO wave at Cz while Responders, in contrast, do show much smaller amplitude of the P3 NOGO wave in comparison with healthy controls. An opposite pattern was obtained for the parietally distributed P3 cue wave: the Non-Responders showed suppression of the P3 cue amplitude in comparison to healthy subjects, while responders showed relatively intact P3 cue. In addition, the responders demonstrated more EEG power in the theta band. So, the Responders and Non-Responders can be **separated** at the baseline on the basis **of QEEG/ERP measures**. It appears that localization of the

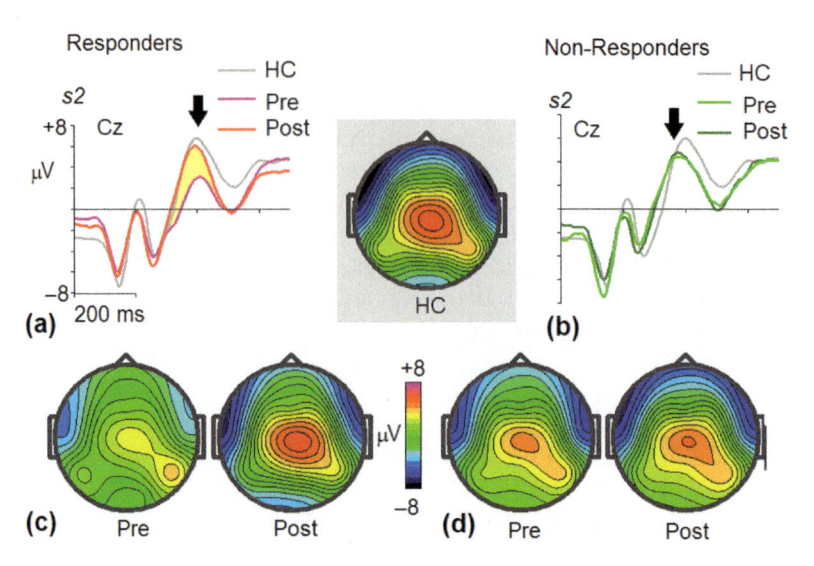

FIGURE 5.1.15 **Predicting response to psychostimulants.** 82 medication-naive ADHD patients were separated into Responders (a, c) and non-Responders (b, d) on the basis of a 4-week trial on stimulant medication. ERPs in the cued GO/NOGO task were recorded twice: at the baseline *(Pre)* and 1 hour after stimulant medication *(Post)*. (a and c) ERPs to the NOGO stimulus at Cz and their maps for *Pre* (*pink*) and *Post* (*red*) recordings for Responders. (b and d) ERPs to the NOGO stimulus at Cz and their maps for *Pre* (*pink*) and *Post* (*red*) recordings for non-Responders. *Partly adapted from Ogrim et al. (2014).*

main dysfunction of the cognitive control in a particular patient subject defines the response to Ritalin: "positive" response if the main dysfunction is localized in the frontal lobe, and "negative" response if the main dysfunction is localized in the parietal lobe. **The comparison with the normative** database is mandatory in this way of prediction.

The results of the study (Fig. 5.1.15) also showed that prediction of how the individual patient would respond to a stimulant could be made from a single-dose recording without comparison to the reference database. ERPs simply have to be recorded in the patient twice: **before and after a single dose of medication**. If an individual NOGO wave at Cz changes significantly after medication this indicates high probability of the patient being a Responder. If not, the chances of a positive response to medication are very low.

DOPAMINE HYPOTHESIS

The **dopamine hypothesis of ADHD** is based on the facts: (1) that symptoms of ADHD are reduced by stimulant treatment which blocks the dopamine reuptake mechanism in the striatum; and (2) that some patients with ADHD have abnormalities in genes responsible for dopamine regulation. However, this hypothesis **has been questioned**. For example, François Gonon from the University of Bordeaux in his 2009 critical review of neurochemical, genetic, neuropharmacological, and imaging data underlines the weaknesses of the dopamine deficit hypothesis of ADHD. He warns that the current dominance of the dopamine hypothesis discourages the human and financial investments needed to explore alternative theories. According to the Gonon's view the main drawback of the dopamine deficit theory is that it gives scientific credence to a view that favors psychostimulant medication over other medical, psychological, and social approaches to ADHD treatment.

NEUROFEEDBACK

Two variants of neurofeedback have been applied for modulating the symptoms of ADHD: **TBR neurofeedback** (or better to say, beta/theta ratio up-training) and training **slow cortical potentials (SCPs)**.

The first protocol was suggested by Joel Lubar and coworkers in the 1970s and got experimental support in the observation that the EEG **TBR** discriminated ADHD children from healthy controls (Lubar & Shouse, 1976). This measure was used as a target for neurofeedback protocols in which children were required to voluntarily decrease excessive theta and increased beta EEG activity at frontocentral electrodes. In the following 20 years several studies found that the effects of TBR neurofeedback on inattention and impulsivity in the ADHD population were comparable with stimulant medication. In recent years several well-controlled studies have been performed to assess the effects of neurofeedback on ADHD symptoms and several studies of metaanalysis have been published. The 2009 metaanalysis on neurofeedback by Martijn Arns and coworkers reported **large effect sizes for inattention and impulsivity** and medium effect sizes for hyperactivity.

The second protocol of neurofeedback was developed by Niels Birbaumer on the basis of the observation that subjects could **voluntarily change SCP** while activation of brain resources in anticipation of motor or cognitive action was accompanied by a negative shift of SCP (for details see chapter: Neurofeedback). It was shown that proconvulsive procedures produced negative shifts of cortical potentials while anticonvulsive therapy worked in the opposite direction. These results served as the basis for application of SCP neurofeedback for treating drug refractory **epilepsy**. As far as ADHD is concerned, the application of SCP neurofeedback for its treatment was governed by the observation of **reduced contingent negative variation (CNV) in ADHD**. In 2004, Hartmut Heinrich and coworkers from the University Hospital of Erlangen published a study on application of SCP neurofeedback in ADHD and showed that reductions in the severity of ADHD were accompanied by an **increase in CNV**.

At least two issues in ADHD neurofeedback research remain to be resolved. First, the fact of heterogeneity of ADHD in which only 30% of patients are characterized by excessive TBR requires **individual neurofeedback protocols** different from conventional theta beta neurofeedback. Although many studies suggest that a personalized approach could improve the efficacy of neurofeedback, future well-controlled multicenter studies with large sample sizes are needed. Second, Martijn Arns and his coworkers in a 2014 review suggested that a possible explanation of positive effects of TBR neurofeedback on all patients irrespective of their QEEG endophenotype could be that in the conventional TBR approach a **compensatory mechanism** is actually trained—rather than an individual neural dysfunction being addressed. By training to increase the relative beta frontally, children could learn to voluntarily reach an attentive state, thereby strengthening attention neuronal networks. This issue has to be clarified in future studies. Finally, the issue of **neurofeedback specificity** has to be clarified in future studies. Indeed, the multisesson neurofeedback procedure requires a substantial amount of client–therapist interaction and produces nonspecific effects such as **cognitive training** by focusing on a computer screen for 30–40 sessions, positive feedback and verbal reinforcements from the therapist, etc.

tDCS

As shown in chapter: Transcranial Direct Current Stimulation, tDCS is a powerful tool for activating cortical areas. To my knowledge, there has only been one double-blind placebo-controlled study of the testing effect of tDCS in ADHD. In 2015, Camila Cosmo and coworkers from the Federal University of Bahia, Brazil used a single 20-min anodal tDCS (1 mA) of the left dorsolateral prefrontal cortex (F3 anode, F4 cathode) to improve inhibitory control in a group of adult ADHD patients as measured by behavioral parameters in the GO/NOGO task. They were not able to demonstrate statistically significant differences between this group and the sham group. There might be several reasons for this. First, the protocol of tDCS needs be individually tailored depending on what ERP component (roughly frontal or parietal) is dysfunctional in the particular patient (recall that the stimulant effect strongly depends on this ADHD subtype). Second, the locations of latent components impaired in ADHD as shown earlier in the chapter are not consistent with the F3–F4 locations of the anodal–cathodal electrodes used in this tDCS study. Third, a 20-min single session might not be enough to produce significant modulations of cortication activity. Further research in this direction is definitely needed.

TRANSCRANIAL MAGNETIC STIMULATION

Attempts at using **TMS in ADHD are rare and inconsistent**. Yuval Bloch and coworkers from Shalvata Mental Health Center in Israel in their 2010 crossover double-blind randomized, sham-controlled pilot study showed positive effects of high-frequency repetitive TMS (rTMS) on attention in adult ADHD patients. In contrast, Laurel Weaver and coworkers from the University of Pennsylvania, Philadelphia did not find the difference between active TMS applied at 10 Hz to the right prefrontal cortex and sham.

CHAPTER 5.2

Schizophrenia

HISTORICAL INTRODUCTION

In contrast to ADHD schizophrenia (SZ) is a less common psychiatric condition (**0.4–0.6% of the world's population**) and is characterized by some symptoms that are not present in the healthy population of corresponding age. Accounts of SZ are believed to be rare until 1809 when the first case reports were published. In 1893, the term dementia praecox (premature dementia) was broadly introduced by German psychiatrist Emil Kraepelin as a **rapid cognitive impairment** beginning in the late teens or early adulthood. The current term SZ (splitting of mind in Greek) was coined by Swiss psychiatrist Eugen Bleuler in 1908. Bleuler realized that the illness was not a dementia as some of his patients improved and described the main symptoms as 4 A's: flattened **Affect**, **Autism**, impaired **Association** of ideas and **Ambivalence**.

About 10% of those diagnosed with SZ eventually commit suicide. Most of SZ population experiences a lifetime disability such as long-term unemployment, poverty, and homelessness. There is emerging consensus about the need to decompose SZ into etiopathologically meaningful and clinically identifiable dimensions.

SYMPTOMS

SZ disorders are characterized by a **diverse set of signs and symptoms**. They include specific distortions in **sensory, motor, executive, and affective systems**. These symptoms are generally classified into the following dimensions: positive and negative symptoms from one perspective, and cognition, disorganization, mood, and motor dimensions from the other perspective. These dimensions are differentially expressed across patients and through the course of the illness.

Positive symptoms involve **reality distortion** and include delusions (false beliefs) as well as hallucinations (perceptions in the absence of external stimuli which look as if they are real). Several kinds of **delusions** can occur including delusions of persecution, reference, control, thought insertion, withdrawal, broadcasting, and some others depending on the individual life setting.

Hallucinations can occur in all sensory modalities, although **auditory hallucinations** are the most common. Auditory hallucinations include threatening or accusatory voices speaking to the patient. Positive symptoms mark the **onset** of the illness. **Dopaminergic hyperactivity** seems to underlie positive symptoms because they are suppressed by **blocking dopaminergic receptors** produced by antipsychotic medications.

Negative symptoms include impairments in the affective system such as abulia (loss of motivation), alogia (poverty of speech), anhedonia (inability to experience pleasure), avolition (lack of initiative), apathy (lack of interest), and reduced sexual drive. It is important to distinguish between primary and secondary negative symptoms. Primary negative symptoms are intrinsic to SZ whereas secondary negative symptoms are caused by "extrinsic" factors such as environmental deprivation, neuroleptic treatment, and depression. **No medication** is known to reduce negative symptoms and no pathophysiological theory explaining negative symptoms exists. In contrast with negative symptoms, SZ patients frequently manifest **mood symptoms** which include increased emotional arousal and reactivity in conjunction with positive symptoms, a phenomenon termed "the emotional paradox" of SZ.

Formal thought disorder refers to fragmentation of the logical, progressive, and goal-directed nature of normal thought processes. It seems to index the process of **loosening of associations** in SZ and involve poverty of thought content and production of neologisms.

Depression is common in SZ. Several mechanisms may contribute to depression in SZ: it could be a part of the illness, it could be a result of developing insight, it could be a comorbid condition, or it could be an adverse effect of antipsychotic medications.

Abnormalities in **psychomotor activity** include slowing associated with negative and depressive symptoms, excessive apparently purposeless motor activity often associated with positive symptoms, and can range from simple isolated movements of posturing, mannerisms, and **stereotypes** to more complex patterns of motion as observed in **catatonic states.** It should be stressed here that catatonic states were described in SZ long before the introduction of antipsychotic agents in the 1950s and were observed in a quarter of drug-naive patients. The full-blown catatonic syndrome is characterized by echolalia, echopraxia, automatic obedience, waxy flexibility, and extreme negativism.

Cognitive deficits are consistently reported in SZ. Recall that dementia praecox literally means cognitive decline with onset in youth. During the last few decades the research into the neurophysiological basis of cognitive deficits has been fueled by advances in **neuropsychological and neuroimaging** methods. Cognitive impairment observed in neuropsychological tasks is universal in patients with SZ and distinguishes them from healthy controls (HCs). Cognitive deficits are present in the **premorbid phase** of SZ illness. A similar pattern of lesser extent is present in nonpsychotic relatives so that cognitive deficits could be considered as candidates for endophenotypes in SZ.

Although **anxiety** in SZ is considered as a comorbidity, it represents a prominent symptom early in the course. Anxiety disorders such as comorbid social phobia, obsessive–compulsive, and panic disorders are common in SZ.

Lack of insight is a cardinal feature of SZ. Many patients with SZ either believe that they do not have any disorder or misattribute their symptoms to other causes.

Sensory distortions are spontaneously reported by patients during early stages of SZ. These sensory deficits cannot be explained by attention and emotion dysfunctions and are objectively documented.

PREVALENCE

SZ affects men and women equally. It occurs at similar rates in all ethnic groups around the world.

TIMECOURSE

The development of SZ is quite different from the development of ADHD (Fig. 5.2.1) Positive symptoms such as **hallucinations and delusions** usually start between the **ages of 16 and 30**. Men tend to experience symptoms a little earlier than women. Most of the time, people do not get SZ after age 45. SZ rarely occurs in children, but awareness of childhood-onset SZ is increasing. Diagnosing SZ in teenagers is difficult because the first signs can include behaviors common among them such as a change of friends, a drop in grades, sleep problems, and irritability.

SZ develops in **sequential stages** including (1) a **prodromal** phase characterized by attenuated positive symptoms and declining cognitive functions, (2) the **first psychotic episode** indicating formal onset of SZ, (3) the **initial decade of illness** with repeated episodes of psychosis with variable degrees of remission and with the most prominent decline of functions during the first 5 years, and finally, and (4) **a stable phase or plateau,** when psychotic symptoms are less prominent and negative symptoms

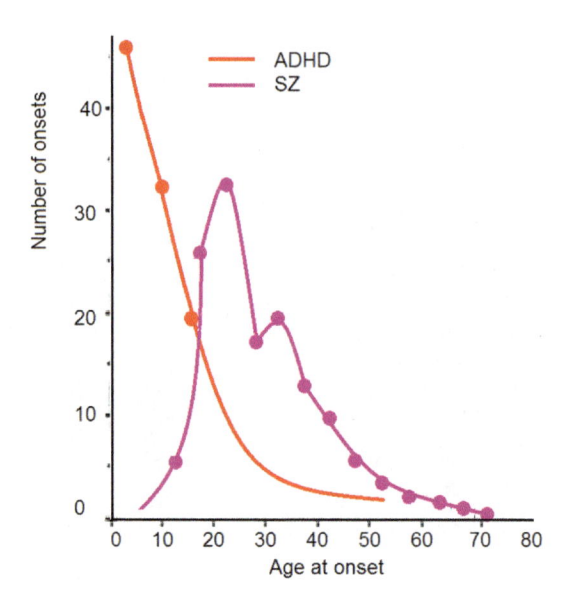

FIGURE 5.2.1 Age onset of SZ in comparison to that of ADHD. *Adapted from Sham et al. (1994) and Peyre et al. (2014).*

and stable cognitive deficits increasingly predominant. **Recovery of varying degrees** can occur at any stage of the illness.

Defining **neuromarkers of the prodromal state** during the period of late adolescence and early adulthood is of major importance. The **idea is that early intervention** during this critical phase of the pathogenetic process could attenuate or perhaps even prevent the onset of SZ. Recent research on functional neuromarkers shows optimism in this respect.

The period of adolescence is a time of profound maturation changes in the highest order cognitive functions, such as reasoning, abstract thinking, and planning. These maturation changes in behavior reflect maturation of the prefrontal cortex via extensive **pruning** of dendritic spines on pyramidal neurons and axon terminals of pyramidal neurons. The neuromarkers that exhibit the most dramatic changes during this maturation process might be of the most importance for SZ. The fact that SZ typically emerges during late adolescence and early adulthood has long led to the neurodevelopmental hypothesis of SZ.

NEURODEVELOPMENT

In general, neuroscience indicates two types of postnatal neurodevelopmental events: **progressive phenomena**, such as cell proliferation, dendritic arborization, and myelination; and **regressive phenomena**, such as apoptosis, synaptic and axonal pruning, and atrophic processes.

As mentioned in chapter: Attention Deficit Hyperactivity Disorder, cell growth, arborization, and accompanying **synaptogenesis** contribute to **increases in cortical gray matter volume** during the first 5–10 years of life. During this stage synapses are **overproduced**. Then the process of synaptic pruning starts which is reflected in a **reduction of gray matter volume**. This process results in the normal elimination of about 40% of cortical synapses.

Katherine Karlsgodt and coworkers from University of Southern California in 2008 proposed the **developmental pruning hypothesis of SZ** which presumes that pruning occurs in excessive degree in patients, placing them below a threshold for developing psychotic symptoms (Fig. 5.2.2). The neurodevelopmental hypothesis of SZ is supported by the 2001 study of Paus Thompson and coworkers from University of California School of Medicine. They made MRI scans at 2-year intervals at three time points in early-onset SZs and HCs and found **accelerated gray matter loss** in SZ with the earliest deficits in parietal brain regions which 5 years later progressed anteriorly into temporal lobes and dorsolateral prefrontal

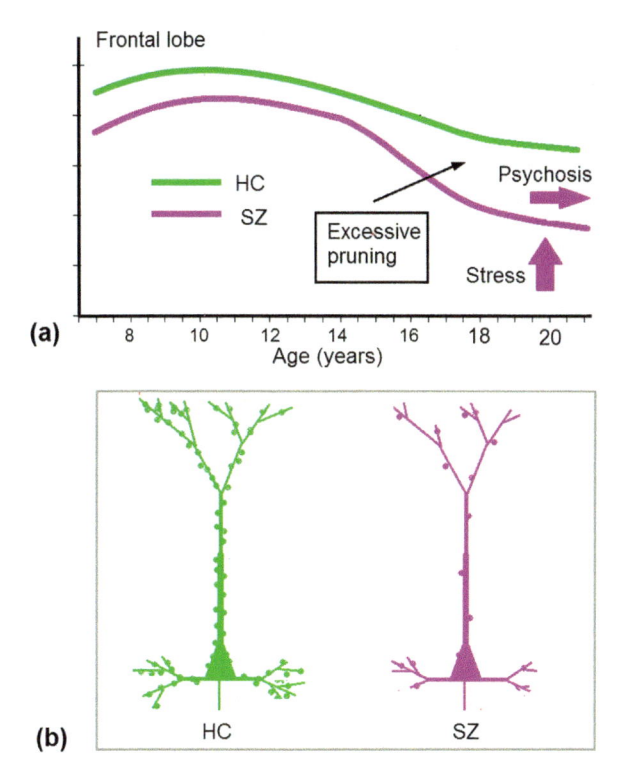

FIGURE 5.2.2 **A schematic diagram of synaptic pruning.** (a) Synaptic pruning as reflected in the size of the frontal lobes for schizophrenia (SZ) and healthy controls (HC). (b) The result of excessive pruning in schizophrenia as reflected in lower number of synapses on a pyramidal neuron in schizophrenia. *Adapted from Karlsgodt et al. (2008).*

cortices. These emerging patterns correlated with **psychotic symptom severity** and mirrored the motor, auditory, visual, and frontal executive impairments in the disease.

HETEROGENEITY

There is significant **heterogeneity** in the etiopathology, symptomatology, and course of SZ. Because of the qualitative diversity of symptoms there is debate regarding whether the diagnosis of SZ represents a single disorder or a number of discrete syndromes. This was noticed at the very beginning when Eugen Bleuler termed the disease *the schizophrenias* (plural). Although SZ may not represent a single disease, alternative approaches in better defining this syndrome and its components so far have been unsuccessful.

The attempt to describe heterogeneity is made by subclassifying SZ into **classical subtypes:** (1) **catatonic type** characterized by marked psychomotor disturbance involving stupor, negativism, rigidity, excitement, and posturing; (2) **disorganized type** characterized by loosening of associations, disorganized behavior, and flat or inappropriate affect; (3) **paranoid type** characterized by preoccupation with systematized delusions or presence of hallucinations related to a single theme; (4) **schizoaffective type** characterized by prominent mood disturbance in conjunction with psychotic symptomatology; (5) **undifferentiated type** characterized by psychotic symptoms that meet criteria for SZ but not for any specific subtype; and (6) **residual type** characterized by the occurrence of at least one prior episode of florid phase of SZ with a current clinical picture free from psychotic symptoms, but with minimal residual cognitive and negative symptoms.

Clinically, however, these subtypes of SZ are **unstable** over the course of the illness, **do not indicate specific treatment or prognosis**, and do not help to explain the pathophysiological heterogeneity of SZ.

HERITABILITY

Genetic factors play important roles in developing SZ. Twin studies have suggested **heritability is as large as 80%**. However, if an identical twin develops SZ, the probability of the other twin having SZ is only 50%. Several **susceptibility genes** for SZ have been identified with a small value of the risk factor for each gene. For most genes the biological basis of the increased risk for the illness is unclear. All these facts indicate that genetic liability alone is not sufficient to cause the clinical features of the illness. It is likely that SZ is a condition of complex inheritance, with several genes interacting to generate risk for SZ. Recent work has suggested that genes of the risk for SZ are nonspecific, and may also raise the risk of developing other psychotic disorders such as bipolar disorder.

ENVIRONMENTAL RISK FACTORS

An example of twin studies shows that the **interaction of genes with the environment** plays a critical role in SZ. Social adversity, racial discrimination, family dysfunction, unemployment, and poor housing conditions have been proposed as contributing factors. For this reason, being a first- or second-generation immigrant creates a strong risk factor for SZ. Childhood experiences of abuse or trauma have also been implicated as risk factors for SZ. There is also evidence that prenatal exposure to infections increases the risk for developing SZ later in life.

TREATMENT

Drugs that are used to treat psychosis are called **antipsychotic agents.** They have been in use for more than 50 years. The discovery of chlorpromazine in the 1950s dramatically reduced the number of patients held in mental institutions. Antipsychotic medication remains the mainstream for treatment of SZ. It is effective in suppressing the psychotic features of the illness but has no effect on negative symptoms and does not improve the social or occupational outcome. Most medications take around **7–14 days** to have an antipsychotic effect.

The two classes of antipsychotics are generally thought to be equally effective for the treatment of positive symptoms but atypical antipsychotics have fewer extrapyramidal side-effects. Typical antipsychotics have an **affinity to D_2 dopamine receptors and block them**. The affinity of different antipsychotic agents strongly correlates with the reduction of psychotic symptoms produced by these drugs. This correlation was first reported by Philip Seeman from University of Toronto and coworkers in 1975.

NEUROPSYCHOLOGICAL ASSESSMENT

When we think of SZ, we think of people who hear voices and have false beliefs about reality. However, clinicians are aware that the key elements of SZ **are cognitive dysfunctions** which determine the quality of life of SZ in more degree than just the severity of hallucinations and delusions.

Comprehensive cognitive assessments using standard neuropsychological tests have suggested a relatively **homogeneous profile of**

impairments across virtually all cognitive ability areas. The best-fitting model, obtained with confirmatory factor analysis, was a **single-factor model** (Keefe, Poe, Walker, Kang, & Harvey, 2006). In other words, impairments on neuropsychological tests are global in nature and do not describe regional brain dysfunctions affecting only a few cognitive abilities. As a consequence, the overall severity of impairment can be indexed with a few measures. For example, the processing speed measured in various coding and continuous performance tasks is slower for SZ than for HCs and is the best predictor of total performance scores on extended batteries of tests.

VOLUMETRIC STUDIES

As Daniel Mamah from Washington University Medical School and coworkers noted in their 2007 study, the volumes of the **basal ganglia** are **enlarged** in SZ patients in comparison with HCs, but with small effect sizes. This is in contrast with ADHD in which the basal ganglia are reduced.

MOTOR ABNORMALITIES

In general, patients with SZ are **slower in motor response** and perform more **omission and commission errors** in continuous performance tasks indicating impairment of the motor system. The effect sizes of discriminating a group of SZ patients from HCs is 0.6–1.1 for reaction time and 0.5–3.0 for omission/commission errors.

There are also signs of impairment of the ocular motor system. A group of SZ patients differs significantly from a group of HCs in viewing pictures or pursuing target movements. For example, Philip Benson and coworkers from King's College, University of Aberdeen (United Kingdom) in their 2012 study showed that a group of SZ patients could be discriminated from HCs with near-perfect accuracy at 98% on almost all eye movement tests (Fig. 5.2.3).

SPONTANEOUS ELECTROENCEPHALOGRAPHY

Research of **QEEG** in SZ has been quite **inconsistent**. While some studies have reported increase of beta and reduction of alpha EEG power, other research has shown no differences and even opposite results. This inconsistency is due to several interacting factors: (1) heterogeneity of the population of patients with SZ, (2) small number of patients in comparison groups, (3) low size effect of differences between the groups, and

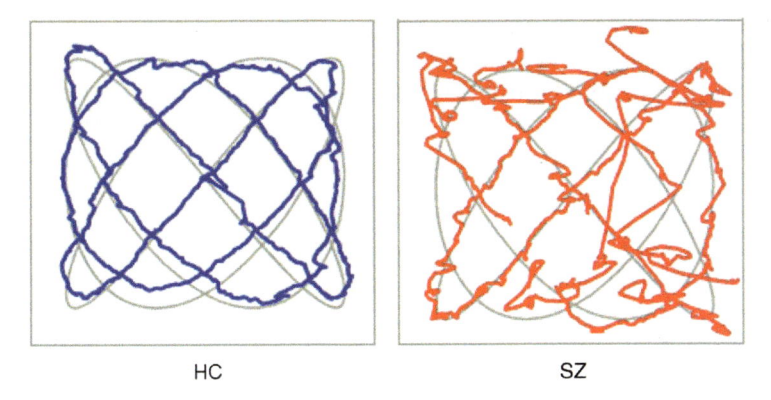

FIGURE 5.2.3 **Simple viewing tests distinguish patients with SZ from HCs with exceptional accuracy.** Scanpaths superposed on the stimulus trajectory for a HC and a patient with SZ. *Adapted with permission from Benson et al. (2012).*

(4) possible change of QEEG pattern in medicated patients in comparison with nonmedicated.

Fig. 5.2.4 shows EEG spectra for group independent midline components for patients with SZ in comparison with healthy control subjects. Increase of **theta and beta EEG power** for the frontal widely distributed

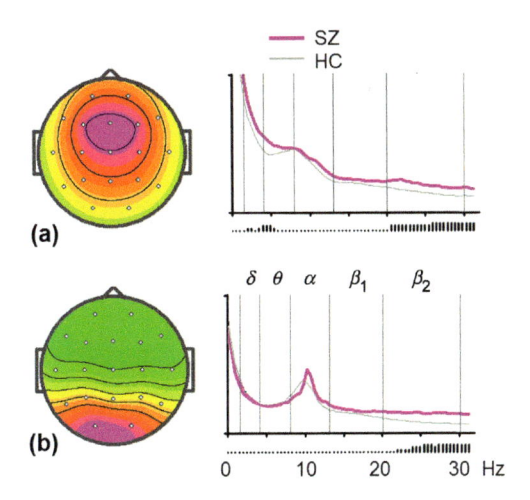

FIGURE 5.2.4 **QEEG in SZ.** (a,b) Topographies and grand average spectra of the two midline group independent components for schizophrenic patients *(pink line)* and healthy subjects *(gray line)* in eyes-open condition. On curves: X-axis—frequency in Hz, Y-axis—square root of power in conventional units. The black bars under the X-axis indicate p values of the difference between schizophrenic and healthy control groups ($p < 0.0005$, short bar; $p < 0.0001$, middle bar; $p < 0.00001$, long bar). Figure obtained by Vera Grin-Yatsenko and Valery Ponomarev from N.P. Bechtereva Institute of the Human Brain from Saint Petersburg on the basis of data from the HBI database.

component is clearly seen both under the eyes-open and eyes-closed conditions (not shown in figure).

Although **no reliable index of background EEG** for SZ has been found, EEG correlates of psychosis as a certain state in SZ have been observed. For example, hallucinations are associated with increase of beta power in EEG over the left temporal areas (Lee et al., 2006).

A systematic review by Galderisi and Maj (2009) from the University of Naples focused on the QEEG and the P300 ERP wave as diagnostic tools in SZ. For QEEG, most studies qualified as level 4 (**case control study with poor reference standard**), and only 24% as level 3b or better. An increase in slow activity in patients is reported by most of these studies while gamma band activity seems to be suppressed in SZ. In contrast, for **P300 amplitude reduction** as a diagnostic index, 63% of the studies qualified as level 3b or better while metaanalysis of the results of 52 studies (60 independent samples) demonstrated a **large effect size**.

SENSORY-RELATED NEUROMARKERS

At the beginning of the 20th century, Bleuler considered sensory processing to be intact in SZ. Accumulated knowledge at the end of 20th century clearly showed that Bleuler was wrong. Psychological studies in the 1960s documented **sensory distortions** spontaneously reported by individuals experiencing early symptoms of SZ. Subsequent electrophysiological studies objectively demonstrated neuronal correlates of sensory deficits.

In the auditory modality, one of the robust sensory-related neuromarkers in SZ is **P50 gating failure**. By definition, **P50** is a midlatency cortical response to a brief auditory stimulus. When two identical clicks are sequentially presented to healthy subjects, the P50 to the second stimulus is **inhibited**. This inhibitory effect is considered an index of sensory gating—an automatic suppression of response to redundent stimuli. In SZ patients the sensory gating effect is substantially smaller than in HCs. The effect size, as reported in a 2004 metaanalysis by **Elvira Bramon** and coworkers from Institute of Psychology in London, is quite large and varies between 1.3 and 1.6 in different studies. The P50 gating effect is enhanced by nicotinic cholinergic mechanisms indicating that smoking in SZ patients might be a form of self-medication. P50 gating failure is observed in all patients with history of psychosis demonstrating its **nonspecificity** for SZ.

The other sensory-related waves in the auditory modality such as **N1 and MMN** are shown to be decreased in SZ. For example, in a 2011 study by John Foxe and coworkers from the Albert Einstein College of Medicine, N1 deficit was shown in first-episode and chronic patients with SZ, as well in first-degree unaffected relatives, suggesting that N1 deficit could be an **endophenotype of** SZ. MMN deficit in SZ is a well-documented phenomenon. The largest effect size is found for MMN in response to **duration deviance**.

MMN has also been used for predicting conversion to psychosis in clinically at risk mental state individuals. The literature on this topic was reviewed in a 2015 paper by an international team headed by Risto Nataanen who was the first to discover the MMN in 1978.

As shown in chapter: Sensory Systems and Attention Modulation, if the deviation in deviant rare auditory stimuli is large enough, MMN is followed by two frontocentrally distributed waves—the novelty **P3 (P3a) and reorienting negativity (RON)**. In a 2012 study by Carol Jahshan and coworkers from the University of California the MMN, P3a, and RON in a duration-deviant auditory oddball paradigm were assessed in 118 subjects across 4 groups: (1) prodromal patients, (2) recent-onset patients, (3) chronic patients, and (4) normal controls. Deficits in sensory-related waves were present in all patient groups. The reduction of MMN and P3a in schizophrenia have been shown in many studies and have been proven to exhibit qualities of endophenotypes, including heritability, test–retest reliability, and trait-like stability. In 2015 the feasibility of adding MMN and P3a to the ongoing study by the Consortium on the Genetics of Schizophrenia (COGS) was tested in a large-scale multicenter study by Gregory Light from University of California San Diego and his coworkers from other universities in the USA (Fig. 5.2.5). Schizophrenia (SZ) 824 healthy control subjects (HC) and 966 patients were tested at five geographically distributed COGS laboratories. The duration increment MMN was measured in the 20-min auditory oddball paradigm in which standard binaural tones of 50 ms duration ($P = 0.9$, 1-kHz, 85-dB, 500 ms stimulus onset-to-onset asynchrony) were randomly interspersed with deviant stimuli of 100 ms duration ($P = 0.1$). Valid ERP recordings were obtained from 91% of HC and 91% of SZ patients. Highly significant MMN ($d = 0.96$) and P3a ($d = 0.93$) amplitude reductions were observed in SZ patients with no significant differences across laboratories. Demographic characteristics (including age, race, and sex) accounted for 26 and 18% of the variance in MMN and P3a amplitudes, respectively.

Auditory-related neuromarkers appear to be **nonspecific for** SZ as shown in another 2012 study by Carol Jahshan and coworkers. The authors recorded the duration of MMN and P3a waves in bipolar disorder and SZ patients in comparison with HCs and found that patients with bipolar disorder also revealed deficits in preattentive auditory processing.

Similarly, in the visual modality, obligatory waves such as **C1, P1, and N1** are decreased in SZ. However, in the case of the visual modality, processing in the **magnocellular visual system** appears to have the largest dysfunction. For example, in a 2007 study by Pamela Butler and coworkers from New York University School of Medicine visual ERPs were

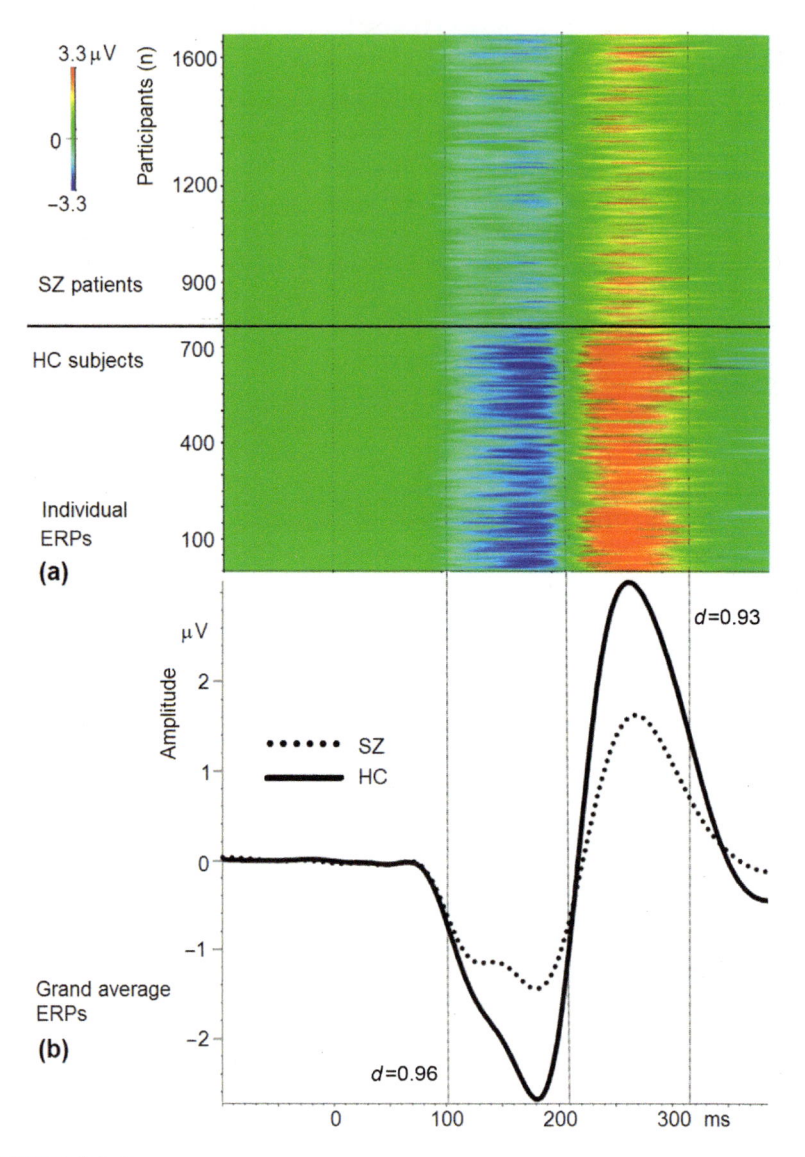

FIGURE 5.2.5 **Auditory MMN and P3a as neuromarkers in SZ.** (a) Amplitude color coded individual Deviant–Standard difference ERP waves for Healthy Control Subjects (HC, N = 753) and schizophrenia patients (SZ, N = 877). (b) Group grand average waveforms for SZ and HC groups. *Adapted with permission from Light et al. (2015).*

recorded to stimuli biased toward the magnocellular system using low versus high spatial frequency sinusoidal gratings. SZ patients showed differences from HCs in C1, P1, and N1, when stimuli were biased toward the magnocellular system, but intact ERP generation when stimuli were biased toward the parvocellular system.

AUTOMATIC PREDICTING ABILITY FAILURE

As we know from chapter: Sensory Systems and Attention Modulation, auditory MMNs reflect brain mechanisms for extracting and maintaining regularities in the auditory and visual environments. In other terms, the brain builds up an internal model of environments which allows healthy subjects to make predictions, facilitating processing of unexpected events and distinguishing between expectations and reality. According to the MMN deficit, people with SZ lack this **automatic predicting ability and reality–expectation discriminability** which might contribute to the cardinal symptoms of the illness, such as auditory hallucinations and delusions.

This hypothetical predicting ability of the human brain was explicitly measured in the three-stimulus auditory oddball paradigm by Judith Ford and Daniel Mathalon from the University of California, San Francisco by contrasting ERPs in response to the first and fourth stimuli in a sequence of four standards reviewed in their 2012 paper (Fig. 5.2.6). It is critical to emphasize that no other features of the fourth standard made it deviant or salient other than violation of the prediction that it was time for a change after three standards in a row. Thus, while the salience of targets is typically based on their different pitch and global probability, the salience of the fourth standard in a row is based only on the low local probability of this pattern. This ability is measured by **the stimulus preceding negativity** elicited before the fourth standard stimulus and by the **novelty N2b/P3**

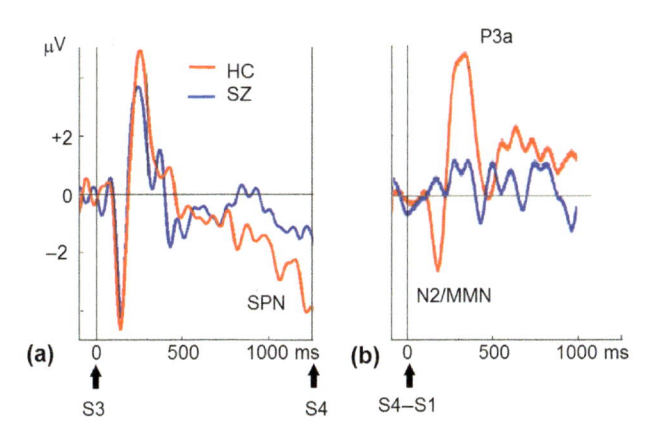

FIGURE 5.2.6 **Reduced predicting ability in SZ.** (a) Grand-averaged ERPs from Cz in the auditory oddball task following the third standard stimulus in a row (S3). Clear stimulus preceding negativity can be seen in healthy subjects but not in SZpatients; and (b) grand-averaged ERPs from vertex (Cz) resulting from the subtraction of the ERP to Standard 1 from Standard 4 (S4 − S1). This subtraction allows visualization and quantification of the N2b/*MMN* and P3a components elicited by Standard 4. *Adapted with permission from Ford & Mathalon (2012).*

component elicited in response to the fourth stimulus. These components are intact in the healthy brain but significantly reduced in SZ patients.

OBJECT RECOGNITION DEFICIT IN N170

As discussed in chapter: Sensory Systems and Attention Modulation, ERP waves are associated with familiar object processing, in general, and face processing, in particular—**N170 and N250** waves. In a 2014, meta-analysis Amanda McCleery and coworkers from UCLA reported that N170 and N250 were suppressed in SZ with the medium effect size, reflecting the well-established behavioral deficit in face emotion processing in SZ patients.

Fig. 5.2.7 demonstrates the N170 deficit in a group of 100 patients with SZ in comparison to a group of 258 HC subjects. Note that the extracted N170 component is generated in the fusiform gyrus—the area of the ventral visual stream.

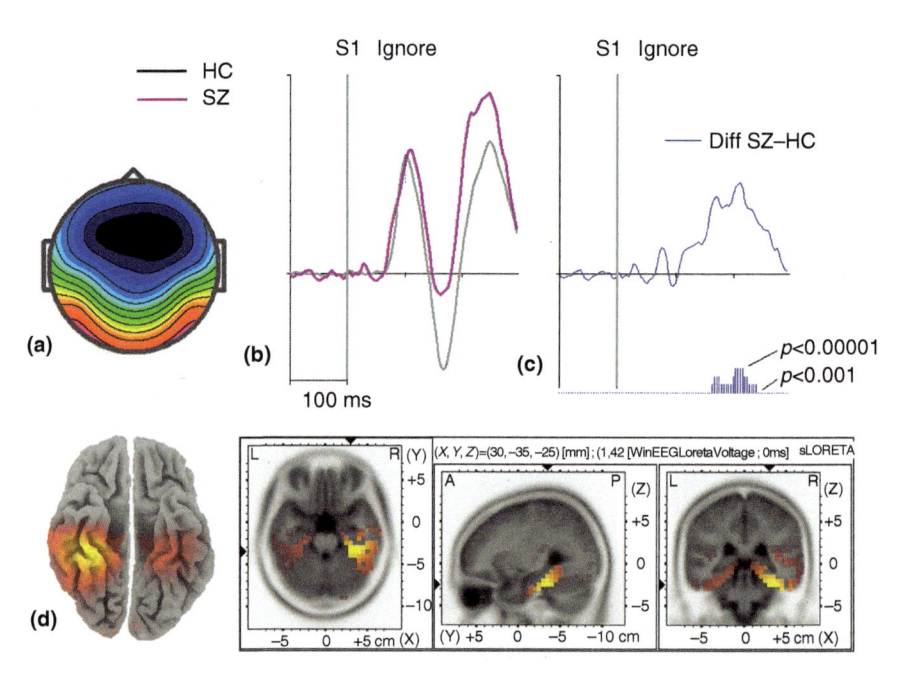

FIGURE 5.2.7 **N170 deficit in SZ.** (a) Topographic map of N170 latent component extracted by the joint diagonalization method in response to Ignore stimuli (images of plants); (b) timecourses of N170 component for a large group of HCs and patients with SZ; (c) the difference etween SZ and HC groups with p values depicted as vertical bars of different lengths; and (d) sLORETA images of the latent component. Note the localization of the component in the fusiform gyrus with maximum on the right side. *Data from the HBI database.*

P3b AS ENDOPHENOTYPE

One of the key signs of SZ is cognitive dysfunction which must be reflected in impairment of the executive ERP components such as P3b, N2, and P3 NOGO. From the cognitive neuroscience perspective, ERP cognitive components are independent of sensory components. As far as SZ research is concerned, this general inference was explicitly confirmed by a 2009 study by Bruce Turetsky and coworkers from the University of Pennsylvania, Philadelphia. In this study 9 auditory ERP measures were acquired in a single testing session from 23 SZ patients and 22 healthy subjects. Hierarchical oblique factor analysis revealed that these measures aggregated into four factors while patient deficits were observed for **two independent factors: N100/MMN and P3a/P3b**. The N100/MMN abnormalities were associated with symptoms of alogia and formal thought disorder. The P3a/P3b abnormalities were associated with **avolition, attentional disturbances, and delusions**.

The target P300 wave has been a focus of research in SZ since the pioneering work of Roth and Cannon in 1972 which in turn was done a few years after discovery of this component. In a 2004, metaanalysis of P300, Bramon and coworkers showed that the amplitude of P300 robustly decreased in SZ with a large effect size of **0.9**. This is in agreement with a 2003 review by Yang-Whan Jeon from the Catholic University of Korea and John Polich from the Scripps Research Institute, La Jolla, California who conducted a metaanalysis of the early literature on P300 in SZ (Fig. 5.2.8). It should be noted that although P300 latency was found delayed in SZ its averaged effect size was smaller than the amplitude. Moreover, note in Fig. 5.2.8 the large variability in computed values of effect size for small-sample sizes (numbers of participants in a study). This visibly demonstrates the critical issue of getting large numbers of patients and HCs for obtaining robust and reliable results.The P3b ERP wave is a trait marker of SZ: it is reliably reduced in patients with SZ including treated patients whose symptoms have improved or largely remitted, first-episode patients, and unaffected first-degree relatives of SZ patients. In the above mentioned 2004 metaanalysis by Elvira Bramon and coworkers metaanalysis study by Bramon and coworkers the target P300 was found to fit all requirements for being **an endophenotype**.

Less consistently cross-sectional studies showed that patients with more severe negative symptoms have smaller auditory P300s. This inconsistency may arise because the variance in P300 amplitude at a single time point reflects both trait and state influences. To avoid this drawback of cross-sectional studies a 2000 longitudinal study by Daniel Mathalon and coworkers from Stanford University directly tested the state aspects of P300 by measuring ERPs and the Brief Psychiatric Rating Scale (BPRS) in three behavioral paradigms on multiple occasions in SZ patients in

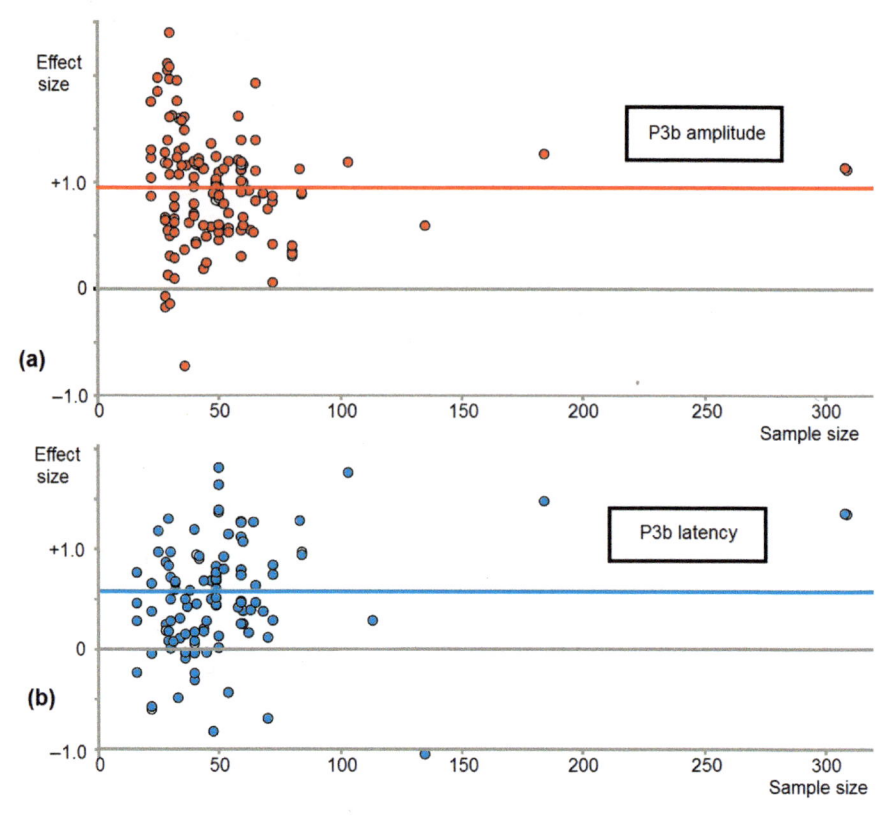

FIGURE 5.2.8 P3b deficit in SZ. Funnel graphs of effect sizes from studies employing the oddball paradigm for P3b (a) amplitude and (b) latency. *Adapted with permission from Jeon & Polich (2003).*

comparison with HCs. The authors showed that P300 amplitude, regardless of elicitation method, tracked BPRS total and positive symptom scores over time, decreasing with symptom exacerbations and increasing with improvements. In addition, target auditory and visual P3b amplitudes tracked negative symptoms, and automatic auditory novelty P3a tracked depression–anxiety symptoms.

P3b AS A PREDICTOR OF PSYCHOSIS

In a 2010 study by van Tricht and coworkers from the University of Amsterdam the target P300 was found to be the best **predictor for subsequent psychosis** in a group of subjects at ultrahigh risk for developing psychosis (Fig. 5.2.9).

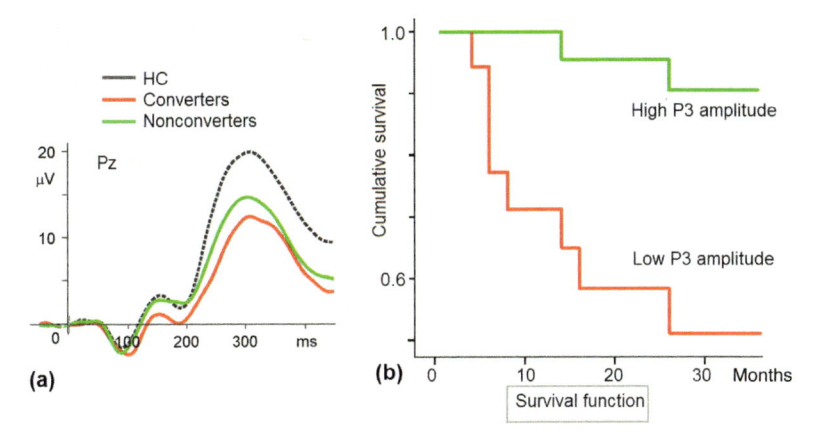

FIGURE 5.2.9 **P3b is a predictor for conversion to psychosis in SZ.** (a) Grand-averaged waveforms for HCs *(dashed gray line)*, for ultrahigh-risk subjects without transition to psychosis (Non-converters, *green* line), for ultrahigh-risk subjects with transition to psychosis (Converters, *red* lines); and (b) the survival function differed significantly between the groups with above *(green)* or below *(blue)* the mean amplitude of P300. *Adapted with permission from van Tricht et al. (2010).*

PROACTIVE COGNITIVE CONTROL DEFICIT

SZ patients were found to have deficit in ERP correlates of voluntary preparatory activity. SZ patients reveal **reduction in Contingent negative variation (CNV) amplitude** and enhancement of so-called post-imperative negative variation—another slow wave that followed the imperative stimulus. This was shown as early as the 1970s by Brigitte Rockstroh and coworkers from the University of Constanz and reviewed in their 1989 book.

Recently, a 2010 study by Jonathan Wynn and coworkers from the University of California, Los Angeles used a cued reaction time–contingent picture-viewing task to demonstrate that SZ subjects produced **lower CNV** and smaller **stimulus preceding negativity (SPN)**—a preparatory activity in the sensory domain (Fig. 5.2.10).

Fig. 5.2.11 demonstrates three latent components associated with proactive cognitive control in a group of patients with SZ ($N = 100$) and a group of healthy control subjects ($N = 258$) from the HBI database. The middle components are associated correspondingly with orienting CNV (CNV-O) and expectancy CNV (CNV-E) (see chapter: Executive System and Cognitive Control) whereas the temporal component might be associated with contralateral delayed activity (CDA) (see chapter: Memory Systems)—a neuromarker of working memory. CNV-O remained intact in the group of patients with SZ. The two other components were substantially reduced with large effect sizes.

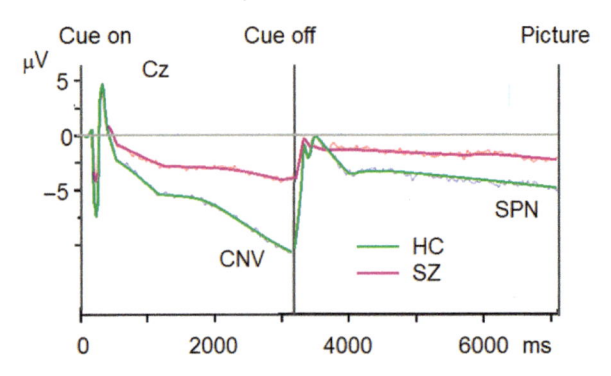

FIGURE 5.2.10 **Reduced CNV and Stimulus Preceding Negativity (SPN) in SZ.** A cued reaction-time-contingent picture viewing task was used to assess ERPs during two types of anticipatory intervals: one preceding a cued motor response (ie, CNV) and a second preceding presentation of an expected emotional stimulus without a motor response (ie, SPN). Green line, HCs); pink line, SZ patients. *Adapted with permission from Wynn et al. (2010).*

REACTIVE COGNITIVE CONTROL IN SCHIZOPHRENIA

NOGO ERP waves as neuromarkers of cognitive control were shown to be decreased in SZ. In a 2008 study by Groom and coworkers, an ERP visual GO/NOGO task and an oddball task were recorded in adolescent groups: (1) early-onset SZ patients, (2) nonpsychotic siblings of SZ patients (SZ-SIB); (3) ADHD, and (4) HCs. Results revealed **reduced P3 amplitude in** SZ patients and nonpsychotic siblings in an auditory oddball and a visual GO/NOGO task.

Fig. 5.2.12 demonstrates three latent components associated with reactive cognitive control in a group of patients with SZ ($N = 100$) and a group of healthy control subjects ($N = 258$) from the HBI database. Middle components are associated correspondingly with conflict monitoring (the frontally distributed component) and action inhibition or action overriding (the two centrally distributed components) (see chapter: Executive System and Cognitive Control). The P3-like activation patterns in these three components are **suppressed** in the SZ group with large effect sizes.

HYPOFRONTALITY—fMRI STUDIES

As shown in chapter: Executive System and Cognitive Control, NOGO trials in the GO/NOGO paradigm require activation of the cognitive control system including the frontal lobes. This activation pattern

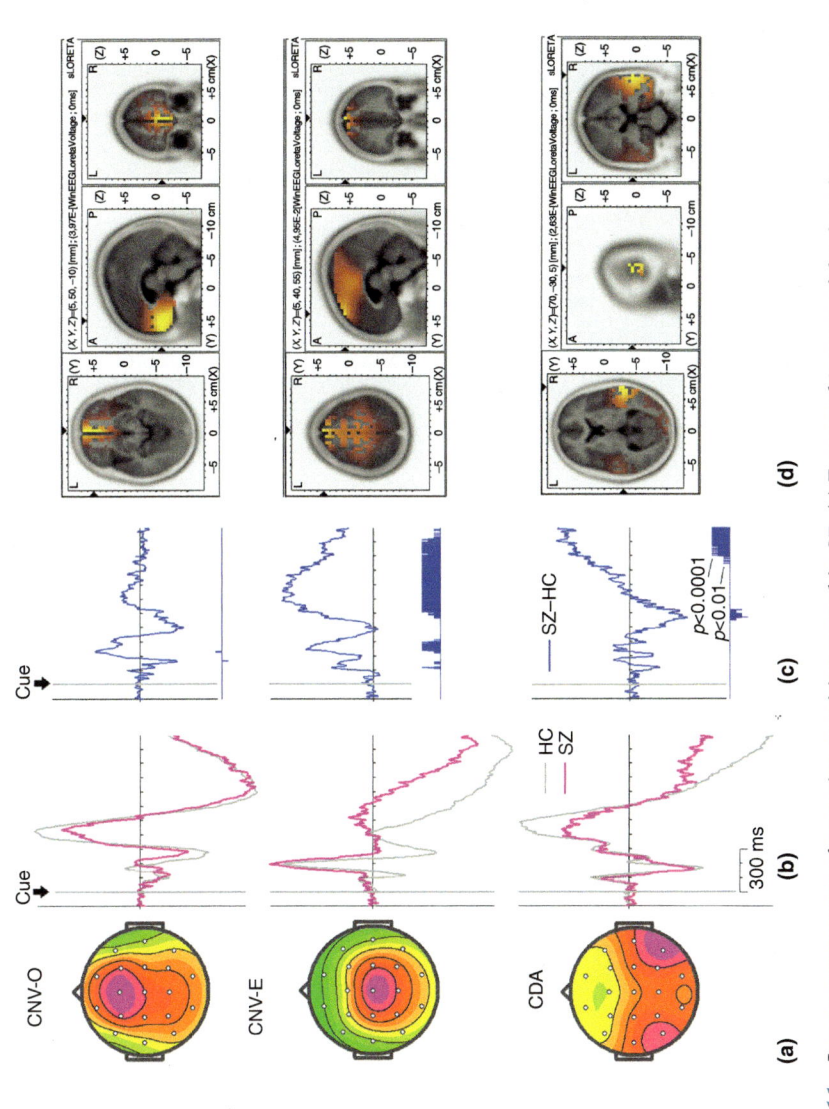

FIGURE 5.2.11 **Latent components of proactive cognitive control in SZ.** (a) Topographic maps of the three latent components in the cued GO/NOGO task in response to Cue; (b) time course of the latent components for SZ (*pink*) patients and HCs; (c) the difference SZ – HC with *p* values below. (d) sLORETA image of the topography. *Data from the HBI database.*

FIGURE 5.2.12 **Latent components of reactive cognitive control in SZ.** (a) Topographic maps of the three latent components in the cued GO/NOGO task in response to NOGO stimuli; (b) timecourse of the latent components for SZ patients (*pink*) and HCs; (c) the difference SZ − HC with *p* values here; and (d) sLORETA image of the topography. *Data from the HBI database.*

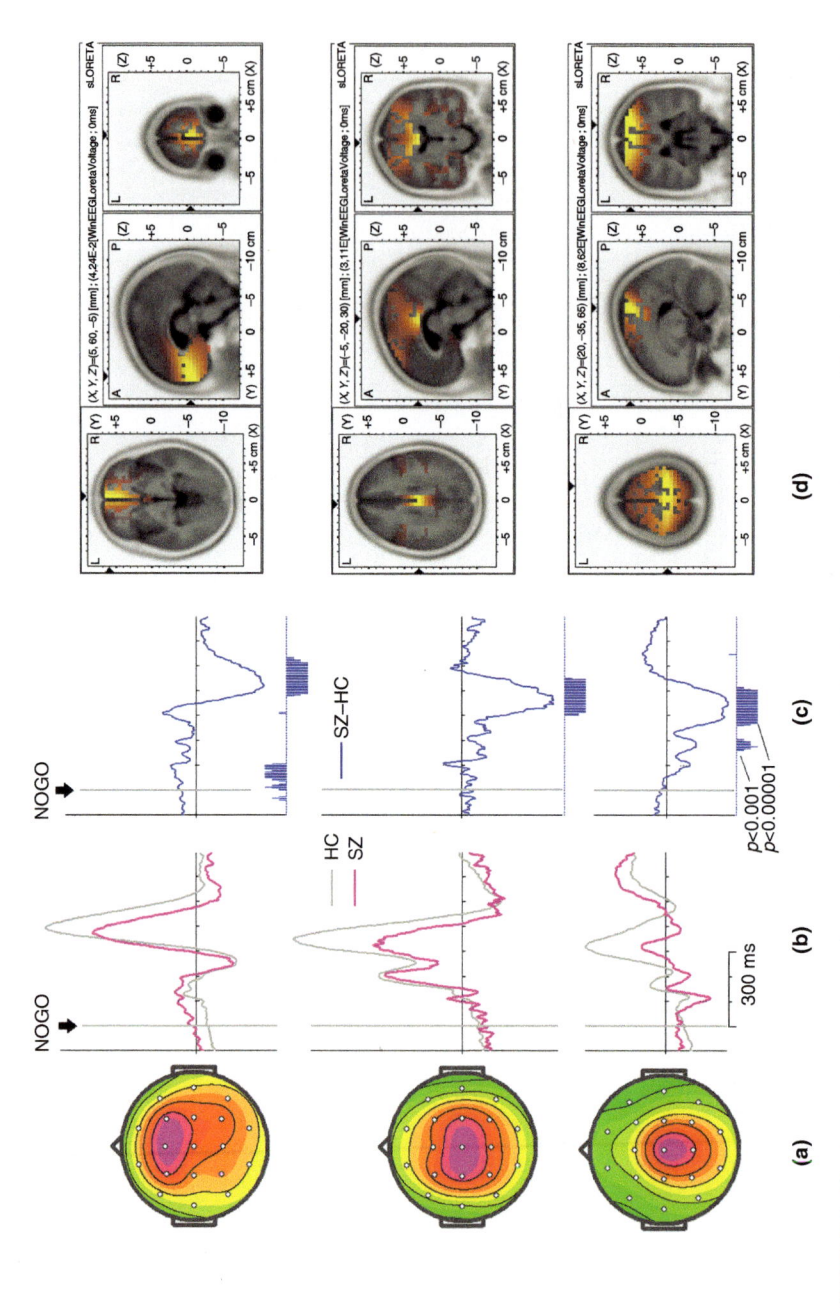

measured by fMRI was shown to be smaller in SZ patients in comparison to HCs. In a 2004 study by Judith Ford and coworkers from Stanford University School of Medicine, ERP and fMRI data were collected in a NOGO task in a group of patients with SZ and a group of healthy control subjects. The results enabled the authors to suggest that patients with schizophrenia established a **weaker prepotent model** in comparison to healthy subjects and responded with a **weaker** frontal activation pattern to NOGO trials. In sum, the differences between GO and NOGO conditions in terms of electrical and metabolic activation were smaller in schizophrenia.

HYPOFRONTALITY AS PREDICTOR OF RESPONSE TO MEDICATION

The NOGO-GO ERP differences were used for differentiation groups of patients with schizophrenia and healthy controls as well as for predicting response to medication. Andreas Fallgatter and Werner Strik from University Hospital of Wuerzburg, Germany in 1999 suggested an ERP parameter that reflected the cognitive control and was named the NoGo anteriorization (NGA). The NGA reflects of a more anterior center of gravity (centroid) of the P3 NOGO wave in contrast to a more posterior center of gravity of the P3GO wave. In 2005 the Fallgatter's team (Zielasek, Ehlis, Herrmann, and Fallgatter, 2005) showed a significant reduction of the NGA in patients with systematic schizophrenias.

In 2012 the same team (Ann-Christine Ehlis and coworkers) aimed at establishing the NGA as a predictor of the treatment response to first- and second-generation antipsychotics. They showed that the patients with low level of NGA (an index of hypo-frontality) responded to atypical medication which effects more frontal lobe function, while the patients high on NGA responded to typical antipsychotics.

NEUROTRANSMITTERS

At least two major neurotransmitters appear to be involved in the pathogenesis of SZ. They are reflected in the **dopamine and glutamate hypotheses** of SZ. The historically older **dopamine hypothesis** of SZ presumes **subcortical dopamine hyperfunction**. Facts that support the dopamine hypothesis are: the clinical doses of antipsychotic medications are related to their affinities for the dopamine D2 receptor, the majority of SZ patients are supersensitive to dopamine activation drugs such as amphetamine and cocaine.

The **glutamate hypothesis** postulates disruption of excitatory neural pathways through **NMDA receptor hypofunction.** Facts that support the glutamatergic hypothesis of SZ are: NMDAR antagonists uniquely reproduce both positive and negative symptoms of SZ, and induce SZ-like cognitive deficits and neurophysiological dysfunction.

Recent research shows that the hypotheses are not exclusive. Indeed, there is a strong interaction between both neurotransmitter systems at the level of single cells and neural networks. In particular, dopamine and glutamate interactions at the frontal–basal ganglia thalamocortical pathways are involved in modulation of **signal-to-noise ratios.**

A 2010 paper by Guillermo Gonzalez-Burgos and coworkers from the University of Pittsburgh suggested that **deficit in production of cortical inhibitory GABA** neurotransmission might be a central element in the pathology of SZ. The hypothesis emerged from a series of postmortem studies. The authors argue that parvalbumin-containing GABA neurons are involved in production of gamma oscillations that are critical for cortical information processing and that their deficit might explain alterations in GABA as well as a decrease of gamma oscillations and impairment of cognitive function in SZ.

NEURONAL MODEL

The prodromal period of SZ coincides with the period of adolescence which is a time of profound changes in the brain structure (**pruning**) and **maturation of the highest-order cognitive functions**, such as reasoning, abstract thinking, and planning. This pruning process is time-locked to the maturation process in the capacity of neuronal networks to process information and to synchronize activity for performing corresponding operations (Fig. 5.2.13).

ERPs as indexes in information processing in neuronal networks dramatically change with age. Fig. 5.2.14 shows the age dynamics of one of the **cognitive control components.** This component is evoked in response to NOGO stimuli in the cued GO/NOGO task, generated in the presupplementary motor area, associated with inhibition or overriding of the prepotent response, and correlates with the neuropsychological index of the energization domain (see chapter: Executive System and Cognitive Control). This is one of many components that are suppressed in SZ. One can see that the late positive fluctuation of the component is of low amplitude and delayed at age 7–10. It reaches its maximum at age 11–17 and reaches minimal latency at age 21–30. After this age the component decreases in amplitude and increases in latency. Between the ages of 51 and 60 it becomes similar to the one at age 7–10, and deteriorates both in amplitude and latency between the ages of 61 and 90. In a way the component seems to reflect changes in the synaptic connectivity of neurons of the inhibitory control circuit.

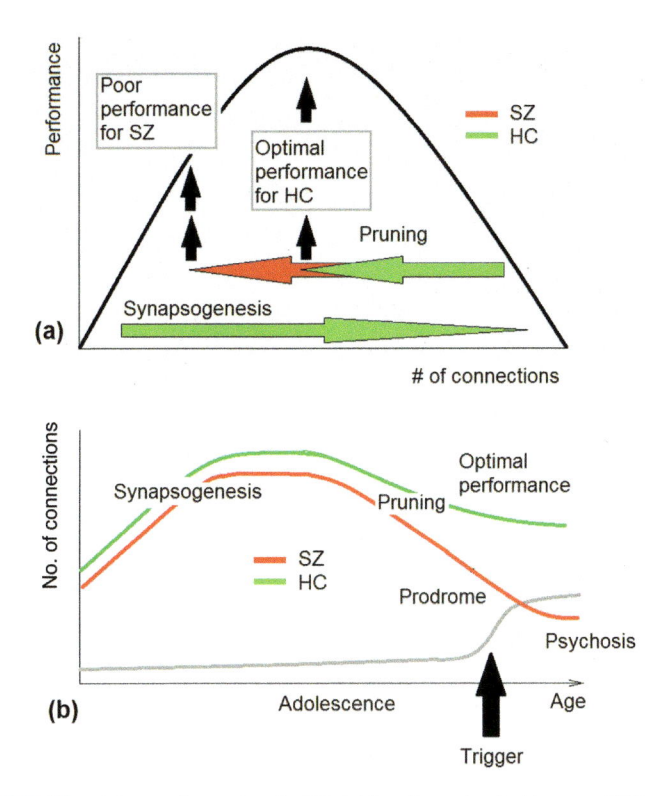

FIGURE 5.2.13 **Enhanced pruning in SZ.** (a) In a hypothetical inverted U law the medium level on connectivity within the neuronal network corresponds to optimal performance. Too much connectivity due to intensive synapsogenesis in childhood is associated with poor performance. In adolescence the pruning process in healthy *(HC)* subjects reaches the optimal number of connections corresponding to optimal performance; and (b) time dynamics of interaction of synapsogenesis and pruning in healthy subjects *(green)* and patients with SZ *(red)*. Any event that temporally enhances the dopamine sensitivity of the brain (Trigger on the *gray* curve) may induce psychosis.

The challenge we face in understanding the mechanisms of cognitive impairments in SZ is that all components of cognitive control (as well as sensory-related components) are suppressed in this illness. This fact might indicate some common sources which become dysfunctional during development.

According to our theory of action selection, within basal ganglia thalamocortical circuits (see chapter: Executive System and Cognitive Control) the striatum serves as a map of all programs of actions that the subject can perform. The representations of different actions are mapped into separate locations of the striatum. The selection of activated representations is done within the striatum by means of inhibitory processes including

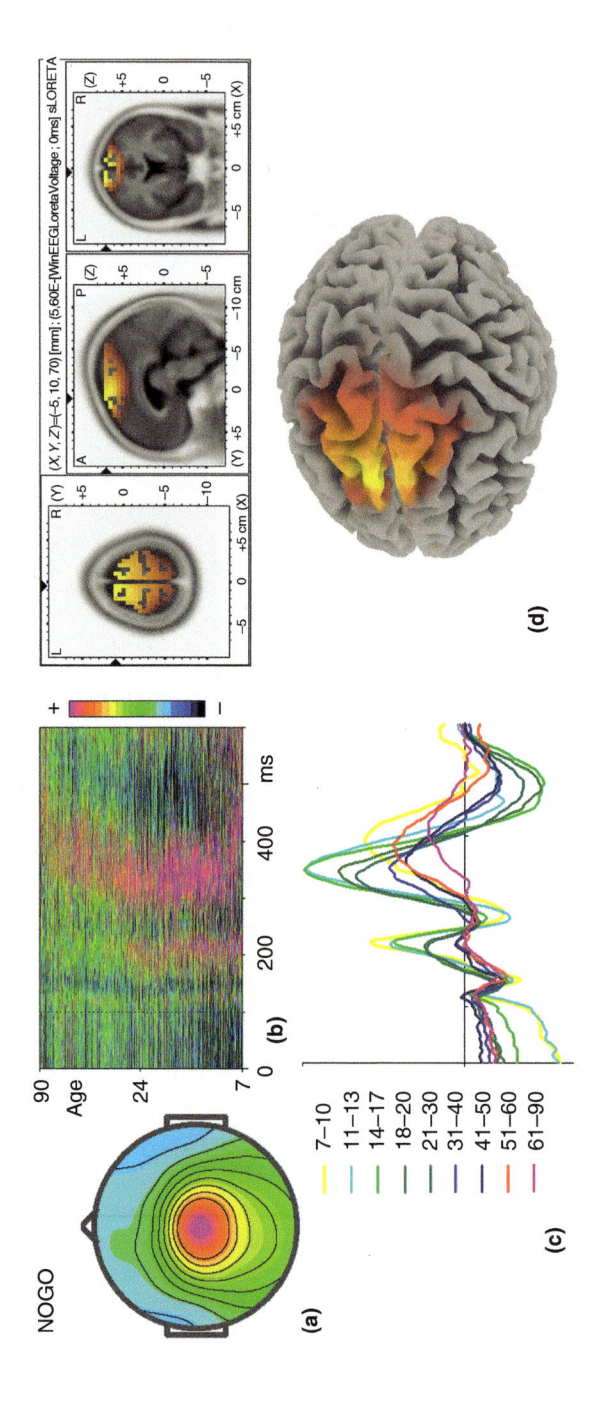

FIGURE 5.2.14 **Age dynamics of a NOGO latent component generated in the presupplementary motor area.** (a) Topography of the component; (b) age dynamics of the individual component. Each line in the graph represents a color-coded component for a single subject from the whole group of healthy subjects ($N = 880$) aged 7 (bottom) to 90 (top) years; (c) Age dynamics of grand averages of the component computed for different age groups such as 7–10 (*yellow*), 11–13 (*blue*), 14–17 (*light green*), 18–20 (*dark green*), etc.; and (d) sLORETA images of the topographic map of the component. *Data from the HBI database.*

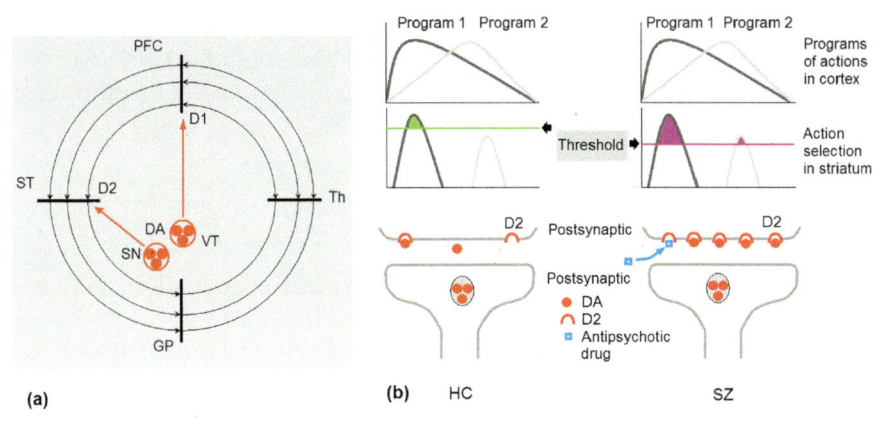

FIGURE 5.2.15 **Impairment of action selection in frontal basal ganglia thalamocortical circuits.** (a) Parallel basal ganglia thalamocortical loops with dopaminergic modulatory input from the substantia nigra *(SN)* to the striatum *(ST)* and from the ventral tegmental *(VT)* area to the prefrontal cortex *(PFC)*; and (b) two representations of potential actions are overlapped in the cortex, but segregated at the striatal level. Increased dopaminergic sensitivity of striatal neurons in SZ (eg, due to the excess density of D2 receptors) sets the effective threshold of these neurons at a lower level than that of HC subjects. This produces a SZ-related condition in which the irrelevant action (depicted as a lower amplitude bell shape) is selected together with the relevant action (giving the largest input to the striatum)—a state called "split of consciousness." *DA*, dopamine; *D-receptors*, postsynaptic striatal receptors to DA; *Antipsychotic drug*, schematic representation of a molecular of an antipsychotic drug that blocks the D-receptors and lessens the symptoms of SZ.

long-distance lateral inhibition. In line with the model, representation of an action is selected if its activation exceeds some **threshold** (Fig. 5.2.15).

The threshold of activation of striatal neurons is set by the mediator dopamine. The threshold is defined as the minimal amount of excitatory postsynaptic potentials coming from the activated cortical representation to trigger firing of striatal neurons. This differs from the classically defined threshold of a neuron (which is constant) because in this context input to the neuron is divided into two parts: fast-acting activation from cortical neurons via glutamatergic excitatory connections and long-lasting modulatory activation from dopaminergic neurons in the substantia nigra. Consequently, increase of the excitatory modulatory dopaminergic input would decrease the amount of the cortical input needed to activate striatal neurons.

The effect of dopamine is defined, in particular, by the density of dopamine receptors at the postsynaptic membrane of striatal neurons. In healthy subjects the threshold is high enough so that only currently important actions are selected while others are suppressed. The selected action is performed and comes into consciousness. In SZ patients the density of D2 receptors is high and consequently the threshold is low. The low threshold enables selection of several actions simultaneously. The selected actions compete with each other and create "the split of consciousness." Blocking D2 receptors by antipsychotic drugs increases the threshold for action selection and restrains psychosis.

tDCS

Attempts at applying tDCS for SZ are rare but almost all of them are quite promising. Only a few randomized controlled studies have been conducted so far. Jerome Brunelin and coworkers from the University of Lyon (France) in 2012 demonstrated the efficacy of a protocol consisting of two daily sessions (20 min each) of 2-mA tDCS for 5 consecutive days in **reducing hallucinatory symptoms** of 30 refractory patients (Fig. 5.2.16a). The anode was placed between F3 and Fp1and the cathode located over a point midway between T3 and P3. The authors also reported improvements in negative symptoms, which they associated with partial reversal of hypoactivity of the left dorsolateral prefrontal cortex.

Robert Smith and coworkers from Nathan S. Kline Institute for Psychiatric Research in 2015 conducted a randomized double-blind, sham-controlled study of the effects of 5 sessions of tDCS (2 mA for 20 min, F3 anode, Fp2 cathode) on 37 outpatients with SZ or schizoaffective disorder. Active compared with sham tDCS subjects showed significant improvements after the fifth tDCS session on **cognitive performance** (Fig. 5.2.16b). However, in contrast with Brunelin's study the authors were not able to demonstrate statistically significant effects on secondary outcome measures of psychiatric symptoms (PANSS scores), hallucinations, and cigarette craving. The difference between the outcomes of the two studies might be explained by differences in the protocols including differences in electrode placement. In general, no commonly accepted tDCS protocol for SZ exists at the moment. Individually tailored protocols on the basis of topographies of latent components might be a good solution. Further research is definitely needed.

TRANSCRANIAL MAGNETIC STIMULATION

In contrast with tDCS, many studies to assess the effectiveness of TMS for patients with SZ have been conducted worldwide. In 2015, Nadine Dougall from the University of Stirling (United Kingdom) conducted a systematic review of all randomized controlled trials comparing TMS with sham or standard treatment. Forty-one trials (1473 participants) survived eligibility criteria. The authors found significant differences in favor of temporoparietal TMS compared with sham TMS for global state and

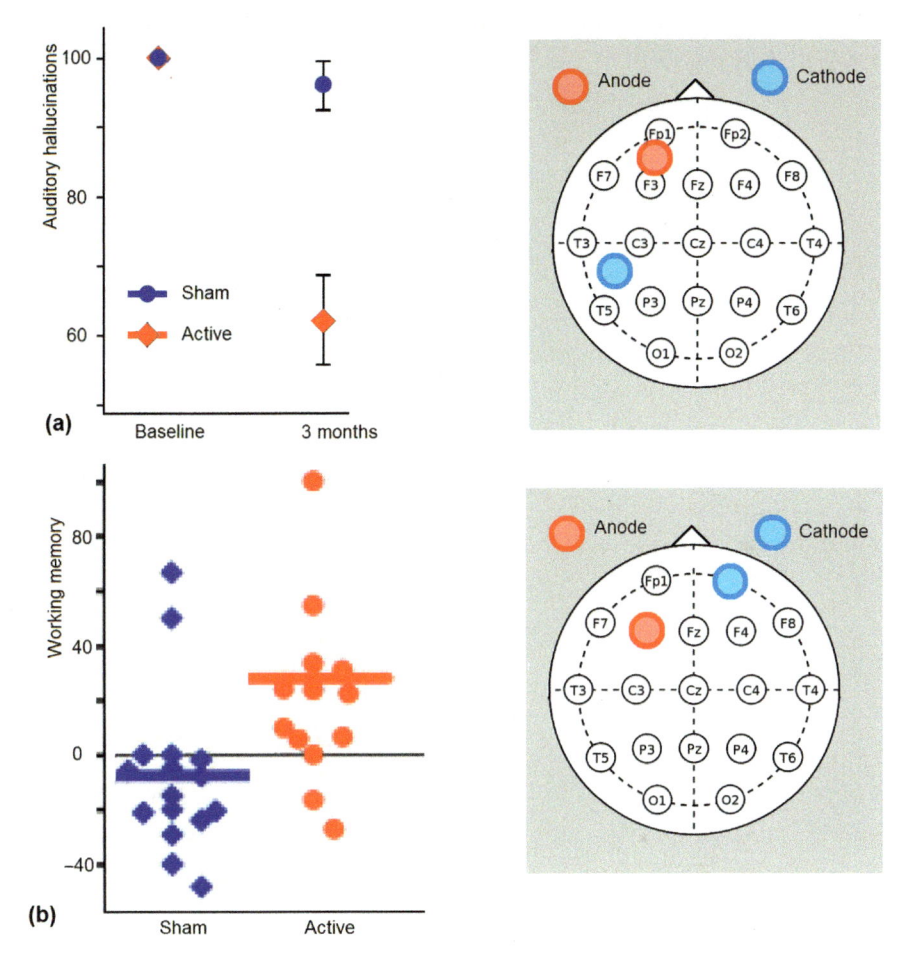

FIGURE 5.2.16 **Effect of tDCS on hallucinations and working memory in SZ.** (a) Effect of tDCS (*red*) in comparison with sham (*blue*) on severity of auditory hallucinations during four different assessments: at baseline, 3 months after tDCS (error bars, standard errors); the placement of anodal and cathodal electrodes is schematically presented in a gray box; and (b) Effect of tDCS (*red*) in comparison with sham (*blue*) on working memory parameters (y-axis, percent change; each dot corresponds to a participant); the placement of anodal and cathodal electrodes is schematically presented in a gray box. *Part a: adapted from Brunelin et al. (2012). Part b: adapted with permission from Smith et al. (2015).*

positive symptoms. However, the authors also found that the quality of trial reporting was frequently suboptimal and the risks of bias were strong or undiscoverable for many trial aspects. On that basis the authors were unable to definitively support or refute the routine use of TMS in clinical practice. High-quality and large-sample TMS studies in SZ research are needed.

CHAPTER
5.3

Obsessive–Compulsive Disorder

HISTORICAL INTRODUCTION

By definition, **obsessions** are impulses in the form of thoughts, ideas, or images that uncontrollably and persistently intrude on consciousness, which in turn causes significant anxiety and stress. **Compulsions** are repetitive motor or mental actions that a subject feels must be performed. Although in DSM-IV **OCD** was referred to anxiety disorders because of the distressful result of obsessional thoughts, it differs from other anxiety disorders and in this book is considered as a separate entity.

In the medical world, compulsive washing, repetitive checking, sexual and religious obsessions have been known for centuries. In the 1700s many OCD patients were considered insane and kept in asylums. Only at the end of the 19th century did the OCD medical world start to think about OCD as a neurosis rather than a psychosis and produced some neurological models of the disorder. Nineteenth-century physicians treated OCD with a variety of medications including opium and morphine.

Freud's **symbolic interpretation** of obsessions and compulsions produced a definite shift in the medical paradigm and continued to be accepted up to the middle of the 20th century. Pavlov, in contrast, related human obsessions and compulsions to pathological processes in the cerebral cortex. **Behaviorism** looked at OCD from the viewpoint of conditioned reflexes, fear, avoidance and suggested a variety of **desensitization techniques**. An exposure-and-response-prevention therapy was developed in the 1960s and 1970s. It is still the primary psychological treatment for OCD today.

There were a number of decades when few pharmaceuticals were available for OCD. The most common **antianxiet**y drugs like benzodiazepines **were found to be unhelpful**. In the 1980s it was discovered that **antidepressants helped to relieve symptoms of OCD**. These drugs increase the level of serotonin in the whole brain but show significant side-effects including drowsiness and loss of sexual desire. In addition, a substantial number of OCD patients do not respond to selective serotonin reuptake inhibitors (SSRIs).

OCD is a relatively common disorder and was described in the lives of well-known figures including Howard Hughes, the 20th-century American engineer, whose story was told in the 2004 film *The Aviator* starring Leonardo DiCaprio.

SYMPTOMS

The name obsessive–compulsive disorder stems from the symptoms of the disease. Obsessions typically include thoughts of harm or death occurring to a loved one, chronic doubting, fears of contamination, blasphemous or socially unacceptable thoughts or impulses, counting, and a preoccupation with symmetry. Compulsions include excessive hand washing, placing objects symmetrically, repeatedly checking (eg, that lights are off), or following set routines. Usually, particular compulsive motor or mental actions are carried out in response to a particular obsession to **neutralize the anxiety associated with that obsession**.

Carrying out a particular compulsion in an OCD patient may temporarily relieve the anxiety brought on by the obsession. For example, hand-washing may reduce the anxiety produced by thoughts of contamination. The **excessive nature of the compulsion**, however, creates its own distress and it appears that the individual may be caught up in a kind of **negative reinforcement loop**. Some patients with OCD report that they must engage in a compulsion a certain number of times, whereas others will repeat the behavior until it satisfies some sensory–perceptual criterion such as feeling "just right." Perfectionism and scrupulosity are also features of the disorder that may involve doing things in a certain order, or a certain number of times; but more often these behaviors are associated with **obsessive–compulsive personality disorder (OCPD)**.

The intrusive, troubling thoughts in OCD are perceived as the product of one's own mind and are distinguished from thought insertions in patients with **schizophrenia**. The majority of patients are aware of the irrationality of their thoughts and behaviors but have **limited control** over them.

PREVALENCE

OCD has a **lifetime prevalence of 2–3%**, more than twice that of schizophrenia. Even when the most aggressive medical and behavioral therapies are applied, an estimated 10% of the OCD population remains severely affected.

DEVELOPMENT

Ritualistic and repetitive behaviors sometimes observed in the healthy population are particularly prominent in early childhood. Children 2–3 years old exhibit marked compulsive behavior including strong preferences for wholeness and symmetry in the environment, rigidity of likes and dislikes, ritualized behaviors. The interval between 3 and 5 years of age is associated with the development of executive functions and maturation of the frontal lobes. Like individuals with OCD, young children have difficulty inhibiting unnecessary responses as well as shifting and maintaining cognitive set. The compulsive-like behaviors in children positively correlate with fears and phobias and seem to produce anxiety reduction. These compulsive-like activities in healthy subjects wane by the age of 8.

As shown in a 2011 systematic review by Taylor from the University of British Columbia two categories of OCD could be separated: **early onset OCD** with mean onset of about 11 years (76% of cases) and **late-onset OCD** with mean onset about 23 years. Metaanalyses indicated that early onset OCD is (1) more likely to occur in males, (2) associated with greater OCD global severity, (3) more likely to be comorbid with tics, and (4) associated with a greater prevalence of OCD in first-degree relatives.

HETEROGENEITY

The nature of obsessions and compulsions **varies greatly** between patients. It seems that a **specific negative thought** (like fear of being infected from dirty hands) in childhood or adolescence for some reason takes over other thoughts and becomes **dominant and intrusive**. Actually, it is not uncommon that some specific thought becomes dominant in human life. Many scientists, sportsmen, artists are preoccupied by great ideas to discover a new phenomenon, to become an Olympic champion, to create a masterpiece. The difference is that in the case of OCD the thought is **negative** and produces **anxiety** and is so strong that it **cannot be controlled**.

HERITABILITY

Twin studies strongly suggest that vulnerability to OCD is **inherited**. Concordance rates for OCD symptoms are significantly higher for monozygotic versus dizygotic, and the disorder is often demonstrably familial, with risk to first-degree relatives estimated at 3–12 times greater than for the wider population. But, in many OCD patients a positive family history is absent and only 10% of the parents of children with OCD have the disorder themselves.

COMORBIDITY

Patients with OCD often suffer from other (comorbid) disorders, including **depression, anxiety, motor tics and Tourette's syndrome, body dysmorphic disorder**. **Major depression** has a lifetime prevalence of 60–70% in OCD patients. On the other hand, compulsive–repetitive behavior is seen in some other psychiatric conditions. Some of these conditions like Tourette syndrome and compulsive skin picking are united under the umbrella of the **obsessive-compulsive spectrum of disorders**, while others like autism and Asperger syndrome fall into the category of **autism-spectrum disorders**.

NEUROPSYCHOLOGICAL PROFILE

In contrast with schizophrenia, individuals with OCD **do not suffer** devastating cognitive impairment in all domains. The neuropsychological profile of OCD reveals two local deficits: **inhibiting** and **planning** motor and cognitive actions. This profile fits well with OCD symptomatology which may be best characterized in terms of **failures to inhibit**, or **shift from**, intrusive, troubling thoughts and repetitive, compulsive motor actions.

Initial evidence for **response inhibition** deficits in OCD came from studies using oculomotor tasks that required the suppression of eye movements. For example, failures of inhibition were identified in treatment-naive children and adults with OCD in a 1997 study by Rosenberg and coworkers. In a computerized GO/NOGO task, Bannon and coworkers in 2002 found that OCD patients made significantly more commission errors than matched control subjects. In a 2005 review by Chamberlain and coworkers from the University of Cambridge the failure in cognitive and behavioral inhibition was considered a **candidate of endophenotype** of OCD. The authors propose that it may be useful to differentiate between two types of inhibition processes: (1) cognitive inhibition, representing control over internal cognitions (eg, intrusive thoughts, mental rituals, or inappropriate strategies); and (2) behavioral inhibition, representing control over externally manifested motoric activities (eg, ritualistic checking behavior).

Some personality traits may differentiate OCD from healthy controls (HCs). One of these traits is **chronic doubting.** Most individuals with OCD describe a feeling of uncertainty in their own behavior. OCD patients demonstrate poor confidence in "reality testing" or the ability to recall whether a behavior was actually performed or merely imagined. **Need for perfection** as an often cited trait of OCD has found inconsistent support in the literature. **Responsibility** as a trait of overestimating the importance of one's own actions and thoughts might be another mark of OCD.

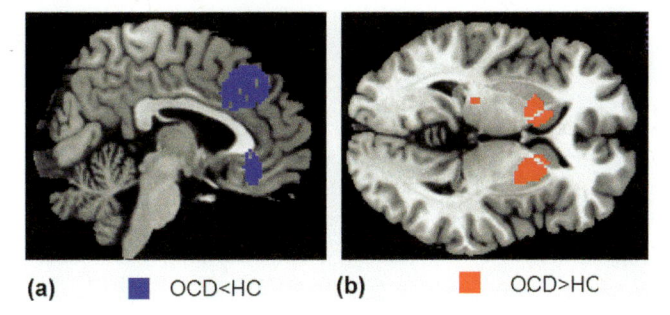

FIGURE 5.3.1 Gray matter differences between OCD patients and HCs. (a) Blue indicates brain regions with decreased gray matter in OCD; and (b) red indicates regions with increased gray matter in OCD patients compared with HCs. *Adapted with permission from Peng et al. (2012).*

LESIONS

Traumatic brain injury to local regions of the brain can cause OCD to develop in adults with no prior history of symptoms. In severe cases, focal lesions are found in **the orbitofrontal cortex and associated basal ganglia structures.**

STRUCTURAL MAGNETIC RESONANCE IMAGING

The results of lesion studies fit well the results of structural MRI studies. In a 2012 metaanalysis by ZiWen Peng and coworkers from the Chinese Academy of Sciences showed that OCD patients had reduced gray matter volume in comparison with HCs in the **medial frontal, anterior cingulated, and orbitofrontal cortical** areas accompanied by enlargement of the **striatum** (Fig. 5.3.1).

fMRI IN SYMPTOM PROVOCATION

As mentioned earlier, the signs of OCD are very individual such that each OCD symptom is triggered by specific sensory stimuli. In a 2013 study by Ali Baioui and coworkers from Justus Liebig University, Giessen fMRI was used to study the neural correlates of contamination/washing-related OCD by using a highly individualized symptom provocation paradigm. The results clearly demonstrated that only confrontation with highly **individual triggers** evokes activation patterns that are **common** for the whole OCD group: activation of the striatum (including caudate and nucleus accumbens) and pallidum (Fig. 5.3.2).

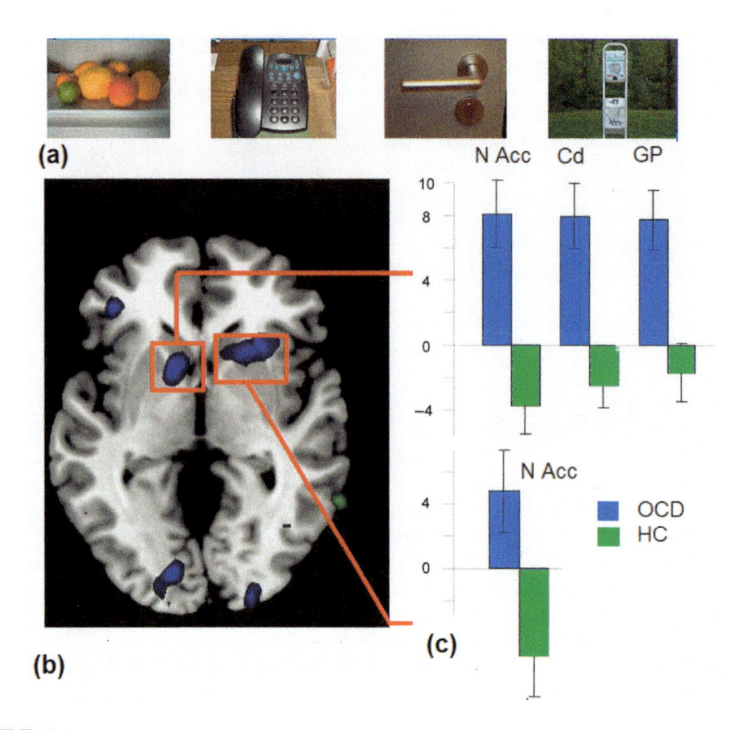

FIGURE 5.3.2 **Hyperactivation of the basal ganglia in OCD during symptom provocation.** (a) Individual stimuli used for symptom provocation; (b) 2D map; and (c) statistical results of comparing neural activation of OCD patients *(blue)* in comparison to HCs *(green)*. *Adapted with permission from Baioui et al. (2013).*

fMRI IN CONFLICT CONDITIONS

In contrast with ADHD and schizophrenia where metabolic activation in response to cognitive demands is reduced in the prefrontal–thalamo-cortical circuits, OCD patients consistently show **hyperactivation of specific cortical areas** especially in tasks requiring conflict monitoring.

According to a 2005 study by Nicholas Maltby and coworkers from the Institute of Living/Hartford Hospital the areas of this hyperactivation during correctly rejected NOGO trials occupy the rostral and caudal anterior cingulate cortex, lateral prefrontal cortex, lateral orbitofrontal cortex, the basal ganglia and posterior regions including the posterior cingulate.

Hyperactivation of the **anterior cingulate** cortex as a key node of the conflict-monitoring brain system has led to a commonly accepted hypothesis of OCD that **excessive error** signals generated by the anterior cingulate cortex produce a subjective sense **that something is wrong** and that some behavior change is needed to correct the problem. In essence, these error signals convey the feeling that things are "not just right," even when no actual error has been made. Thus, exaggerated or false error signals may explain a wide range of compulsive behaviors that involve an **intense, irrational need to repeat a behavior** because it was "not just right."

fMRI research shows exaggerated activation of the anterior cingulate cortex in response to errors. However, in reality not errors but **correctly performed actions** force OCD patients to repeat the action over and over.

QUANTITATIVE ELECTROENCEPHALOGRAPHY

Leslie Sherlin and Marco Congedo from Novatech in 2005 using LORETA imaging found an **excess of current source density of beta activity in the cingulate cortex** in OCD patients compared with healthy subjects.

In a 2011 study Marco Congedo, now at Grenoble University, together with Jana Koprivova and coworkers from the Prague Psychiatric Centre used normative group independent component analysis on spontaneous EEG and found the evidence of hyperactivation of the medial prefrontal cortex in OCD.

In the study of our group the most striking result of QEEG/ERS/ERD analysis in OCD in comparison with HCs was **excessive frontal beta synchronization** in response to Cue in the cued GO/NOGO (Fig. 5.3.3).

Taking into account that frontal beta rhythm is an activation rhythm which is present under conditions requiring engagement of the frontal lobes (see chapter: Beta and Gamma Rhythms) we can speculate that adult OCD patients activate their frontal lobes more strongly than HC subjects.

All the previously mentioned results in the QEEG field show **hyperfrontality** in OCD which fits fMRI data and which is an opposite pattern to hypofrontality in schizophrenia (see chapter: Schizophrenia).

ERROR-RELATED NEGATIVITY AND N2 EVENT-RELATED POTENTIAL WAVES

As shown in chapter: Executive System and Cognitive Control, there are two ERP waves that are considered indexes of **conflict monitoring**: (1) error-related negativity (**ERN**) in response to erroneous trials, and (2) **N2 waves** in response to conflict in NOGO trials. These waves are generally localized in the anterior cingulate cortex.

Studies in OCD patients report **enhancement of ERN** waves in the disorder. For example, in a 2013 study Melisa Carrasco and coworkers from

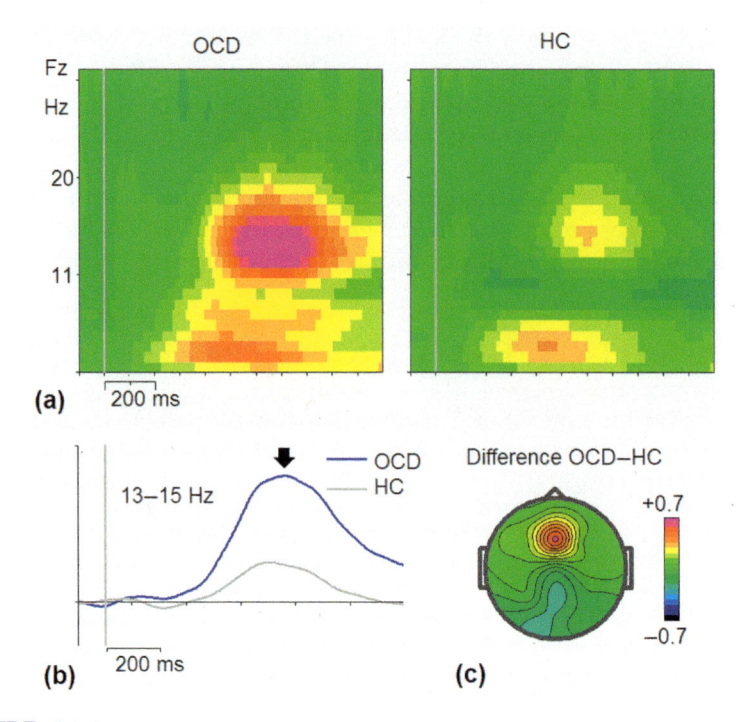

FIGURE 5.3.3 **Excessive frontal beta synchronization in OCD.** (a) Grand-averaged power–time frequency representations (wavelet decomposition) of EEG responses to Cue in the cued GO/NOGO task in the group of OCD patients ($N = 66$) and HCs ($N = 312$); (b) grand average of Event-Related Synchronization (ERS) in the 13- to 15-Hz band for OCD and HC; and (c) topographic map of the ERS differences OCD – HC at time interval indicated by *arrow* at (b). *Data from the HBI database.*

the University of Michigan recorded ERN in three groups: OCD, non-OCD anxiety disorder, and HCs. The authors found significant increase of ERN amplitude in patients with either OCD or a non-OCD anxiety disorder (Fig. 5.3.4). Scores from the Child Behavior Checklist DSM-oriented anxiety problem scale had a significant correlation with ERN amplitude in all subjects. The results provide further evidence that the pathophysiology of OCD and some non-OCD anxiety disorders involves **increased anterior cingulate cortex (ACC)** activity and that ERN may serve as a **quantitative phenotype** in genetic and longitudinal studies of these complex traits.

The results of the N2 NOGO wave in conflict tasks are **inconsistent**. Some studies, such as the 2007 study by Martin Ruchsow and coworkers from Department of Psychiatry, Christophsbad (Germany), reported an increase of N2 NOGO, while a 2007 study by Myung-Sun Kim and coworkers from Sungshin Women's University, Seoul (South Korea) reported a decrease of N2 NOGO.

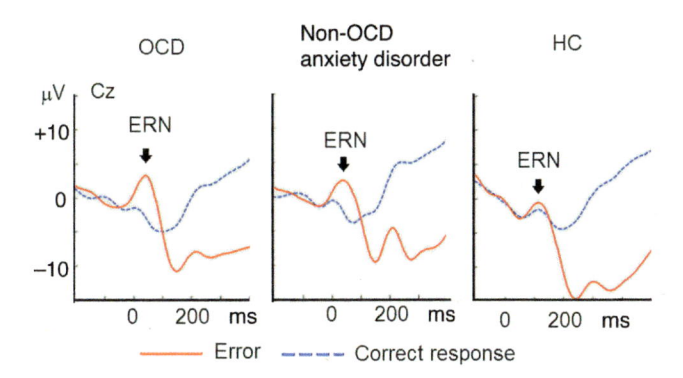

FIGURE 5.3.4 Enhancement of ERN in OCD and anxiety disorder in comparison to HC. *Adapted with permission from Carrasco et al. (2013).*

LATENT COMPONENTS OF COGNITIVE CONTROL

In our own study with latent components of cognitive control, we were not able to show any deviations in the frontal N2 NOGO component (see the frontally distributed component in Fig. 5.3.5, top), although the **preparing activity** and the late N600 waves in this component differ significantly between the two groups (OCD vs. HC) with large effect sizes.

The most striking difference in our studies was found in the component associated with **action inhibition** (Fig. 5.3.5, bottom). This component was found to be twice as large in OCD than in HCs with large effect size. This observation is in clear contrast with the one found in the group of patients with schizophrenia where this component is almost absent. So again we demonstrate the hyperactivation pattern in OCD in contrast with the hypoactivation pattern in schizophrenia.

NEURONAL MODEL

At least two operations can be considered for maintenance of obsessions and compulsions: first, the urge attributed to some sensory conditions, such as the urge to wash the hands because of the fear of being contaminated. A neutral for a healthy subject visual image such as dirty hands may induce abnormally exaggerated emotional reaction in an OCD patient. This is the feature of the affective system.

The second operation that might be theoretically implicated in OCD is the ability to suppress an inappropriate action such as repetitive hand washing by reasoning that the hands are clean enough. This is a feature of the executive system of the brain. Any misbalance between these

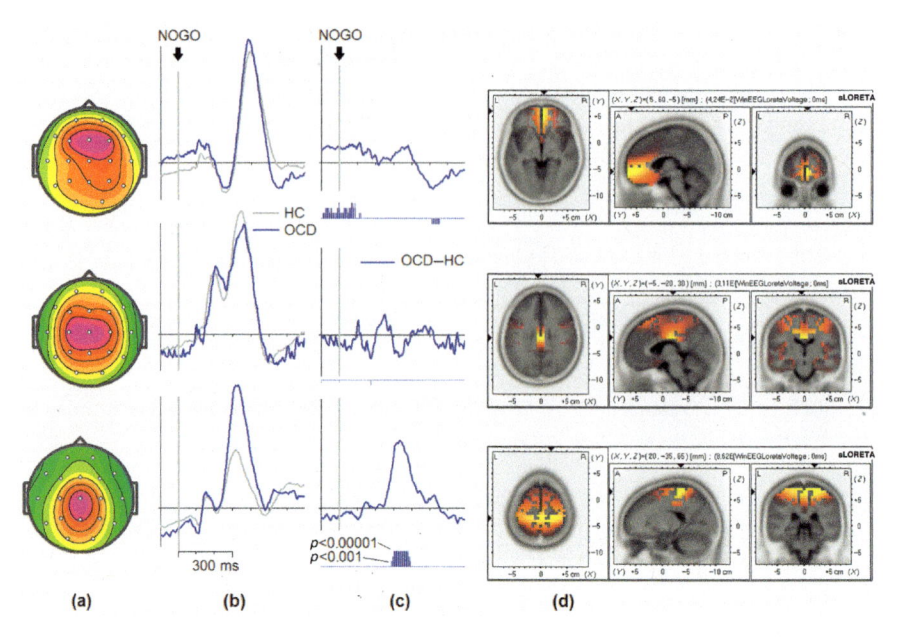

FIGURE 5.3.5 **Latent components of reactive cognitive control in OCD.** (a) Topographic maps of the three latent components in the cued GO/NOGO task in response to NOGO stimuli; (b) time course of the latent components for OCD patients *(blue)* and HCs; (c) the difference OCD – HC with *p* values below; and (d) sLORETA image of the topography. *Data from the HBI database.*

operations of affective and executive systems theoretically could produce symptoms of OCD.

Usually an argument in favor of inhibitory control failure in OCD is the fact that patients with OCD make more commission errors than normal controls in GO/NOGO paradigms that require inhibitory control. Action-monitoring operations are associated with detecting mismatch between expectation and reality and changing/adapting behavior in order to minimize the mismatch.

Another executive operation that was theoretically associated with OCD is the set-shifting operation. Recall that set-shifting refers to the ability to learn and respond to a sorting rule, and to subsequently adapt when the rule is changed. Although some research suggests that individuals with OCD make significantly more perseverative errors than normal controls on set-shifting tasks, other studies have failed to observe differences in set-shifting performance between normal controls and individuals with OCD.

Accumulating data show that hyperactivity in a specific prefrontal–basal ganglia–thalamocortical system is the core of OCD. The system includes the orbitofrontal cortex and the ventral part of the anterior cingulate cortex as the cortical parts of the affective loop via the nucleus accumbens. The system also includes the presupplementary motor cortex and the dorsal part of the anterior cingulate cortex as the cortical parts of the motor loop through the putamen and caudate nucleus. At the ERP level these hyperactivations are reflected in enhancement of **action-monitoring and action-inhibiting** latent components. As far as the ERPs in the cued GO/NOGO task concern, following are the possible neuromarkers of impairment in cognitive control:

1. The presence of N600 pattern in the latent component generated in the ventral cingulate cortex and associated with action monitoring operation. This ERP pattern seems to reflect reactivation of conflict-detection operation in OCD patients—the pattern which is completely absent in healthy population.
2. The enhancement of the P3 NOGO fluctuation in the latent component generated over the Rolandic fissure and associated with action inhibition/overriding operation. This ERP pattern seems to reflect hyperactivation in the medial prefrontal cortex of OCD patients. Recall that this part of the cortex receives strong input from the basal ganglia (which are hyperactive in OCD) through the ventral thalamus.

NEUROTRANSMITTERS

There is strong evidence that the **serotonergic system** modulates OCD symptoms. Serotonin reuptake inhibitors are unique among antidepressants in producing at least some clinical benefit in most patients with OCD. Interestingly, both the serotonin transporter and some serotonin receptor subtypes implicated in OCD are highly expressed in the **nucleus accumbens** where they could influence functioning of the affective basal ganglia thalamocortical circuit. In theory, **dopamine** as another neuromodulator of this circuit may play a supportive role in the SSRI treatment of OCD. Indeed, neuroleptics, ineffective as monotherapy in OCD, are beneficial when added to ongoing serotonin reuptake inhibitor treatment.

FIRST-LINE TREATMENT

The most effective treatments for OCD based on strong evidence in the literature are **cognitive behavioral therapy such as the Exposure and Response Prevention method and** pharmacological therapy by **serotonin reuptake inhibitors**. Taken together these methods are considered "first-line" treatments for OCD.

PSYCHOSURGERY AND DEEP-BRAIN STIMULATION

However, in 10–30% of cases **severe OCD turns out to be resistant to these conventional treatments and invasive methods such as psychosurgery and deep-brain stimulation (DBS) with electrical currents can be used**.

John Fulton in 1951 was the first to suggest that the **anterior congulate cortex** would be an appropriate target for psychosurgical intervention. Thomas Ballantine and coworkers from Harvard Medical School in the 1980s demonstrated the safety and effectiveness of **cingulotomy** in a large number of patients. Independent analysis of patients who underwent cingulotomy demonstrated **no significant intellectual deficits** as a result of the cingulate lesions themselves.

Cingulotomy is used for treatment of OCD in the Neurosurgery Clinic of the Institute of the Human Brain in Saint Petersburg. In our laboratory we are using ERPs for assessment localization of cortical hyperactivation in OCD patients as well as for **assessing the treatment outcome**. Fig. 5.3.6 shows the effect of cingulotomy on NOGO ERP components in a group of patients with obsessions. Note **selective suppression of the independent component** located by sLORETA in the anterior cingulate cortex.

In 2010 Damiaan Denys and coworkers from the University of Amsterdam reported successful application of **DBS of the nucleus accumbens** in 9 of 16 patients. In 2013, Patric Blomstedt and coworkers from Umeå University and the Institute of Neurology, London reviewed the results of DBS in OCD presented in 25 reports with 130 patients. Sixty-eight of these patients underwent implantation in the region of the internal capsule/ventral striatum, including the nucleus accumbens, and showed a 50% reduction in OCD scores, depression, and anxiety.

NEUROFEEDBACK

Until recently reports on successful **neurofeedback treatment** in OCD were rare because most of the attempts used a simplistic treatment approach of alpha enhancement training. A 2014 study by Jana Koprivova and coworkers from Prague Psychiatric Center, Prague, Marco Congedo and coworkers from Grenoble University used **independent component neurofeedback** in 20 patients with OCD. The neurofeedback group showed a significantly higher percentage reduction of compulsions compared with the sham feedback group.

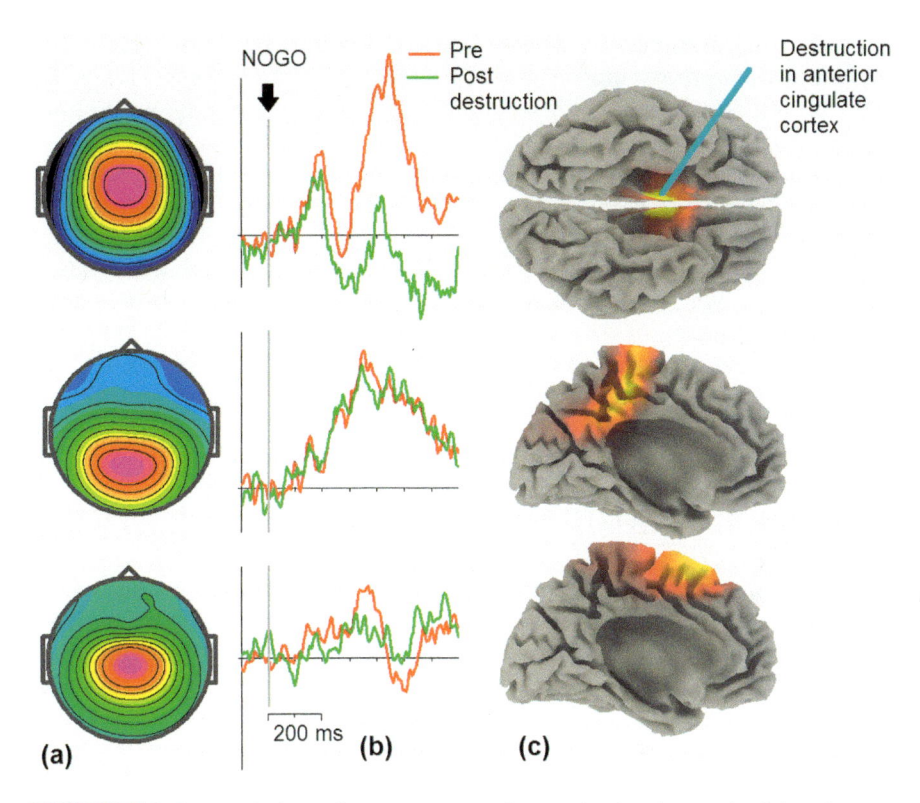

FIGURE 5.3.6 ERP independent components for monitoring the result of cingulotomy in OCD. ERPs in the cued GO/NOGO task were recorded in OCD patients before (pre destruction) and after (post destruction) a stereotactic cingulotomy. (a) Topographies of independent components extracted from NOGO ERPs; (b) time course of the components pre *(red)* and post *(green)* cingulotomy; and (c) sLORETA images of the topographies. *Data from the HBI database.*

TRANSCRANIAL MAGNETIC STIMULATION

In 2013, Marcelo Berlim and coworkers from Douglas Mental Health University Institute, Montreal searched the literature for randomized and sham-controlled trials using rTMS for treating OCD. Data were obtained from 10 studies, involving 282 subjects with OCD. Response rates were 35 and 13% for patients receiving active and sham rTMS, respectively. **Low-frequency (inhibitory) rTMS protocols targeting the orbitofrontal cortex or the supplementary motor area** seem to be the most promising. Indeed, Antonio Mantovani from Columbia University/New York State Psychiatric Institute in 2010 published the results of a randomized sham-controlled double-blind study in which medication-resistant OCD patients were assigned for 4 weeks to either active or sham 1-Hz rTMS to the supplementary motor area bilaterally. After 4 weeks the response rate in the completer sample was 67% with active and 22% with sham rTMS.

In a 2014 review on neuromodulation methods in OCD, Kyle Lapidus and coworkers from Icahn School of Medicine at Mount Sinai, New York showed that multiple sham-controlled studies of rTMS to the dorsolateral prefrontal cortex at 1 or 10 Hz produced no benefit of active treatment over placebo, yet the **supplementary motor area** appeared to be a very promising target.

Both reviews concluded that future research should include larger sample sizes and be more homogeneous in terms of clinical symptoms, stimulation parameters, and brain targets.

TRANSCRANIAL DIRECT CURRENT STIMULATION

The supplementary and presupplementary motor areas, as shown in the chapter, are hyperactive on OCD. So, theoretically, inhibition of these areas by cathodal tDCS could lessen the urge for repetitive actions. This was proved in a 2015 study by Giordano D'Urso and coworkers from the University Hospital of Naples Federico II on a patient with resistant OCD. In this study, anodal stimulation worsened the severity of symptoms whereas **cathodal stimulation induced a dramatic clinical improvement**.

PART 6

ASSESSING FUNCTIONAL NEUROMARKERS

CHAPTER 6.1

Working Hypothesis

REASONS FOR ASSESSMENT

Theoretically, there could be several reasons functional neuromarkers must be assessed in a particular person.

One reason comes from **disability insurance companies** that want to know if the applicant indeed has the psychiatric condition he or she claims. In this approach obtaining the objective measures of brain functioning is important for making a decision about whether a claimant is malingering or not. **Malingering**, the intentional production of false or grossly exaggerated physical or psychological symptoms, has been estimated to occur in 7.5–33% of disability claimants. Malingering is motivated by such reasons as avoiding military duty or work, obtaining financial compensation, escaping criminal prosecution. Because malingering is difficult to detect solely on the basis of unstructured interviews, clinicians should use as many sources of information as possible including QEEG and ERP parameters.

A second reason comes from **forensic psychiatry and forensic neuropsychology** that want to know what is happening in the brains of current or potential criminals. On the one hand, a forensic psychiatrist has to answer to the court two questions about **competency of a client to stand trial** and about **his/her mental state** at the time of the offence. On the other hand, forensic psychiatric teams supervise past criminals and suspected or potential offenders with mental health problems to promote both the welfare of their clients and the safety of the public. In all cases information about the degree to which the QEEG/ERP parameters of the client deviate from the reference average could be of vital importance.

A third reason comes from **sport psychologists** who want to know how the brain performance of their clients can be enhanced. In this case the concept of **peak performance** is introduced. Peak performance does not depend on physical strength alone but rather on combination with emotional and cognitive states. Athletes need to relax (stress management) and control their mental state (focus enhancement) if they want to perform at the peak of their abilities. Commonly defined neurofeedback procedures have proved to be helpful in achieving peak performance. **Individually tailored neurofeedback protocols** are definitely more efficient because they provide the trainer and trainee with knowledge about their baseline brain functioning and how optimal functioning of brain circuits may be achieved.

A fourth reason comes from **biologically oriented psychiatrists** who want to obtain objective measures of brain dysfunction in order (1) to get an aid in making correct diagnosis, (2) to construct the optimal treatment protocol including pharmacological and neuromodulation techniques, and (3) to monitor the outcome of treatment.

Each of the reasons for assessment of functional neuromarkers needs to answer specific questions and consequently requires a specific protocol of the assessment procedure. In this book we focus on the assessment protocol for psychiatry.

CONVENTIONAL DIAGNOSTIC CATEGORIES AS A STARTING POINT

Although most psychiatric conditions are heterogeneous brain disorders, at the moment we do not have anything better for their classification than DSM or ICD diagnostic manuals. Biological bases of the conventional diagnostic categories like ADHD, schizophrenia, OCD have been intensively studied in the **last 50 years** by means of **QEEG, ERPs, MRI/fMRI, and PET**. Numerous behavioral paradigms and data-processing methodological approaches have been applied. Various hypothetical neuronal models of these conventional diagnostic categories have been suggested and tested. Several functional neuromarkers of brain dysfunctioning have been discovered.

For example, EEG for the assessment of ADHD has been applied (according to Pubmed) in 1170 published studies from 1968 through 2015, fMRI for the assessment of ADHD has been published in 1131 papers since 1990. Schizophrenia as a diagnostic category has been studied even more intensively: fMRI studies are reflected in 5441 papers and EEG studies are reflected in 3605 papers according to Pubmed. It is important to stress that the number of publications per year **increases dramatically**. For example, in schizophrenia research only 2 papers on ERPs were published in 1963 while in 2014 the number increased to 121 papers.

So the starting point for assessment must be a real or suspected **conventional diagnosis** or a set of possible diagnoses.

MULTIPLE CAUSES OF ADHD

As was discussed in the previous part of this book, a single diagnostic category does not have a single brain pathophysiology but rather is considered as a heterogeneous condition in which similar symptoms are caused by impairments in different parts of the same brain system or even by impairments in different systems. Let us consider for example **ADHD** as a suspected diagnosis. Let us imagine a boy coming to the door of a psychiatrist. The boy shows all the symptoms of ADHD—**inattention, impulsivity, and hyperactivity**.

A biologically oriented psychiatrist, especially one who has read the previous chapters of the book, knows that there are several causes for the boy to manifest this behavioral pattern. These causes are briefly listed here:

1. **epifocus**—a patient might have a focus near the Rolandic fissure of his cortex, which without any overt symptoms of epilepsy may impair operations of cognitive control and, consequently, may mimic attention deficit or hyperactivity;
2. **hypoarousal**—a patient might have a lack of overall cortical activation due to dysfunction of the ascending reticular system of the brainstem;
3. **hyperactivation of the frontal lobe**—a patient might have hyperactive frontal lobes which according to the inverted U-law produces poor performance on executive functions associated with a high level of distractibility;
4. **prefrontostriatothalamic dysfunction** ("cool" dysfunction)—a patient might have dysfunction of the prefrontostriatothalamic system due to structural, functional abnormality including increase of dopamine reuptake transporters in the striatum;
5. **limbic system dysfunction** ("hot" dysfunction)—a patient might have dysfunctioning of the affective system which may produce anxiety, emotional instability, and hyperactivity;
6. **maturation lag**—a patient might have a lag in brain maturation such that older children behave as younger ones;
7. **default mode dysfunction**—a patient might have the default mode system of the brain dysfunctional such that it interferes with task systems and produces large intraindividual variability in performance reflected in increased reaction time variability;
8. **delay aversion**—the choice of smaller immediate reward over large delayed reward in ADHD is motivated by the desire to escape from delay to avoid the negative emotional states associated with delay.

THESES TO TEST

A primary question to ask in ADHD objective assessment is: **does the patient have any of the previously mentioned brain dysfunctions?** The evidence presented in the previous chapters of the present book enables us to claim that recording multichannel EEG in two resting state (eyes open and eyes closed) and in one task condition (in the case of ADHD the best option is the cued GO/NOGO task) can answer (at least partly) this question. Each of the eight hypothetical suggestions listed earlier can be tested separately with specific methods adjusted for this particular working hypothesis. So, when the diagnostic category is approximately defined and the potential reasons for this categorical disease are formulated, each of the possible causes can be tested as a separate **working hypothesis.**

PROGNOSTIC POWER

Knowing the cause of a particular behavioral pattern helps not only in making the correct diagnosis but also in providing the prognosis of how to treat the cause. Depending on the biological source of the behavioral pattern the psychiatrist has in his/her arsenal the following treatment procedures:

1. medication using **dopamine reuptake inhibitors**
2. medication using **noradrenalin reuptake inhibitors**
3. medication using **GABA agonists** which may "shut down" the cortical focus
4. **neurofeedback**
5. **tDCS**
6. **TMS**.

Answering the question about the course of the behavioral pattern in a particular individual will enable the psychiatrist to select the appropriate treatment protocol.

CHAPTER
6.2

Technical Implementation

ARRANGEMENT OF THE WORKING SPACE

In contrast with the 20th century psychiatrist whose working space was a room with two armchairs (one for the patient and one for the psychiatrist), the working space of the 21st century psychiatrist may look like that presented in Fig. 6.2.1. It includes the **EEG/ERP system for recording spontaneous EEG and event-related potentials**.

The electrical activity of the patient's brain is recorded during the first visit of the psychiatrist and the recorded EEG data are sent to the **report service center**. The report made by an expert is further sent to the psychiatrist who uses it for making a decision about individual treatment. The possible content of the report depends on the goals of the psychiatrist. If, for example, the goal is to define the source of an inattention/hyperactivity behavioral pattern, the content of the report is discussed as given in the following sections.

There are many EEG/ERP systems in the world. However, there are not many QEEG databases and just a **few ERP databases**.

QEEG/ERP DATABASES

The first commercially available normative database was developed in the 1970s and 1980s by **Roy John** and coworkers from the University of New York. The term "neurometrics" was first used by this group to describe an analogy to psychometric assessment, commonly used in clinical psychology. The Neurometrics database uses the following EEG features: absolute power, relative power, coherence mean frequency within band, and symmetry (left–right and front–back) extracted from approximately 2 min of EEG recorded under the eyes-closed condition and selected for being minimally contaminated by artifact. The analyzed EEG frequency range extends from 0.5 to 25 Hz. The **NxLink** database includes measures from 782 "normal" individuals aged between 6 and 90 years. The limitations of the database are as follows: (1) only recordings made with the eyes closed at rest were analyzed and normed, (2) comparisons were made

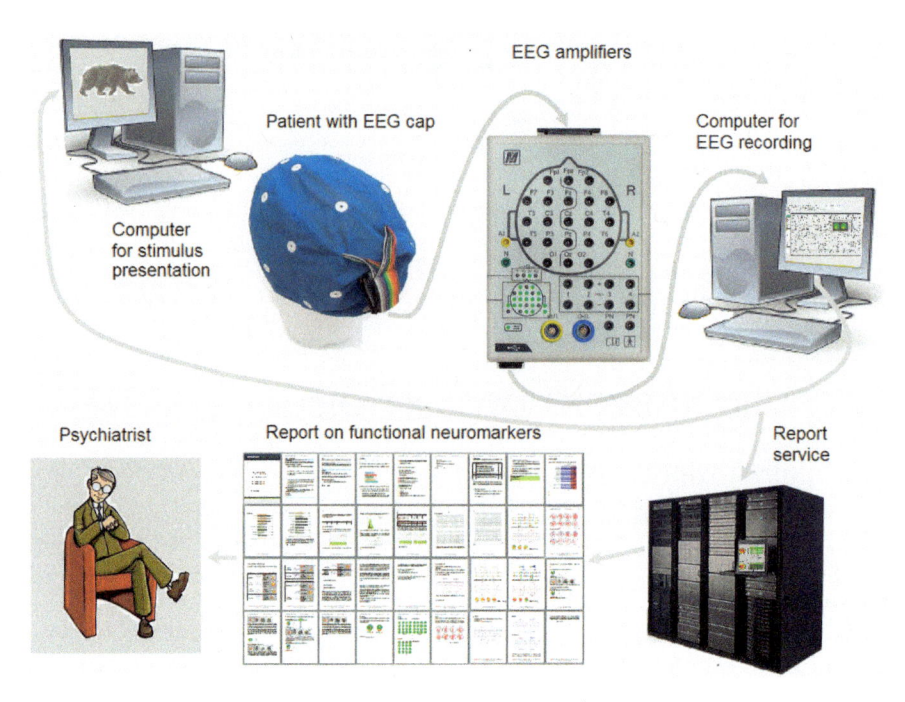

FIGURE 6.2.1 **Working space of the biologically oriented psychiatrist.** The psychiatrist orders the assessment of functional neuromarkers which includes (a) recording of EEG of a patient under resting state and task conditions; (b) processing the data; (c) comparing the results of processing with the reference database; and (d) compiling the report.

only with banded EEG characteristics such as delta, theta, alpha, and low-frequency beta bands.

The **Neuroguide** database developed by **Robert Thatcher** and co-workers is widely used in the field of neurofeedback. The database includes QEEG data from 625 healthy individuals, covering the age range 2 months to 82.6 years. Nine hundred and forty-three variables are computed for each subject including measures of absolute and relative power, coherence, phase, asymmetry, and power ratios. Z-score transforms are available in single-hertz bins. Sliding averages are used to compute age-appropriate norms. The principles of the database and its application for neurofeedback are presented in the 2012 book by R. Thatcher. The main limitation of the database is the absence of recordings in task conditions.

The first standardized **international brain database** was developed by the **Brain Resource Company** (BRC). It overcomes the limitations of the previously mentioned databases. A consortium of leading neuroscientists was consulted to come up with the optimal choice of tests to assess the brain's major networks in the shortest time. Six sites were set up with identical equipment and software under the auspices of the BRC. Hundreds of normative subjects and patients from different diagnostic categories were acquired. The BRC database involves data collection not only of EEG and ERP parameters in a battery of tasks (such as the auditory and visual oddball tasks), but also a comprehensive psychological test battery undertaken using a touchscreen monitor. Structural and functional MRI data are also obtained for many selected individuals. Further, genetic information will be systematically collected for comparison with neuroanatomical, neurophysiological, and psychometric measures. The principles of the database are presented in a 2005 paper by Evian Gordon and coworkers.

The present book is based on application of the HBI database. The principles of the database are described in Kropotov (2009). The current state of the database is described in the Introduction of the book.

ARSENAL OF THE 21ST CENTURY PSYCHIATRIST

In contrast with the 20th century psychiatrist the modern psychiatrist has many treatment options which include **electrical modulation** of brain neuronal networks. The options are listed in Fig. 6.2.2. The list is not complete and will definitely be expanded in the future.

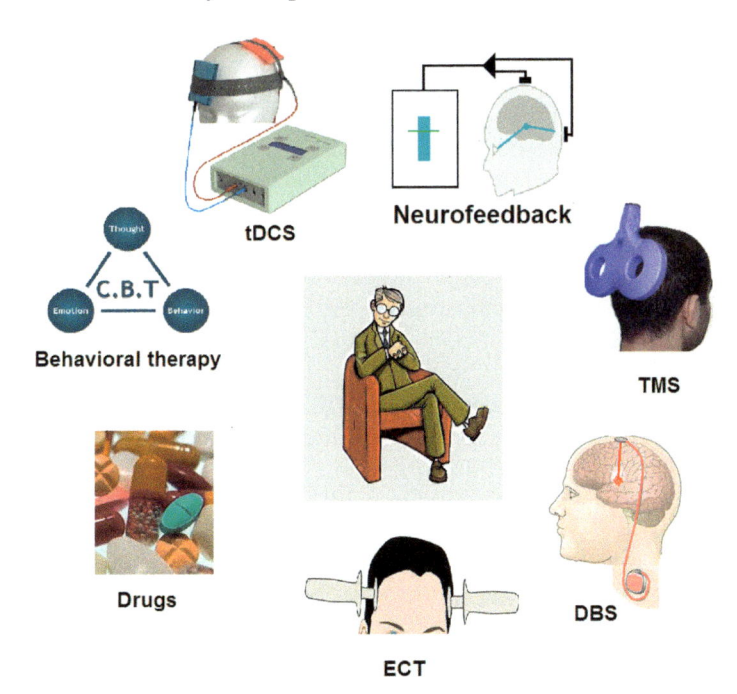

FIGURE 6.2.2 Treatment methods available for the biologically oriented psychiatrist. *tDCS*, transcranial direct current stimulation. *TMS*, transcranial magnetic stimulation; *ECT*, electro convulsive therapy; *DBS*, deep brain stimulation.

SELECTING THE BEHAVIORAL PARADIGM

As shown in Part 3 of the present book, assessing a certain brain system such as the sensory, executive, memory, or affective system, requires a specific set of behavioral paradigms involving the main operations in these systems. As far as ADHD is concerned, this dysfunction involves attention and cognitive control brain circuits. The 10-year experience of our laboratory and the review of the literature in chapter: Attention Deficit Hyperactivity Disorder enable us to conclude that the **cued GO/NOGO task** is one of best ERP behavioral approaches for assessing brain dysfunctions in ADHD.

CORRECTING ARTIFACTS

EEG is contaminated by various **nonbrain sources** of electrical potentials—**artifacts**. Eye movement artifacts such as blinking and saccades are quite often the largest. They can be orders of magnitude larger than the EEG and can propagate across much of the scalp, masking and distorting brain signals. Eye movements are frequent and vigorous because the patient is constantly stimulated by being presented with visual images such as in the cued GO/NOGO task.

The cornea of the eye is electrically positive relative to the back of the eye, it is not affected by the presence or absence of light, is considered a resting potential, and is in the range 0.4–1.0 mV. This source behaves as if it were **a single dipole oriented from the cornea to the retina**. Eye blink results in reflexive upward vertical eye movement that produces positive deflection at frontal areas with maximum at the Fp1, Fp2 electrodes. Horizontal eye movements (saccades) produce positive–negative fluctuation of potentials at the F7, F8 electrodes. Examples of topographies for these two types of artifacts are presented in Fig. 6.2.3.

Eye blink artifacts are corrected by zeroing the activation curves of individual independent components corresponding to blinking. These components are obtained by application of ICA to raw EEG fragments. The method was previously described by Ricardo Vigário from Helsinki University of Technology in 1997.

Besides the eye movement artifact, EEGs may be contaminated by mechanical movements of electrodes, poor electrode contacts, and some other uncontrollable artifacts. So in addition to the eye movement correction procedure, epochs with excessive amplitude of filtered EEG and/or excessive faster and/or slower frequency activity are automatically marked and excluded from further analysis.

In the HBI database software, exclusion thresholds are set as follows:

1. 100 μV for filtered EEGs
2. 50 μV for slow waves in the 0- to 1-Hz band
3. 35 μV for fast waves filtered in the 20- to 35-Hz band (Fig. 6.2.4).

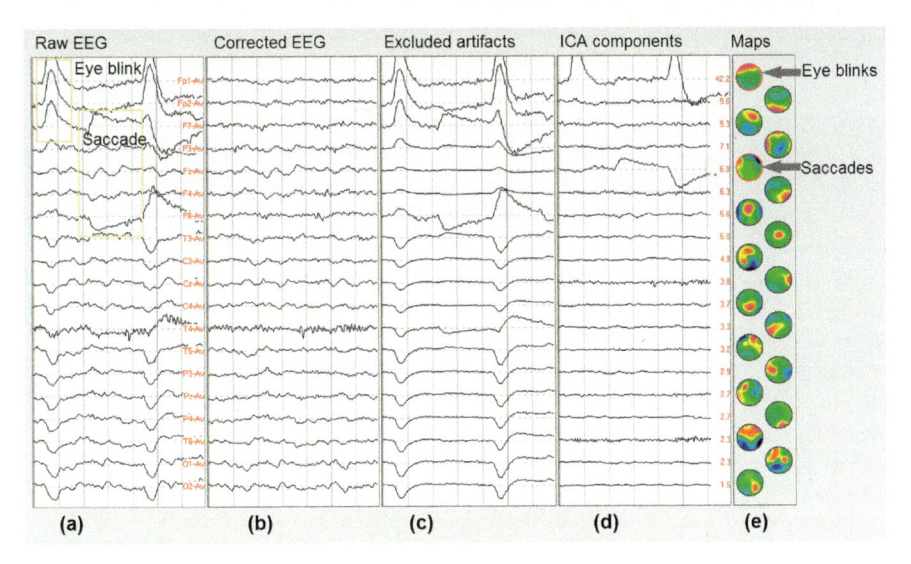

FIGURE 6.2.3 **Eye movement correction by spatial filtration.** The raw EEG in the resting state with the eyes open (a) is decomposed by ICA with the topographic maps presented (e) and timecourses presented in (d). Maps of the components corresponding to eye blinks and saccades are selected (e) and corresponding components (c) are excluded from the raw EEG to obtain the corrected EEG (b). *From HBI software.*

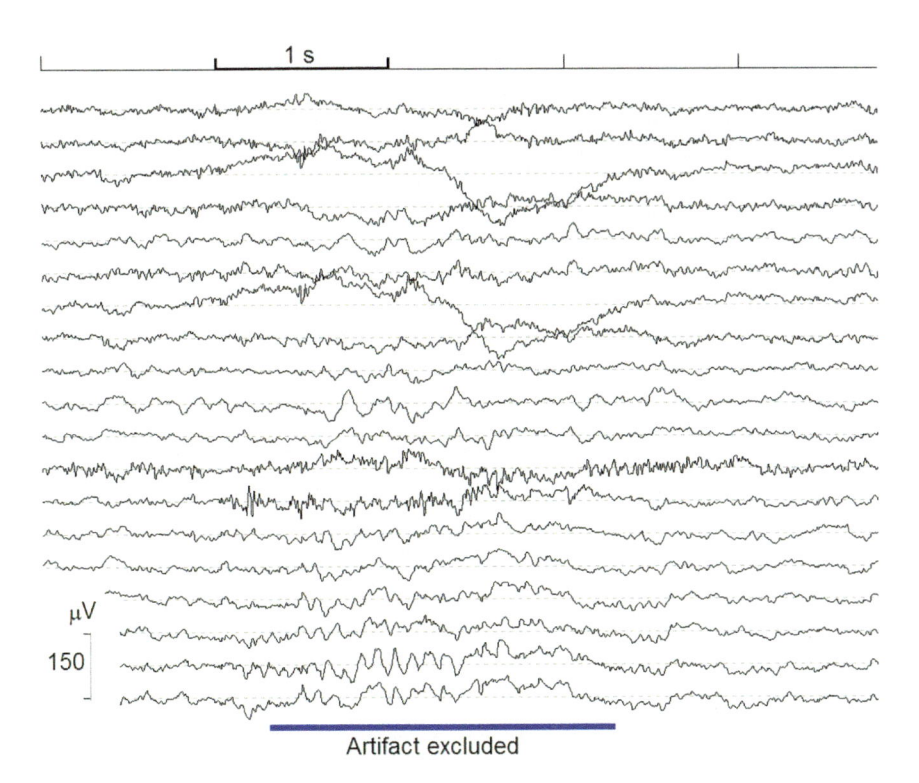

FIGURE 6.2.4 **Removing artifacts by setting thresholds for amplitude/spectral parameters of EEG.** *From HBI software.*

CHAPTER

6.3

Testing Working Hypotheses: Spontaneous EEG

Cortical self-regulation (see Part 2) may be impaired in the diseased brain. Since normal mechanisms of cortical self-regulation are reflected in different types of **brain oscillations**, different types of EEG **dysrhythmia** are a priori expected in brain disorders. ADHD is characterized by a specific pattern of cortical dysregulation (chapter: Attention Deficit Hyperactivity Disorder). The chapter describes the steps of analysis of spontaneous EEG oscillations in patients with symptoms of ADHD in an attempt to answer some of the questions regarding the possible cause of the observed behavioral pattern from the point of view of cortical dysregulation.

ROLANDIC SPIKES

As we learned from chapter: Attention Deficit Hyperactivity Disorder a **focus over the Rolandic fissure** is considered a risk factor of ADHD. The focus may disturb normal information processing in **premotor areas** of the cortex and may produce symptoms of hyperactivity. Indeed, about 6% of children with ADHD without epilepsy show Rolandic spikes in their EEG. Children with Rolandic spikes tended to exhibit more hyperactive–impulsive symptoms.

Spikes in raw EEG can be detected by visual inspection of raw EEG. There are three characteristics that define a spike in EEG. They are (1) paroxysmal character, (2) high degree of sharpness, and (3) short duration. The paroxysmal character of a spike is associated with its large amplitude: the spike pops up from background activity. The amplitude of a spike in turn depends on several factors: (1) position of the focus within the brain with lower amplitudes generated by deeper sources; (2) the size of the focus with lower amplitudes generated by smaller foci; and (3) orientation of the electric dipole corresponding to the spike with tangential dipoles generating smaller amplitudes than radially oriented dipoles. The degree of spike sharpness can be estimated by the second derivative of the EEG signal.

These and some other parameters of spikes can be used for automatic spike detection. This procedure is implemented in HBI software. The results of application of the spike detection procedure for resting-state EEG in a 7-year-old ADHD patient are presented in Fig. 6.3.1.

Theoretically, there are at least two options in treating this subtype of ADHD. They are **anticonvulsive medication and neurofeedback**.

In a 2012 study by Nira Schneebaum-Sender and coworkers from Dana Children's Hospital in Tel Aviv the efficacy of antiepileptic drugs in patients with Rolandic (centrotemporal) spikes was assessed. The authors claimed that a reduction of pathologic activity in some patients was accompanied by improvement in attention; however, this improvement was either temporary or not significant enough to discontinue methylphenidate.

At least two recent reviews showed the efficacy of neurofeedback in epilepsy. One was conducted by Barry Sterman from the School of Medicine, UCLA and Tobias Egner from Columbia University in 2006, and the other was done by Gabriel Tan and coworkers from the Michael E. DeBakey Veterans Affairs Hospital and Baylor College of Medicine in 2009. Both reviews showed that **training sensorimotor rhythm (SMR) over the Rolandic fissure** (or mu-rhythm in the terminology of the book) represents a viable alternative to anticonvulsant pharmacotherapy.

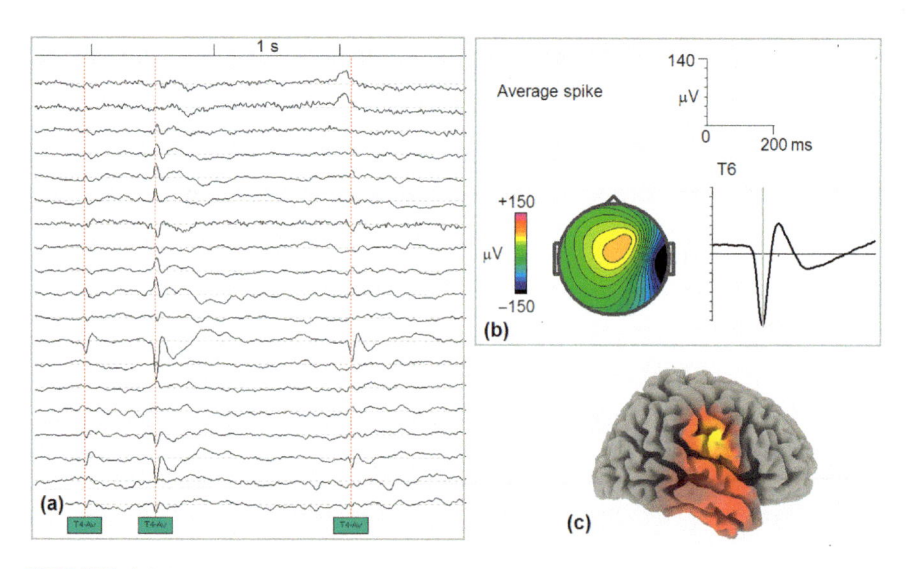

FIGURE 6.3.1 **Rolandic spikes in ADHD.** EEG was recorded under the eyes-open condition of a 7-year-old patient with diagnosis of ADHD. (a) Spikes are automatically detected in a fragment of raw EEG; (b) topography and time dynamics of the average spike; and (c) sLORETA image of the topography.

A 2012 case study by Maria Pachalska and coworkers from Andrzej Frycz Modrzewski Cracow University, Cracow, Poland showed that neurofeedback training in an ADHD child with Rolandic spikes improved his cognitive performance accompanied by significant decrease in the number of spikes.

EXCESSIVE THETA/BETA RATIO

Around 30% of ADHD children and adolescents show an elevated theta/beta ratio (TBR). This proportion is substantially larger than that of ADHD patients with Rolandic spikes. Decomposition of EEG in these patients with excessive TBR quite often shows the source of the elevated theta activity at the supplementary motor cortex (Brodmann area, BA 6) (Fig. 6.3.2d).

Fig. 6.3.2 demonstrates the case of an ADHD patient with excessive (in comparison with healthy controls) TBR (Fig. 6.3.2c). In this case, excessive

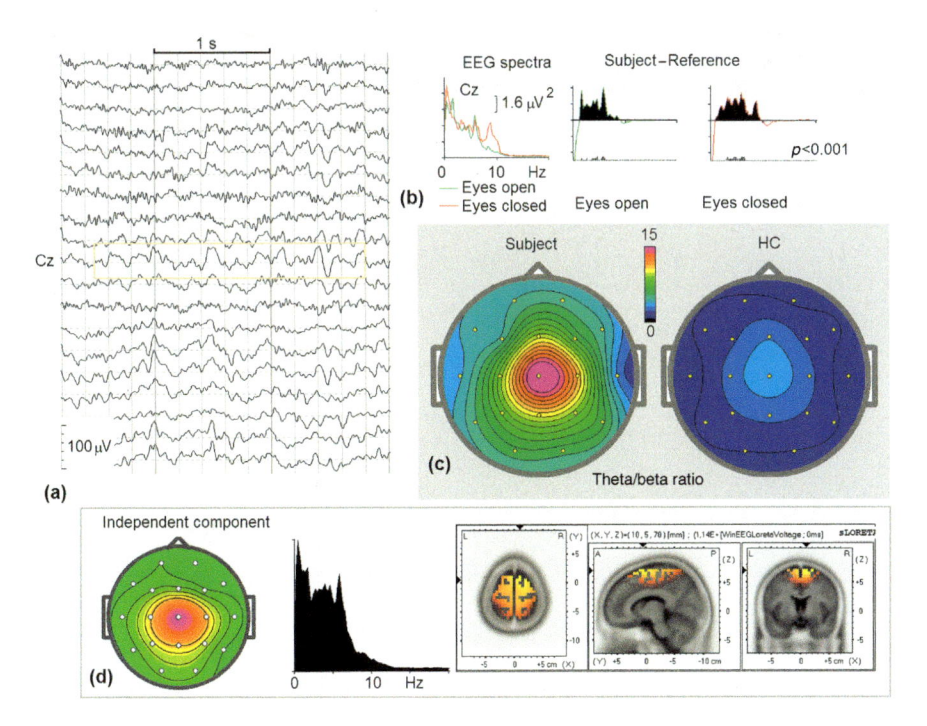

FIGURE 6.3.2 Elevated TBR in ADHD. EEG was recorded in the resting-state condition with the eyes open (a) and the eyes closed in an ADHD adolescent. EEG spectra were computed (b, left) and compared with the HBIdb (b, right). Note the excessive theta activity at Cz. The TBR measured at Cz was twice as high in the patient in comparison with the average of the healthy group both in eyes open (c) and eyes-closed (not shown) conditions. Independent component corresponding to the excessive theta was extracted by ICA. (d) Topography, spectra, and sLORETA image of the independent component.

TBR is associated with elevated theta activity at central electrodes (Fig. 6.3.2b). In this particular case "EEG slowing" at Cz can even be seen with the naked eye in the EEG record (Fig. 6.3.2a); however, in the majority of cases comparison with the database is needed. Note that in some cases excessive TBR is associated with decreased beta activity.

The experience of our team shows that ADHD patients with elevated TBR respond positively to theta/beta training or relative beta neurofeedback training. These patients also respond to psychostimulants.

EXCESS OF FRONTAL BETA ACTIVITY

About 20% of the whole ADHD sample show **elevated frontal beta activity**. As shown above, frontal beta activity increases with task load in healthy subjects while ADHD children show lower levels of beta synchronization in cognitive tasks than healthy controls.

An example of this QEEG subtype of ADHD is presented in Fig. 6.3.3. One can see a burst of low beta rhythm at the frontal electrodes (Fig. 6.3.3a). This beta rhythmicity is seen as a peak on the EEG spectrum and is significantly higher than in the averaged spectra of healthy controls (Fig. 6.3.3b). The independent component extracted from the spontaneous EEG and

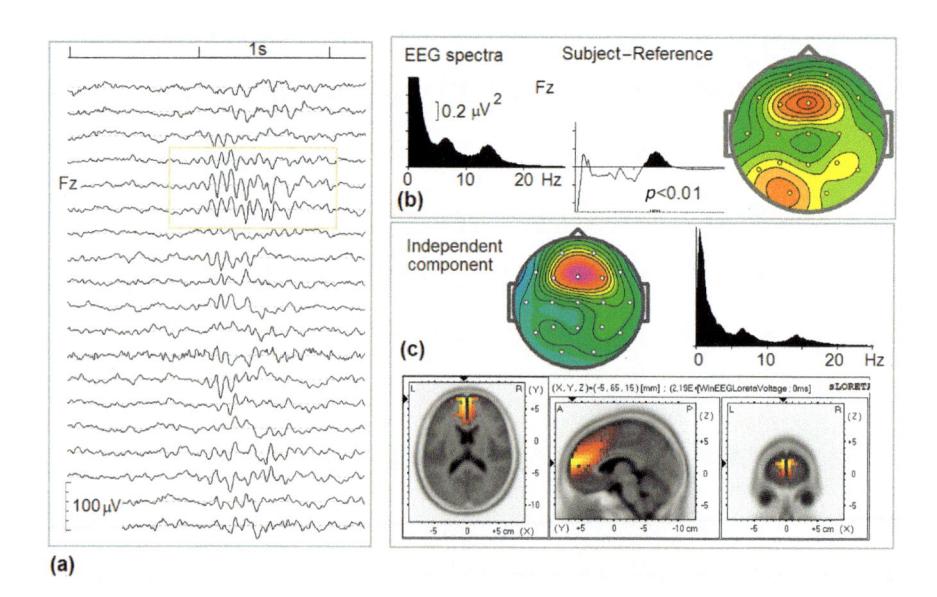

FIGURE 6.3.3 **Elevated frontal low beta activity in ADHD.** (a) EEG fragment in eyes-open condition in a 14-year-old ADHD patient; (b, left to right) EEG spectrum at Fz, deviation from the reference, and spectra of the difference at 14 Hz; (c) independent component corresponding to the excessive beta: topography, spectra, and sLORETA image.

corresponding to excessive beta rhythm is generated in the prefrontal cortex according to sLORETA (Fig. 6.3.3c).

The excessive frontal beta subtype of ADHD was specifically studied by **Adam Clark** and coworkers from the University of Wollongong. The studies of this team, presented in consecutive 2001, 2003, 2007, and 2013 papers, demonstrated that the excessive frontal beta ADHD subtype responded to stimulant medications without any significant changes in EEG. Importantly, it was also shown that this subtype was not associated with hyperarousal as previously suggested. It does not exclude the hyperactivation pattern of the frontal lobes in this subtype though. Recall the inverted U-law according to which overactivation of the frontal lobes may lead to poor performance in inhibitory control rather than in tasks requiring sustained attention. This fits the observation by Clark and coworkers that the elevated beta subtype is associated with the impulsivity and/or hyperactivity aspects of the disorder but not with inattention.

As shown in chapter: Beta and Gamma Rhythms, beta rhythms come in various overlapping ranges and are associated with different functions. In some rare cases of ADHD taken from the HBI database, excessive frontal beta appeared at high frequency (21–30 Hz) (Fig. 6.3.4). The medication effect, particularly of the use of sedatives, is excluded in this case. One can see spindles of high beta rhythms in frontal leads (Fig. 6.3.3a). EEG spectra

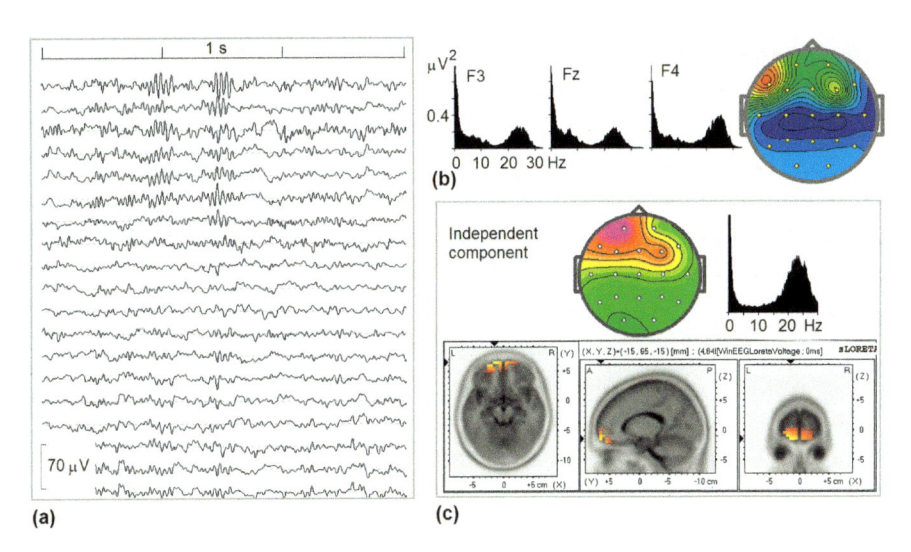

FIGURE 6.3.4 **Elevated frontal high beta activity in ADHD.** EEG was recorded under the resting-state condition with the eyes open (a) and eyes closed in an ADHD adult. EEG spectra were computed (b, left). The map of beta activity in the frequency band 20–30 Hz is presented in (b, right). Note the high-amplitude beta activity frontally which deviates from the reference data significantly at $p < 0.001$. Independent component corresponding to the excessive beta was extracted by ICA. (c) Topography, spectra, and sLORETA image of the independent component.

show clear peaks around 25 Hz (Fig. 6.3.3b) while the independent component corresponding to the spindling beta is generated in the superior frontal gyrus according to sLORETA.

According to a 2005 paper by Jack Johnstone, Jay Gunkelman from Q-metrix and J. Lunt, the spindling excessive beta QEEG endophenotype is probably best considered a nonspecific sign of dysfunction or encephalopathy. The experience of the authors with **neurofeedback** shows that this pattern responds very poorly to any higher frequency beta enhancement training, exacerbating the symptom complex. **Beta suppression** directly in the area of dysfunction has shown good clinical response. The band of frequencies to be suppressed should be selected based on individual profiles—not by standard bands.

According to the experience of our team, another neurofeedback option for the excessive beta subtype is any relaxation protocol including the protocol of **uptraining the mu-rhythm**. Recall that the mu-rhythm is under control of the premotor and motor cortical areas.

EXCESSIVE FRONTAL MIDLINE THETA RHYTHM

Excessive **frontal midline theta** is found in relatively small (less than 10%) subgroup of ADHD. When found the theta rhythm at Fz appears in **long (up to 10 s) bursts** which are in strong contrast with the short bursts of the healthy frontal midline theta rhythm (Fig. 6.3.5a). The frequency of

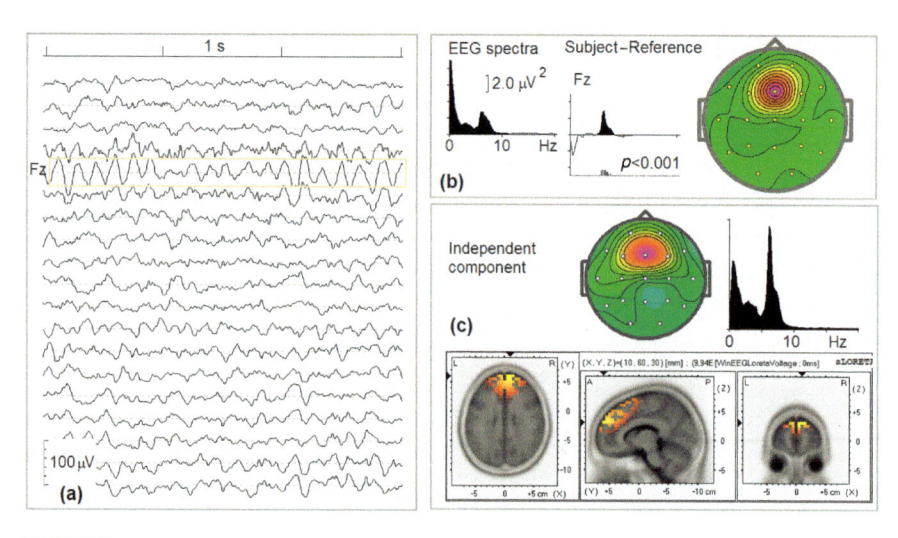

FIGURE 6.3.5 Elevated frontal midline theta rhythm in ADHD. (a) EEG fragment under the eyes-open condition in a 10-year-old ADHD patient; (b, left to right) EEG spectrum at Fz, deviation from the reference of the spectra in eyes open and map of the subject–reference difference at 6 Hz; (c) independent component corresponding to the excessive frontal midline theta rhythm: topography, spectra, and sLORETA image.

such excessive theta rhythms is around **6 Hz** and the topography shows a sharp maximum at **Fz** (Fig. 6.3.5b). Generators of excessive theta rhythms according to sLORETA are usually located in the medial frontal/dorsal part of the anterior cingulate cortex (Fig. 6.3.5c).

Agatha Lenartowicz and coworkers from the University of California San Diego in a 2014 study separated the frontal midline theta component in EEG by independent component analysis and demonstrated its enhancement in ADHD during the working memory task.

This QEEG subtype of ADHD is associated with **emotional dysregulation**, that is, the inability to properly modulate and regulate emotions. The experience of the HBI team shows that this group of patients might respond to downtraining of the theta rhythm at Fz. The group also responds well to **behavioral therapy**.

EXCESSIVE ALPHA ACTIVITY

In contrast with the frontal midline theta subtype, the alpha subtype is more commonly found in ADHD. It is characterized by excessive alpha rhythms in posterior parts of the cortex. In contrast with healthy subjects, this rhythm **is not fully suppressed by opening the eyes** and is clearly seen in the raw spontaneous EEG under the eyes-open condition not only at posterior electrodes but also at **frontal electrodes** (Fig. 6.3.6a). In EEG spectra at the frontal electrodes the rhythm is seen as a sharp peak in the

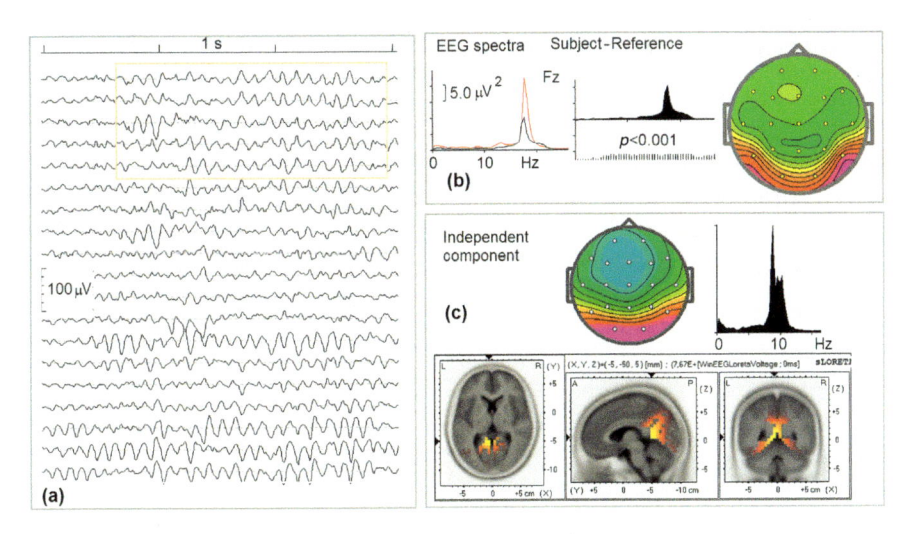

FIGURE 6.3.6 **Elevated alpha activity in ADHD.** (a) EEG fragment in eyes-open condition in an adult ADHD patient; (b, left to right) EEG spectrum at Fz in eye-open *(black)* and eyes-closed *(red)* conditions, deviation from the reference, and map of the difference subject–reference at 9 Hz; (c) independent component corresponding to the excessive frontal midline theta rhythm: topography, spectra, and sLORETA image.

alpha frequency band (Fig. 6.3.6b). Nevertheless, sLORETA images consistently show **generators in posterior parts of the brain** (Fig. 6.3.6c).

Jack Johnstone and coworkers in their 2006 paper named this QEEG endophenotype as "persistent alpha—with eyes open" and associated it with underarousal. According to the experience of the team, **downtraining of the posterior alpha rhythm** at the individual frequency in combination with beta enhancement is considered a neurofeedback option for this ADHD subtype. This conclusion fits the experience of the HBI team. In general, all techniques directed for elevation of the arousal level including drinking coffee are beneficial for the patient.

INDIVIDUAL INDEPENDENT COMPONENTS FOR NEUROMODULATION PROTOCOLS

When the EEG spectra of the tested subject are compared with the grand average of healthy controls, the statistically significant deviations ($p < 0.05$) from the reference are taken into account at the beginning of assessment. These statistically significant deviations must be present in all resting (eyes open, eyes closed) and in a task (eg, the cued GO/NOGO task) conditions. Moreover, these deviations must explain the subject's complaints either on the basis of the clinical–physiological evidence obtained in the previous studies of the same diagnostic category or on the basis of the functional meaning of the deviations known from basic neuroscience.

Figs. 6.3.2–6.3.6 illustrate the first situation, when the EEG spectra deviations found in different subgroups of ADHD are supported by numerous studies in ADHD population. In all these cases independent components corresponding to the clinically relevant deviations from the reference can be extracted from spontaneous EEG. The topographies of these independent components can be used for defining electrodes placement in neurofeedback and tDCS protocols.

The case presented on Fig. 6.3.7 illustrates the second situation when no a priori information is available and the user has to rely on the basic information known from neuroscience literature. This is a case of a 7 year old boy with symptoms of ADHD and autistic spectrum disorder (ASD). It should be stressed here that nowadays there is almost no research on cooccurring ADHD and ASD because of exclusion criteria of DSM-IV. The new DSM 5 allows a dual diagnosis and enables the researchers to study this broader phenotype in future, but at the moment we need to relay what is known in basic neuroscience.

As shown in Fig. 6.3.7a strong deviations from the reference in EEG spectra of this subject are observed over temporal areas with maximums at T5 and T6. The deviations are expressed in excessive theta rhythmicities (around 5Hz) at the left and right temporal areas. These rhythmicities

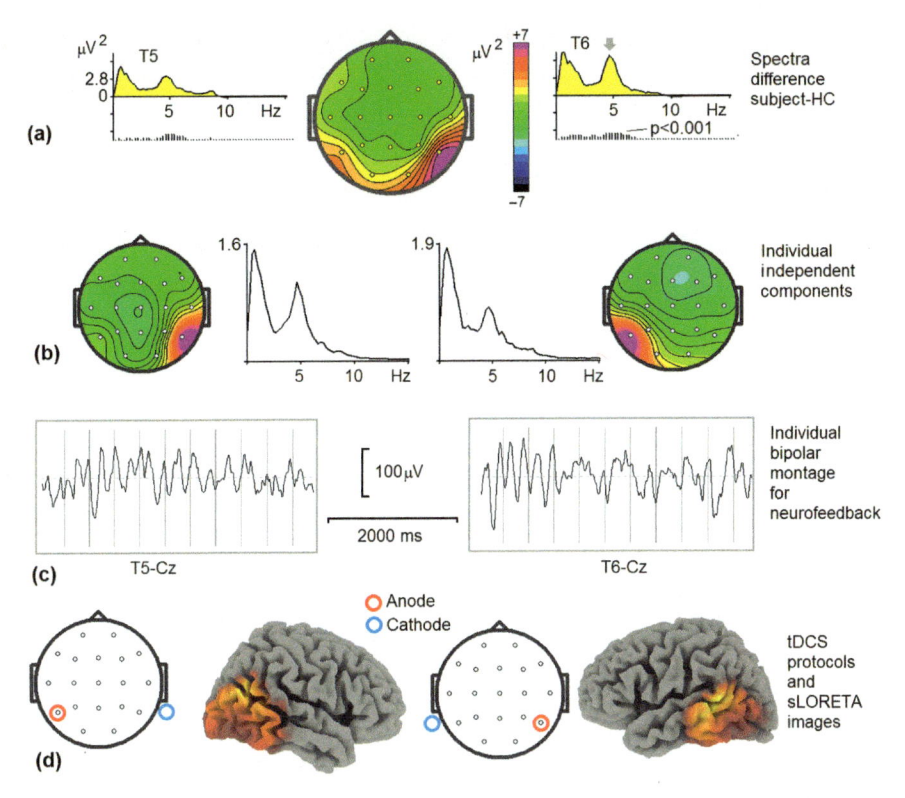

FIGURE 6.3.7 **Elevated temporal theta rhythmicity in a boy with symptoms of ADHD and autistic spectrum disorder.** (a) Difference spectra (individual-grand average of healthy controls—HC) at T5 and T6 electrodes with p-values of statistical significance and map (middle) at peak frequency of the difference *(indicated by gray arrow)*. (b) Maps and EEG spectra of two independent components extracted from the spontaneous EEG of the individual. (c) Two bipolar montages for neurofeedback protocols on the basis of the maps of individual independent components. (d) Location of anodal and cathodal electrodes for the individually tailored tDCS protocol depicted together with sLORETA images of the corresponding independent components. *Data from the HBI database.*

are extracted from the spontaneous EEG in two independent components (Fig. 6.3.7b). The topographies of the independent components allow us to construct neurofeedback protocols (Fig. 6.3.7c) and protocols of tDCS (Fig. 6.3.7d).

The basic neuroscience tells us that theta rhythmicity in the temporal areas must be considered as a sign of severe inhibition (functional suppression) of these parts of the cortex (see Fig. 2.2.13c). Recall that temporal areas of the ventral visual stream are responsible for extraction of object-based features from the visual images so that inhibition of this area might produce difficulties of visual recognition and, as consequence, difficulties in face recognition—a hallmark of ASD.

CHAPTER
6.4

Testing Working Hypotheses: Event-Related Potentials

Information flow in neuronal networks of the brain (see Part 3) may be impaired in the diseased brain. Since normal mechanisms of information flow are reflected in different types of ERP waves and components, different types of ERP impairment are a priori expected in brain disorders. ADHD is characterized by a specific pattern of abnormal information flow in neuronal networks of cognitive control (chapter: Attention Deficit Hyperactivity Disorder). The chapter describes the steps taken for ERP analysis of a patient with symptoms of ADHD in an attempt to answer some of the questions regarding the possible cause of the observed behavioral pattern from the point of view of cortical information flow.

INDEPENDENCE FROM OTHER FUNCTIONAL NEUROMARKERS

Recall that there are at least **three types of functional neuromarkers**: (1) those that reflect spontaneous or induced EEG oscillations, (2) those that reflect behavioral parameters such as reaction time and reaction time variability, and (3) those that reflect the stages of information flow as measured in the ERP components (Fig. 6.4.1). These functional neuromarkers reflect quite different products of the brain functioning and theoretically are independent. Practically, some correlations between ERP components and behavioral parameters are observed in healthy subjects. However, these correlations quite often are weakened or even reversed in mental illness. The experience of the HBI team indicates that many psychiatric patients do not show any deviations of EEG spectra from healthy controls. Moreover, in some cases even behavioral parameters could be within normal limits.

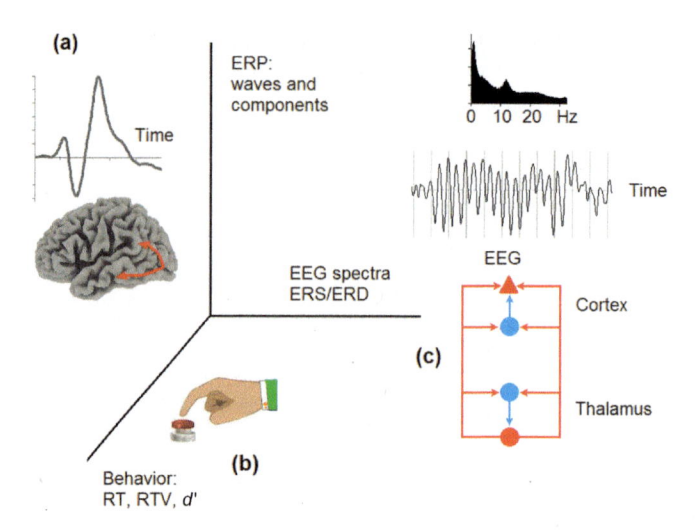

FIGURE 6.4.1 Three independent dimensions in the assessment of functional neuromarkers. (a) Event-related potentials *(ERPs)* reflect stages of information flow within neuronal networks of the brain. (b) Behavioral parameters—such as reaction time *(RT)*, RT variability *(RTV)*, and the d' index (sensitivity index)—are the final results of motor output. (c) EEG reflects mechanisms of cortical self-regulation.

SELECTIVE DEFICIT OF COGNITIVE CONTROL

As noticed previously, each diagnostic category of mental illness requires a special set of hypotheses to test. In case of ADHD in relation to ERPs the primary hypotheses would be impairment of operations of cognitive control. Recall from the chapter: Attention Deficit Hyperactivity Disorder that in ADHD group ERP differences from healthy controls are related to specific operations of proactive and reactive cognitive control. Two different localizations of the disturbed operations in ADHD can be separated: parietal and frontal. The parietal operations are associated with P3cue and P3GO conditions while the frontal operations are associated with P3 NOGO conditions. Recall, that the response to psychostimulants depends critically on the location of the operation which is particularly impaired in a given individual.

Two steps in the ERP assessment are separated:

1. Individual ERPs are compared with the grand averaged ERPs computed for a group of healthy subjects of the corresponding age. P-values of the statistical significance for the difference waves are computed and the candidates for neuromarkers (P3 cue, P3 GO, P3 NOGO, and CNV) are tested;
2. ERPs are decomposed into individual independent components by means of spatial filtration on the basis of group independent

components and the obtained components of cognitive control are compared with the averaged independent components computed for a group of healthy subjects of the corresponding age. Again, p-values of the statistical significance for the components' differences are computed and candidates for neuromarkers (P3 cue parietal, P3 cue temporal, CNV subcomponents, P3b GO, P3 NOGO early, P3 NOGO late, P3 novelty) are tested.

Fig. 6.4.2a and b illustrates the two steps of ERP assessment for a young man diagnosed with ADHD. The EEG spectra in this patient did not show any statistically significant deviations from the reference in all three conditions (eyes open, eyes closed, and during the cued GO/NOGO task). The theta/beta ratio was absolutely within normal limits. No differences in behavioral parameters were found: reaction time (RT) was 70 ms faster than the average of healthy controls, variance of RT was 20% smaller than the corresponding average of healthy controls, there were only 1% of omission errors, and no commission errors were made by the patient. So, the patient looks absolutely normal from the point of view of behavioral and EEG spectral functional neuromarkers.

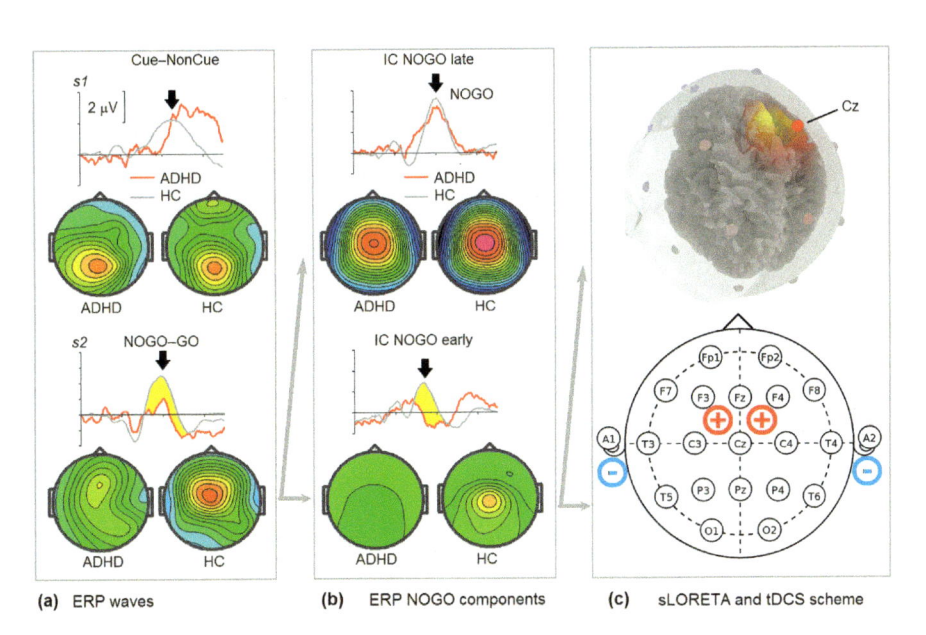

FIGURE 6.4.2 Selective impairment of the P3 NOGO early independent component (IC) in an individual ADHD. (a) ERP difference waves Cue–NonCue (top) and NOGO–GO (bottom) for a 20-year-old ADHD patient compared with the group average of healthy controls *(HC)*. Statistically significant differences are shown in *yellow*. (b) ERP NOGO (IC) for the ADHD patient compared with HC. (c) sLORETA image of the IC NOGO early (top) and a potential tDCS protocols.

In step 1 the ERP waves such as P3 cue, P3 NonCue, P3 GO, and P3 NOGO are compared with the grand averages of healthy controls of the corresponding age. Significant differences are found only for P3 NOGO wave. Fig. 6.4.2a illustrates the finding by depicting the ERP difference waves (Cue-NonCue and NOGO-GO). In sum, no impairment of the proactive cognitive control is found while the reactive cognitive control is disturbed.

In step 2 (Fig. 6.4.2b) the individual P3 NOGO wave is decomposed into two independent components by means of spatial filtration based on the group independent components. Note selective decrease of P3 NOGO early component and intact P3 NOGO late component. So, for this patient we can determine the specific decrease of the P3NOGO early component—an index of reactive cognitive control in ADHD.

According to sLORETA imaging the component is localized in the supplementary motor area. Recall from Chapter: Executive System and Cognitive Control that this area represents a cortical node of the complex basal ganglia–thalamo–cortical neuronal network responsible for the operation of inhibition/overriding in the cognitive control. TMS and tDCS are the potential neuromodulation techniques that could be used for activating this area in this particular case.

Information about the use of tDCS to modulate cognitive control in ADHD is limited. Paulo Boggio and coworkers from the University of Sao Paulo (Brazil) under the supervision of Felipe Fregni from Harvard Medical in a 2007 study showed that anodal stimulation of the left dorsolateral prefrontal cortex significantly improved task performance in the GO/NOGO task in patients with major depression (Boggio et al., 2007). We can presume that ADHD patients could also benefit from this tDCS protocol (anode, F3; cathode, over the left supraorbital area; current <2 mA).

According to the topography of P3 NOGO early component associated with action inhibition/overriding and impaired in the patient from Fig. 6.4.2 we could suggest two symmetrical protocols of placement of electrodes (Fig. 6.4.2c bottom). The first protocol (anode—between Fz and C3, cathode—the right mastoid) is intended to activate the left supplementary motor cortex. The second protocol (anode—between Fz and C4, cathode—the left mastoid) is intended to activate the right supplementary motor cortex.

CHAPTER
6.5

Monitoring Treatment Effects

As described in chapters: Spontaneous Electroencephalogram and Event-Related Potentials (ERPs), the QEEG/ERP parameters are rather **stable** with good-to-excellent test–retest reliability. The author of the book was recorded in the cued GO/NOGO task seven times during the last 10 years. Although the time of the recording, the emotional or physical state, and the laboratory equipment were not controlled, in all cases the ERPs were almost identical and the EEG spectra were similar. The stability of QEEG/ERP parameters opens an opportunity to use corresponding functional neuromarkers to monitor the effects of treatment.

PHARMACO-ELECTROENCEPHALOGRAPHY

The idea that QEEG may be applied in pharmacology for **monitoring the effect of pharmacological treatment** was born in the early 1960s. At the beginning the goal was quite modest: to find a new functionally oriented method that would classify drug effects as alternative to structural methods based on chemical similarities between substances.

In 1980 the International Pharmaco-EEG Society (IPEG), a nonprofit organization, was established. The society includes people involved in electrophysiological brain research in preclinical and clinical pharmacology, personalized medicine, and neurotoxicology. It has an official journal and holds biannual meetings.

The main result of research in the field of pharmaco-EEG is demonstration that different classes of pharmaceutical agents differently affect **spatiotemporal parameters (in the form of spectral maps) of background EEG.** However, individual QEEG profiles for different classes of drugs often overlap with each other due to large interindividual differences, sharing the effects of distinct classes of drugs, and the unspecific nature of spontaneous EEG.

The use of pharmaco-EEG in psychiatry was reviewed in a 2006 paper by Armida Mucci and coworkers from the University of Naples

(Italy). The authors focused on two attempts to transfer pharmaco- EEG methods to psychiatric clinic: (1) monitoring psychotropic drug toxicity, and (2) predicting clinical response to treatment with psychotropic drugs. However, in spite of progress pharmaco-EEG remained until recently an empirical method with a poor theoretical background and **without any significant effect** on the drug industry or on clinical application.

PHARMACO-EVENT RELATED POTENTIALS

Cases in which **ERPs** have been applied in psychopharmacology are quite rare and restricted mostly by N1/P2 and P3 waves in the auditory modality. For example, the serotoninergic system in depression was assessed by the loudness dependence of the auditory N1/P2 component (see chapter: Sensory Systems and Attention Modulation) and was used to predict the response to SSRI.

Recently, Geir Ogrim from the HBI team conducted a study to test whether changes in ERPs induced by a single-dose stimulant medication could predict long-term outcome (see chapter: Attention Deficit Hyperactivity Disorder). Based on data from daily ratings of parents and teachers during the 4-week trial, patients were divided into two groups: responders and non-Responders. Fig. 6.5.1 shows the ERPs of responders to GO stimuli (Fig. 6.5.1a) and NOGO stimuli (Fig. 6.5.1b) before (Pre) and after (Post) taking one dose medication. Note that only ERPs to **NOGO stimuli** show large difference between premedication and postmedication conditions. The effect size of the difference *post − pre* is quite large with an 80% increase in P3 NOGO amplitude after taking one dose of psychostimulant. sLORETA images made on the maps of difference waves (Fig. 6.5.1d) indicate generators in the medial prefrontal cortex (Fig. 6.5.1e). This example illustrates the **power of the ERP approach** in the pharmacological field, in general, and in the predicting response to psychostimulants, in particular.

NEUROFEEDBACK

There have not been many **systematic attempts** to study the **QEEG** changes induced by neurofeedback. **Linda and Michael Thompson** from the ADD Centre, Mississauga, (Canada) in a 2010 paper together with A. Ried reviewed their own experience of training 150 clients with Asperger's syndrome and 9 clients with Autistic Spectrum Disorder over a 15-year period (1993–2008) in a clinical setting. Together with improvement of symptoms and cognitive measures, they demonstrated a decrease in relevant EEG ratios such as the **theta/beta ratio**.

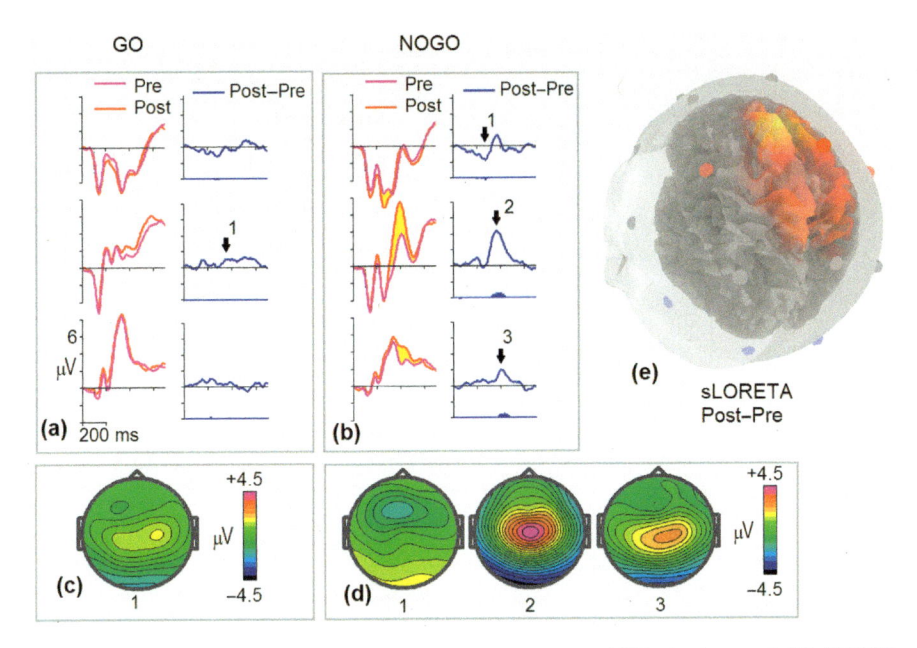

FIGURE 6.5.1 **Effect of stimulant medication on ERPs.** ERPs in the cued GO/NOGO task were recorded in 58 ADHD children (responders to psychostimulants) without medication before the trial (*pink, pre*) and 1 h after taking medication (*red, post*). (a) Pre- and post-ERPs and their differences post–pre for the GO condition with a map of the maximal difference (c); (b) pre- and post-ERPs and their differences post–pre for the NOGO condition with maps at three latencies marked by *arrows* (d); and (e) the most probable sLORETA image of difference for the NOGO condition. Note that in contrast with Fig. 5.1.16 the prestimulus base is corrected (Data from Geir Ogrim).

ERPs provide a good tool for monitoring changes in brain functioning induced by neurofeedback sessions. The procedure can be done at two levels: (1) the **group level** to assess the effect of neurofeedback training on a group of patients who underwent **the same neurofeedback protocol** during multiple sessions, (2) the **single-subject level** to assess the effect of neurofeedback training on a single subject who performed an **individually tailored protocol** for improvement of a certain brain dysfunction.

Fig. 6.5.2 illustrates the group comparison taken from our study of the effect of relative beta training on ERP correlates of cognitive control.

An example of application of ERPs for monitoring the effects of **an individually tailored neurofeedback protocol** is presented in Fig. 6.5.3. It illustrates the case of an ADHD 14-year-old boy with an **excessive frontal beta phenotype** who performed 20 sessions of relaxation protocol of **mu-rhythm training**. The ERPs in the cued GO/NOGO task were recorded pre and post neurofeedback training. One can see a clear increase in the P3 NOGO wave (Fig. 6.5.3b) but not in the P3 GO wave (Fig. 6.5.3a) after neurofeedback training.

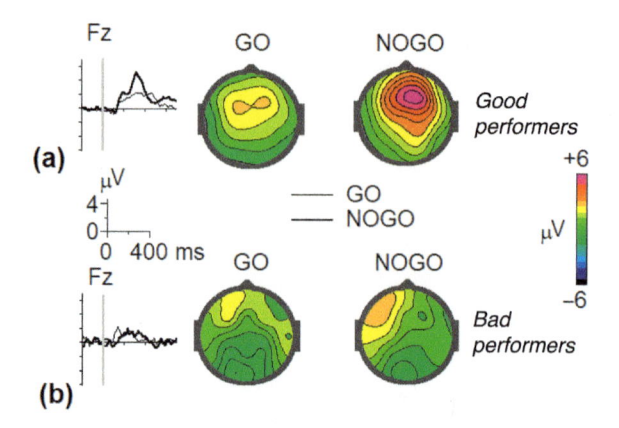

FIGURE 6.5.2 **Effect of 20 sessions of relative beta neurofeedback on ERPs.** ERPs in the auditory cued GO/NOGO task were recorded before *(Pre)* and after *(Post)* 20 sessions of relative beta training neurofeedback. On the basis of quality of performance during neurofeedback sessions the patients were divided into two groups: bad performers and good performers. (a) Post–pre ERP differences at Fz and separate maps for GO and NOGO stimuli for good performers; and (b) post–pre ERP differences at Fz and separate maps for GO and NOGO stimuli for bad performers. *Adapted with permission from Kropotov et al. (2005).*

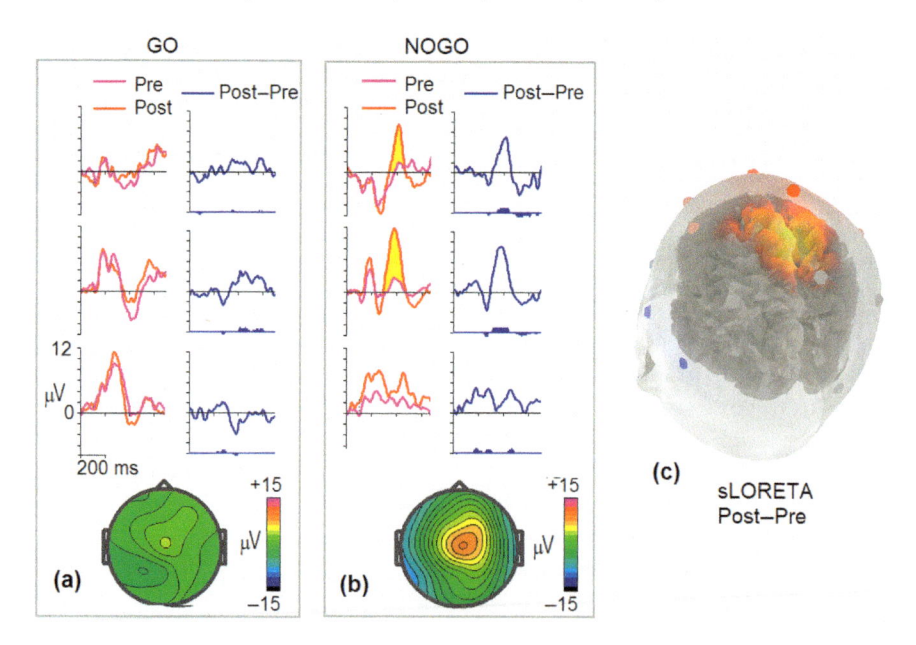

FIGURE 6.5.3 **Effect of 20 sessions of neurofeedback on individual ERPs.** ERPs in the cued GO/NOGO task were recorded in an ADHD adolescent before 20 sessions of neurofeedback *(pink, pre)* and after intervention *(red, post)*. (a) Pre- and post-ERPs and their differences post–pre for the GO condition with a map of the maximal difference (c); (b) pre- and post-ERPs and their differences post–pre for the NOGO condition with a map; and (c) the sLORETA image of difference for the NOGO condition.

One thing must be mentioned in conclusion. Examining changes in ERP induced by psychostimulants and different types of neurofeedback, we can see a common pattern: increase in the P3NOGO wave with maximum at Cz. Recall that this part of the P3NOGO wave is reflected in the P3 early component which in turn is associated with the energization function of the brain. This observation suggests that all the methods of neuromodulation described lead to the same output—enhancement of the energization function of the brain.

PART 7

THE STATE OF THE ART: OVERVIEW

CHAPTER
7.1

Objective Measures of Human Brain Functioning

Human brain functioning can be directly and objectively measured by neuroscience methods such as fMRI, PET, EEG, and ERPs (Fig. 7.1.1). Indirectly, human brain functioning is measured by behavioral indexes of speed and quality of performance such as reaction time, reaction time variability, omission, and commission errors.

In magnetic resonance imaging (MRI), a measured parameter is amplitude of radio waves reemitted by hydrogen protons of the brain located in a strong magnetic field (Fig. 7.1.1a, right). Functional MRI is developed for studies of vascular reactions of the brain tissue in response to different tasks. The primary contrast mechanism exploited for fMRI is blood oxygenation level dependent (BOLD) contrast. Task-specific BOLD signal changes are expressed as a small (0.5–5%) percentage of signal change in regional image intensity which slowly develops over 3–8 s following task initiation (Fig. 7.1.1b, right). Because of low temporal resolution, indirect relation to neuronal activity, and low intrasubject reliability the fMRI indexes are quite limited in psychiatric practice. PET is based on physical properties of isotopes to emit positrons. In the brain, neurons consume the radioactive substance or neuronal receptors bind the corresponding ligand. The radioactive substance, when it is accumulated in a certain area of the brain, emits positrons which emit two gamma-quantums when encounter with electrons. The source parameter for PET is number of such events. To restore the three-dimensional pattern of the radioactive substance density, mathematical reconstruction methods (similar to those used for MRI) are applied. However the spatial resolution of PET is significantly lower than that of the MRI. In PET many radioligands have been synthesized for studying receptor systems of the brain. The quantitative imaging for several receptors showed potential clinical importance.

EEG is electric field measured from scalp electrodes (Fig. 7.1.1a, left). The main feature of EEG is oscillatory nature of potential dynamics. The oscillations are seen at different frequency bands: infralow (0.01–0.1 Hz),

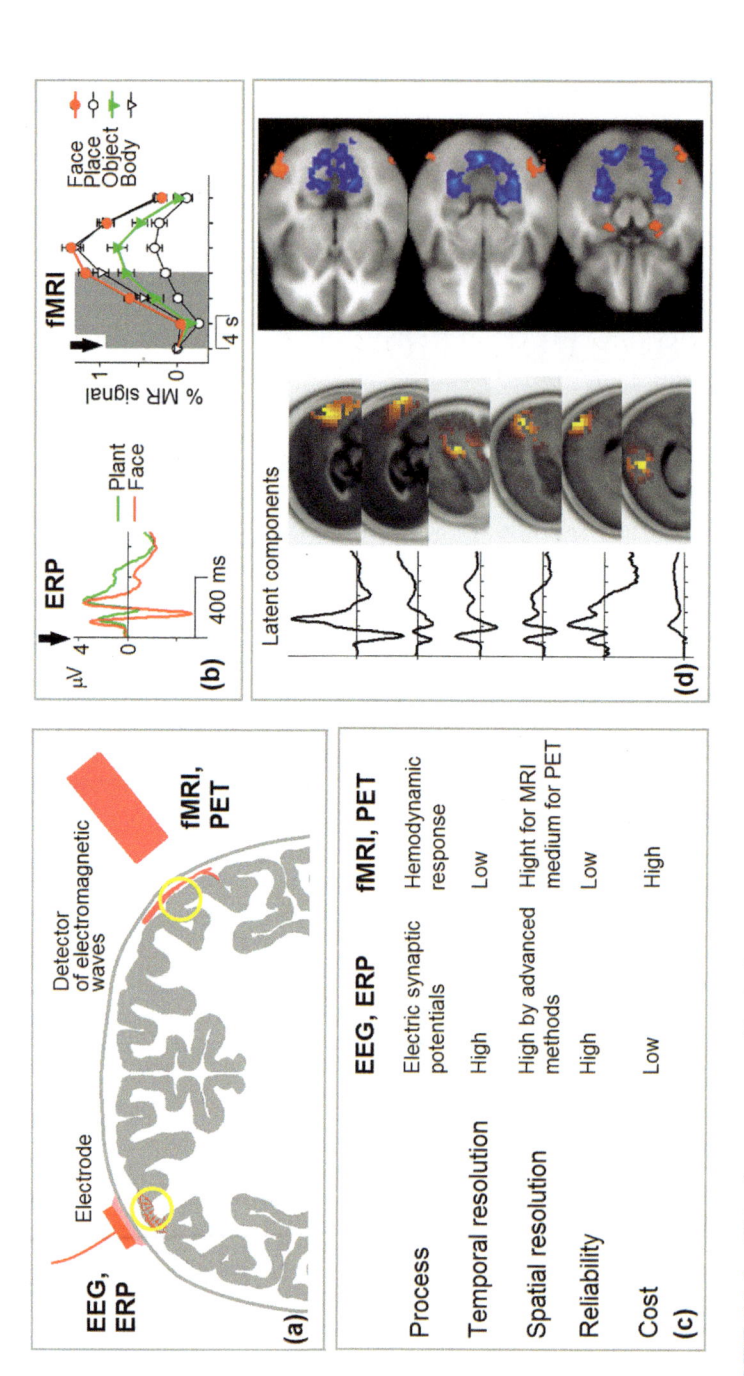

FIGURE 7.1.1 Objective measures of brain functioning. (a) Left: EEG and ERP are recorded by scalp electrodes and pick up averaged synaptic activity of cortical neurons. Right: fMRI and PET are recorded by detectors of electromagnetic waves and measure hemodynamic response of brain tissue under the detector. (b) Left: ERPs at T6 in response to a brief presentation of images of plants (*green*) and faces (*red*). Note time scale of 400 ms and increase of N170 wave in ERP to faces. Right: Delayed hemodynamic response of an area in the fusiform gyrus to a brief presentation of different categories of images including objects (*green*) and faces (*red*). Note time scale of 4 s and increase of hemodynamic response to faces. (c) Comparison of the basic properties of the measured parameters with the Left for EEG/ERP and the Right for fMRI/PET. (d) Left: decomposition of group ERPs in response to visual images of different categories into latent components for which operations of category discrimination and comparison to working memory are spatially and temporally segregated. Right: Category selectivity to intact and scrambled images in the ventral stream with face-selective (*red*) and place-selective (*blue*) activation areas. Parts b, d Left: adapted with permission from Kropotov and Ponomarev (2015). Parts b, d right: adapted with permission from Andrews et al. (2010).

low (0.1–1 Hz), delta (1–4 Hz), theta (4–8 Hz), alpha (8–12 Hz), and beta (>13 Hz) bands. The generators of these rhythms are neuronal networks with different membrane and synaptic mechanisms based on interplay between excitatory and inhibitory processes. This interplay leads to oscillations which in turn control the neuronal network functioning through voltage gated channels. Methods of extracting and compressing the information about rhythmicity and its dynamics are given by Fourier and wavelet analyses. EEG as electrical field is volume-conducted so that a given single current dipole generates a distribution of positive and negative potentials on the whole scalp. This feature is often used as an indicator of low spatial resolution of EEG. However, application of distributed source models (such as LORETA) and blind source separation methods (such as independent component analysis) improves spatial resolution of EEG substantially.

Event-related potentials (ERPs) are scalp recorded voltage fluctuations that are time-locked to an event. Depending on the nature of functional task, ERPs reflect stages of information processing in the sensory-related hierarchical neuronal networks, in the networks of cognitive control as well as in the memory and affective systems of the brain. The ERP amplitude is usually smaller than the amplitude of background EEG so that the reliable ERP is obtained by averaging EEG fragments in multiple trials. Each ERP wave represents a sum of potentials generated in widely distributed cortical sources. The functionally distinct sources of ERPs are called components and are associated with discrete hypothetical psychological operations (Fig. 7.1.1d, left). Recently, methods of blind source separation have been successfully applied in ERP research for component extraction. Many ERP components are quite reliable measures of brain functioning with good to excellent test–retest reliability. Although ERPs are quite stable in time they vary substantially from subject to subject reflecting interindividual fundamental differences in information processing.

To sum up, EEG, ERP on one hand and fMRI, PET on the other hand reflect quite different parameters of brain functioning (Fig. 7.1.1c). The parameters of EEG/ERP reflect averaged fast synaptic potentials of cortical neurons and, as consequence, show high temporal resolution and high test–retest reliability. The parameters of fMRI/PET reflect slow hemodynamic responses and show low temporal resolution and test–retest reliability. When advanced methods of blind source separation are used, the two sets of parameters can have similar spatial resolution. The cost of EEG machines is at least 100 times lower than that of fMRI/PET machines.

Any objective measure of brain functioning is made with errors. Errors in neuromarker measures arise from variations in the subject state, variations of the environment as well as spontaneous fluctuations in the measured parameter. Test–retest reliability as the correlation between the scores at two different times on a selected population of subjects is often used for estimating the error. To be used in clinical practice the reliability must be good or excellent.

Most of functional neuromarkers are normally or log-normally distributed. Z-scores are used to measure the individual deviation from the mean value of the population. When a neuromarker is used for diagnosis, the test outcome can be positive or negative for a tested subject and may or may not match the subject's actual status. Sensitivity of neuromarker is defined as probability of the positive test given that the subject is ill. Specificity is probability of the negative test given that the subject is healthy. For any test, there is a trade-off between sensitivity and specificity represented graphically in a receiver operating characteristic curve. The other measure of the difference between the group of patients and the group of healthy controls is given by effect size—the difference between means in terms of the standard deviation. For being a real help for psychiatry, the neuromarker must show high sensitivity and specificity and correspondingly large effect size.

CHAPTER 7.2

Rhythms of the Healthy Brain

In EEG recordings rhythmicities are usually pop out from the background and coexist with irregular, arrhythmic patterns (Fig. 7.2.1). The amplitude of arrhythmic oscillations decreases with frequency according to the power–law function. In infralow frequency band, there are at least two types of spontaneous oscillations: (1) periodic oscillations with a peak frequency around 0.1 Hz and (2) arrhythmic fluctuations with no clear peak at EEG spectrograms. Strong experimental evidence suggests that the 0.1 Hz oscillations are associated with local hemodynamic regulatory oscillations in the human brain. The neuronal source of the aperiodic infraslow spontaneous fluctuations is not so clear. Functional meaning of the infraslow oscillations is based on two facts: the phase of these oscillations correlates with (1) human psychophysical performance, and (2) amplitude of oscillations in higher frequency bands. Although these findings are correlational, there is a temptation to suggest that the infraslow potentials of the human brain reflect fluctuations in cortical excitability which in turn modulate faster fluctuations in the brain state and behavioral performance.

German psychologist Hans Berger was the first to observe electrical alpha rhythms from the scalp of human subjects in 1929 (Fig. 7.2.1, top). It took more than 50 years for scientists to discover neuronal mechanisms of alpha rhythms. Of specific importance is the discovery of a novel form of rhythmic burst firing, termed high-threshold bursting, which occurs in a subset of thalamocortical neurons under a depolarized state. Now we know that during relaxed wakefulness the human brain exhibits several types of rhythmic electrical activity in the alpha frequency band (8–13 Hz) in occipital, parietal, and central areas. These rhythms differ in topography, frequency, and sensitivity to tasks. Despite these differences alpha oscillations share a general function: they provide active and adequate inhibition of the irrelevant sensory pathways. If we consider the thalamus as the gate to the cortex, this function can be named as closing the sensory gate to the cortex. Fig. 7.2.1a, top, schematically represents thalamocortical network for generation of mu-rhythms. Note inhibitory feedback to

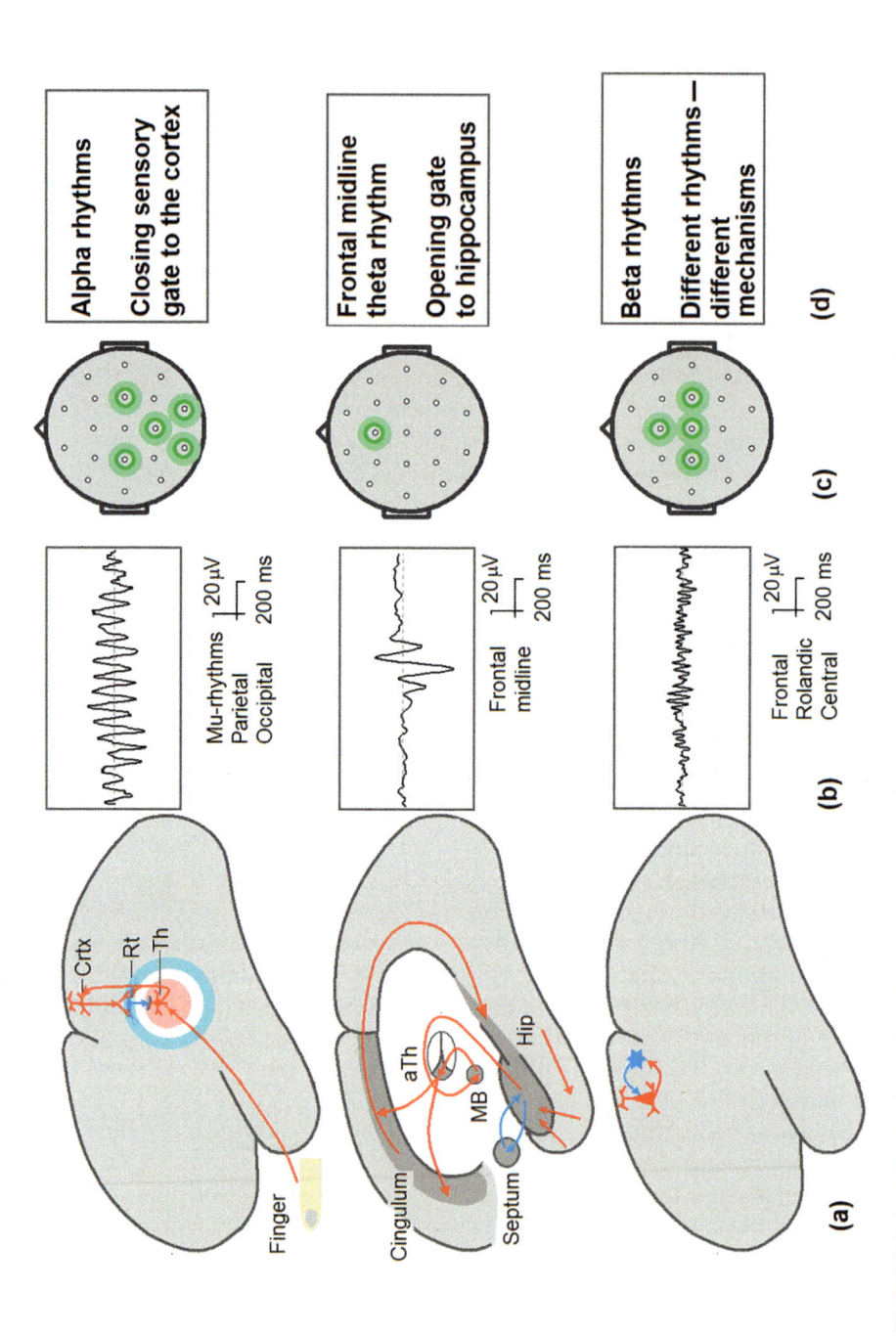

FIGURE 7.2.1 **EEG rhythms of cortical self-regulation: alpha (top), frontal midline theta (middle), and beta (bottom) rhythms.** (a) Hypothetical neuronal network for rhythm generation. Excitatory neurons and pathways are marked by *red*. Inhibitory neurons and pathways are marked by *blue*. (b) An example of rhythm recorded from the human scalp. (c) Locations of maximums of the rhythm power. (d) Functional meaning.

the thalamocortical neurons from the neurons in the reticular nucleus of the thalamus. The frequency of alpha rhythms reaches the highest values at around 20 years of age and slowly declines with aging. The absence of alpha rhythms is found in 10% of population with prevalence in anxiety disorders. Frontal alpha asymmetry may serve a neuromarker for depression. At behavioral level, alpha oscillations are responsible for maintaining the optimal level of sensory processing.

Beta rhythms (Fig. 7.2.1, bottom) come in ranges: beta 1 (13–20 Hz), beta 2 (21–30 Hz), and gamma (30–60 Hz). Beta rhythms of the basal ganglia are reflected in scalp recorded Rolandic beta rhythms which are inversely correlated with the fMRI-BOLD signals in the precentral cortex. The frontal beta rhythm, in contrast, synchronizes in response to activation of the frontal lobes. The vertex beta rhythm is induced in response in unexpected situations. Existence of several beta rhythms with different frequencies, topographies, and functional properties presumes no single neuronal mechanism for their generation. The scheme in Fig. 7.2.1a, bottom, reflects locality of beta rhythms rather than a single mechanism.

In healthy brain there is only one rhythm in the theta band during wakefulness: the frontal midline theta rhythm (Fig. 7.2.1, middle). Because this rhythm appears in short bursts (of few seconds) with long and varied interburst intervals in a small (10–40%) group of healthy subjects, and is enhanced by task load; it can be reliably measured by spectral analysis in highly demanding cognitive tasks. Appearance of the frontal midline theta rhythm is more likely in less neurotic and less anxious subjects. The frontal midline theta rhythm in humans is often associated with the hippocampal theta rhythms in animal research. From this research the rhythm can be associated with opening the sensory gate to the hippocampus for the intermediate storage of episodic information. The scalp topography of the rhythm power is frontal with a maximum at Fz. sLORTEA localizes the rhythm in the medial part of the prefrontal cortex including the anterior gyrus cingulate. The frequency of frontal midline theta varies from 5 to 7.5 Hz with average around 6 Hz. The rhythm is associated with working memory, episodic encoding, and retrieval. It also appears during hypnosis and deep meditation.

CHAPTER 7.3

Information Flow in the Healthy Brain

The brain regions where neurons respond to stimulation of a certain type of receptor are called sensory systems. There are different sensory modalities that give us sensations of images, sounds, and body movements. The information flow in the sensory systems is modulated by attention which in turn is associated with enhancement of relevant sensory information and suppression of irrelevant sensory information. A canonical ERP in a sensory modality includes P1 and N1 waves. In the auditory modality, the P1/N1 complex is modulated by intensity of auditory stimulation whereas its dependence on loudness is defined by the cortical serotonin level. In the visual modality, the N1 is associated with discrimination operation. When stimuli are repeatedly presented in a particular order, the brain forms the neuronal model of sensory stimulation so that when a new stimulus violates the model hypothetical "change," detectors are activated. The activation of these change detectors is reflected in visual and auditory mismatch negativity ERP waves. If the deviancy is large, it produces orienting response which is reflected in the Novelty P3 wave. Readout from a personal memory is attributed to the visual N170 wave generated in the fusiform gyrus. This area is also involved in generation of N250 wave which appears in response to stimulus repetition. When the stimulus is mismatched with the template in working memory the temporally generated P2 (or P250) wave emerges.

The operations of sensory processing are separated by different methods of blind source separation approach. Fig. 7.3.1b represents grand average ERPs in response to GO, NOGO, and Ignore stimuli in the cued GO/NOGO task. Note that a visual stimulus presentation in Ignore condition elicits a modest P1/N1 response which is much smaller than GO and NOGO P3 waves. Two sensory-related latent components are shown in Fig. 7.3.1c. The time course of these components is modulated by operations of categorization and comparison to working memory as seen in the

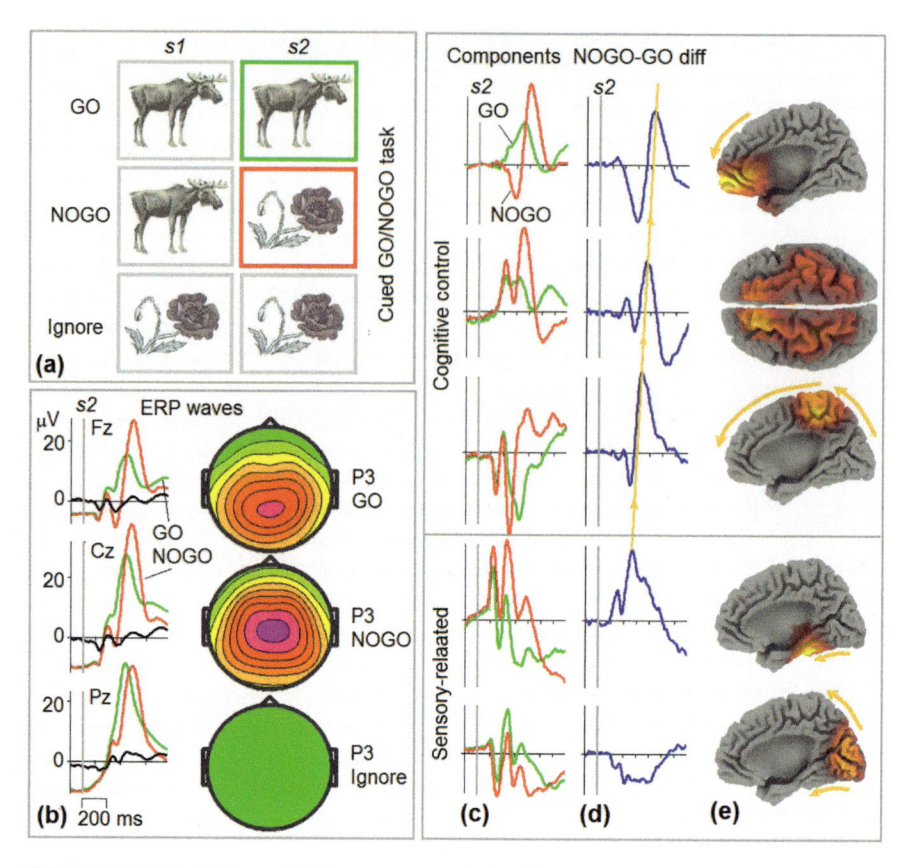

FIGURE 7.3.1 **ERPs and ERP components in healthy brain.** (a) Stimuli in the cued GO/NOGO task (examples). (b) Left: grand average ERPs in referential montage for a group of 114 healthy subjects of 18–23 years old for GO (*green*), NOGO (*red*), and Ignore (*black*) conditions. Right: maps of ERPs for the three conditions at maximums around 300 ms. (c) Latent components extracted from the collection of ERPs for GO (green) and NOGO (red) conditions. (d) Difference NOGO-GO for the corresponding components. Note a progressive delay of the P3 peak latency from occipital to frontal locations. (e) sLORETA images with *arrows* indicating hypothetical information flow.

differences in Fig. 7.3.1d. sLORETA localizes the first component in the prestriate cortex and the second component in the fusiform gyrus.

One of the basic goals in human behavior is selection from a large repertoire of possible behavioral options those actions that are more likely to promote human well-being. The hypothetical operations of cognitive control include: action preparation and action selection, working memory, shift from one action to another, suppression of prepared but irrelevant action, inhibition of ongoing activity, detection of conflict, and adjusting behavior in order to avoid conflicts. Many behavioral paradigms such as willed selection, task switching, inhibition paradigms, delayed and

Stroop tasks are used for studying different aspects of cognitive control. Two modes of cognitive control are separated: proactive mode reflecting the sustained and anticipatory maintenance of internal goals, and reactive control reflecting transient mechanism in response to conflict detection.

The ERP correlates of proactive cognitive control include readiness potential, the contingent negative variation, and the stimulus preceding negativity which are associated with preparatory activities in motor, cognitive, and sensory systems. Concept of prepotent response (or prepotent model of behavior) is used as a theoretical background. The prepotent automaticity makes our life efficient because it frees limited cognitive resources from the numerous routine requirements.

When conflict is detected, the automatic response is shifted to the reactive cognitive control mode. This is reflected in NOGO P3 wave (Fig. 7.3.1b). One can see that when contrasted to GO P3 wave the NOGO ERP reveals the N2 NOGO and P3 NOGO fluctuations. These fluctuations are often associated with conflict detection and action inhibition operations. However the N2/P3 dichotomy does not fit experimental data and is substituted by decomposition of ERPs into functional components on the basis of the blind source separation approach.

Three components of cognitive control obtained by the method of joint diagonalization of cross-covariance matrixes are presented in Fig. 7.3.1c. Time course of these components is differently modulated by operations of conflict detection, inhibition of ongoing activity, and suppression/overriding of the prepared action. The functions of cognitive control are implemented by the cortical basal ganglia thalamocortical system with the frontal and parietal lobes as nodes of the system. Recent studies have highlighted the role of the noradrenergic system in engagement operations as well as the role of the dopaminergic system in working memory and operations of cognitive control.

The affective system is a network of cortical and subcortical anatomical structures which maps sensory stimuli into rewards and punishments, expresses emotions, and is responsible for feelings of those emotional reactions. Thus the affective system provides two new dimensions to behavior: negative and positive affective reactions and states. One of the hypotheses suggests asymmetric involvement of prefrontal cortical regions in positive affect (dominance of the left hemisphere) and negative affect (dominance of the right hemisphere). The left–right frontal asymmetry in human subjects is expressed in the anterior alpha asymmetry index of QEEG. The key structure of the affective system is the amygdala which receives sensory information, extracts affective memories, and sends the results to the prefrontal cortex. The amygdala reacts with increased activation to fearful stimuli and via feedback connections enhances early visual ERPs components. As a result, anxiety is often associated with enhancement of visual P1/N1 waves.

The other key structure of the affective system is the ventral part of the anterior cingulate cortex which receives strong input from amygdala and is in position of regulating the emotional state from restless anxiety to focused relaxation. The affective and cognitive control systems are mutually interconnected so that the dysfunction of the affective system can lead to impairment of cognitive control. The loss of cognitive control during stress exposure leads to a number of maladaptive behaviors, such as drug addiction, smoking, drinking alcohol, and overeating. Prolonged stress is a major risk factor for depression, and exposure to traumatic stress can cause posttraumatic stress disorder. Among all neuromodulators the highest concentration of serotonin is found in the affective system.

There are different types of memory. In relation to timing immediate, short-term, and long-term memories are separated. The contralateral delayed activity in the posterior brain areas represents an ERP correlate of short-term (working) memory. In relation to content, the long-term memory is divided into declarative and procedural memories. The hippocampus serves as a key element in declarative memory by mapping representations of the current episode into an intermediate form. These intermediate representations are relatively resistant to interference and serve as references to the memories. The encoding of episode is accompanied by a burst of hippocampal theta rhythm and controlled by acetylcholine.

Procedural memory is based on learning and recalling motor and cognitive skills. The representations of actions are stored in the frontal–parietal networks and are mapped into striatum. Procedural memory, unlike episodic memory, does not need a separate system for encoding and consolidating events. An engram of procedural memory represents slowly changing synaptic connections in the executive system. There are two main neuromodulators of procedural memory: dopamine that is transported to the striatum from the substantia nigra and acetylcholine that is produced by specific cholinergic cells within the striatum itself.

CHAPTER
7.4

Current Treatment Options in Psychiatry

The modern psychiatry has in its arsenal different treatment options including pharmacotherapy, neurotherapy, and neuromodulation techniques. The basic idea of the pharmacotherapy is that the core of the brain functioning is chemical so that mental illness is the result of imbalances among neurotransmitters. Consequently, the main goal for pharmacology is to find a neurotransmitter system involved in a given psychopathology. Taken into account high heterogeneity of functional proteins (receptors, transporters, ion channels, enzymes) in the brain, the number of selective pharmaceutical agents used by mental health care is supposed to be big. However during the last years, the psychopharmacology faced a certain crisis and shrinked neuroscience research projects. Neurotherapy and neuromodulation options emerged as alternative approaches for curing mental illness.

Neurofeedback (NF) is a technique of a self-regulation in which current parameters of EEG recorded from the subject's head are presented to a subject through visual, auditory, or tactile modality while the subject is supposed to alter these parameters to reach a more efficient mode of brain functioning. According to the two types of electrical brain phenomena there are two main types of neurofeedback: conventional EEG biofeedback (including LORETA NF) and infralow frequency (ILF) neurofeedback. The conventional NF uses the spectral characteristics of EEG in 0.5–50 Hz frequency band whereas the ILF NF uses either amplitude itself or phase of the voltage fluctuations below 0.1 Hz. ILF NF is done in discrete and continuous forms. In contrast to the "one size fits all," discrete protocol the continuous ILF protocols are different for different symptoms. The protocols of the conventional NF can be divided into activation and relaxation protocols. The neurofeedback of at the beginning (middle of 20th century) was driven by the theory of operant conditioning, but recently a bulldozer principle has been suggested with the aim to normalize a pathologically abnormal EEG pattern. NF should not be applied without relevant diagnostic procedures of QEEG and ERPs.

Electroconvulsive therapy (ECT) is the oldest nonpharmacological therapy currently available for psychiatric disease. Electrodes are placed either unilaterally or bilaterally with electric pulses of 500–800 mA. Mechanisms of the therapeutic effect of ECT are not known. ECT is frequently associated with retrograde and anterograde amnesia. Relapse after a successful ECT course is a major limitation of the therapy.

In transcranial direct current stimulation (tDCS) a small amount of direct electric current (1–2 mA) is applied to the skin of the head by two relatively large electrodes. The electric current flows according to the Ohm's law and depolarizes/hyperpolarizes pyramidal cells at their basal membrane depending on direction of the current. tDCS is fundamentally different from ECT by inducing much smaller currents which do not evoke action potentials but change overall neuronal activity due to collective effects. tDCS long-term aftereffect is a function of the intensity and duration of tDCS application and occurs through NMDA-dependent mechanisms similar to long-term synaptic potentiation and depression. When applied according to standard requirements, the tDCS is a safe procedure.

Transcranial magnetic stimulation (TMS) is based on the law of electromagnetic induction. In TMS, a pulse of electric current passing through a coil placed over a subject's head generates a rapidly changing magnetic pulse that penetrates the scalp and skull to reach the cortex with negligible attenuation. The pulse of magnetic field in turn induces a secondary ionic current in the brain which can trigger action potentials in cortical neurons. TMS in clinical practice is applied in form of continuous trains and is named repetitive TMS (rTMS). rTMS can induce changes in neuronal excitability that persist beyond the time of stimulation. These neuromodulatory effects of TMS are used in patients with neurologic and psychiatric diseases to maintain or restore brain functions. rTMS is reasonably safe with mild side effects when performed in compliance with the recommended safety guidelines.

At the beginning of 20th century, psychosurgery for severe psychiatric conditions aimed in destruction of large portions of the brain. In the mid of 20th century, it was replaced by stereotactic local lesions. Currently deep brain stimulation (DBS) substitutes these ablation techniques. Effects of DBS are fully reversible and can be adjusted to patients' symptoms.

CHAPTER
7.5

Functional Neuromarkers in Diseased Brain

During the last 20 years, intensive research on functional neuromarkers in different psychiatric conditions such as ADHD, schizophrenia, OCD provided a vast amount of empirical knowledge with clear indication that at least some of those neuromarkers are reliable and powerful tools in discriminating patients from healthy controls. Nowadays we are facing a decade of translation that focuses on application of functional neuromarkers for providing an early detection of brain dysfunction as well as a personalized care for patients with the dysfunction. The early detection of neuromarkers of mental illness in turn requires development of preventive interventions.

Attention-deficit hyperactivity disorder (ADHD) is a highly prevalent neuropsychiatric condition with onset in childhood characterized by a persistent and age-inappropriate pattern of descriptive symptoms of inattention, hyperactivity, and impulsivity. ADHD puts children at risk for other psychiatric and substance abuse disorders. Although symptoms decline with age, ADHD persists into adulthood in some cases. Neuropsychological parameters of variability of performance and response inhibition differentiate ADHD from healthy controls. Methylphenidate, by blocking dopamine reuptake in the striatum, has a positive therapeutic effect in approximately 65–70% of patients. On the basis of EEG, several subgroups of ADHD are separated including those characterized by presence of Rolandic spikes, excessive theta/beta ratio, excessive frontal beta, and persistent alpha in eyes open condition. On the basis of ERPs in GO/NOGO paradigm, at least two subgroups of ADHD with selective decrease of parietal and frontal ERP components are separated. Patients with the frontal ERP deficit respond to psychostimulants. The specific pattern of ERP change in response to a single dose stimulant medication predicts a positive response to stimulant medication. There are numerous studies on application of neurofeedback in treatment symptoms of ADHD while studies on application of tDCS and TMS are still in their infancy.

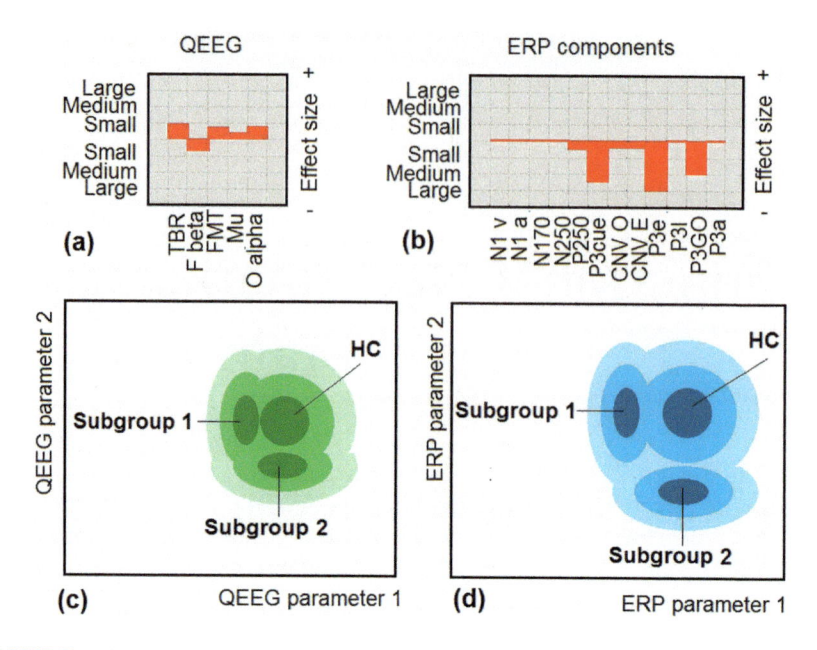

FIGURE 7.5.1 **Functional neuromarkers in diseased brain.** A hypothetical diagnostic category (eg, ADHD) on the basis of QEEG and ERP components is decomposed in several subgroups (only two subgroups are depicted for simplicity). (a) Profile of effect size of the deviation from the mean of the healthy control group in parameters of QEEG. TBR, theta beta ratio; F beta, frontal beta in the GO/NOGO task; FMT, frontal midline theta; O alpha, occipital alpha in eyes open condition. Y-axis: effect size. (b) Profile of effect size of the deviation from the mean of the healthy control group in parameters of ERPs. Y-axis: effect size. X-axis: amplitudes of ERP latent components. (c) Schematic presentation of distribution of QEEG parameters in two-dimensional space for two subgroups of patients (Subgroup 1 and Subgroup 2) versus healthy controls (HC). Note an overlap between ADHD and HC so that a substantial part of ADHD patients is not discriminated from HC. (d) Schematic presentation of distribution of ERP latent components in two-dimensional space for two subgroups of patients (Subgroup 1 and Subgroup 2) versus healthy controls (HC). Note a clear separation between ADHD and HC so that a given individual can be ascribed to one of the groups with high specificity and sensitivity.

Figure 7.5.1 schematically represents a summary of the main results of research on QEEG and ERP neuromarkers in ADHD. Note that ADHD includes several subgroups of patients with specific QEEG/ERP profiles. The heterogeneity is especially evident in QEEG profiles reflecting different subtypes of cortical self-regulation and presenting both in ADHD and healthy control groups but at different probabilities. Because of such heterogeneity, the effect sizes for discriminating the whole group from similarly heterogeneous group of healthy controls is small for all QEEG parameters (Fig. 7.5.1a). However, discrimination power increases substantially in multidimensional space of multiple neuromarkers (Fig. 7.5.1c).

A different picture is seen in ERP profiles (Fig. 7.5.1b). These profiles describe impairments in specific operations of sensory processing and cognitive control. As one can see in this particular example, no impairment is seen in sensory-related components as well as in some executive component suggesting specific ERP profiles for ADHD. This makes effect sizes for ERP components higher than for QEEG parameters while discriminability in the multidimensional ERP space could be reliably used in clinical practice for individual assessment (Fig. 7.5.1d). Moreover, defining to which subgroup does a particular patient belongs enables the psychiatrist to predict how the patient will respond to a particular form of treatment.

In contrast to ADHD, schizophrenia (SZ) is a less common psychiatric condition characterized by a diverse set of symptoms including specific distortions in sensory, motor, executive, and affective systems. SZ is a result of complex inheritance in which interaction of genes with environment plays a critical role. Most of SZ patients experience a lifetime disability with 10% eventually committing suicide. Positive symptoms (delusions and hallucinations) start between ages 16 and 30, are associated with dopaminergic hyperactivity and are suppressed by antipsychotic drugs via blocking dopaminergic receptors. Negative symptoms include impairments in the affective domain such abulia, anhedonia, and apathy as well as impairments in cognitive domain with no medication known to reduce them. SZ is developing during prodromal phase so that defining neuromarkers of this state is of big importance for SZ prevention. No consistent changes in QEEG have been so far reported in SZ; however, ERP research shows reliable decrease of many ERP components in SZ with large effect sizes. The ERP neuromarkers reflect sensory-related deficits such as failure of P50 gating effect, decrease of auditory/visual N1 waves, MMN, and novelty P3, as well as cognitive related deficits such as reduction of the target P3, N2 and P3 NOGO waves, CNV and SPN. Some of these waves predict conversion to psychosis whereas the others predict response to antipsychotic medication. Rare attempts of applying tDCS for schizophrenia report reduction of hallucinations and improvement of cognitive functions. TMS has been used more widely but still needs larger multicenter studies.

Obsessions include socially unacceptable thoughts or impulses, chronic doubting, fears of contamination, preoccupations with symmetry, and so forth. Compulsions include excessive hand washing, placing objects symmetrically, repeatedly checking and so forth. The nature of the obsessions and compulsions varies greatly between patients with obsessive–compulsive disorder (OCD). The intrusive thoughts in OCD are perceived as the product of one's own mind and are different from thought insertions in SZ. The neuropsychological profile of OCD reveals local deficits in inhibiting and planning motor and cognitive actions. In contrast to ADHD and schizophrenia, OCD patients consistently show hyperactivation of the medial prefrontal cortex including the anterior cingulate area. The error-related negativity and conflict-related N2 ERP waves generated in the anterior cingulum are elevated in OCD and correlate of the feeling that things are "not just right." The serotonergic system modulates OCD symptoms with SSRI producing clinical benefit. The other effective treatment is the exposure and response prevention method of cognitive behavior therapy. In 10–30% of cases, severe OCD is resistant to conventional treatments so that psychosurgery and deep brain stimulation remain as the only options. Until recently reports on successful neurofeedback treatment in OCD were rare but individually tailored protocols could be beneficial. Inhibition of the presupplementary motor area by TMS or tDCS might be a promising protocol.

CHAPTER
7.6

Implementation in Clinical Practice

Motives for assessment functional neuromarkers in a particular person come from insurance companies, sport psychologists, forensic psychiatrists, as well as from biologically oriented psychiatrists. A single diagnostic category represents a heterogeneous condition in which similar symptoms may be caused by different sources. For example, ADHD patient might have a focus in the Rolandic fissure, maturation lag, hypoarousal, limbic dysfunction, or cortical–striatal–thalamic–cortical dysfunction with either frontal or parietal nodes. Knowledge about the cause of the particular behavioral pattern helps not only in making the correct diagnosis but also in providing the prognosis of how to treat the cause. For example, depending on the biological source of the behavioral pattern of ADHD, the psychiatrist may use dopamine reuptake or noradrenalin reuptake inhibitors, GABA agonists, neurofeedback, tDCS, TMS.

The working space of the 21st century psychiatrist includes the EEG/ERP system for recording spontaneous EEG and event-related potentials (Fig. 7.6.1a). A possible content of the report made by an expert depends on the goals of the psychiatrist and includes recommendations for the individual treatment. There are many EEG/ERP systems in the world but there are few QEEG/ERP databases.

Since normal mechanisms of cortical self-regulations are reflected in brain oscillations in several frequency bands, different kinds of EEG dysrhythmia are expected in brain disorders. For example, ADHD is characterized by a specific pattern of cortical dysregulation including Rolandic spikes (Fig. 7.6.1b, top), excess of theta–beta ratio and/or central theta (Fig. 7.6.1b, middle), increase of frontal beta oscillations and frontal midline theta rhythm activity, persistent alpha rhythms with eyes open. Individual ERP independent components for the deviations from the reference might be used for constructing neurofeedback and tDCS protocols (Fig. 7.6.1b, bottom).

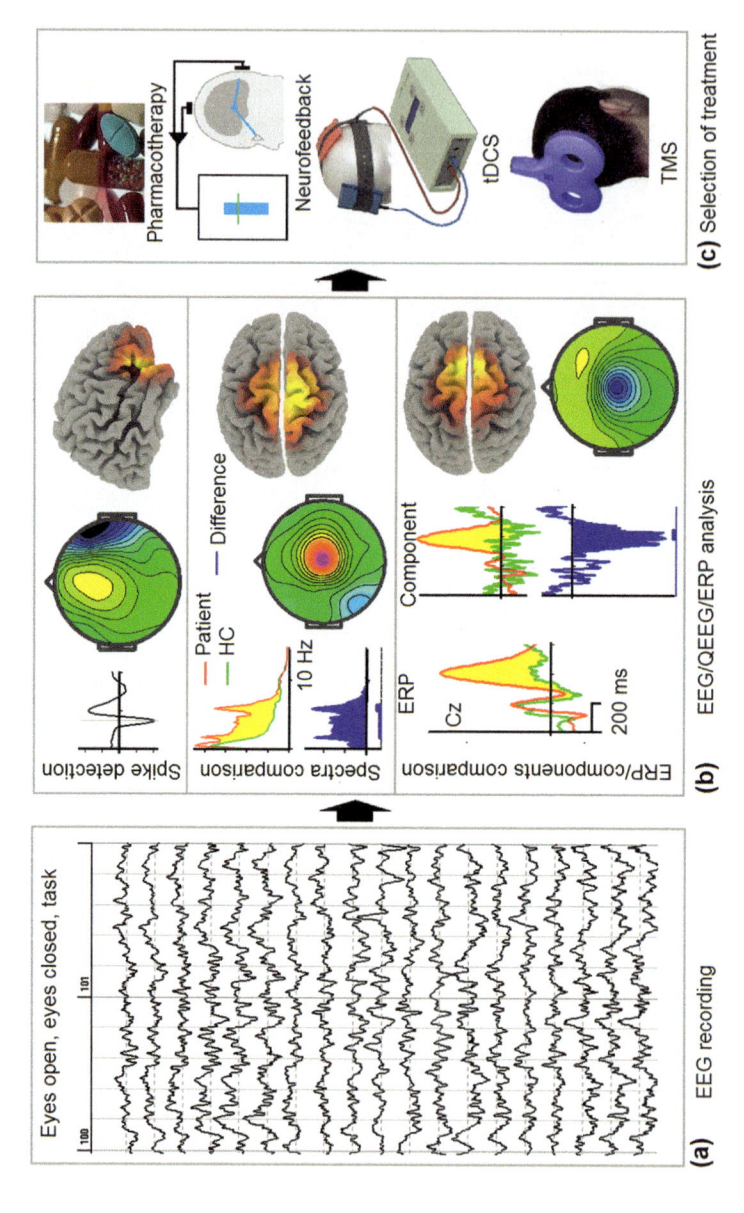

FIGURE 7.6.1 **Implementing functional neuromarkers into psychiatric practice.** (a) EEG recording in resting and task conditions. (b) Top: automatic spike detection and averaging, mapping, and constructing sLORETA image. Middle: computing EEG spectra, comparing with the normative data, extracting the corresponding independent component, and constructing sLORETA image. Bottom: computing ERPs and ERP components by spatial filtration, comparing with the normative data, and constructing sLORETA image for the deviant component. (c) Selection of appropriate treatment on the basis of the results of assessment as well as on the basis of the obtained research knowledge.

Since psychological operations are reflected in ERP components, different types of ERP impairment are expected in brain disorders. In substantial number of psychiatric patients, the ERP abnormalities occur without any significant deviations from the reference in spectral EEG and even behavioral characteristics. ADHD in particular is characterized by impairments in parietal and frontal nodes of neuronal networks of cognitive control. The ERP analysis includes comparing raw ERPs and ERP components with the parameters obtained in healthy controls of the same age. An example at Fig. 7.6.1b, bottom, shows ERP in NOGO condition of an ADHD patient without any impairment in EEG spectra and behavioral characteristics. One can see statistically significant decrease of P3 NOGO wave and selective decrease of the P3 early component associated with operation of response inhibition. The topography and sLORETA image of the component clarify the source of dysfunction and possible neuromodulation techniques for activating this part of the prefrontal cortex (Fig. 7.6.1c).

Attempts of applying QEEG for monitoring effects of pharmacological treatment so far were not successful. The cases of application of ERPs in psychopharmacology are quite rare and restricted by N1/P2 and P3 waves in the auditory modality. Recently ERPs in the cued GO/NOGO task were successfully applied for predicting response to stimulant medication of the basis on a single-dose trial. ERPs provide a reliable tool for monitoring changes in brain functioning induced by neurotherapeutic and neuromodulation techniques.

Postscriptum

The German psychiatrist Emil Kraepelin, the founder of modern scientific psychiatry, in opposition to the dominating theories at the beginning of the 20th century did not believe that certain symptoms were characteristic of specific illnesses. On the basis of his experience he suggested that the main source of psychiatric disease is biological and genetic dysfunction. His ideas were oppositional to the psychoanalytic ideas of the Austrian psychiatrist Sigmund Freud.

At the same time in Russia the psychiatrist Vladimir Bechterev and the physiologist Ivan Pavlov—both started their carriers at the Military Academy of Medicine in Saint Petersburg—laid down the physiological foundations of mental illness. On the basis of his research, Bechterev suggested that each specific "zone" of the brain had a specific function and that psychiatric conditions occurred as a consequence of dysfunctions of those zones. He also believed that there was no definite distinction between "nervous" and mental disorders because they may occur in conjunction with each other.

Thirty years after the death of Vladimir Bechterev, his granddaughter and the supervisor of the author of the present book Natalia Bechtereva introduced into clinical practice the methods of psychosurgery and deep-brain stimulation for treating neurological and psychiatric conditions. She and her pupils at the Institute of Experimental Medicine in Saint Petersburg in 1960–1980 carried out research on neuronal mechanisms of neurofeedback and tDCS (electrical microstimulation in the Russian transcription). In our laboratory, which is located 200 m from the former Pavlov laboratory, we continue research into the neurophysiological mechanisms of mental illness and into implementation of the methods of neuromodulation in clinical practice.

This book was written in memory of the giants who laid down the foundations of biological psychiatry. It has been intended to show that we now facing the decade of translation in which the accumulated knowledge about functional neuromarkers can and must be used in clinical practice. If this book inspires at least one psychiatrist, I will be happy.

References

Accornero, N., Li Voti, P., La Riccia, M., & Gregori, B. (2006). Visual evoked potentials modulation during direct current cortical polarization. *Experimental Brain Research, 178*(2), 261–266.

Acosta, M. T., Castellanos, F. X., Bolton, K. L., Balog, J. Z., Eagen, P., Nee, L., Jones, J., Palacio, L., Sarampote, C., Russell, H. F., Berg, K., Arcos-Burgos, M., & Muenke, M. (2008). Latent class subtyping of Attention-Deficit/Hyperactivity Disorder and comorbid conditions. *Journal of the American Academy of Child & Adolescent Psychiatry, 47*, 797–807.

Ahveninen, J., Jaaskelainen, I. P., Raij, T., Bonmassar, G., Devore, S., Hamalainen, M., Bentin, S., Allison, T., Puce, A., Perez, E., & McCarthy, G. (1996). Electrophysiological studies of face perception in humans. *Journal of Cognitive Neuroscience, 8*(6), 551–565.

Alexander, G. E., & Crutcher, M. D. (1990). Functional architecture of basal ganglia circuits: neural substrates of parallel processing. *Trends in Neurosciences, 13*(7), 266–271.

Andrews, T. J., Clarke, A., Pell, P., & Hartley, T. (2010). Selectivity for low-level features of objects in the human ventral stream. *Neuroimage, 49*(1), 703–711.

Anokhin, A. P. (2014). Genetic psychophysiology: advances, problems, and future directions. *International Journal of Psychophysiology, 93*(2), 173–197.

Arns, M., Gunkelman, J., & Breteler, M. (2008). EEG phenotypes predict treatment outcome to stimulants in children with ADHD. *Journal of Integrative Neuroscience, 7*(3), 421–438.

Arns, M., De Ridder, S., Strehl, U., Breteler, M., & Coenen, A. (2009). Efficacy of neurofeedback treatment in ADHD: the effects on inattention, impulsivity and hyperactivity: a meta-analysis. *Clinical EEG and Neuroscience Official Journal of the EEG and Clinical Neuroscience Society ENCS, 40*, 180–189.

Baddeley, A. D., & Hitch, G. (1974). Working memory. In G. H. Bower (Ed.), *The Psychology of learning and motivation: Advances in research and theory* (pp. 47–89). (8). New York: Academic Press.

Bai, S., Loo, C., Dokos, S., 2010. A computational model of direct brain stimulation by electroconvulsive therapy. 32nd Annual International Conference of the IEEE EMBS, Buenos Aires, Argentina, August 31–September 4.

Bai, S., Loo, C., Al Abed, A., & Dokos, S. (2012). A computational model of direct brain excitation induced by electroconvulsive therapy: comparison among three conventional electrode placements. *Brain Stimulation, 5*(3), 408–421.

Baioui, A., Pilgramm, J., Merz, C. J., Walter, B., Vaitl, D., & Stark, R. (2013). Neural response in obsessive–compulsive washers depends on individual fit of triggers. *Frontiers in Human Neuroscience, 7*, 134.

Barkley, R. A. (1997a). *ADHD and the Nature of Self-Control*. New York: Guilford Press.

Barkley, R. A. (1997b). Behavioral inhibition, sustained attention, and executive functions: constructing a unifying theory of ADHD. *Psychology Bulletin, 121*, 65–94.

Barry, R. J., Johnstone, S. J., & Clarke, A. R. (2003). A review of electrophysiology in attention-deficit/hyperactivity disorder: II. Event-related potentials. *Clinical Neurophysiology, 114*(2), 184–198.

Bechtereva, N. P., Kropotov, J. D., Ponomarev, V. A., & Etlinger, S. C. (1990). In search of cerebral error detectors. *International Journal of Psychophysiology, 8*(3), 261–273.

Benson, P. J., Beedie, S. A., Shephard, E., Giegling, I., Rujescu, D., & St. Clair, D. (2012). Simple viewing tests can detect eye movement abnormalities that distinguish schizophrenia cases from controls with exceptional accuracy. *Biological Psychiatry, 72*, 716–724.

Boggio, P. S., Bermpohl, F., Vergara, A. O., Muniz, A. L., Nahas, F. H., Leme, P. B., Rigonatti, S. P., & Fregni, F. (2007). Go–no-go task performance improvement after anodal transcranial DC stimulation of the left dorsolateral prefrontal cortex in major depression. *Journal of Affective Disorders, 101*(1–3), 91–98.

Braver, T. S. (2012). The variable nature of cognitive control: a dual mechanisms framework. *Trends in Cognitive Sciences, 16*, 106–113.

Brunner, J. F., Hansen, T. I., Olsen, A., Skandsen, T., Haberg, A., & Kropotov, J. (2013). Long-term test–retest reliability of the P3 NoGo wave and two independent components decomposed from the P3 NoGo wave in a visual Go/NoGo task. *International Journal of Psychophysiology, 89*, 106–114.

Brunner, J. F., Olsen, A., Aasen, I. E., Lohaugen, G. C., Haberg, A. K., & Kropotov, J. (2015). Neuropsychological parameters indexing executive processes are associated with independent components of ERPs. *Neuropsychologia, 66*, 144–156.

Buchsbaum, M.S., Silverman, J. (1968). Stimulus intensity control and the cortical evoked response. *Psychosomatic Medicine, 30*(1), 12–22.

Cantero, J. L., Atienza, M., Stickgold, R., Kahana, M. J., Madsen, J. R., & Kocsis, B. (2003). Sleep-dependent theta oscillations in the human hippocampus and neocortex. *The Journal of Neuroscience, 23*(34), 10897–10903.

Carrasco, M., Hong, C., Nienhuis, J. K., Harbin, S. M., Fitzgerald, K. D., Gehring, W. J., & Hanna, G. L. (2013). Increased error related brain activity in youth with obsessive–compulsive disorder and other anxiety disorders. *Neuroscience Letters, 541*, 214–218, http://dx.doi.org/10.1016/j.neulet.

Cavanagh, J. F., & Shackman, A. J. (2014). Frontal midline theta reflects anxiety and cognitive control: meta-analytic evidence. *Journal of Physiology, 109*, 3–15.

Cerletti, U. (1940). L'Elettroshock. *Rivista Sperimentale di Frenatria, 1*, 209–231.

Chang, W. P., Lu, H. C., & Shyu, B. C. (2015). Treatment with direct-current stimulation against cingulate seizure-like activity induced by 4-aminopyridine and bicuculline in an in vitro mouse model. *Experimental Neurology, 265*, 180–192.

Cohen, J. (1988). *Statistical power analysis for the behavioral sciences* (2nd ed.). New Jersey: Lawrence Erlbaum.

Cook, I. A., O'Hara, R., Uijtdehaage, S. H., Mandelkern, M., & Leuchter, A. F. (1998). Assessing the accuracy of topographic EEG mapping for determining local brain function. *Electroencephalography and Clinical Neurophysiology, 107*(6), 408–414.

Cooley, J. W., & Tukey, J. W. (1965). An algorithm for machine calculation of complex Fourier series. *Mathematics of Computation, 19*, 297–301.

Damasio, A. (1999). *The feeling of what happens: Body and emotion in the making of consciousness.* New York: Harcourt Brace.

Davidson, R. J. (1993). Cerebral asymmetry and emotion: methodological conundrums. *Cognition and Emotion, 7*, 115–138.

Deecke, L., Grözinger, B., & Kornhuber, H. H. (1976). Voluntary finger movement in man: cerebral potentials and theory. *Biological Cybernetics, 23*(2), 99–119.

Dostrovsky, J. O., Levy, R., Wu, J. P., Hutchison, W. D., Tasker, R. R., & Lozano, A. M. (2000). Microstimulation-induced inhibition of neuronal firing in human globus pallidus. *Journal of Neurophysiology, 84*(1), 570–574.

Doyle, A., Biederman, J., Seidman, L. J., Weber, W., & Faraone, S. V. (2000). Diagnostic efficiency of neuropsychological test scores for discriminating boys with and without attention-deficit/hyperactivity disorder. *Journal of Consulting and Clinical Psychology, 68*, 477–488.

Ehlis, A. C., Pauli, P., Herrmann, M. J., Plichta, M. M., Zielasek, J., Pfuhlmann, B., Stober, G., Ringel, T., Jabs, B., & Fallgatter, A. J. (2012). Hypofrontality in schizophrenic patients and its relevance for the choice of antipsychotic medication: an event-related potential study. *The World Journal of Biological Psychiatry, 13*, 188–199.

Fair, D. A., Bathula, D., Nikolas, M. A., & Nigg, J. T. (2012). Distinct neuropsychological subgroups in typically developing youth inform heterogeneity in children with ADHD. *Proceedings of the National Academy of Sciences, 109*(17), 6769–6774.

Feige, B., Scheffler, K., Esposito, F., Di Salle, F., Hennig, J., & Seifritz, E. (2005). Cortical and subcortical correlates of electroencephalographic alpha rhythm modulation. *Journal of Neurophysiology, 93*(5), 2864–9872.

Fell, J., Ludowig, E., Rosburg, T., Axmacher, N., & Elger, C. E. (2008). Phase-locking within human mediotemporal lobe predicts memory formation. *NeuroImage, 43*, 410–419.

Fernandez, G., Effern, A., Grunwald, T., Pezer, N., Lehnertz, K., Dumpelmann, M., Van Roost, D., & Elger, C. E. (1999). Real-time tracking of memory formation in the human rhinal cortex and hippocampus. *Science, 285*, 1582–1585.

Ford, J. M., & Mathalon, D. H. (2012). Anticipating the future: automatic prediction failures in schizophrenia. *International Journal of Psychophysiology, 83*, 232–239.

Frodl, T., & Skokauskas, N. (2012). Meta-analysis of structural MRI studies in children and adults with attention deficit hyperactivity disorder indicates treatment effects. *Acta Psychiatrica Scandinavica, 125*, 114–126.

Galderisi, S., & Maj, M. (2009). Deficit schizophrenia: an overview of clinical, biological and treatment aspects. *European Psychiatry, 24*(8), 493–500.

Gordon, E., Cooper, N., Rennie, C., Hermens, D., & Williams, L. M. (2005). Integrative neuroscience: the role of a standardized database. *Clinical EEG and Neuroscience, 36*(2), 64–75.

Greenberg, P. A., & Wilson, F. A. W. (2004). Functional stability of dorsolateral prefrontal neurons. *Journal of Neurophysiology, 92*, 1042–1055.

Gretchin, V. B. & Kropotov, Y. D. (1979). Slow non-electric rhythms of the human brain. Leningrad: Nauka.

Grossmann, A., & Morlet, J. (1984). Decomposition of hardy functions into square integrable wavelets of constant shape. *SIAM Journal on Mathematical Analysis, 15*(4), 723–736.

Hämäläinen, M. S., & Ilmoniemi, R. J. (1994). Interpreting magnetic fields of the brain: minimum norm estimates. *Medical & Biological Engineering & Computing, 32*(1), 35–42.

Hart, H., Radua, J., Mataix, D., & Rubia, K. (2013). Meta-analysis of fMRI studies of inhibition and attention in ADHD: exploring task-specific, stimulant medication and age effects. *JAMA Psychiatry, 70*, 185–198.

He, B. J., Zempel, J. M., Snyder, A. Z., & Raichle, M. E. (2010). The temporal structures and functional significance of scale-free brain activity. *Neuron, 66*, 353–369.

Hegerl, U., Gallinat, J., & Juckel, G. (2001). Event-related potentials: do they reflect central serotonergic neurotransmission and do they predict clinical response to serotonin agonists? *Journal of Affective Disorders, 62*, 93–100.

Herskovits, E. H., Megalooikonomou, V., Davatzikos, C., Chen, A., Bryan, R. N., & Gerring, J. P. (1999). Is the spatial distribution of brain lesions associated with closed-head injury predictive of subsequent development of attention-deficit/hyperactivity disorder? Analysis with brain-image database. *Radiology, 213*(2), 389–394.

Hopf, J. -M., Vogel, E., Woodman, G., Heinze, H. -J., & Luck, S. J. (2002). Localizing visual discrimination processes in time and space. *Journal of Neurophysiology, 88*(4), 2088–2095.

Howells, F. M., Stein, D. J., & Russell, V. A. (2012). Synergistic tonic and phasic activity of the locus coeruleus norepinephrine (LC-NE) arousal system is required for optimal attentional performance. *Metabolism Brain Disease, 27*, 267–274.

Hsieh, L. T., & Rangnath, C. (2014). Frontal midline theta oscillations during working memory maintenance and episodic encoding and retrieval. *Neuroimage, 85*, 721–729.

Hughes, S. W., & Crunelli, V. (2007). Just a phase they're going through: the complex interaction of intrinsic high-threshold bursting and gap junctions in the generation of thalamic alpha and theta rhythms. *International Journal of Psychophysiology, 64*(1), 3–17.

Hughes, S. W., Lorincz, M. L., Parri, H. R., & Crunelli, V. (2011). Infraslow (<0.1 Hz) oscillations in thalamic relay nuclei basic mechanisms and significance to health and disease states. *Progress in Brain Research, 193*, 145–262.

Inase, M., Li, B. M., & Tanji, J. (1997). Dopaminergic modulation of neuronal activity in the monkey putamen through D1 and D2 receptors during a delayed Go/Nogo task. *Experimental Brain Research, 117*(2), 207–218.

Jensen, O., Goel, P., Kopell, N., Pohja, M., Hari, R., & Ermentrout, B. (2005). On the human sensorimotor-cortex beta rhythm: sources and modeling. *Neuroimage, 26*(2), 347–355.

Jeon, Y. W., & Polich, J. (2003). Meta-analysis of P300 and schizophrenia: patients, paradigms, and practical implications. *Psychophysiology, 40,* 684–701.

Ji, J., Porjesz, B., & Begleiter, H. (1998). ERP components in category matching tasks. *Electroencephalography and Clinical Neurophysiology, 108*(4), 380–389.

John, E. R. (1977). *Neurometrics: Clinical applications of quantitative electrophysiology.* New Jersey: Lawrence Erlbaum Associates.

John, E. R. (1990). Principles of neurometrics. *American Journal of EEG Technology, 30,* 251–266.

Johnson, M. D., & Ojemann, G. A. (2000). The role of the human thalamus in language and memory: evidence from electrophysiological studies. *Brain and Cognition, 42*(2), 218–230.

Johnstone, S. J., Barry, R. J., & Anderson, J. W. (2001). Topographic distribution and developmental timecourse of auditory event-related potentials in two subtypes of attention-deficit hyperactivity disorder. *International Journal of Psychophysiology, 42,* 73–94.

Johnstone, J., Gunkelman, J., & Lunt, J. (2005). Clinical database development: characterization of EEG phenotypes. *Clinical EEG and Neuroscience, 36*(2), 99–107.

Karlsgodt, K. H., Sun, D., Jimenez, A. M., Lutkenhoff, E. S., Willhite, R., Van Erp, T. G. M., & Cannon, T. D. (2008). Developmental disruptions in neural connectivity in the pathophysiology of schizophrenia. *Development and Psychopathology, 20,* 1297–1327.

Kastner, S., & Ungerleider, L. G. (2001). The neural basis of biased competition in the human visual cortex. *Neuropsychologia, 39,* 1263–1276.

Kimura, M. (2012). Visual mismatch negativity and unintentional temporal-context-based prediction in vision. *International Journal of Psychophysiology, 83*(2), 144–155.

Klimesch, W. (1999). EEG alpha and theta oscillations reflect cognitive and memory performance: a review and analysis. *Brain Research Reviews, 29,* 169–195.

Krolak-Salmon, P., Henaff, M. A., Vighetto, A., Bertrand, O., & Mauguière, F. (2004). Early amygdala reaction to fear spreading in occipital, temporal, and frontal cortex: a depth electrode ERP study in humans. *Neuron, 42,* 665–676.

Kropotov, J. D. (2009). *Quantitative EEG, event-related potentials and neurotherapy.* Amsterdam, London: Elsevier, Academic Press.

Kropotov, J. D., & Etlinger, S. C. (1999). Selection of actions in the basal ganglia-thalamocortical circuits: review and model. *International Journal of Psychophysiology, 31*(3), 197–217.

Kropotov, J. D., & Ponomarev, V. A. (2015). Differentiation of neuronal operations in latent components of event-related potentials in delayed match-to-sample tasks. *Psychophysiology, 52*(6), 826–838.

Kropotov, J. D., Etlinger, S. C., & Ponomarev, . V. A. (1997). Human multiunit activity related to attention and preparatory set. *Psychophysiology, 34,* 495–500.

Kropotov, J. D., Ponomarev, V. A., Kropotova, O. V., Anichkov, A. D., & Nechaev, V. B. (2000). Human auditory-cortex mechanisms of preattentive sound discrimination. *Neuroscience Letters, 280,* 87–90.

Kropotov, J. D., Etlinger, S. C., & Ponomarev, V. A. (1997). Human multiunit activity related to attention and preparatory set. *Psychophysiology, 34,* 495–500.

Kropotov, J. D., Grin-Yatsenko, V. A., Ponomarev, V. A., Chutko, L. S., Yakovenko, E. A., & Nikishena, I. S. (2005). ERP correlates of EEG relative beta training in ADHD children. *International Journal of Psychophysiology, 55,* 23–34.

Kropotov, J. D., Poliakov, Iu. I., Ryzhenkova, Iu. Iu., Konenkov, S. Iu., Ponomarev, V. A., Anichkov, A. D., & Pronina, M. V. (2007). Changes in the late positive component of evoked potentials in the GO/NOGO test after cryocingulotomy. *Fiziol Cheloveka, 33*(2), 16–22 (Russian).

Lang, P.J., Bradley, M.M., Cuthbert, B.N., 2005. International affective picture system (IAPS): Affective ratings of pictures and instruction manual. Technical Report A-6. University of Florida, Gainesville, FL.

Linkenkaer-Hansen, K., Nikulin, V. V., Palva, S., Ilmoniemi, R. J., & Palva, J. M. (2004). Prestimulus oscillations enhance psychophysical performance in humans. *Journal of Neuroscience, 24*, 10186–10190.

Light, G. A., Swerdlow, N. R., Thomas, M. L., Calkins, M. E., Green, M. F., Greenwood, T. A., Gur, R. E., Gur, R. C., Lazzeroni, L. C., Nuechterlein, K. H., Pela, M., Radant, A. D., Seidman, L. J., Sharp, R. F., Siever, L. J., Silverman, J. M., Sprock, J., Stone, W. S., Sugar, C. A., Tsuang, D. W., Tsuang, M. T., Braff, D. L., & Turetsky, B. I. (2015). Validation of mismatch negativity and P3a for use in multi-site studies of schizophrenia: characterization of demographic, clinical, cognitive, and functional correlates in COGS-2. *Schizophrenia Research, 163*(1–3), 63–72.

Lubar, J. F., & Shouse, M. N. (1976). EEG and behavioral changes in a hyperkinetic child concurrent with training of the sensorimotor rhythm (SMR): a preliminary report. *Biofeedback Self Regulation, 3*, 293–306.

McLoughlin, G., Makeig, S., & Tsuang, M. T. (2014). In search of biomarkers in psychiatry: EEG-based measures of brain function. *American Journal of Medical Genetics, 165B*(2), 111–121.

Monto, S., Palva, S., Voipio, J., & Palva, J. M. (2008). Very slow EEG fluctuations predict the dynamics of stimulus detection and oscillation amplitudes in humans. *Journal of Neuroscience, 28*(33), 8268–8272.

Neumann, M. F., & Schweinberger, S. R. (2008). N250r and N400 ERP correlates of immediate famous face repetition are independent of perceptual load. *Brain Research, 1239*, 181–190.

Nikulin, V. V., Fedele, T., Mehnert, J., Lipp, A., Noack, C., Steinbrink, J., & Curio, G. (2014). Monochromatic ultra-slow (~0.1 Hz) oscillations in the human electroencephalogram and their relation to hemodynamics. *NeuroImage, 97*, 71–80.

Nitsche, M. A., Liebetanz, D., Lang, N., Antal, A., Tergau, F., & Paulus, W. (2003). Safety criteria for transcranial direct current stimulation (tDCS) in humans. *Clinical Neurophysiology, 114*(11), 2220–2222.

Nitsche, M. A., & Paulus, W. (2001). Sustained excitability elevations induced by transcranial DC motor cortex stimulation in humans. *Neurology, 57*(10), 1899–1901.

Norman, D. A., & Shallice, T. (1986). Attention to action: willed and automatic control of behaviour. In R. J. Davidson, G. E. Schwartz, & D. Shapiro (Eds.), *Consciousness and self-regulation: advances in research and theory* (Volume 4). NY: Plenum Press.

Ogrim, G., Kropotov, J., Brunne, J. F., Candrian, G., Sandvik, L., & Hestad, K. A. (2014). Predicting the clinical outcome of stimulant medication in pediatric attention-deficit/hyperactivity disorder: data from quantitative electroencephalography, event-related potentials, and a go/no-go test. *Journal of Neuropsychiatric Disease and Treatment, 10*, 231–242.

Onton, J., & Makeig, S. (2006). Information-based modeling of event-related brain dynamics. *Progress in Brain Research, 159*, 99–120.

Othmer, S., & Othmer, S. F. (2006). Efficacy of neurofeedback for pain management. In M. V. Boswell, & B. E. Cole (Eds.), *Weiner's pain management: A practical guide for clinicians* (7th ed., pp. 719–739). Boca Raton, FL: CRC Press.

Pascual-Marqui, R. D., Michel, C. M., & Lehmann, D. (1994). Low resolution electromagnetic tomography: a new method for localizing electrical activity in the brain. *International Journal of Psychophysiology, 18*(1), 49–65.

Penfield, W. (1952). Memory mechanisms. *AMA Archives of Neurology and Psychiatry, 67*, 178–198.

Peng, Z., Lui, S. S., Cheung, E. F., Jin, Z., Miao, G., Jing, J., & Chan, R. C. K. (2012). Brain structural abnormalities in obsessive–compulsive disorder: converging evidence from white matter and grey matter. *Asian Journal of Psychiatry, 5*, 290–296.

Perrin, F., Pernier, J., Bertrand, O., & Echallier, J. F. (1989). Spherical splines for scalp potential and current density mapping. *Electroencephalography and Clinical Neurophysiology, 72*(2), 184–187.

Pessoa, L., Kastner, S., & Ungerleider, L. G. (2002). Attentional control of the processing of neutral and emotional stimuli. *Cognitive Brain Research, 15*, 31–45.

Peyre, H., Hoertel, N., Cortese, S., Acquaviva, E., De Maricourt, P., Limosin, F., & Delorme, R. (2014). Attention-deficit/hyperactivity disorder symptom expression: a comparison of individual age at onset using item response theory. *Journal of Clinical Psychiatry, 75*(4), 386–392.

Pliszka, S. R., Liotti, M., & Woldorff, M. G. (2000). Inhibitory control in children with attention-deficit/hyperactivity disorder: event-related potentials identify the processing component and timing of an impaired right-frontal response-inhibition mechanism. *Biological Psychiatry, 48*(3), 238–246.

Ponomarev, V. A., Mueller, A., Candrian, G., Grin-Yatsenko, V. A., & Kropotov, J. D. (2014). Group independent component analysis (gICA) and current source density (CSD) in the study of EEG in ADHD adults. *Clinical Neurophysiology, 125*(1), 83–97.

Raghavachari, S., Kahana, M. J., Rizzuto, D. S., Caplan, J. B., Kirschen, M. P., Bourgeois, B., Madsen, J. R., & Lisman, J. E. (2001). Gating of human theta oscillations by a working memory task. *Journal of Neuroscience, 21*(9), 3175–3183.

Reato, D., Rahman, A., Bikson, M., & Parra, L. C. (2010). Low-intensity electrical stimulation affects network dynamics by modulating population rate and spike timing. *Journal of Neuroscience, 30*, 15067–15079.

Rellecke, J., Sommer, W., & Schacht, A. (2012). Does processing of emotional facial expressions depend on intention? Time-resolved evidence from event-related brain potentials. *Biological Psychology, 90*, 23–32.

Ritter, P., Moosmann, M., & Villringer, A. (2009). Rolandic alpha and beta EEG rhythms' strengths are inversely related to fMRI-BOLD signal in primary somatosensory and motor cortex. *Human Brain Mapping, 30*, 1168–1187.

Rosvold, H. E., Mirsky, A. F., Sarason, I., Bransome, E. D., Jr., & Beck, L. H. (1956). A continuous performance test of brain damage. *Journal of Consulting Psychology, 20*, 343–350.

Schmidt, B., & Hanslmayr, S. (2009). Resting frontal EEG alpha-asymmetry predicts the evaluation of affective musical stimuli. *Neuroscience Letters, 460*, 237–240.

Schmolesky, M. T., Wang, Y., Hanes, D. P., Thompson, K. G., Leutgeb, S., Schall, J. D., & Leventha, A. G. (1998). Signal timing across the macaque visual system. *Journal of Neurophysiology, 79*(6), 3272–3278.

Sebastian, A., Jung, P., Krause-Utz, A., Lieb, K., Schmahl, C., & Tüscher, O. (2014). Frontal dysfunctions of impulse control – a systematic review in borderline personality disorder and attention-deficit/hyperactivity disorder. *Frontiers in Human Neuroscience, 8*, 698.

Sham, P. C., MacLean, C. J., & Kendler, K. S. (1994). A typological model of schizophrenia based on age at onset, sex and familial morbidity. *Acta Psychiatrica Scandinavica, 89*(2), 135–141.

Sherlin, L., & Congedo, M. (2005). Obsessive–compulsive dimension localized using low-resolution brain electromagnetic tomography (LORETA). *Neuroscience Letters, 387*, 72–74.

Siegel, M., Kording, K. P., & Konig, P. (2000). Integrating top-down and bottom-up sensory processing by somato-dendritic interactions. *Journal of Computational Neuroscience, 8*, 161–173.

Singh, K. D., Barnes, G. R., Hillebrand, A., Forde, E. M., & Williams, A. L. (2002). Task related changes in cortical synchronization are spatially coincident with the hemodynamic response. *NeuroImage, 16*, 103–114.

Smith, R. C., Boules, S., Mattiuz, S., Youssef, M., Tobe, R. H., Sershen, H., Lajtha, A., Nolan, K., Amiaz, R., & Davis, J. M. (2015). Effects of transcranial direct current stimulation (tDCS) on cognition, symptoms, and smoking in schizophrenia: a randomized controlled study. *Schizophrenia Research, 168*(1–2), 260–266, doi: 10.1016/j.schres.2015.06.011.

Sperling, G. (1963). A model for visual memory tasks. *Human Factors, 5*, 19–31.

Sreenivasan, K. K., Gratton, C., Vytlacil, J., & D'Esposito, M. (2014). Evidence for working memory storage operations in perceptual cortex. *Cognitive, Affective, & Behavioral Neuroscience, 14*(1), 117–128.

Stuss, D. T., Shallice, T., Alexander, M. P., & Picton, T. W. (1995). A multidisciplinary approach to anterior attentional functions. *Annals NY Academy of Science, 769*, 191–211.

Sugase, Y., Yamane, S., Ueno, S., & Kawano, K. (1999). Global and fine information coded by single neurons in the temporal visual cortex. *Nature, 400*(6747), 869–873.

Sutton, S., Braren, M., Zubin, J., & John, E. R. (1965). Evoked-potential correlates of stimulus uncertainty. *Science, 150*, 1187–1188.

Takahashi, T., Murata, T., Hamada, T., Omori, M., Kosaka, H., Kikuchi, M., Yoshida, H., & Wada, Y. (2005). Changes in EEG and autonomic nervous activity during meditation and their association with personality traits. *International Journal of Psychophysiology, 55*, 199–207.

Uhlhaas, P. J., & Singer, W. (2015). Oscillations and neuronal dynamics in schizophrenia: the search for basic symptoms and translational opportunities. *Biological Psychiatry, 77*(12), 1001–1009.

Ulbert, I., Heit, G., Madsen, J., Karmos, G., & Halgren, E. (2004). Laminar analysis of human neocortical interictal spike generation and propagation: current source density and multiunit analysis in vivo. *Epilepsia, 45*(Suppl. 4), 48–56.

Vanhatalo, S., Voipio, J., & Kaila, K. (2005). Full-band EEG (fbEEG): a new standard for clinical electroencephalography. *Clinical EEG and Neuroscience, 36*(4), 311–317.

van Tricht, M. J., Nieman, D. H., Koelman, J. H., van der Meer, J. N., Bour, L. J., de Haan, L., & Linszen, D. H. (2010). Reduced parietal P300 amplitude is associated with an increased risk for a first psychotic episode. *Biological Psychiatry, 68*(7), 642–648.

Vogel, E. K., & Machizawa, M. G. (2004). Neural activity predicts individual differences in visual working memory capacity. *Nature, 428*(6984), 748–751.

Weinberg, A., & Hajcak, G. (2011). Electrocortical evidence for vigilance-avoidance in generalized anxiety disorder. *Psychophysiology, 48*(6), 842–885.

Whittle, S., Allen, N. B., Lubman, D. I., & Yucel, M. (2006). The neurobiological basis of temperament: towards a better understanding of psychopathology. *Neuroscience & Biobehavioral Reviews, 30*(4), 511–525.

Wolfers, T., Buitelaar, J. K., Beckmann, C. F., Franke, B., & Marquand, A. F. (2015). From estimating activation locality to predicting disorder: A review of pattern recognition for neuroimaging-based psychiatric diagnostics. *Neuroscience and Biobehavioral Reviews, 57*, 328–349.

Woodman, G. F., & Arita, J. T. (2011). Direct electrophysiological measurement of attentional templates in visual working memory. *Psychological Science, 22*, 212–215.

Wutzler, A., Winter, C., Kitzrow, W., Uhl, I., Wolf, R. J., Heinz, A., & Juckel, G. (2008). Loudness dependence of auditory evoked potentials as indicator of central serotonergic neurotransmission: simultaneous electrophysiological recordings and in vivo microdialysis in the rat primary auditory cortex. *Neuropsychopharmacology, 33*, 3176–3181.

Wynn, J. K., Horan, W. P., Kring, A. M., Simons, R. F., & Green, M. F. (2010). Impaired anticipatory event-related potentials in schizophrenia. *International Journal of Psychophysiology, 77*, 141–149.

Zhang, X., Zhaoping, L., Zhou, T., & Fang, F. (2012). Neural activities in v1 create a bottom-up saliency map. *Neuron, 73*(1), 183–192.

Zielasek, J., Ehlis, A. C., Herrmann, M. J., & Fallgatter, A. J. (2005). Reduced prefrontal response control in patients with schizophrenias: a subgroup analysis. *Journal of Neural Transmission (Vienna), 112*(7), 969–977.

Further Readings

Aladzhalova, N. A., Kol'tsova, A.V. (1971). Infralow frequency spectrum of electrical phenomena in the brain. *Doklady Akademii Nauk SSSR, 197*, 973–976.

Alexander, G. E., DeLong, M. R., & Strick, P. L. (1990). Parallel organization of functionally segregated circuits linking basal ganglia and cortex. *Annual Review of Neuroscience, 9*, 357–381.

Allan, K., & Rugg, M. D. (1997). An event-related potential study of explicit memory on test of word–stem cued recall and recognition memory. *Cognitive Brain Research, 4*, 251–262.

Allen, J. J. B., Harmon-Jones, E., & Cavender, J. H. (2001). Manipulation of frontal EEG asymmetry through biofeedback alters self-reported emotional responses and facial EMG. *Psychophysiology, 38*, 685–693.

Allman, J. M., Hakeem, A., Erwin, J. M., Nimchinsky, E., Hof, P., & Hixon, F. P. (2001). The anterior cingulate cortex: the evolution of an interface between emotion and cognition. *Annals of the New York Academy of Sciences, 935*, 107–117.

Anderson, B., & Sheinberg, D. L. (2008). Effects of temporal context and temporal expectancy on neural activity in inferior temporal cortex. *Neuropsychologia, 46*, 947–957.

Arcos-Burgos, M., & Acosta, M. T. (2007). Tuning major gene variants conditioning human behavior: the anachronism of ADHD. *Current Opinion in Genetics & Development, 17*(3), 234–238.

Arns, M., Conners, C. K., & Kraemer, H. C. (2012). A decade of EEG theta/beta ratio research in ADHD: a meta-analysis. *Journal of Attention Disorders, 17*(5), 374–383.

Arns, M., Heinrich, H., & Strehle, U. (2014). Evaluation of neurofeedback in ADHD: the long and winding road. *Biological Psychology, 95*, 108–115.

Arsalidou, M., & Taylor, M. J. (2011). Is 2 + 2 = 4? Meta-analyses of brain areas needed for numbers and calculations. *NeuroImage, 54*, 2382–2393.

Aston-Jones, G., & Cohen, J. D. (2005). Adaptive gain and the role of the locus coeruleus–norepinephrine system in optimal performance. *Journal of Comparative Neurology, 493*(1), 99–110.

Atkinson, R. C., & Shiffrin, R. M. (1971). The control of short-term memory. *Scientific American, 225*, 82–90.

Baddeley, A. D. (1986). *Working memory*. Oxford: Clarendon Press.

Baddeley, A. D. (2012). Working memory: theories, models, and controversies. *Annual Review of Psychology, 63*, 1–29.

Bakhtadze, S., Beridze, M., Geladze, N., Khachapuridze, N., & Bornstein, N. (2016). Effect of EEG biofeedback on cognitive flexibility in children with Attention Deficit Hyperactivity Disorder with and without epilepsy. *Applied Psychophysiology and Biofeedback, 41*(1), 71–79.

Banich, M. T. (2009). Executive function: the search for an integrated account. *Current Directions in Psychological Science, 18*(2), 89–94.

Bannon, S., Gonsalvez, C. J., Croft, R. J., & Boyce, P. M. (2002). Response inhibition deficits in obsessive–compulsive disorder. *Psychiatry Research, 110*(2), 165–174.

Barker, A. T., Jalinous, R., & Freeston, I. L. (1985). Non-invasive magnetic stimulation of human motor cortex. *Lancet, 325*(8437), 1106–1107.

Barlow, D. H. (2000). Unraveling the mysteries of anxiety and its disorders from the perspective of emotion theory. *American Psychologist, 55*(11), 1247–1263.

Bauer, H., & Pllana, A. (2014). EEG-based local brain activity feedback training–tomographic neurofeedback. *Frontiers in Behavioral Neuroscience, 8,* 1005.

Benabid, A. L., Pollak, P., Louveau, A., Henry, S., & de Rougemont, J. (1987). Combined (thalamotomy and stimulation) stereotactic surgery of the VIM thalamic nucleus for bilateral Parkinson disease. *Applied Neurophysiology, 50*(1–6), 344–346.

Benabid, A. L., Pollak, P., Louveau, A., Henry, S., & de Rougemont, J. (1987). Combined (thalamotomy and stimulation) stereotactic surgery of the VIM thalamic nucleus for bilateral Parkinson disease. *Applied Neurophysiology, 50,* 344–346.

Berlim, M. T., Neufeld, N. H., & Van den Eynde, F. (2013). Repetitive transcranial magnetic stimulation (rTMS) for obsessive–compulsive disorder (OCD): an exploratory meta-analysis of randomized and sham-controlled trials. *Journal of Psychiatric Research, 47*(8), 999–1006.

Birbaumer, N., Elbert, T., Canavan, A., & Rockstroh, B. (1990). Slow potentials of the cerebral cortex and behavior. *Physiological Reviews, 70,* 1–41.

Bitsakou, P., Psychogiou, L., Thompson, M., & Sonuga-Barke, E. J. (2009). Delay aversion in attention deficit/hyperactivity disorder: an empirical investigation of the broader phenotype. *Neuropsychologia, 47,* 446–456.

Bleuler, E. (1908). Die Prognose der Dementia praecox (Schizophreniegruppe). *Allgemeine ZeitschriftfürPsychiatrie und psychisch-gerichtliche Medicin, 65,* 436–464.

Bloch, Y., Harel, E. V., Aviram, S., Govezensky, J., Ratzoni, G., & Levkovitz, Y. (2010). Positive effects of repetitive transcranial magnetic stimulation on attention in ADHD subjects: a randomized controlled pilot study. *World Journal of Biological Psychiatry, 11*(5), 755–758.

Blumenfeld, H., McNally, K. A., Ostroff, R. B., & Zubal, I. G. (2003a). Targeted prefrontal cortical activation with bifrontal ECT. *Psychiatric Research, 123*(3), 165–170.

Blumenfeld, H., Westerveld, M., Ostroff, R. B., Vanderhill, S. D., Freeman, J., Necochea, A., Uranga, P., Tanhehco, T., Smith, A., Seibyl, J. P., Stokking, R., Studholme, C., Spencer, S. S., & Zubal, I. G. (2003b). Selective frontal, parietal and temporal networks in generalized seizures. *NeuroImage, 19,* 1556–1566.

Boonstra, A. M., Kooij, J. J. S., Oosterlaan, J., Sergeant, J. A., & Buitelaar, J. K. (2010). To act or not to act, that's the problem: primarily inhibition difficulties in adult ADHD. *Neuropsychology, 24,* 209–221.

Bose, A., Shivakumar, V., Narayanaswamy, J. C., Nawani, H., Subramaniam, A., Agarwal, S. M., Chhabra, H., Kalmady, S. V., & Venkatasubramanian, G. (2014). Insight facilitation with add-on tDCS in schizophrenia. *Schizophrenia Research, 156*(1), 63–65.

Bramon, E., Rabe-Hesketh, S., Sham, P., Murray, R. M., & Frangou, S. (2004). Meta-analysis of the P300 and P50 waveforms in schizophrenia. *Schizophrenia Research, 70,* 315–329.

Braver, T. S., & Barch, D. M. (2002). A theory of cognitive control, aging cognition and neuromodulation. *Neuroscience & Biobehavioral Reviews, 26,* 809–817.

Braver, T. S., Gray, J. R., & Burgess, G. C. (2007). Explaining the many varieties of working memory variation: dual mechanisms of cognitive control. In A. Conway et al. (Ed.), *Variation in working memory* (pp. 76–106). Oxford: Oxford University Press.

Bresnahan, S. M., Anderson, J. W., & Barry, R. J. (1999). Age-related changes in quantitative EEG in Attention- Deficit/Hyperactivity Disorder. *Biological Psychiatry, 46,* 1690–1697.

Brewer, J. B., & Moghekar, A. (2002). Imaging the medial temporal lobe: exploring new dimensions. *Trends in Cognitive Sciences, 6,* 217–223.

Broadbent, D. E. (1958). *Perception and communication.* London: Pergamon Press.

Brozoski, T. J., Brown, R. M., Rosvold, H. E., & Goldman, P. S. (1979). Cognitive deficit caused by regional depletion of dopamine in prefrontal cortex of rhesus monkey. *Science, 205*(4409), 929–932.

Brunelin, J., Mondino, M., Gassab, L., Haesebaert, F., Gaha, L., Suaud-Chagny, M. F., Saoud, M., Mechri, A., & Poulet, E. (2012). Examining transcranial direct-current stimulation (tDCS) as a treatment for hallucinations in schizophrenia. *The American Journal of Psychiatry, 169,* 719–724.

Brunoni, A. R., & Vanderhasselt, M. A. (2014). Working memory improvement with non-invasive brain stimulation of the dorsolateral prefrontal cortex: a systematic review and meta-analysis. *Brain and Cognition, 86*, 1–9.

Buckner, R. L., & Wheeler, M. E. (2001). The cognitive neuroscience of remembering. *Nature Reviews Neuroscience, 2*, 624–634.

Butler, P. D., Martinez, A., Foxe, J. J., Kim, D., Zemon, V., Silipo, G., Mahoney, J., Shpaner, M., Jalbrzikowski, M., & Javitt, D. C. (2007). Subcortical visual dysfunction in schizophrenia drives secondary cortical impairments. *Brain, 130*, 417–430.

Buzsáki, G., Anastassiou, C. A., & Koch, C. (2012). The origin of extracellular fields and currents—EEG, ECoG, LFP and spikes. *Nature Reviews Neuroscience, 13*(6), 407–420.

Cade, J. (1949). Lithium salts in the treatment of psychotic excitement. *Medical Journal of Australia, 2*(36), 349–352.

Chabot, R. J., & Serfontein, G. (1996). Quantitative electroencephalographic profiles of children with attention deficit disorder. *Biological Psychiatry, 40*(10), 951–963.

Chamberlain, S. R., Blackwell, A. D., Fineberg, N. A., Robbins, T. W., & Sahakian, B. J. (2005). The neuropsychology of obsessive–compulsive disorder: the importance of failures in cognitive and behavioral inhibition as candidate endophenotypic markers. *Neuroscience & Biobehavioral Reviews, 29*, 399–419.

Chernigovskaya, N. (1984). Biofeedback control in epilepsy and neuroses. In T. Elbert, B. Rockstroh, W. Lutzenberger, & N. Birbaumer (Eds.), *Self-Regulation of the brain and behavior*. Berlin: Springer-Verlag.

Chi, R. P., Fregni, F., & Snyder, A. W. (2010). Visual memory improved by non-invasive brain stimulation. *Brain Research, 1353*, 168–175.

Cisek, P., & Kalaska, J. F. (2005). Neural correlates of reaching decisions in dorsal premotor cortex: specification of multiple direction choices and final selection of action. *Neuron, 45*(5), 801–814.

Clarke, A., Barry, R., McCarthy, R., & Selikowitz, M. (2001a). Excess beta in children with attention-deficit/hyperactivity disorder: an atypical electrophysiological group. *Psychiatric Research, 103*, 205–218.

Clarke, A., Barry, R., McCarthy, R., & Selikowitz, M. (2001b). EEG-defined subtypes of children with attention-deficit/hyperactivity disorder. *Clinical Neurophysiology, 112*, 2098–2105.

Clarke, A. R., Barry, R. J., McCarthy, R., & Selikowitz, M. (2001c). Electroencephalogram differences in two subtypes of attention-deficit/hyperactivity disorder. *Psychophysiology, 38*(2), 212–221.

Clarke, A. R., Barry, R. J., McCarthy, R., Selikowitz, M., Clarke, D. C., & Croft, R. J. (2003). Effects of stimulant medications on children with attention-deficit/hyperactivity disorder and excessive beta activity in their EEG. *Clinical Neurophysiology, 114*(9), 1729–1737.

Clarke, A. R., Barry, R. J., McCarthy, R., Selikowitz, M., & Johnstone, S. J. (2007). Effects of stimulant medications on the EEG of girls with Attention-Deficit/Hyperactivity Disorder. *Clinical Neurophysiology, 118*(12), 2700–2708.

Clarke, A. R., Barry, R. J., Dupuy, F. E., McCarthy, R., Selikowitz, M., & Johnstone, S. J. (2013). Excess beta activity in the EEG of children with attention-deficit/hyperactivity disorder: a disorder of arousal? *International Journal of Psychophysiology, 89*(3), 314–319.

Congedo, M., Lubar, J. F., & Joffe, D. (2004). Low-resolution electromagnetic tomography neurofeedback. *IEEE Transactions on Neural Systems and Rehabilitation Engineering, 12*(4), 387–397.

Congedo, M., John, R. E., De Ridder, D., & Prichep, L. (2010). Group independent component analysis of resting state EEG in large normative samples. *Clinical Neurophysiology, 78*(2), 89–99.

Cools, R., & D'Esposito, M. (2011). Inverted-U-shaped dopamine actions on human working memory and cognitive control. *Biological Psychiatry, 69*, 113–125.

Corbetta, M., & Shulman, G. L. (2011). Spatial neglect and attention networks. *Annual Review of Neuroscience, 34,* 569–599.

Cosmo, C., Ferreira, C., Miranda, J. G. V., do Rosário, R. S., Baptista, A. F., Montoya, P., & de Sena, E. P. (2015). Spreading effect of tDCS in individuals with Attention-Deficit/Hyperactivity Disorder as shown by functional cortical networks: a randomized, double-blind, sham-controlled trial. *Frontiers in Psychiatry, 6.*

Cui, H., & Andersen, R. A. (2007). Posterior parietal cortex encodes autonomously selected motor plans. *Neuron, 56*(3), 552–559.

Czigler, I. (2010). *Representation of regularities in visual stimulation: Event-related potentials reveal the automatic acquisition. unconscious memory representation in perception* (pp. 107–132). Amsterdam: John Benjamins.

D'Urso, G., Brunoni, A. R., Anastasia, A., Micillo, M., de Bartolomeis, A., & Mantovani, A. (2015). Polarity-dependent effects of transcranial direct current stimulation in obsessive-compulsive disorder. *Neurocase, 22*(1), 60–64.

deCharms, R. C., Maeda, F., Glover, G. H., Ludlow, D., Pauly, J. M., Soneji, D., Gabrieli, J. D. E., & Mackey, S. C. (2005). Control over brain activation and pain learned by using real-time functional MRI. *Proceedings of the National Academy of Sciences, 102,* 18626–18631.

Deffke, I., Sander, T., Heidenreich, J., Sommer, W., Curio, G., Trahms, L., & Lueschow, A. (2007). MEG/EEG sources of the 170-ms response to faces are co-localized in the fusiform gyrus. *NeuroImage, 35*(4), 1495–1501.

DeLong, M. R., & Wichmann, T. (2007). Circuits and circuit disorders of the basal ganglia. *Archives of Neurology, 64,* 20–24.

Demirtas-Tatlidede, A., Vahabzadeh-Hagh, A. M., & Pascual-Leone, A. (2013). Can noninvasive brain stimulation enhance cognition in neuropsychiatric disorders? *Neuropharmacology, 64,* 566–578, doi: 10.1016/j.neuropharm.2012.06.020.

Denys, D., Mantione, M., Figee, M., van den Munckhof, P., Koerselman, F., Westenberg, H., Bosch, A., & Schuurman, R. (2010). Deep brain stimulation of the nucleus accumbens for treatment-refractory obsessive–compulsive disorder. *Archives of General Psychiatry, 67,* 1061–1068.

D'Esposito, M. (2007). From cognitive to neural models of working memory. *Philosophical Transactions of the Royal Society B, 362,* 761–772.

Dmochowski, J. P., Datta, A., Huang, Y., Richardson, J. D., Bikson, M., Fridriksson, J., & Parra, L. C. (2013). Targeted transcranial direct current stimulation for rehabilitation after stroke. *NeuroImage, 75,* 12–19.

Donchin, E. (1981). Surprise! Surprise! *Psychophysiology, 18*(5), 493–513.

Donchin, E., & Coles, M. G. (1988). Is the P300 component a manifestation of context updating. *Behavioral and Brain Science, 11,* 357–374.

Dosenbach, N. U. F., Fair, D. A., Cohen, A. L., Schlaggar, B. L., & Petersen, S. E. (2008). A dual-networks architecture of top-down control. *Trends in Cognitive Sciences, 12,* 99–105.

Dougall, N., Maayan, N., Soares-Weiser, K., McDermott, L. M., & McIntosh, A. (2015). Transcranial magnetic stimulation for schizophrenia. *Schizophrenia Bulletin, 41*(6), 1220–1222.

Douglas, R. J., & Martin, K. A. C. (2007). Mapping the matrix: the ways of neocortex. *Neuron, 56,* 226–238.

Durand, D. M. (2003). Electric field effects in hyperexcitable neural tissue: a review. *Radiation Protection Dosimetry, 106*(4), 325–331.

Dux, P. E., & Marois, R. (2009). How humans search for targets through time: a review of data and theory from the attentional blink. *Attention, Perception, & Psychophysics, 71,* 1683–1700.

Ehlis, A. C., Pauli, P., Herrmann, M. J., Plichta, M. M., Zielasek, J., Pfuhlmann, B., Stober, G., Ringel, T., Jabs, B., & Fallgatter, A. J. (2012). Hypofrontality in schizophrenic patients and its relevance for the choice of antipsychotic medication: an event-related potential study. *World Journal of Biological Psychiatry, 13,* 188–199.

Enriquez-Geppert, S., Huster, R. J., & Herrmann, C. S. (2013). Boosting brain functions: improving executive functions with behavioral training, neurostimulation, and neurofeedback. *International Journal of Psychophysiology, 88*, 1–16.

Enriquez-Geppert, S., Huster, R. J., Figge, C., & Herrmann, C. S. (2014). Self-regulation of frontal–midline theta facilitates memory updating and mental set shifting. *Frontiers in Behavioral Neuroscience, 8*, 420.

Escera, C., Alho, K., Winkler, I., & Naatanen, R. (1998). Neural mechanisms of involuntary attention to acoustic novelty and change. *Journal of Cognitive Neuroscience, 10*, 590–604.

Eysenck, H. J. (1967). *The biological basis of personality*. Springfield, IL: Thomas.

Faraone, S. V., Sergeant, J., Gillberg, C., & Biederman, J. (2003). The worldwide prevalence of ADHD: is it an American condition? *World Psychiatry, 2*(2), 104–113.

Faraone, S. V., Perlis, R. H., Doyle, A., Smoller, J. W., Goralnick, J. J., & Holmgren, M. A. (2005). Molecular genetics of attention-deficit/hyperactivity disorder. *Biological Psychiatry, 57*, 1313–1323.

Fiske, D. W. (1949). Consistency of the factorial structures of personality ratings from different sources. *Journal of Abnormal Psychology, 44*(3), 329–344.

Flöel, A. (2014). tDCS-enhanced motor and cognitive function in neurological diseases. *NeuroImage, 85*(3), 934–947.

Ford, J. M., Gray, M., & Whitfield, S. L. (2004). Acquiring and inhibiting prepotent responses in schizophrenia: event-related brain potentials and functional magnetic resonance imaging. *Archives of General Psychiatry, 61*, 119–129.

Foxe, J. J., Yeap, S., Snyder, A. C., Kelly, S. P., Thakore, J. H., & Molholm, S. (2011). The N1 auditory evoked potential component as an endophenotype for schizophrenia: high-density electrical mapping in clinically unaffected first-degree relatives, first-episode, and chronic schizophrenia patients. *European Archives of Psychiatry and Clinical Neuroscience, 261*(5), 331–339.

Fulton, J. E. (1951). *Frontal Lobotomy and Affective Behavior—A Neurophysiological Analysis*. New York: WW Norton.

Garcia, L., D'Alessandro, G., Bioulac, B., & Hammond, C. (2005). High-frequency stimulation in Parkinson's disease: more or less? *Trends in Neuroscience, 28*(4), 209–216.

Gemba, H., & Sasaki, K. (1989). Potential related to no-go reaction of go/no-go hand movement task with color discrimination in human. *Neuroscience Letter, 101*, 263–268.

Giurgea, C. E. (1989). Kupalov's concept of shortened conditional reflexes: psychophysiological and psychopharmacological implications. *Pavlovian Journal of Biological Science, 24*(3), 81–89.

Gonon, F. (2009). The dopaminergic hypothesis of attention-deficit/hyperactivity disorder needs re-examining. *Trends Neuroscience, 32*(1), 2–8, doi: 10.1016/j.tins.2008.09.010.

Gonzalez-Burgos, G., Hashimoto, T., & Lewis, D. A. (2010). Alterations of cortical GABA neurons and network oscillations in schizophrenia. *Current Psychiatry Reports, 12*(4), 335–344, doi:10.1007/s11920-010-0124-8.

Gordon, E. (2007). Integrating genomics and neuromarkers for the ear of brain-related personalized medicine. *Personalized Medicine, 4*(2), 201–217.

Gottesman, I. I., & Shields, J. (1973). Genetic theorizing and schizophrenia. *British Journal of Psychiatry, 122*, 15–30.

Gray, J. A. (1982). *The Neuropsychology of Anxiety* (1st ed.). Oxford: Oxford University Press.

Haber, S. N. (2008). *Functional anatomy and physiology of the Basal Ganglia: Non-motor functions. current clinical neurology: Deep brain stimulation in neurological and psychiatric disorders.* (pp. 33-62). Totowa, NJ: Humana Press.

Hajcak, G., MacNamara, A., & Olvet, D. M. (2010). Event-related potentials, emotion, and emotion regulation: an integrative review. *Developmental Neuropsychology, 35*(2), 129–155.

Hegerl, U., & Hensch, T. (2014). The vigilance regulation model of affective disorders and ADHD. *Neuroscience & Biobehavioral Reviews, 44*, 45–57.

Heimrath, K., Sandmann, P., Becke, A., Müller, N. G., & Zaehle, T. (2012). Behavioral and electrophysiological effects of transcranial direct current stimulation of the parietal cortex in a visuo-spatial working memory task. *Frontiers in Psychiatry, 3*, 56.

Heinrich, H., Gevensleben, H., Freisleder, F. J., Moll, G. H., & Rothenberger, A. (2004). Training of slow cortical potentials in attention-deficit/hyperactivity disorder: evidence for positive behavioral and neurophysiological effects. *Biological Psychiatry, 55*(7), 772–775.

Heiss, W. D., & Herholz, K. (2006). Brain receptor imaging. *Journal of Nuclear Medicine, 47*(2), 302–312.

Hikosaka, O. (2008). Decision-making and learning by cortico-basal ganglia network. *Brain Nerve, 60*, 799–813.

Hillyard, S. A., Anllo-Vento, L., 1998. Event-related brain potentials in the study of visual selective attention. *Proceedings of the National Academy of Sciences USA, 95*,781–787.

Hillyard, S. A., Squires, K. C., Bauer, J. W., & Lindsay, P. H. (1971). Evoked potential correlates of auditory signal detection. *Science, New York, 172*(3990), 1357–1360.

Hillyard, S. A., Vogel, E. K., & Luck, S. J. (1998). Sensory gain control (amplification) as a mechanism of selective attention: electrophysiological and neuroimaging evidence. *Philosophical Transactions of the Royal Society, 353*, 1257–1270.

Hinterberger, T., Veit, R., Strehl, U., Trevorrow, T., Erb, M., Kotchoubey, B., Flor, H., & Birbaumer, N. (2003). Brain areas activated in fMRI during self-regulation of slow cortical potentials (SCPs). *Experimental Brain Research, 152*, 113–122.

Holtmann, M., Becker, K., Kentner-Figura, B., & Schmidt, M. H. (2003). Increased frequency of rolandic spikes in ADHD children. *Epilepsia, 44*, 1241–1244.

Holtzheimer, P. E., 3rd, Kosel, M., & Schlaepfer, T. (2012). Brain stimulation therapies for neuropsychiatric disease. *Neurobiology of Psychiatric Disorders, 106*, 681–695.

Hopfinger, J. B., Woldorff, M. G., Fletcher, E. M., & Mangun, G. R. (2001). Dissociating top-down attentional control from selective perception and action. *Neuropsychologia, 39*(12), 1277–1291.

Hubel, D. H., & Wiesel, T. N. (1979). Brain mechanisms of vision. *Scientific American, 241*, 150–162.

Huster, R. J., Enriquez-Geppert, S., Lavallee, C. F., Falkenstein, M., & Herrmann, C. S. (2013). Electroencephalography of response inhibition tasks: functional networks and cognitive contributions. *International Journal of Psychophysiology, 87*, 217–233.

Jahshan, C., Wynn, J. K., Mathis, K. I., Altshuler, L. L., Glahn, D. C., & Green, M. F. (2012). Cross-diagnostic comparison of duration mismatch negativity and P3a in bipolar disorder and schizophrenia. *Bipolar Disorders, 14*, 239–248.

Iannaccone, R., Hauser, T. U., Staempfli, P., Walitza, S., Brandeis, D., & Brem, S. (2015). Conflict monitoring and error processing: new insights from simultaneous EEG–fMRI. *NeuroImage, 105*, 395–407.

Ikkai, A., McCollough, A. W., & Vogel, E. K. (2010). Contralateral delay activity provides a neural measure of the number of representations in visual working memory. *Journal of Neurophysiology, 103*, 1963–1968.

Jackson, D. C., Mueller, C. J., Dolski, I., Dalton, K. M., Nitschke, J. B., Urry, H. L., Rosenkranz, M. A., Ryff, C. D., Singer, B. H., & Davidson, R. J. (2004). Making a life worth living: neural correlates of well-being. *Psychological Science, 15*, 367–372.

Jefferys, J. G., Deans, J., Bikson, M., & Fox, J. (2003). Effects of weak electric fields on the activity of neurons and neuronal networks. *Radiation Protection Dosimetry, 106*(4), 321–323.

Kamiya, J. (1968). Conscious control of brain waves. *Psychology Today, 1*, 56–60.

Keefe, R. S., Poe, M., Walker, T. M., Kang, J. W., & Harvey, P. D. (2006). The Schizophrenia Cognition Rating Scale: an interview-based assessment and its relationship to cognition, real world functioning, and functional capacity. *American Journal of EEG Technology, 163*, 426–432.

Keiji, T. (2003). Columns for complex visual object features in the inferotemporal cortex: clustering of cells with similar but slightly different stimulus selectivities. *Cerebral Cortex, 13*(1), 90–99.

Kenemans, J. L., Bekker, E. M., Lijffijt, M., Overtoom, C. C. E., Jonkman, L. M., & Verbaten, M. N. (2005). Attention deficit and impulsivity: selecting, shifting, and stopping. *International Journal of Psychophysiology, 58*, 59–70.

Kim, M. S., Kim, Y. Y., Yoo, S. Y., & Kwon, J. S. (2007). Electrophysiological correlates of behavioral response inhibition in patients with obsessive–compulsive disorder. *Depression and Anxiety, 24*(1), 22–31.

Koberda, J. L., Moses, A., Koberda, P., & Koberda, L. (2013). Clinical advantages of quantitative electroencephalogram (QEEG)-electrical neuroimaging application in general neurology practice. *Clinical EEG and Neuroscience, 44*(4), 273–285.

Kolb, B., & Whishaw, I. Q. (2000). *An Introduction to Brain and Behavior*. New York: Worth.

Kopřivová, J., Congedo, M., Raszka, M., Praško, J., Brunovský, M., & Horáček, J. (2013). Prediction of treatment response and the effect of independent component neurofeedback in obsessive–compulsive disorder: a randomized, sham-controlled, double-blind study. *Neuropsychobiology, 67*(4), 210–223, doi: 10.1159/000347087.

Kornhuber, H. H., & Deecke, L. (1965). Hirnpotentialanderungen bei Willkurbewegungen und passiven Bewegungen des Menschen. Bereitschaftspotential und reafferente Potential. *Pflugers Arch. Gesamte Physiol. Menschen Tiere, 10*(284), 1–17.

Kornhuber, H. H., & Deecke, L. (1990). Readiness for movement—the Bereitschaftpotential story. *CC/Life Science, 33*(4), 14.

Kraepelin, E. (1983). *Lebenserinnerungen*. Heidelberg: Springer, 290p.

Krause, K. -H., Dresel, S. H., Krause, J., Kung, H. F., & Tatsch, K. (2000). Increased striatal dopamine transporter in adult patients with attention deficit hyperactivity disorder: effects of methylphenidate as measured by single photon emission computed tomography. *Neuroscience Letters, 285*, 107–110.

Kringelbach, M. L., & Rolls, E. T. (2004). The functional neuroanatomy of the human orbitofrontal cortex: evidence from neuroimaging and neuropsychology. *Progress in Neurobiology, 72*(5), 341–372.

Kropotov, J. D. (1989). Brain organization of perception and memory: hypothesis for action programming. *Human Physiology, 15*(3), 19–27.

Kuhn, R. (1958). The treatment of depressive states with G 22355 (imipramine hydrochloride). *The American Journal of Psychiatry, 115*(5), 459–464.

Kuo, M. F., Paulus, W., & Nitsche, M. A. (2014). Therapeutic effects of non-invasive brain stimulation with direct currents (tDCS) in neuropsychiatric diseases. *NeuroImage, 85*(3), 948–960.

Lapidus, K. A. B., Stern, E. R., Berlin, H. A., & Goodman, W. K. (2014). Neuromodulation for obsessive–compulsive disorder. *Neurotherapeutics, 11*(3), 485–495.

Lashley, K., 1950. In search of the engram. *Society of Experimental Biology Symposium, 4*, 454–482.

Lazarsfeld, P. F. (1950). The obligations of the 1950 pollster to the 1984 historian. *Public Opinion Quarterly, 14*, 618–638.

Lee, S. H., Wynn, J. K., Green, M. F., Kim, H., Lee, K. J., Nam, M., Park, J. K., & Chung, Y. C. (2006). Quantitative EEG and low resolution electromagnetic tomography (LORETA) imaging of patients with persistent auditory hallucinations. *Schizophrenia Research, 83*(2–3), 11–119.

Lehtela, L., Salmelin, R., & Hari, R. (1997). Evidence for reactive magnetic 10-Hz rhythm in the human auditory cortex. *Neuroscience Letters, 222*(2), 111–114.

Lenartowicz, A., Delorme, A., Walshaw, P. D., Cho, A. L., Bilder, R. M., McGough, J. J., McCracken, J. T., Makeig, S., & Loo, S. K. (2014). Electroencephalography correlates of spatial working memory deficits in attention-deficit/hyperactivity disorder: vigilance, encoding, and maintenance. *Journal of Neuroscience, 34*(4), 1171–1182.

Leslie, G., Ungerleider, G., & Haxby, J. V. (1994). "What" and "where" in the human brain. *Haxby Current Opinion in Neurobiology, 4*, 157–165.

Levanen, S., Lin, F. H., Sams, M., Shinn-Cunningham, B. G., Witzel, T., & Belliveau, J. W. (2006). Task-modulated "what" and "where" pathways in human auditory cortex. *Proceedings of the National Academy of Sciences of USA, 103*(39), 14608–14613.

Levy, B. J., & Wagner, A. D. (2011). Cognitive control and right ventrolateral prefrontal cortex: reflexive reorienting, motor inhibition, and action updating. *Annals of the New York Academy of Sciences, 1224*, 40–62.

Liddell, B. J., Brown, K. J., Kemp, A. H., Barton, M. J., Das, P., Peduto, A., & Williams, L. M. (2005). A direct brainstem–amygdala–cortical 'alarm' system for subliminal signals of fear. *NeuroImage, 24*, 235–243.

Liechti, M. D., Valko, L., Müller, U. C., Döhnert, M., Drechsler, R., Steinhausen, H. C., & Brandeis, D. (2013). Diagnostic value of resting electroencephalogram in attention-deficit/hyperactivity disorder across the lifespan. *Brain Topography, 26*(1), 135–151.

Liu, Y., Murray, S. O., & Jagadeesh, B. (2009). Time course and stimulus dependence of repetition-induced response suppression in inferotemporal cortex. *Journal of Neurophysiology, 101*, 418–436.

López-Muñoz, F., Alamo, C., Cuenca, E., Shen, W. W., Clervoy, P., & Rubio, G. (2005). History of the discovery and clinical introduction of chlorpromazine. *Annals of Clinical Psychiatry, 17*(3), 113–135.

Lubar, J. F. (1991). Discourse on the development of EEG diagnostics and biofeedback treatment for attention-deficit/hyperactivity disorders. *Biofeedback and Self-Regulation, 16*, 201–225.

Luck, S. J., Mathalon, D. H., O'Donnell, B. F., Hämäläinen, M. S., Spencer, K. M., Javitt, D. C., & Uhlhaas, P. J. (2011). A roadmap for the development and validation of event-related potential biomarkers in schizophrenia research. *Biological Psychiatry, 70*, 28–34.

Luria, A. R. (1978). The human brain and conscious activity. In G. E. Schwartz, & D. Shapiro (Eds.), *Consciousness and Self-Regulation* (pp. 1–35). New York: Wiley.

Luria, A. R. (1980). *Higher cortical functions in man* (2nd ed.). New York: Basic Books.

Lynch, J. J., Paskewitz, D. A., & Orne, M. T. (1974). Inter-session stability of human alpha rhythm densities. *Electroencephalography and Clinical Neurophysiology, 36*, 538–540.

MacKay, D. J. C. (1992). Bayesian interpolation. *Neural Computation, 4*, 415–447.

MacLean, P. D. (1949). Psychosomatic disease and the visceral brain; recent developments bearing on the Papez theory of emotion. *Psychosomatic Medicine, 11*(6), 338–353.

Makeig, S., Jung, T. -P., Ghahremani, D., Bell, A. J., & Sejnowski, T. J. (1997). Blind separation of event-related brain responses into independent components. *Proceedings of the National Academy of Sciences of the United States of America, 94*, 10979–10984.

Maltby, N., Tolin, D. F., Worhunsky, P., O'Keefe, T. M., & Kiehl, K. A. (2005). Dysfunctional action monitoring hyperactivates frontal–striatal circuits in obsessive–compulsive disorder: an event-related fMRI study. *NeuroImage, 24*(2), 495–503.

Mamah, D., Wang, L., Barch, D., de Erausquin, G. A., Gado, M., & Csernansky, J. G. (2007). Structural analysis of the basal ganglia in schizophrenia. *Schizophrenia Research, 89*(1–3), 59–71.

Mantovani, A., Simpson, H. B., Fallon, B. A., Rossi, S., & Lisanby, S. H. (2010). Randomized sham-controlled trial of repetitive transcranial magnetic stimulation in treatment-resistant obsessive–compulsive disorder. *International Journal of Neuropsychopharmacology, 13*, 217–227.

Martinez, A., Di Russo, F., Annlo-Vento, L., & Hillyard, S. A. (2001). Electrophysiological analysis of cortical mechanisms of selective attention to high and low spatial frequencies. *Clinical Neurophysiology, 112*(11).

Mathalon, D. H., Ford, J. M., & Pfefferbaum, A. (2000). Trait and state aspects of P300 amplitude reduction in schizophrenia: a retrospective longitudinal study. *Biological Psychiatry, 47*, 434–449.

May, P. J., & Tiitinen, H. (2010). Mismatch negativity (MMN), the deviance-elicited auditory deflection, explained. *Psychophysiology, 47*(1), 66–122.

Mayer, K., Wyckoff, S. N., & Strehl, U. (2013). One size fits all? Slow cortical potentials neurofeedback: a review. *Journal of Attention Disorders, 17*(5), 393–409.

McCleery, A., Lee, J., Joshi, A., Wynn, J. K., Hellemann, G. S., & Green, M. F. (2014). Meta-analysis of face processing event-related potentials in schizophrenia. *Biological Psychiatry, 77*(2), 116–126.

McNaughton, B. L. (1989). The neurobiology of spatial computation and learning. In D. J. Stein (Ed.), *Lectures on Complexity, Santa Fe Institute Studies in the Sciences of Complexity* (pp. 389–437). Redwood, CA: Addison-Wesley.

Meduna, L. J. (1937). *Die Konvulsionstherapie der Schizophrenie*. Halle am Saale, Germany: Carl Marhold.

Miller, G. A. (1956). The magical number seven plus or minus two: some limits on our capacity for processing information. *Psychological Review, 63*(2), 81–97.

Miller, R. (2007). *A theory of the Basal Ganglia and their disorders*. Boca Raton, FL: CRC Press.

Miller, E. K., & Cohen, J. (2001). An integrative theory of prefrontal cortex function. *Annual Review of Neuroscience, 24*, 167–202.

Miller, G. A., Galanter, E., & Pribram, K. H. (1960). *Plans and the structure of behavior*. New York: Holt, Rinehart & Winston.

Miyachi, S., Hikosaka, O., Miyashita, K., Karadi, Z., & Rand, M. K. (1997). Differential roles of monkey striatum in learning of sequential hand movement. *Experimental Brain Research, 151*, 1–5.

Moser, E. I., Kropff, E., & Moser, M. -B. (2008). Place cells, grid cells, and the brain's spatial representation system. *Annual Review of Neuroscience, 31*, 69–89.

Motohiro, K. (2012). Visual mismatch negativity and unintentional temporal-context-based prediction in vision. *International Journal of Psychophysiology, 83*(2), 144–155.

Mucci, A., Volpe, U., Merlotti, E., Bucci, P., & Galderisi, S. (2006). Pharmaco-EEG in psychiatry. *Clinical EEG and Neuroscience, 37*(2), 81–98.

Mueller, A., Candrian, G., & Kropotov, J. (2011). *ADHS—Neurodiagnostik in der Praxis*. Berlin: Springer Verlag.

Munz, M. T., Prehn-Kristensen, A., Thielking, F., Mölle, M., Göder, R., & Baving, L. (2015). Slow oscillating transcranial direct current stimulation during non-rapid eye movement sleep improves behavioral inhibition in attention-deficit/hyperactivity disorder. *Frontiers in Cellular Neuroscience, 9*, 307–315.

Näätänen, R., Gaillard, A. W., & Mantysalo, S. (1978). Early selective-attention effect on evoked potential reinterpreted. *Acta Psychologica, 42*(4), 313–329.

Näätänen, R., Shiga, T., Asano, S., & Yabe, H. (2015). Mismatch negativity (MMN) deficiency: A break-through biomarker in predicting psychosis onset. *International Journal of Psychophysiology, 95*, 338–344.

Neisser, U. (1978). Memory: What are the important questions? In M. M. Gruneberg, P. E. Morris, & R. N. Sykes (Eds.), *Practical aspects of memory* (pp. 3–24). London: Academic Press.

Neisser, U. (1979). Images, models, and human nature. *Behavioral and Brain Sciences, 2*(4), 561.

Nitsche, M. A., Cohen, L. G., Wassermann, E. M., Priori, A., Lang, N., Antal, A., Paulus, W., Hummel, F., Boggio, P. S., Fregni, F., & Pascual-Leone, A. (2008). Transcranial direct current stimulation: state of the art 2008. *Brain Stimulation, 1*(3), 206–223.

Noesselt, T., Hillyard, S. A., Woldorff, M. G., Schoenfeld, A., Hagner, T., Jancke, L., Tempelmann, C., Hinrichs, H., & Heinze, H. J. (2002). Delayed striate cortical activation during spatial attention. *Neuron, 35*(3), 575–587.

Ogrim, G., Kropotov, J., & Hestad, K. (2012). The QEEG theta/beta ratio in ADHD and normal controls: sensitivity, specificity, and behavioral correlates. *Psychiatry Research, 198*(3), 482–488.

O'Neill, B. V., Croft, R. J., & Nathan, P. J. (2008). The loudness dependence of the auditory evoked potential (LDAEP) as an in vivo biomarker of central serotonergic function in humans: rationale, evaluation and review of findings. *Human Psychopharmacology, 23*(5), 355–370.

Othmer, S., Othmer, S. F., Kaiser, D. A., & Putman, J. (2013). Endogenous neuromodulation at infra-low frequencies. *Seminars in Pediatric Neurology, 20*(4), 246–257.

Pąchalska, M., Kropotov, I. D., Mańko, G., Lipowska, M., Rasmus, A., Łukaszewska, B., Bogdanowicz, M., & Mirski, A. (2012). Evaluation of a neurotherapy program for a child with ADHD with benign partial epilepsy with Rolandic spikes (BPERS) using event-related potentials. *Medical Science Monitor, 8*(11), CS94–104.

Panksepp, J. (1992). A critical role for "affective neuroscience" in resolving what is basic about basic emotions. *Psychological Review, 99*, 554–560.

Papez, J. W. (1937). A proposed mechanism of emotion. *Archives of Neurology and Psychiatry, 38*, 725–743.

Pavlov, I. P., 1927. *Conditioned reflexes: An investigation of the physiological activity of the cerebral cortex* (translated by G.V. Anrep). London: Oxford University Press.

Peniston, E., & Kulkosky, P. (1989). Alpha-theta brainwave training and beta-endorphin levels in alcoholics. *Alcoholism: Clinical and Experimental Research, 13*(2), 271–279.

Pennick, M. R., & Kana, R. K. (2012). Specialization and integration of brain responses to object recognition and location detection. *Brain Behavior, 2*(1), 6–14.

Petersen, S. E., & Posner, M. I. (2012). The attention system of the human brain: 20 years after. *Annual Review of Neuroscience, 35*, 73–89.

Pierce, L., Scott, L. S., Boddington, S., Droucker, D., Curran, T., & Tanaka, J. T. (2011). The N250 brain potential to personally familiar and newly learned faces and objects. *Frontiers in Neuroscience, 5*(111), 1–13.

Pliszka, S. R., Lopez, M., Crismon, M. L., Toprac, M. G., Hughes, C. W., Emslie, G. J., & Boemer, C. (2003). A feasibility study of the children's medication algorithm project (CMAP) algorithm for the treatment of ADHD. *Journal of the American Academy of Child & Adolescent Psychiatry, 42*, 279–287.

Pogarell, O., Padberg, F., Karch, S., Segmiller, F., Juckel, G., Mulert, C., Hegerl, U., Tatsch, K., & Koch, W. (2011). Dopaminergic mechanisms of target detection—P300 event related potential and striatal dopamine. *Psychiatry Research, 194*(3), 212–218.

Polich, J., & Criado, J. R. (2006). Neuropsychology and neuropharmacology of P3a and P3b. *International Journal of Psychophysiology, 60*, 172–185.

Posner, M. I., & Petersen, S. E. (1990). The attention system of the human brain. *Annual Review of Neuroscience, 13*(1), 25–42.

Posner, M. I., Snyder, C. R. R., 1975. Attention and cognitive control. In Solso, R.L. (Ed.), *Information processing and cognition: The Loyola symposium*, Hillsdale (pp. 55–85). Lawrence Erlbaum Associates, Mahwah, NJ.

Prasad, K. M., & Keshavan, M. S. (2008). Structural cerebral variations as useful endophenotypes in schizophrenia: do they help construct "extended endophenotypes"? *Schizophrenia Bulletin, 34*(4), 774–790.

Preskorn, S. H., Ross, R., & Stanga, C. Y. (2004). Selective serotonin reuptake inhibitors. In S. H. Preskorn, H. P. Feighner, C. Y. Stanga, & R. Ross (Eds.), *Antidepressants: past, present and future* (pp. 241–262). Berlin: Springer.

Radman, T., Su, Y., An, J. H., Parra, L. C., & Bikson, M. (2007). Spike timing amplifies the effect of electric fields on neurons: implications for endogenous field effects. *The Journal of Neuroscience, 27*(11), 3030–3036.

Rae, C. L., Hughes, L. E., Weaver, C., Anderson, M. C., & Rowe, J. B. (2014). Selection and stopping involuntary action: a meta-analysis and combined fMRI study. *NeuroImage, 86*, 381–391.

Raichle, M. E., MacLeod, A. M., Snyder, A. Z., Powers, W. J., Gusnard, D. A., & Shulman, G. L. (2001). A default mode of brain function. *Proceedings of the National Academy of Sciences of the United States of America, 98*(2), 676–682.

Raiteri, M. (2006). Functional pharmacology in human brain. *Pharmacological Reviews, 58*(2), 162–193.

Rajkowski, J., Majczynski, H., Clayton, E., & Aston-Jones, G. (2004). Activation of monkey locus coeruleus neurons varies with difficulty and behavioral performance in a target detection task. *Journal of Neurophysiology, 92*, 361–371.

Rampersad, S. M., Janssen, A. M., Lucka, F., Aydin, Ü., Lanfer, B., Lew, S., Wolters, C. H., Stegeman, D. F., & Oostendorp, T. F. (2014). Simulating transcranial direct current stimulation with a detailed anisotropic human head model. *IEEE Transactions on Neural Systems and Rehabilitation Engineering, 22*(3), 441–452.

Randall, W. M., & Smith, J. L. (2011). Conflict and inhibition in the cued-Go/NoGo task. *Clinical Neurophysiology, 122*, 2400–2407.

Ranganath, C., & Rainer, G. (2003). Neural mechanisms for detecting and remembering novel events. *Nature Reviews Neuroscience, 4*, 193–202.

Rayshubskiy, A., Wojtasiewicz, T. J., Mikell, C. B., Bouchard, M. B., Timerman, D., Youngerman, B. E., McGovern, R. A., Otten, M. L., Canoll, P. D., McKhann, G. M., 2nd, & Hillman, E. M. (2014). Direct, intraoperative observation of approx. 0.1 Hz hemodynamic oscillations in awake human cortex: implications for fMRI. *NeuroImage, 87*, 323–331.

Reynolds, J. H., Chelazzi, L., & Desimone, R. (1999). Competitive mechanisms subserve attention in macaque areas V2 and V4. *Journal of Neuroscience, 19*(5), 1736–1753.

Rice, M. E., & Harris, G. T. (2005). Comparing effect sizes in follow-up studies: ROC area, Cohen's d, and r. *Law and Human Behavior, 29*(5), 615–620.

Ritter, W., Simson, R., & Vaughan, H. G., Jr. (1983). Event-related potential correlates of two stages of information processing in physical and semantic discrimination tasks. *Psychophysiology, 20*(2), 168–179 Albert Einstein College of Medicine.

Rizzolatti, G., & Luppino, G. (2001). The cortical motor system. *Neuron, 31*, 889–901.

Rockstroh, B., Elbert, T., Canavan, A. G. M., Lutzenberger, W., & Birbaumer, N. (1989). *Slow cortical potentials and behavior* (2nd ed.). Munchen: Urban and Schwarzenberg.

Rockstroh, B., Elbert, T., Birbaumer, N., Wolf, P., Düchting-Röth, A., Reker, M., & Dichgans, J. (1993). Cortical self-regulation in patients with epilepsies. *Epilepsy Research, 14*, 63–72.

Rosenberg, D. R., Dick, E. L., O'Hearn, K. M., & Sweeney, J. A. (1997). Response-inhibition deficits in obsessive–compulsive disorder: an indicator of dysfunction in frontostriatal circuits. *Journal of Psychiatry & Neuroscience, 22*(1), 29–38.

Roth, W. T., & Cannon, E. H. (1972). Some features of the auditory evoked response in schizophrenics. *Archives of General Psychiatry, 27*, 466–547.

Ruchsow, M., Reuter, K., Hermle, L., Ebert, D., Kiefer, M., & Falkenstein, M. (2007). Executive control in obsessive–compulsive disorder: event-related potentials in a Go/Nogo task. *The Journal of Neural Transmission, 114*(12), 1595–1601.

Rugg, M. D., & Wilding, E. L. (2000). Retrieval processing and episodic memory. *Trends in Cognitive Sciences, 4*(3), 108–115.

Rugg, M. D., & Yonelinas, A. P. (2003). Human recognition memory: a cognitive neuroscience perspective. *Trends in Cognitive Sciences, 7*(7), 313–319.

Sadeh, T., Ozubko, J. D., Winocur, G., & Moscovitch, M. (2014). How we forget may depend on how we remember. *Trends in Cognitive Sciences, 18*, 26–36.

Salmi, J., Rinne, T., Koistinen, S., Salonen, O., & Alho, K. (2009). Brain networks of bottom-up triggered and top-down controlled shifting of auditory attention. *Brain Research, 1286*, 155–164.

Sammer, G., Blecker, C., Gebhardt, H., Bischoff, M., Stark, R., Morgen, K., & Vaitl, D. (2007). Relationship between regional hemodynamic activity and simultaneously recorded EEG-theta associated with mental arithmetic-induced workload. *Human Brain Mapping, 28*(8), 793–803.

Sarter, M., Hasselmo, M. E., Bruno, J. P., & Givens, B. (2005). Unraveling the attentional functions of cortical cholinergic inputs: interactions between signal-driven and top-down cholinergic modulation of signal detection. *Brain Research Reviews, 48*, 98–111.

Sarter, M., Gehring, W. J., & Kozak, R. (2006). More attention must be paid: the neurobiology of attentional effort. *Brain Research Reviews, 51*, 145–160.

Saville, C. W. N., Dean, R. O., Daley, D., Intriligator, J., Boehm, S., Feige, B., & Klen, C. (2011). Electrocortical correlates of intra-subject variability in reaction times: average and single-trial analyses. *Biological Psychology, 87*(1), 74–83.

Saville, C. W. N., Pawling, R., Trullinger, M., Daley, D., Intriligator, J., & Klen, C. (2011). On the stability of instability: optimising the reliability of intra-subject variability of reaction times. *Personality and Individual Differences, 51*(2), 148–153.

Schildkraut, J. (1965). The catecholamine hypothesis of affective disorders: a review of supporting evidence. *American Journal of Psychiatry, 122*(5), 509–522.

Schneebaum-Sender, N., Goldberg-Stern, H., Fattal-Valevski, A., & Kramer, U. (2012). Does a normalizing electroencephalogram in benign childhood epilepsy with centrotemporal spikes abort attention deficit hyperactivity disorder? *Pediatric Neurology, 47*(4), 279–283.

Schulz, R., Gerloff, C., & Hummel, F. C. (2013). Non-invasive brain stimulation in neurological diseases. *Neuropharmacology, 64*, 579–587.

Seamans, J. K., & Yang, C. R. (2004). The principal features and mechanisms of dopamine modulation in the prefrontal cortex. *Progress in Neurobiology, 74*(1), 1–58.

Seeman, P., Chau-Wong, M., Tedesco, J., & Wong, K. (1975). Brain receptors for antipsychotic drugs and dopamine: direct binding assays. *Proceedings of the National Academy of Sciences, 72*(11), 4376–4380.

Sergeant, J. A. (2000). The cognitive-energetic model: an empirical approach to attention-deficit hyperactivity. *Neuroscience & Biobehavioral Reviews, 24*, 7–12.

Shallice, T. (1988). *From neuropsychology to mental structure*. Cambridge, UK: Cambridge University Press.

Shaw, P., Eckstrand, K., Sharp, W., Blumenthal, J., Lerch, J. P., Greenstein, D., & Clasen, L. (2007). Attention-deficit/hyperactivity disorder is characterized by a delay in cortical maturation. *Proceedings of the National Academy of Sciences, 104*(49), 19649–19654.

Shaw, P., Malek, M., Watson, B., Sharp, W., Evans, A., & Greenstein, D. (2012). Development of cortical surface area and gyrification in attention-deficit/hyperactivity disorder. *Biological Psychiatry, 72*(3), 191–197.

Siever, D. (2013). Transcranial DC stimulation. *NeuroConnections, Spring Issue*, 33–40.

Simson, R., Vaughan, H. G., Jr., & Ritter, W. (1977). The scalp topography of potentials an auditory and visual go-nogo tasks. *Electroencephalography and Clinical Neurophysiology, 43*, 864–875.

Singer, B. H., & Davidson, R. J. (2003). Now you feel it, now you don't: frontal brain electrical asymmetry and individual differences in emotion regulation. *Psychological Science, 14*(6), 612–617.

Sokolov, E. N. (1990). The orienting response, and future directions of its development. *The Pavlovian Journal of Biological Science, 25*(3), 142–150.

Solanto, M. V., Abikoff, H., Sonuga-Barke, E., Schachar, R., Logan, G. D., Wigal, T., Hechtman, L., Hinshaw, S., & Turkel, E. (2001). The ecological validity of delay aversion and response inhibition as measures of impulsivity in AD/HD: a supplement to the NIMH multimodal treatment study of AD/HD. *The Journal of Abnormal Child Psychology, 29*, 215–228.

Sonuga-Barke, E. J. S., & Castellanos, F. X. (2007). Spontaneous attentional fluctuations in impaired states and pathological conditions: a neurobiological hypothesis. *Neuroscience & Biobehavioral Reviews, 31*, 977–986.

Spearman, C. E. (1904). General intelligence, objectively determined and measured. *American Journal of Psychology, 15*, 201–293.

Spratling, M., & Johnson, M. H. (2004). A feedback model of visual attention. *Journal of Cognitive Neuroscience, 16*(2), 219–237.

Spruston, N. (2008). Pyramidal neurons: dendritic structure and synaptic integration. *Nature Reviews Neuroscience, 9*, 206–221.

Squire, L. R., Stark, C. E. L., & Clark, R. E. (2004). The medial temporal lobe. *Annual Review of Neuroscience, 27*, 279–306.

Squires, N. K., Squires, K. C., & Hillyard, S. A. (1975). Two varieties of long-latency positive waves evoked by unpredictable auditory stimuli in man. *Electroencephalography and Clinical Neurophysiology, 38*(4), 387–401.

Sreenivasan, K. K., Sambhara, D., & Jha, A. P. (2011). Working memory templates are maintained as feature-specific perceptual codes. *Journal of Neurophysiology, 106*, 115–121.

Steele, V. R., Aharoni, E., Munro, G. E., Calhoun, V. D., Nyalakanti, P., Stevens, M. C., Pearlson, G., & Kiehl, K. A. (2013). A large scale (N = 102) functional neuroimaging study of response inhibition in a Go/NoGo task. *Behavioural Brain Research, 256*, 529–536.

Sterman, M. B., & Egner, T. (2006). Foundation and practice of neurofeedback for the treatment of epilepsy. *Applied Psychophysiology and Biofeedback, 31*(1), 21–35.

Sterman, M. B., & Friar, L. (1972). Suppression of seizures in an epileptic following sensorimotor EEG feedback training. *Electroencephalography and Clinical Neurophysiology, 33*, 89–95.

Sterman, M. B., & Kaiser, D. (2000). A new QEEG analysis metric for assessment of structural and functional disorders of the central nervous system. *Journal of Neurotherapy, 4*(3), 73–83.

Sterman, M. B., Wyrwicka, W., & Roth, S. R. (1969). Electrophysiological correlates and neural substrates of alimentary behavior in the cat. *Annals NY Academy of Science, 157*, 723–739.

Sternbach, L. H. (1979). The benzodiazepine story. *Journal of Medicinal Chemistry, 22*(1), 1–7.

Steven, J. L., Geoffrey, F. W., & Edward, K. V. (2000). Event-related potential studies of attention. *Trends in Cognitive Sciences, 4*(11), 432–440.

Stokes, M. G., Atherton, K., Patai, E. Z., & Nobre, A. C. (2012). Long-term memory prepares neural activity for perception. *Proceedings of the National Academy of Sciences, 109*(6), 360–367.

Stroop, J. R. (1935). Studies of interference in serial verbal reactions. *Journal of Experimental Psychology, 18*(6), 643–662.

Sulzer, J., Haller, S., Scharnowski, F., Weiskopf, N., Birbaumer, N., Blefari, M. L., Bruehl, A. B., Cohen, L. G., DeCharms, R. C., Gassert, R., Goebel, R., Herwig, U., LaConte, S., Linden, D., Luft, A., Seifritz, E., & Sitaram, R. (2013). Real-time fMRI neurofeedback: progress and challenges. *NeuroImage, 76*, 386–399.

Swanson, J., Baler, R. D., & Volkow, N. D. (2011). Understanding the effects of stimulant medications on cognition in individuals with attention-deficit hyperactivity disorder: a decade of progress. *Neuropsychopharmacology, 36*, 207–226.

Talairach, J., & Tournoux, P. (1988). *Co-planar stereotaxic atlas of the human brain: Three-dimensional proportional system—an approach to cerebral imaging.* New York: Thieme Medical Publishers.

Talairach, J., Hecaen, H., & David, M. (1949). Lobotomies prefrontal limitée par electrocoagulation des fibres thalamo-frontales à leur emergence du bras anterieur de la capsule interne. *Revue Neurologique, IV Congrès Neurologique International.* Paris: Masson.

Tan, G., Thornby, J., Hammond, D. C., Strehl, U., Canady, B., Arnemann, K., & Kaiser, D. A. (2009). Meta-analysis of EEG biofeedback in treating epilepsy. *Clinical EEG and Neuroscience, 40*(3), 173–179.

Taylor, S. (2011). Early versus late onset obsessive–compulsive disorder: evidence for distinct subtypes. *Clinical Psychology Review, 31*, 1083–1100.

Tharyan, P., & Adams, C. E. (2005). Electroconvulsive therapy for schizophrenia. *Cochrane Database System Review.* CD000076.

Thatcher, R. (2012). *Handbook of quantitative electroencephalography and EEG biofeedback.* St. Petersburg, FL: Anipublishing Co.

Thibault, R. T., Lifshitz, M., Birbaumer, N., & Raz, A. (2015). Neurofeedback, self-regulation, and brain imaging: clinical science and fad in the service of mental disorders. *Psychotherapy and Psychosomatics, 84*(4), 193–207.

Thome, J., Ehlis, A. C., Fallgatter, A. J., Krauel, K., Lange, K. W., Riederer, P., Romanos, M., Taurines, R., Tucha, O., Uzbekov, M., & Gerlach, M. (2012). Biomarkers for attention-deficit/hyperactivity disorder (ADHD). A consensus report of the WFSBP task force on biological markers and the World Federation of ADHD. *World Journal of Biological Psychiatry, 13*(5), 379–400.

Thompson, P. M., Vidal, C., Giedd, J. N., Gochman, P., Blumenthal, J., Nicolson, R., Toga, A. W., & Rapoport, J. L. (2001). Mapping adolescent brain change reveals dynamic wave of accelerated gray matter loss in very early-onset schizophrenia. *Proceedings of the National Academy of Sciences, 98*(20), 11650–11655.

Thurstone, L. L. (1934). The vectors of mind. *Psychological Review, 41*(1), 1–32.

Treue, S. (2001). Neural correlates of attention in primate visual cortex. *Trends in Neuroscience, 24*, 295–300.

Turetsky, B. I., Bilker, W. B., Siegel, S. J., Kohler, C. G., & Gur, R. E. (2009). Profile of auditory information-processing deficits in schizophrenia. *Psychiatry Research, 165*(1–2), 27–37.

UK ECT Review Group. (2003). Efficacy and safety of electroconvulsive therapy in depressive disorders: a systematic review and meta-analysis. *Lancet, 361*(9360), 799–808.

Ullsperger, M., Danielmeier, C., & Jocham, G. (2014). Neurophysiology of performance monitoring and adaptive behavior. *Physiological Reviews, 94*, 35–79.

Ulric, N. (1978). Anticipations, images, and introspection. *Cognition, 6*(2), 169–174.

Ungerleider, L. G., & Mishkin, M. (1982). Two cortical visual systems. In D. J. Ingle, M. A. Goodale, & R. J. W. Mansfield (Eds.), *Analysis of visual behavior*. Cambridge, MA: MIT Press.

van Tricht, M. J., Nieman, D. H., Koelman, J. T., Mensink, A. J., Bour, L. J., van der Meer, J. N., van Amelsvoort, T. A., Linszen, D. H., & de Haan, L. (2012). Sensory gating in subjects at ultra high risk for developing a psychosis before and after a first psychotic episode. *World Journal of Biological Psychiatry, 16*(1), 12–21.

Vasilevsky, N. (1973). Adaptive self-control function and its relationship with the dynamic control of the endogenous biorhythms. *Journal of Evolutionary Biochemistry and Physiology, 9*(4), 374–382.

Verleger, R. (1988). Event related potentials and cognition: a critique of context updating hypothesis and alternative interpretation of P3. *Behavioral and Brain Sciences, 11*, 343–427.

Vigário, R. N. (1997). Extraction of ocular artefacts from EEG using independent component analysis. *Electroencephalography and Clinical Neurophysiology, 103*, 340–395.

Von Helmholtz, H. (1866). *Handbuch der Physiologischen Optik*, V.3.

Walter, W. G., Cooper, R., Aldridge, V. J., McCallum, W. C., & Winter, A. L. (1964). Contingent negative variation: an electric sign of sensorimotor association and expectancy in the human brain. *Nature, 203*(4943), 380–384.

Wang, W. J., Wu, X. H., & Li, L. (2008). The dual-pathway model of auditory signal processing. *Neuroscience Bulletin, 24*, 173–182.

Wheeler, M. E., Petersen, S. E., & Buckner, R. L. (2000). Memory's echo: vivid remembering reactivates sensory-specific cortex. *Proceedings of the National Academy of Sciences of the USA, 97*, 11125–11129.

Wickelgren, W. A. (1979). Chunking and consolidation: a theoretical synthesis of semantic networks, configuring in conditioning, S–R versus congenitive learning, normal forgetting, the amnesic syndrome, and the hippocampal arousal system. *Psychological Review, 86*(1), 44–60.

Wiecki, T. V., & Frank, M. J. (2013). A computational model of inhibitory control in frontal cortex and basal ganglia. *Psychological Review, 120*(2), 329–355.

Wiener, N. (1948). *Cybernetics, or control and communication in the animal and the machine*. Cambridge, MA: MIT Press.

Willcutt, E. G., Doyle, A. E., Nigg, J. T., Faraone, S. V., & Pennington, B. F. (2005). Validity of the executive function theory of attention-deficit/hyperactivity disorder: a meta-analytic review. *Biological Psychiatry, 57*, 1336–1346.

William, J. (1890). *The principles of psychology*. New York: H. Holt & Company.

Williams, B. R., Hultsch, D. F., Strauss, E. H., Hunter, M. A., & Tannock, R. (2005). Inconsistency in reaction time across the life span. *Neuropsychology, 19*(1), 88–96.

Williams, L. M., Palmer, D., Liddell, B. J., Song, L., & Gordon, E. (2006). The when and where of perceiving signals of threat versus non-threat. *NeuroImage, 31*, 458–467.

Woodman, G. F., Carlisle, N. B., & Reinhart, R. M. G. (2013). Where do we store the memory representations that guide attention? *Journal of Vision, 13*, 1–17.

Yeredor, A. (2010). Blind separation of Gaussian sources with general covariance structures: bounds and optimal estimation. *IEEE Transactions on Signal Processing, 58*(10), 5057–5068.

Subject Index

A

Abnormal beta rhythms, 119
　beta activity, asymmetry of, 119
　cortical irritability, 119
　observed reactivity, 119
　scalp distribution, 119
Abnormal theta rhythms, 132
　in closed brain injury, 133
　in patients with ADHD, 133
AC. *See* Anterior commissure (AC)
ACC. *See* Anterior cingulated cortex (ACC)
Acetylcholine, 240
　broad areas of cortex, activation, 240
　theta rhythm amplitude, 240
Acetylcholinesterase, 29
Action-monitoring/action-inhibiting, obsessive-compulsive disorder, 360
Activation, 302
Addiction, 212
ADHD. *See* Attention deficit hyperactivity disorder (ADHD)
Affective neuroscience, 207
Affective operations, 59
Affective system, 60, 207
　anatomy, 208
　emotions, 209
　information flow, 208
　motivations, 209
　neuromodulators, 228
　stages of reactions, 225
　　emotional reaction, 225
　　feeling stage, 225
　　monitoring stage, 226
　　sensation, 225
Age dynamics, 97
　of alpha rhythms, 97
Alpha activity, 383
Alpha rhythms, 90
　in dysfunctional brain, 99
　in inferior frontal/superior temporal lobe, 99
　mu, 90

optimal performance, 103
parietal, 94
posterior, 90
in somatosensory cortex, 90
tau, 90
of visual system, 91
Ambivalence, 323
American Psychiatric Association (APA), 268
AMPA receptors, 44
Amplitude, 41
Amygdala, 207, 409, 410
　detector of fearful stimuli, 218
Amygdalotomy, 285
Anatomical folding patterns, 77
Animal behavioral reactions, 248
Anterior cingulated cortex (ACC), 166, 222
　activity, 212, 357
　affective part, 222
　executive part, 222
Anterior commissure (AC), 17
Anterior congulate cortex, 362
Antianxiety, 351
　neutralize, 352
Anticipatory maintenance, 173
Anticonvulsive medication, 378
Antidepressants, 246
Antipsychotic agents, 329
Anxiety, 129, 211, 354
　disorders, 302
　　hyperactivation, 302
　　hypoactivation, 302
　schizophrenia, 325
　state of preparing to fear, 220
APA. *See* American Psychiatric Association (APA)
Apical dendrites, 36
Arousal, 302
Arrhythmic beta activity, 116
　as index of cortical activation, 116
Arrhythmic electroencephalograms, 81
Arrhythmic fluctuations, 83
Asperger's syndrome, 392

Attentional control operations, 10, 171, 413
    detectability parameter, 11
    energization, 10
    ERP profiles, 414
    facilitates, 10
    false negatives errors, 11
    false positive errors, 11
    hyperactivation of the medial prefrontal
        cortex, 415
    monitoring, 10
    QEEG/ERP neuromarkers, 414
    quality control, 10
    task-setting, 10
Attention deficit hyperactivity disorder
        (ADHD), 13, 84, 291, 356, 369, 417
    age onset, 295
    birth-related risk factors, 296
        bilirubin, 296
    causes, 369
        default mode dysfunction, 369
        delay aversion, 369
        epifocus, 369
        frontal lobe, hyperactivation of, 369
        hypoarousal, 369
        limbic system dysfunction, 369
        maturation lag, 369
        prefrontostriatothalamic dysfunction,
            369
    comorbidity, 296
    copy number variation (CNV), 319
    default mode, interference, 301
    delay aversion, 300
    dopamine hypothesis, 318
    energization function, 302
    environmental factors, 296
    executive dysfunction theory, 297
    executive functions, 297
    fMRI, 310
    genetic factors, 296
        dopamine receptor, 296
        linkage analysis, 296
        transporter genes, 296
    graph theory, 299
    hyperactivity, 291
    hypoactivation, 302
    hypoarousal hypothesis, 302
    impulsivity, 291
    inattention, 291
    inhibition deficit, 299
    magnetic resonance imaging, 310
    maturation lag, 304
    neurodevelopment, delay in, 303
    neuropsychological assessment, 297
    neuropsychological profile, 298
    objective assessment, 370
    pathogenetic factor, 303
    P3b wave, 310
    persistence in adulthood, 295
    pharmacological treatment, 315
        amphetamine, 315
        atomoxetine, 315
        methylphenidate, 315
    prevalence, 295
    QEEG endophenotypes in, 307
    reaction time variability, 300
    rolandic focus, 297
    state regulation, 302
    symptoms, 292, 369
        hyperactivity, 369
        impulsivity, 369
        inattention, 369
        latent classes, 293
        multidimensional space, 293
    tDCS, 320
    theta/beta ratio, 304
    transcranial magnetic stimulation, 321
Attention mechanisms, 60
Attention modulation, 140
    bottom–up processes, 142
    top–down processes, 142
Auditory mismatch negativity, 162
Auditory modality, 159
Auditory N1/P2 wave, 160
    loudness dependence, 168
Autism, 323
Autistic spectrum disorder, 247, 354, 392
Automatic predicting ability, 335
    failure, schizophrenia (SZ), 335
Averaging, 3, 61

B

Background emotions, 210
Baddeley's model, 177
Barkley's model, 177
Basal ganglia-thalamo-cortical loops, 183,
        330
    motor loop, 109
Bayesian information criterion (BIC), 75
BCI. See Brain-computer interface (BCI)
Behavioral inhibition system, 218
Behavioral paradigms, 174, 218, 374
    cognitive control operations, 175
    inhibition, 175
    oddball, 176
    task switching, 175
    willed selection, 175

SUBJECT INDEX

Behavioral parameters, 388
  reaction time, 388
  RT variability (RTV), 388
  sensitivity index, 388
Behavioral therapy, 383
Behaviorism, obsessive-compulsive disorder, 351
Behavior, prepotent model, 174
  efficient, 174
  habitual, 174
  innate, 174
  motivational, 174
Benzodiazepine (BNZ), 117
Best estimate diagnosis, 7
Beta rhythms, 109, 381
  abnormal, 119
  frontal, 111
  hypothetical functions, 109
    action – selection network, 109
    coordination, 109
    decision making, 109
    inhibition of movement, 109
    motor planning, 109
    status quo preservation, 109
  multiple, 109
  occipital rebound, 114
  rolandic, 109
  vertex, 113
Beta suppression, 382
Beta waves, 107
BIC. *See* Bayesian information criterion (BIC)
Big 5. *See* Five-factor model (Big 5)
Bimanual motor task, 111
Biological feedback, 249
Biologically oriented psychiatrists, 368, 372
  available treatment methods, 373
  working space, 372
Blind source separation, 52
Blocking dopaminergic receptors, 324
Blood–brain barrier, 36
Blood oxygenation level dependent (BOLD), 19
  blood flow, 19
  blood volume, 19
  fMRI oscillations, 84
  hemoglobin oxygenation, 19
  neuronal activity, 19
  signal, 95, 256
    negative correlation, 95
Blood oxygenation level-dependent (BOLD) signal, 35, 399
BNZ. *See* Benzodiazepine (BNZ)

Body dysmorphic disorder, 354
BOLD. *See* Blood oxygenation level dependent (BOLD)
Bold infralow fluctuations, 20
Bold response, 19
Bottom–up primary sensory information, 46
BRAIN. *See* Brain Research through Advancing Innovative Neurotechnologies (BRAIN)
Brain. *See also* Healthy brain
Brain-behavior interrelations, 9
Brain-computer interface (BCI), 265
  ERP-based, 266
Brain dysfunction, 413
Brain electrical activity, 126
Brain lesions, 10
Brain model, 212
Brain neuronal networks, 373, 387
  electrical modulation, 373
Brain oscillations, 82, 417
Brain Research through Advancing Innovative Neurotechnologies (BRAIN), 246
Brain Resource Company (BRC), 372
Brain rhythms, 108
Brain systems, 231
Bramon, 332
BRC. *See* Brain Resource Company (BRC)
Broca's aphasia, 234

C

Calcium channels, 101
Canonical basal-ganglia thalamo-cortical circuit, 184
Canonical visual event-related potential, 66
Cardioballistic artifact, 39
Catatonic states, 324
Cathodal stimulation, dramatic clinical improvement, 364
Causal function, 102
CDA. *See* Contralateral delay activity (CDA)
Center excitation–surround inhibition model, 104
Central executive, 171
Centroparietal electrodes, 226
Changing conditions, 4
Chemical messengers, 245
Childhood ADHD, frontal beta synchronization, 309
Chronic doubting, 354
Cingulotomy, 362
Classical test theory, 3

Clinical practice, implementation, 417–419
Clinical research, 78
CNV. *See* Contingent negative variation (CNV)
Coefficient of variation (CV), 13
Cognitive behavioral therapy, obsessive-compulsive disorder, 361
Cognitive control systems, 59, 171, 223, 291, 388
   components, 344
   dual mechanisms of, 11
   ERP correlates, 190
   fMRI, 188
   independent components, 192
   latent components of, 359
   models, 177
      Baddeley's, 177
      Barkley's, 177
      Miller and Cohen's, 177
   modes, 173
      maintenance of action, 173
      monitoring, 173
   operations, 172
      action execution, 173
      action preparation, 173
      action selection, 173
      adjusting future behavior, 173
      detection of conflict, 173
      inhibition of ongoing activity, 173
      shift from one action to another, 173
      suppression of prepared action, 173
      working memory, 173
   selective deficit, 388
Cognitive dysfunctions, 325, 329
Cognitive functions, highest-order, 344
Cognitive impairment, 29
Cognitive maps, 9
Cognitive tasks, 23
Coherence, 44
   intracranial electroencephalogram protocols, 254
Comorbidity, 354
Components, 69
Compulsions, 351
Computational neuronal model, 269
Conflict detection hypothesis, 190
Conflict monitoring, 129, 357
   operation, 311
Constraint, 212
Contingent negative variation (CNV), 85, 179, 255
   amplitude, 339

Continuous performance tasks (CPTs), 15
   integrated visual and auditory CPT (IVA), 15
   test of variables of attention (TOVA), 15
Continuous training protocol, 256
Contralateral delay activity (CDA), 233
Conventional diagnosis, 369
Coordinate system, 17
Copy number variation (CNV), 319
Correlational methods, 23
   model-dependent, 23
   model-free, 23
Cortex, 33
   electrical events, measurement, 33
      in neuronal module, 33
      in single neuron, 33
      at systemic level, 33
Cortical circuits, 46
Cortical focus, 47
Cortical potential neurofeedback, 255
Cortical self-regulation, 377
   EEG rhythms of, 404
Cortical spikes, 47
Cortical-striatal-thalamic-cortical dysfunction, 417
Cortico-basal ganglia-thalamic pathways, 172
Cortico-cortical pathways, 172
Covariance matrixes, joint diagonalization of, 74
CPTs. *See* Continuous performance tasks (CPTs)
CSD. *See* Current source density (CSD)
Current source density (CSD), 32, 51, 90
CV. *See* Coefficient of variation (CV)

D

DAT. *See* Dopamine transporter (DAT)
DBS. *See* Deep-brain stimulation (DBS)
DC amplifiers. *See* Direct-coupled (DC) amplifiers
D_2 dopamine receptors, 329
Decision making, 10
Declarative memory, 234
   acetylcholine as neuromodulator, 239
Deep-brain stimulation (DBS), 285, 362, 412
   advantages, 287
   inhibitory effect, 287
   limitations, 287
   neuronal mechanism, 286
   nucleus accumbens, 362
   procedure, 286

Default mode network (DMN), 14, 23, 95, 301
Delayed tasks, 176
Delusions, 324
Dendritic integration, 46
Deoxyhemoglobin, 19
Depolarization, 250
Depression, 324, 354
Desensitization techniques, 351
Detectability parameter, 11
Dichotic listening task, 68
Differential amplifiers, 35
Different neuronal mechanisms, 250
Diffuse tensor imaging (DTI), 24
Dipole orientation, 48
Direct-coupled (DC) amplifiers, 35
Discrepancy effect, 153
Discrimination procedures, 8
Distributed source models, 49
DLPFC. *See* Dorsolateral prefrontal cortex (DLPFC)
DMN. *See* Default mode network (DMN)
Dopamine, 245, 343, 361
   D2 receptors, 29
   levels, 109
   reuptake inhibitors, 370
Dopaminergic hyperactivity, 324
Dopaminergic receptors, 415
Dopaminergic synapses, 310
Dopamine transporter (DAT), 28, 310, 316
Dorsal visual stream, 63
Dorsolateral prefrontal brain lesions, 302
Dorsolateral prefrontal cortex (DLPFC), 212
Double-blind placebo control protocol, 263
DSM-IV, 351
DTI. *See* Diffuse tensor imaging (DTI)
Dysrhythmia, 377

E

Early posterior negativity (EPN), 226
Eastern self-regulation techniques, 263
ECG. *See* Electrocardiogram (ECG)
ECT. *See* Electroconvulsive therapy (ECT)
EEG. *See* Electroencephalogram (EEG)
Effect size (ES), 7
Electrical activity, 36
   alpha waves, 36
   beta waves, 36
   delta waves, 36
   direct current, 36
   electrical potentials, 36
   gamma band, 36
   infraslow fluctuations, 36
   theta waves, 36
Electrical currents, 44
   neuronal sources, 44
Electrical dipole, 45
Electrical modulation, 373
Electrocardiogram (ECG), 39
Electroconvulsive therapy (ECT), 267, 412
   consensus conference, 268
   contraindications, 270
   efficacy, 270
   historical perspective, 267
   mechanisms, 269
      biochemical, 269
      metabolic, 270
      neuronal model, 269
         computational model, 269
   parameters, 268
   relapse, 270
   side-effects, 271
      anterograde amnesia, 271
      retrograde amnesia, 271
Electroencephalogram (EEG), 31, 297
   bipolar derivations, 31
   devices, 3
   dysrhythmia, 417
   EEG–TMS experiments, 102
   ERP systems, 417
   measurement, 31
      differential amplifiers, 31
      potential difference, 31
   nonbrain events, 37
   as reflection of cortical self-regulation, 37
   spectral analysis of, 40
      rhythmicity, 40
   spectra, power-law function of, 81
   volume-conducted, 31
Electrooculogram (EOG), 37
Electrophysiological brain research, 391
Emotional facial expressions, 226
Emotionally competent stimulus, 209
   punishers, 209
   rewards, 209
Emotional reactions, 209
   affective system, stages of reactions, 225
   obsessive-compulsive disorder, 359
Emotions, 207
   classification, 209
      background, 209
      primary, 209
      social, 209
   fMRI, 224
   as habitual responses, 209

Emotions (*cont.*)
  happiness, 224
  left-right asymmetry model, 214
  sadness, 224
  as separate dimension, 209
Encoding paradigm, 237
Endophenotypes, 354
  markers, 78
  QEEG, 307
Energization, 10
Environmental risk factors, schizophrenia, 329
EOG. *See* Electrooculogram (EOG)
Epileptiform activity, 46
Episodic memory, 234
  functional neuromarkers of, 237
    old – new effect, 237
  hippocampus, 235
  neuronal model, 238
  perceptual representations, 236
  ventral stream system, 236
EPN. *See* Early posterior negativity (EPN)
ERD. *See* Event-related desynchronization (ERD)
ERPs. *See* Event-related potentials (ERPs)
Error-related negativity (ERN), 357
  enhancement of
    obsessive-compulsive disorder (OCD), 359
Errors, 3
  random, 3
  systematic, 3
  variance, 4
ERS. *See* Event-related synchronization (ERS)
ES. *See* Effect size (ES)
Event-related desynchronization (ERD), 94
Event-related potentials (ERPs), 16, 59, 388, 392, 401
  beta neurofeedback effect, 394
  EEG systems, 417
  GO/NOGO task, 393
  independent components, obsessive-compulsive disorder, 363
  neuromarkers, 78
  NOGO condition, 401
  paradigms, 68
    multiple sources, 68
  psychological operations, 401
  stimulant medication, 393
Event-related synchronization (ERS), 94
Excessive error signals, obsessive-compulsive disorder, 356
Excessive frontal beta synchronization, 357
  obsessive-compulsive disorder (OCD), 358

Excessive nature of compulsion, obsessive-compulsive disorder, 352
Excitatory neurons, 65
Executive brain system, 291
Executive control, 171
Executive dysfunction theory, 297
ex-Gaussian distribution, 12
Expectancy wave, 180
Exponential parameter, 12
Extraversion, 217
Eye blink, 37
Eysenck's models, 217

F

Facial expressions, 209
Facilitates, 10
False negatives errors, 11
False positive errors, 11
Fast-acting inhibitory receptors, 28
Fast Fourier transform (FFT), 41
Fast-wave activity, 307
fbEEG. *See* Full-band EEG (fbEEG)
fcMRI. *See* Functional connectivity MRI (fcMRI)
Feedback projections, 63
Feedforward processing, 65
FFT. *See* fast Fourier transform (FFT)
Five-factor model (Big 5), 9, 15, 216
  agreeableness, 217
  conscientiousness, 217
  extraversion, 217
  intellect, 217
  neuroticism, 217
fMRI. *See* Functional magnetic resonance imaging (fMRI)
fNIRS. *See* Functional near-infrared spectroscopy (fNIRS)
Forensic neuropsychology, 367
Forensic psychiatry, 367
Formal thought disorder, 324
Fourier analysis, 42
Fourier theorem, 40
Frequency bands, 35
Frontal alpha asymmetry, 98
Frontal beta activity, 380
  in ADHD, 380
Frontal beta rhythms, 111
  frontal leads, 111
Frontal leucotomy, 221
Frontal lobe functions, 183, 238
Frontally distributed component, 313
Frontal midline theta rhythm, 121, 307, 382
  adaptive behavioral, 129

in ADHD, 382
  age dynamics, 125
  conflict detection, 129
  cortical metabolism, 126
  functional features, 122
  genetic factors, 125
  historical perspective, 121
  localization, 123
  personality traits, 126
  prevalence, 124
  task load, 127
  working memory, 127
Frontal-parietal network, 178
Frontocentral electrodes, 304
Full-band EEG (fbEEG), 35
Functional connectivity MRI (fcMRI), 24
Functional imaging, processing steps in, 22
  baseline, 22
  subject movement, 22
  transformation, 22
Functional magnetic resonance imaging (fMRI), 17, 310
  activation maps of, 22
  challenge, 25
    lack of fMRI testing, 25
    lack of standardization, 25
    low signal-to-noise ratio, 25
    uncertainty in BOLD modulation, 25
  clinical application, 25
  in conflict conditions, obsessive-compulsive disorder, 356
  fMRI–BOLD signal, 96
  language lateralization, 25
  motor cortex, localization of, 25
  in neurological practice, 25
  physiology, 22
  in symptom provocation, 355
Functional meaning, 87
Functional near-infrared spectroscopy (fNIRS), 265
Functional neuromarkers, 387, 413, 414
  attention deficit hyperactivity disorder (ADHD), 413
  behavioral parameters, 388
  cortical self-regulation, 388
  event-related potentials (ERPs), 388
  independent dimensions, 388
  obsessions, 415
  schizophrenia (SZ), 415
Functional neuromarkers, implementing into psychiatric practice, 418
Functional reactivity, 91
Fusiform gyri, 96

G

GABA. See Gamma-aminobutyric acid (GABA)
Gamma-aminobutyric acid (GABA), 28
  agonists, 370, 417
  GABAergic mechanisms, 119
  mediated recurrent inhibition, 47
  neurotransmission, 344
  receptors, 44
Gamma rhythms, 117, 381
Ganglia–thalamocortical system, obsessive-compulsive disorder, 360
General factor (g factor), 11
g factor. See General factor
Glutamate, 44
Glutamate hypothesis, 343
Goal-directed flexible behavior, 171
Gold standard, 7
GO/NOGO paradigm, 413
G protein–coupled receptors, 28
Gray's models, 217
Group activation maps, 25
Group independent component analysis
  multiple tasks, 72
  single tasks, 72

H

Habits of mind, 209
Hallucinations, 324
Hanning windows, 41
HBI database, 381
Healthy brain
  information flow, in sensory systems, 407
    cognitive control, 409, 410
    conflict detectiopn, 409
    ERPs and ERP components, 408, 409
    GO P3 wave, 409
    NOGO P3 wave, 409
    P1/N1 waves, 409
    procedural memory, 410
    sLORETA localizes, 407
  rhythms of
    alpha oscillations, 403
    beta rhythms, 405
    cortical self-regulation, EEG rhythms of, 404
    EEG recordings, 403
    electrical alpha rhythms, 403
    infralow frequency band, oscillations, 403
    theta rhythm, frontal midline, 405

Healthy controls (HCs)
  differences, 355
  obsessive-compulsive disorder, 354
Heart rate variability, 20, 249
Hemodynamic oscillations, 20
Hemodynamic regulatory oscillations, 86
Heritability, 101, 216, 328
  obsessive-compulsive disorder (OCD), 353
  schizophrenia (SZ), 328
Heterogeneity, 4
  obsessive-compulsive disorder, 353
  schizophrenia, 328
High-threshold (HT)
  alpha rhythm generation, 103
  calcium potentials, 102
Hippocampal slice preparation, 249
Hippocampal theta rhythms, 121
  episodic memory, 131
  hippocampus integrates, 131
  hypothetical neuronal mechanisms, 130
  information quantum, 132
  memory chunk, 132
  Papez circle, 130
  polysensory signal, 131
  pyramidal cells, 130
  in rats, 130
  reciprocal connections, 131
  septal inhibitory, 131
  theta oscillations, 130
Hippocampus, 207
HKD. *See* Hyperkinetic disorder (HKD)
Homunculus map, 93
5HT3A receptor–expressing interneurons (5HTs), 46
5HTs. *See* 5HT3A receptor-expressing interneurons (5HTs)
Human brain functioning
  magnetic resonance imaging (MRI), 399
  objective measures of, 399–402
Human hippocampus, 121
Human motor cortex, 281
Hyperactivation
  anterior cingulate cortex, 356
  of basal ganglia, obsessive-compulsive disorder, 356
  of specific cortical areas, 356
Hyperactive–impulsive symptoms, 297
Hyperactivity, 302
  symptoms, 413
Hyperfrontality, obsessive-compulsive disorder, 357
Hyperkinetic disorder (HKD), 292
Hyperpolarization, 101

Hypervigilance, 228
Hypoarousal hypothesis, 302
Hypofrontality as response predictor to medication
  schizophrenia, 343
Hypofrontality (fMRI) studies
  schizophrenia, 340
Hypothalamotomy, 285
Hypothalamus, 207, 210, 220
  expression of emotions, 220
Hypothetical operations, 69
  psychological, 69

I

IAPS. *See* International affective picture system (IAPS)
ICA. *See* Independent component analysis (ICA)
ICD. *See* International Classification of Disorder (ICD)
Iconic memory, 231
Identical conditions, 4
Idling, 90
IFs. *See* Infraslow fluctuations (IFs)
IID. *See* Independent and identically distributed (IID)
Implanted electrodes, 130
Impulse activity, 20, 108
Impulsivity, 302, 413
Inattention symptoms, 413
Independent and identically distributed (IID), 53
Independent component analysis (ICA), 23, 53, 71
  clinical practice, decomposition in, 55
  independent and identically distributed (IID), 53
  linear and instantaneous mixing process, 53
  non-Gaussian probability density functions, 53
  sLORETA images, 57
  statistically independent sources, 53
  topographies, 57
Independent component neurofeedback, obsessive-compulsive disorder, 362
Infancy, 216
Information flow, 59, 387
  in local network, 64
  motor-related activity, 59
  sensory-related activity, 59
  two packets, 65
  in visual pathways, 63

Information processing, 10, 77
Infralow frequencies, 20
  neurofeedback, 256
Infraslow electrical oscillations, 83
  arrhythmic fluctuations, 83
  historical perspective, 83
  periodic oscillations, 83
Infraslow fluctuations (IFs), 15, 35, 87, 250
  in performance, 15
Inhibition paradigms, 175
Inhibitory interneurons, 46
  5HT3A receptor–expressing interneurons (5HTs), 46
  parvalbumin-expressing interneurons (PVs), 46
  somatostatin-expressing interneurons (SOMs), 46
Integrated visual and auditory CPT (IVA), 15
Interictal spikes, 48
Interindividual differences, 41
Interindividual variability, 77
Internal representations, 9
International affective picture system (IAPS), 218
International brain database, 372
International Classification of Disorder (ICD), 292
International Pharmaco-EEG Society (IPEG), 391
10–20 International System, 34
Intracortical connectivity, 45
  interneurons, 45
  principal cells, 45
Intracranial EEG recordings, 123
  of monkeys, 123
  of patients, 123
Intracranial recordings, 108
Intraindividual standard deviation (ISD), 14
  inconsistency, life span, 14
Intrasubject reliability, 25
Ionotropic receptors, 28
IPEG. See International Pharmaco-EEG Society (IPEG)
ISD. See Intraindividual standard deviation (ISD)
Isotopes, 27
IVA. See Integrated visual and auditory CPT (IVA)

K

Korsakoff's syndrome, 235
Kullback–Leibler divergence, 54

L

Latent class analysis (LCA), 293
Late positive complex (LPC), 226
LCA. See Latent class analysis (LCA)
Lesions, 355
  orbitofrontal cortex, 355
LFPs. See Local field potentials (LFPs)
Lifetime prevalence, obsessive-compulsive disorder, 352
Ligand-gated receptors, 28
Local field potentials (LFPs), 44, 108
Log-normal distributions, 12
Log value, 5
Long-term memory, 231
  explicit, 234
  implicit, 234
LORETA. See Low resolution tomography (LORETA)
Low resolution tomography (LORETA), 50, 254
Low-voltage fast EEG, 100
LPC. See Late positive complex (LPC)

M

Magnetic resonance imaging (MRI), 18
  obsessive-compulsive disorder (OCD), 355
  physical basis of, 18
Magnetoencephalography (MEG), 90, 116
Magnocellular neurons, 64
Magnocellular visual system, 333
Malingering, 367
Mapping tool, 222
Mathematical algorithms, 18
Maturational lag, 417
  hypothesis, 303
Maximum entropy, 54
Maxwell equations, 51
Mayer waves, 20
Mean value, 5
Medications, 246
MEG. See Magnetoencephalography (MEG)
Membrane depolarization, 37
Memory, 231
  impairment, 29
  related operations, 59
  systems, 60
  temporal aspects, 231
  working representations, 233
Mental health, 98
MEPs. See Motor-evoked potentials (MEPs)
Metabotropic receptors, 28

Methylphenidate, 300, 378, 413
Miller and Cohen's model, 177
Minimal brain dysfunction, 291
Minnesota multiphasis personality inventory (MMPI), 9
Mirror neurons, 209
Mismatch negativity (MMN), 63
MMN. See Mismatch negativity (MMN)
MMPI. See Minnesota multiphasis personality inventory (MMPI)
MNI. See Montreal Neurological Institute (MNI)
MNI305 atlas, 18
Monitoring, 10
Monozygotic vs. dizygotic, 353
Montages, 31, 60
　common average, 31
Montreal Neurological Institute (MNI), 18
Mood symptoms, 324
Morlet functions, 43
Morlet wavelets, 42
Motivational behavior, 210
Motor abnormalities, schizophrenia, 330
Motor cortex, 379
　cortical areas, 110
Motor-evoked potentials (MEPs), 250, 281
Motor response, 110
Motor tics, 354
Movement-related noncortical pathways, 184
Multiple beta rhythms, 109
Multiple research groups, 8
Mu-rhythms, 90, 382
Muscle artifacts, 38

N

NA. See Negative affectivity (NA)
Negative affectivity (NA), 212
Negative facial expression, 226
Negative reinforcement loop, 352
Neurodevelopment, schizophrenia, 326–327
Neurofeedback (NF), 216, 247, 378, 411, 417
　advantages, 263
　　minimizes side-effects, 264
　bulldozer principle, 260
　classification, 252
　conventional electroencephalogram based, 250
　cortical potential neurofeedback, 255
　EEG-based, 252
　　activation, 252
　　biofeedback, 411
　　relaxation, 252

infraslow fluctuation based, 250, 411
　limitations, 263
　LORETA-based, 255
　methodology, 84
　neurophysiological basis, 257
　　relative frontal beta training, 258
　nonelectrical, 265
　nonspecific effects, 263
　protocols, 252, 367
　specific effects, 263
　technique, 35, 248
　treatment, 249, 362
　　obsessive-compulsive disorder, 362
　types of, 411
Neuroguide database, 372
Neuroimaging methods, 325
Neuroleptics drugs, 29
Neuromarkers, 3, 334. See also Functional neuromarkers
　in clinical practice, 8
　　aid psychiatrists, 8
　　brain functioning, specific index of, 8
　　prognostic procedure, 8
　　reliable, 8
　　sensitive, 8
　measures, 401, 402
　quantitative measurement, 3
　standardization, 8
Neurometrics, 250
　database, 371
Neuromodulators, 245
Neuronal impulses reverberation, 238
Neuronal mechanisms, 86
Neuronal model, 157, 359
　network, 82
　obsessive-compulsive disorder (OCD), 359
　schizophrenia (SZ), 344–347
Neurons, 210
Neuron-specific enolase (NSE), 250
Neuropsychological assessment, schizophrenia, 329
Neuropsychological methods, 325
Neuropsychological parameters, 195
Neuropsychological profile, 354
　heterogeneity, 298
　inhibiting and planning motor, 354
　obsessive-compulsive disorder (OCD), 354
Neuropsychological testing, 10
Neuropsychology, 10
Neuroreceptors, 28
　membrane protein, 28

SUBJECT INDEX

Neurotherapy, 250
Neuroticism, 217
Neurotoxicology, 391
Neurotransmitters, 361
   obsessive-compulsive disorder (OCD), 361
   receptors, 28
      G protein – coupled, 28
      ionotropic, 28
      ligand-gated, 28
      metabotropic, 28
   schizophrenia (SZ), 343
N_2 event-related potential waves, 357
NF. See Neurofeedback (NF)
Nicotinic receptors, 29
NMDA receptor hypofunction, 343
N2 NOGO wave, 190, 358
N170, object recognition deficit schizophrenia (SZ), 336
NOGO anteriorization (NGA), 343
NOGO ERP components, 362
NOGO P3 waves, 407
Nonspecific effects, 263
   cognitive training, 263
   positive reinforcement, 263
Noradrenalinergic pathways, 302
Noradrenalin reuptake inhibitors, 370, 417
Norepinephrine (NE), 227, 245
Normative databases, 8
NSE. See Neuron-specific enolase (NSE)
Nucleus accumbens, obsessive-compulsive disorder, 361
N1 wave, 145
N2 wave, 357
N600 wave, 359
NxLink database, 371

O

Observed score, 3
Obsessions, 351, 415
   Freud's symbolic interpretation, 351
Obsessive–compulsive disorder (OCD), 415
   action inhibition, 359
   action-monitoring/action-inhibiting, 360
   anterior cingulated, 355
   behaviorism, 351
   cingulate cortex, 357
   cognitive behavioral therapy, 361
   cognitive control, latent components of, 359
   comorbidity, 354
   development, 353
   DSM-IV, 351
   early/late onset, 353
   emotional reaction, 359
   ERP independent components, 363
   error-related negativity (ERN), 357
      enhancement of, 359
   excessive error signals, 356
   excessive frontal beta synchronization, 358
   excessive nature of compulsion, 352
   first-line treatment, 361
   fMRI
      in conflict conditions, 356
      in symptom provocation, 355
   ganglia–thalamocortical system, 360
   healthy controls (HCs), 354
      differences, 355
   heritability, 353
   heterogeneity, 353
   historical overview, 351
   hyperactivation
      of basal ganglia, 356
      of specific cortical areas, 356
   hyperfrontality, 357
   independent component neurofeedback, 362
   lesions, 355
      orbitofrontal cortex, 355
   lifetime prevalence, 352
   medial frontal, 355
   negative, 353
   neurofeedback treatment, 362
   neuronal model, 359
   neuropsychological profile, 354
      inhibiting and planning motor, 354
   neurotransmitters, 361
   N_2 event-related potential waves, 357
   nucleus accumbens, 361
   orbitofrontal cortical, 355
   preparing activity, 359
   prevalence, 352
   psychosurgery/deep-brain stimulation, 362
   quantitative electroencephalography, 357
   reactive cognitive control, latent components of, 360
   relieve symptoms, 351
   response inhibition deficits, 354
   rTMS protocols, low-frequency (inhibitory), 363
   serotonergic system, 361
   spectrum of disorders, 354
   structural magnetic resonance imaging, 355

Obsessive–compulsive disorder (OCD) (cont.)
  supplementary motor area, 364
  symbolic interpretation, 351
  symptoms, 352
  transcranial direct current stimulation, 364
  transcranial magnetic stimulation, 363
Obsessive–compulsive personality disorder (OCPD), 352
Occipital alpha rhythms, 91
Occipital cortex, 93
Occipital rebound beta rhythms, 114
  occipital–temporal areas, 114
  rebound synchronization, 114
OCD. See Obsessive-compulsive disorder (OCD)
ODD. See Oppositional defiant disorder (ODD)
Oddball paradigms, 176
OFC. See Orbitofrontal cortex (OFC)
Operant conditioning, 248
  clinical application, 260
  theory, 259
Opioid receptors, 29
Oppositional defiant disorder (ODD), 296
Orbitofrontal cortex (OFC), 98, 207, 212
  as map of rewards and punishers, 221
  obsessive-compulsive disorder (OCD), 355
Orienting wave, 180
Oscillations, 37, 81
Oscillatory beta activity, 109
Oxyhemoglobin, 19

P

Papez circle, 130
Parietal alpha rhythm, 94
Parietally distributed component, 313
Parietotemporal lobes, 238
Parkinson's disease, 109, 184
Paroxysmal activity, 86
Parvalbumin-expressing interneurons (PVs), 46
Passive oddball task, 68
Pathological locus coeruleus activity, 302
P3b
  as endophenotype, schizophrenia, 337
  as psychosis predictor, schizophrenia, 338
  wave, 302
PC. See Posterior commissure (PC)
Peak performance, 367
Peniston–Kulkosky protocol, 254
Percentage signal change, 19
Percentiles, 5
Perceptual landscapes, 225
Performance, preportent model, 68
Perinatal hypoxic, 310
Periodic oscillations, 83
Persistent alpha, 384
Personality trait, 216
Personalized medicine, 391
PET. See Positron emission tomography (PET)
PFC. See Prefrontal cortex (PFC)
P50 gating effect, 415
Pharmaco-electroencephalography, 391
Pharmaco-event related potential, 392
Pharmacologically induced convulsive therapy, 267
Pharmacology, 391
  clinical, 391
  preclinical, 391
Pharmacotherapy/neurotherapy/neuromodulation techniques, 411
Phase spectra, 41
Plasticity, 259
P3 NOGO wave, 190, 395
Polarographic method, 20
Population, distribution across, 5
Positron emission tomography (PET), 27, 116, 222
  invasive method, 28
  as neuromarker, 30
  physical basis, 27
  spatial resolution, 27
Positrons, 27
Posterior commissure (PC), 17
Posteriors
  alpha rhythm, 384
  alpha rhythms, 90
Posttraumatic stress disorder (PTSD), 223
Power-law function, 81
  of EEG, 82
Precentral cortex, 111
Preclinical research, 78
Prefrontal cortex (PFC), 177
Preparatory cortical activities, 179
Preparatory slow fluctuations, 85
Preset threshold, 256
Primary emotions, 210
Primary sensory areas, 178
Proactive cognitive control
  deficit, schizophrenia, 339
  schizophrenia (SZ), 341
Proactive control processes, 11

Procedural memory, 234
  frontoparietal networks, 240
  neuromodulators, 242
    acetylcholine, 242
    dopamine, 242
  new actions, learning of, 240
Prognostic power, 370
Prominent mu-rhythms, 91
Pruning, 344, 345
Psychiatrist working space, 371
  arrangement, 371
Psychiatry, treatment options
  deep brain stimulation (DBS), 412
  electroconvulsive therapy (ECT), 412
  pharmacotherapy/neurotherapy/
    neuromodulation techniques, 411
  transcranial direct current stimulation
    (tDCS), 412
  transcranial magnetic stimulation
    (TMS), 412
Psychoactive drugs, 286
Psychomotor activity, 324
Psychopathology, 245
Psychopharmacology, 246
  current crisis, 246
Psychostimulants, 395
Psychosurgery/deep-brain stimulation,
    362
  obsessive-compulsive disorder (OCD),
    362
Psychoticism, 217
Psychotropic drug, 246
PTSD. See Posttraumatic stress disorder
    (PTSD)
Pulsation, 39
PVs. See Parvalbumin-expressing
    interneurons (PVs)
Pyramidal cells, 36

Q
QEEG. See Quantitative EEG (QEEG)
Quality control, 10
Quantitative EEG (QEEG), 307
  changes, 392
  neurofeedback, 392
  parameters, 415
    ERP databases, 371
Quantitative electroencephalography,
    obsessive-compulsive disorder, 357

R
Radioactive substance, 27
Radiofrequency band, 18

Radioligands, 28
  D2, 28
  dopamine transporter (DAT), 28
  5-HT2, 28
  serotonin transporter, 28
Radio waves, amplitude of, 18
Random errors, 3
Rapid cognitive impairment, 323
Reaction time (RT), 10, 12
  variability, 12
Reaction time variability (RTV), 13, 300
Reactive cognitive control, latent
    components of
  obsessive-compulsive disorder (OCD),
    360
Reactive cognitive control, schizophrenia,
    340, 342
Reactive control processes, 11
Readiness potential (RP), 179
Reality–expectation discriminability, 335
Real-time fMRI (rtfMRI), 265
Rebound synchronization, 110
Receiver operating characteristic (ROC)
    curve, 6
Reference electrode, 248
Regional image intensity, 19
Relapse, 270
Relative frontal beta, 253
Relative residual energy (RRE), 49
Relaxation protocols, 263
Relaxation training, 254, 257
Reliability, 4
Relieve symptoms, obsessive-compulsive
    disorder, 351
Repeatability, 4
Repetitive TMS (rTMS), 282, 283, 321, 412
  protocols, low-frequency (inhibitory), 363
  theta burst stimulation, 283
Reproducibility, 4
Responsibility, 354
  response inhibition deficits, obsessive-
    compulsive disorder, 354
  response inhibition performance, 311
  responses to tasks, 85
    contingent negative fluctuation, 85
    red arrows, 85
Retrieval operations, 238
Retrograde amnesia, 267
Rhythmic electrical activity, 403
Rhythmicity, 40
  frontal and central beta rhythms, 40
  frontal midline theta rhythms, 40
  posterior alpha rhythms, 40

ROBBIA. *See* Rotman–Baycrest battery for investigating attention (ROBBIA)
ROC. *See* Receiver operating characteristic (ROC) curve
Rolandic alpha, 90
Rolandic beta rhythms, 109
　mechanism of generation, 118
Rolandic fissure, 297, 378, 417
Rolandic spikes, 377, 413
　in ADHD, 378
Rostral anterior cingulate, 126
Rotman–Baycrest battery for investigating attention (ROBBIA), 10
RP. *See* Readiness potential (RP)
RRE. *See* Relative residual energy (RRE)
RT. *See* Reaction time (RT)
rtfMRI. *See* Real-time fMRI (rtfMRI)
rTMS. *See* Repetitive TMS (rTMS)
RTV. *See* Reaction time variability (RTV)
RT variability (RTV), 388

S

SAS. *See* Supervisory attentional system (SAS) model
Scale invariance, 81
Schizophrenia (SZ), 415
　age onset, 325, 326
　auditory MMN, 334
　automatic predicting ability failure, 335
　developmental pruning hypothesis, 327
　dopamine hypothesis, 343
　endophenotype, 337
　environmental risk factors, 329
　frontal basal ganglia thalamocortical circuits
　　impairment of action selection, 347
　GO/NOGO paradigm, 340, 346
　hallucinations and delusions, 325
　heritability, 328
　heterogeneity, 328
　historical overview of, 323
　hypofrontality (fMRI), 340
　　as response predictor to medication, 343
　　studies, 340
　lack of insight, 325
　2-mA tDCS, 348
　motor abnormalities, 330
　N2b/P3 component, 335
　neurodevelopment, 326–327
　neuromarkers of prodromal state, 326
　neuronal model, 344–347
　neuropsychological assessment, 329
　neurotransmitters, 343
　N100/MMN and P3a/P3b, 337, 338
　　reduced P3 amplitude, 340
　N170/N250, 336
　P300 amplitude reduction, 332
　PANSS scores, 348
　P3b
　　as endophenotype, 337
　　as psychosis predictor, 338
　P50 gating failure, 332
　predictor for subsequent psychosis, 338, 339
　prevalence, 325
　proactive cognitive control, 341
　　deficit, 339
　pruning, 345
　QEEG, 330, 331
　reactive cognitive control, 340, 342
　reduced CNV/stimulus preceding negativity (SPN), 340
　reduced predicting ability, 335
　sensory-related neuromarkers, 332–333
　simple viewing tests, 331
　spontaneous electroencephalography, 330–332
　striatal neurons, threshold of activation, 347
　subclassifying, 328
　symptoms, 323–325
　synaptic pruning, schematic diagram of, 327
　tDCS for, 348
　　hallucinations, 349
　timecourse, 325–326
　transcranial magnetic stimulation (TMS), 348
　treatment, 329
　volumetric studies, 330
SCP. *See* Slow cortical potential (SCP) protocol
SD. *See* Standard deviation (SD)
SE. *See* Standard error (SE)
Segregated reentrant loops, 184
Selective serotonin reuptake inhibitors (SSRIs), 245
Semantic memories, 234
Sensitivity, 6
Sensitivity index, 388
Sensorimotor rhythm (SMR), 249, 378
Sensory distortions, 325, 332
Sensory information flow, 183
Sensory modalities, 22, 137
Sensory-motor interface, 178

Sensory-motor rhythm (SMR), 90
Sensory-related neuromarkers,
    schizophrenia, 332–333
Sensory-related operations, 59
Sensory systems, 60, 137
Separating components, 69
    covariance matrixes, joint
        diagonalization of, 74
    group independent component
        analysis, 72
    in multiple tasks, 72
    in single tasks, 72
    single trial independent component
        analysis, 71
    subtraction approach, 69
Septohippocampal circuit, 218
Serotonergic projections, 229
Serotonergic system, obsessive-compulsive
    disorder, 361
Serotonin (5-HT) receptors, 29, 228, 246
Short-term memory, 231
Signal-to-noise ratios, 61, 344
Single-factor model, 329
Single-photon emission computed
    tomography (SPECT), 316
Situational conditioned reflexes, 248
sLORETA. *See* Standardized low-resolution
    brain electromagnetic tomography
    (sLORETA)
Slow cortical potential (SCP) protocol, 255
SMA. *See* Supplementary motor area (SMA)
Smeared potential topography, 47
SMR. *See* Sensorimotor rhythm (SMR)
Somatosensory cortex, 102
Somatostatin-expressing interneurons
    (SOMs), 46
SOMs. *See* Somatostatin-expressing
    interneurons (SOMs)
Spatial filters, 76
Spatial filtration, eye movement correction,
    375
Spatial resolution, 27
Specificity, 6
    sensitivity measurement, 6
SPECT. *See* Single-photon emission
    computed tomography (SPECT)
Spectrogram, 111
Spherical spline interpolation, 32
SPN. *See* Stimulus preceding negativity
    (SPN)
Spontaneous electroencephalogram, 16
Spontaneous electroencephalography
    schizophrenia (SZ), 330–332

Spontaneous fluctuations, 250
Spontaneous variations, 3
Sport psychologists, 367
SSRIs. *See* Selective serotonin reuptake
    inhibitors (SSRIs)
Standard deviation (SD), 7, 12
Standard error (SE), 7
Standardized low-resolution brain
    electromagnetic tomography
    (sLORETA), 93
    images, 123
    localization, 405, 407
Stereotactic operations, 193
Stereotypes, 324
Stimulus preceding negativity (SPN), 179,
    339
Stroop tasks, 176
Subcerebral projection neurons, 46
Subcortical dopamine hyperfunction, 343
Subtraction approach, 69
Supervisory attentional system (SAS)
    model, 10
Supplementary motor area (SMA), 180
    obsessive-compulsive disorder, 364
Suppress, 68
Symbolic interpretation, obsessive-
    compulsive disorder, 351
Synaptic activity, 44
Synaptic transmissions, 108, 245
Systematic errors, 3
Systemic blood pressure, 20
SZ. *See* Schizophrenia (SZ)

T

Talairach brain, 17
Task-negative networks, 23
Task-positive networks, 14, 23
Task-setting, 10
Task switching paradigms, 175
Tau rhythm, 90
TBR. *See* Theta/beta ratio (TBR)
tDCS. *See* Transcranial direct current
    stimulation (tDCS)
Temperament
    brain model, 212
    dimensions, 211
        monitoring affect, 211
        negative affect, 211
        positive affect, 211
Temporal binding, 117
Temporal dynamics, 231
Temporally distributed component, 313
Test of variables of attention (TOVA), 15

Test–retest reliability, 4, 25, 28, 57, 76
Thalamic midline nuclei, 100
Thalamic neurons, 104
   impulse activity, 83
   infraslow fluctuation in, 83
Thalamocortical neurons, 101, 403
   burst mode, 101
   tonic mode, 101
Theory of action programming, 187
Theta/beta ratio (TBR), 304, 379
   in ADHD, 379
   metaanalysis, 306
Theta burst stimulation, 283
Timecourse, schizophrenia, 325–326
TMS. See Transcranial magnetic stimulation (TMS)
Tonic motor activity, 109
Tonotopic organization, 160
Tourette's syndrome, 354
TOVA. See Test of variables of attention (TOVA)
Transcranial direct current stimulation (tDCS), 247, 412
   attention deficit hyperactivity disorder (ADHD), 320
   historical perspective, 247
   limitations, 257
   long-term post-tdcs, 255
   neurophysiological basis, 252
   NMDA involvement, 256
   nonlinear collective effect, 254
   obsessive-compulsive disorder, 364
   procedure, 247
   protocols, 417
   safety, 257
   for schizophrenia, 348
   side-effects, 257
   vs. electroconvulsive therapy, 250
Transcranial magnetic stimulation (TMS), 214, 250, 281, 348, 412
   model, 283
   obsessive-compulsive disorder, 363
   physical principles, 281
   physiological effect, 281
   safety, 283
   rTMS, 283
   schizophrenia, 348
Translational research, 78
True score, 3
Two-click paradigm, 68

U

UK ECT Review Group, 270
Unilateral electrode placement, 267
Unusual cortical areas, 99

V

Validity, 4
Vanderbilt Assessment Scale for Parents (VAS-P), 294
VAS-P. See Vanderbilt Assessment Scale for Parents (VAS-P)
Ventral visual stream, 63
Veritable–sham–veritable approach, 259
Vertex beta rhythms, 113
   in resting state, 114
Virtual intracranial electroencephalogram protocols, 254
Visual discrimination, 145
Visual mismatch negativity (MMN), 146
Visual modality, decay of information, 232
Visual N170, 148
Visual N250 repetition effect, 151
Visual N1 wave, 228
Visual P2, 153
Voltage-gated channels, 37
   ion channels, 37
Volume conductance, 47

W

WAIS. See Wechsler Adult Intelligence Scale (WAIS)
Wavelets, 42
Wavelet transformation, 42
Wechsler Adult Intelligence Scale (WAIS), 97
   perceptual organization, 97
   processing speed, 97
   verbal comprehension, 97
   working memory, 97
Wernike aphasia, 234
What stream, 160
Where stream, 160
White noise, 81
Willed selection paradigms, 175
Working memory, 128, 178, 233
   representations, 178

Z

Z scores, 5, 402